POLYMER SINGLE CRYSTALS

POLYMER SINGLE CRYSTALS

By PHILLIP H. GEIL

Camille Dreyfus Laboratory
Durham, North Carolina
and
Case Institute of Technology
Cleveland, Ohio

ROBERT E. KRIEGER PUBLISHING COMPANY
Huntington, N. Y.
1973

ORIGINAL EDITION 1963

REPRINT 1973

Printed and Published by

ROBERT E. KRIEGER PUBLISHING CO., INC.
BOX 542, HUNTINGTON, NEW YORK. 11743

© Copyright 1963 by

JOHN WILEY & SONS INC.

Reprinted by arrangement

Library of Congress Card No. 63-19663
SBN #0-88275-088-7

PRINTED IN THE UNITED STATES OF AMERICA.

PREFACE

The years since 1957 have seen the beginnings of the development of a new field of physics: that concerned with the detailed structure and properties of the solid state of crystalline polymers. From the time of the recognition of their macromolecular character, polymers in the liquid and solid state were treated and thought of primarily on a statistical basis, on both the molecular and morphological scales. The application of electron microscopy to crystalline polymers since 1957 has pointed out that a vast amount of detailed morphological information is susceptible to study and description; that which has been reported is the subject of this review. In addition, through the joint application of microscopy and such other techniques as detailed studies of mechanical, dielectric, and nuclear relaxations and x-ray diffraction, the initial steps toward the description of polymers on a molecular scale are occuring—descriptions in terms of features such as twinning, slip, dislocations, defects, and the like, which are familiar as a result of their use in the solid state physics of atomic solids.

Although rapid advances have been and still are being made in our knowledge of the morphology of crystalline polymers, the gaps in our knowledge are at present, as will become obvious, both wide and numerous. In addition, little—in fact essentially nothing— is known of the morphology of noncrystalline polymers. Application and utilization of the knowledge that is available are just beginning. It is hoped that the results presented in this book will not only stimulate further research in the area of morphology but will also serve as a basis to permit the development of the foundations of the solid-state physics of polymers.

This review was originally completed as an E. I. du Pont de Nemours & Co., Inc., internal report in the early part of 1961. The author wishes to express his appreciation to the du Pont Company for permission to include a considerable amount of material from that report which has not previously been published. The large number of papers published between then and the end of 1962 required numerous additions to a number of sections of the initial report, additions that in several cases might better have been handled by revising the format and contents of entire chapters. Experience showed, however, that such revisions would have no end in view of the rapid increase in research in this field and, thus, several recent items of considerable importance have had to be merely inserted in the previously established format.   The various additions and revisions were made while the author was at the Camille Dreyfus Laboratory.   Appreciation is expressed for financial support from the Camille and Henry Dreyfus Foundation during that time.

We have attempted, in this book, to consider all of the recent results pertinent to the general problem of polymer morphology.   Undoubtedly we have overlooked some reports, and apologies are offered to both the authors and the readers.   In several instances, possibly pertinent data has knowingly been omitted because of the impossibility of correlation and comparison with other results.   While we have attempted to critically review the present state of the field, it will be obvious that considerable speculation is still present.   For this, we offer no apologies, only pointing out that speculation must remain until knowledge increases and, in addition, hoping that we have made adequate distinction between observation and speculation in the text.   It should also be pointed out that we have not attempted to evaluate the older work;   that is, observations in such areas as crystal structure and the optical microscopy of spherulites.

The author wishes to express his appreciation to several members of the du Pont Company for their assistance in preparing the original manuscript, to D. H. Reneker for reviewing most of it, to R. P. Schatz for preparing many of the drawings, to J. P. Crumlich for his care in preparing the micrographs, and to C. H. Hamblet for reviewing the contents of the present manuscript.   Considerable indebtedness is also due to numerous workers in the field for kindly furnishing results of their research prior to publication. It would have been impossible to prepare this review in its current form without their generosity.   In this respect, especial appreciation is due to A. Keller and D. C. Bassett for numerous comments and new results, and to R. L. Miller and L. E. Nielsen for permission to include their latest table of crystallographic data as an Appendix.

Case Institute of Technology   P. H. GEIL
Cleveland, Ohio
July 1963

ACKNOWLEDGMENTS

Permission to reproduce copyrighted micrographs and other figures, as listed below, is gratefully acknowledged. Those micrographs for which no reference is given in the caption are from the author's file.

Academic Press Inc., Figure II-67 (*J. Colloid Sci.*); Academie royale de Belgique, Figures IV-56, 64 (*Bull. Acad. roy. Belg. (Classe Sci.)*); American Chemical Society, Figures V-24, 25 (*J. Phys. Chem.*); American Institute of Physics, Figures I-30, II-18, 22, 24, 26, 31, 37, 38, 41, 50, 57, 58, 63, 65, III-4a, 22, 23, 24a, IV-44, V-5, 22, 23, 26, VII-22, 23, 39, 40 (*J. Applied Phys.*), Figures IX-4, 5 (*Phys. Rev. Letters*), Figures VI-4, 5, 6 (*J. Chem. Phys.*), and Figure IX-3 (*Phys. Rev.*); American Telephone & Telegraph Co., Figures II-52, 55 (*Bell Sys. Tech. J.*); Butterworths & Co., Limited, Figures I-2, 3, 4 (*Electron Diffraction* (1953)), and Figures II-62, III-10, IV-46a, 47a (*Polymer*); Dr. Alfred Huthig Verlag, GmbH, Figures II-13a, V-50 (*Makromol. Chem.*); Istituto Lombardo di Scienze e Lettere, Figures III-19, 20 (*Rend. Ist. Lomb. Sci. e Lettere*); Macmillan & Co., Limited, Figure II-43 (*Nature*); National Bureau of Standards, Figures V-54, IX-2 (*J. Research Natl. Bur. Standards*); Philips Electronic Instruments, Figure I-9c, d, e, f (*Norelco Reporter*); Royal Microscopical Society, Figures V-1, 12 (*J. Royal Micro. Soc.*); Royal Society, Figures I-11, 12, 20, 21 (*Proc. Roy. Soc. (London)*); Societa Italiana di Fiscia, Figures I-7, 10, 13, 17, 18, 23, 24, 28, 29, 31, 32 (*Nuovo Cimento*); Dr. Dietrich Steinkopff Verlag, Figures I-9a, b, 19, II-15, III-18, IV-30, 31, 32, 33, 34 (*Kolloid-Z.*); Taylor & Francis, Ltd., Figures I-25, II-11, 35, 42, 66, 78, 81, 82, V-19 (*Phil. Mag.*); Verlag Chemie, GmbH, Figures V-31, 32 (Sym. Macromol., Weisbaden (1959)), and Figures V-3, 4, 6, 7, 33, 49, VII-4, 41, 42 (*Angew. Chem.*); John Wiley & Sons, Inc., Figures I-14, 15, 22, II-1, 2, 4, 9, 32, 33, 44, 45, 46, 47, 48, 64, 68, 70, 74, 75, 84, 88, III-5, 11, 12, IV-2, 10, 11, 17, 20, 41, 43, 45, 63, 69, 70, 71, 72, V-29, 53, VII-1, 3, 8 (*J. Polymer Sci.*), Figures V-10a, 14 (*J. Appl. Polymer Sci.*), and Figures III-13, 16 (*Growth and Perfection of Crystals* (1958)).

CONTENTS

I. INTRODUCTION

Two major advances in polymer science have occurred recently. In the field of polymer chemistry the discovery and use of stereospecific polymerization has created a broad new class of polymers. This work has recently been reviewed (1). Many of the unique physical properties associated with these polymers result from their ability to crystallize. In the field of polymer physics the recognition and understanding of the way polymers crystallize, which is the subject of this book, will, it is believed, lead to better utilization of polymers in their present areas of use as well as to the development of new polymers and new uses for old polymers. In brief, it has been found that most, perhaps all, crystalline polymers can and usually do crystallize from the melt and from solution in the form of thin lamellae, on the order of 100 A thick, in which the molecules are folded back and forth on themselves.

In the remainder of this chapter we shall briefly discuss various aspects of the structure of polymer molecules, inter- and intramolecular forces, the apparent structure of the melt or solution from which the polymer molecules crystallize, and previous ideas of how polymer molecules crystallize. Also included are descriptions of some of the experimental techniques, their limitations, and their applicability to the investigation of polymer morphology. The crystal structures of most of the polymers considered in subsequent chapters are included here. For further details the reader should consult the various references listed. The remaining chapters discuss in detail the morphology of polymers crystallized from solution and the melt, present ideas as to how this morphology develops and some speculations as to how it is related to polymer physical properties.

1. POLYMER MOLECULES AND MOLECULAR FORCES

In the latter half of the 1800's there was evidence that molecules could exist with molecular weights on the order of 10,000 or more. The concept that macromolecules or polymers consist of long chains of atoms held together by covalent bonds was not, however, generally accepted until the 1920's.* It is now recognized that most polymers of commercial interest, in-

* Short histories of the development of the concept of long-chain molecules are given in several polymer chemistry textbooks (2,3).

cluding those with which we shall be concerned, have molecular weights between several thousand and several million. These molecules average about 10 to 20 molecular weight units per angstrom length. In general, the synthetic polymer is built up by the repeated addition to the growing chain of one or more small chemical units, termed monomers. If the monomers are polymerized according to a regular pattern the polymer is known as a "tactic" polymer* and usually is able to crystallize. If, however, the monomers are polymerized in an irregular pattern the resulting "atactic" polymer usually cannot crystallize.

The predominant bond between atoms within a polymer molecule is the covalent bond. These primary bonds are formed when one or more pairs of valence electrons are formed by and shared between two atoms or groups. They are relatively strong bonds, having dissociation energies on the order of 2 to 3 ev. In polymers whose backbone consists of covalently bonded carbon atoms, as is the case in many but not all commercially valuable polymers, the angle between successive single bonds is usually a few degrees larger than the tetrahedral angle, 109° 30'. A change in shape of the molecule occurs by rotation about these bonds in preference to either elongation or bending of the bonds.

Between molecules or portions of a molecule not connected by primary bonds, so-called secondary bonds or van der Waals forces result in attractive forces. These forces vary as the inverse sixth power of the separation of the atoms. They are at least two orders of magnitude weaker than the covalent bond. The cohesive energy per unit chain length is a measure of the strength of these forces. The cohesive energy is that energy required to remove a molecule to a position far from its neighbors. Typical values range from $8 \cdot 10^{-3}$ ev per angstrom chain length for polyethylene to $5 \cdot 10^{-2}$ ev per angstrom chain length for polyamides. A particularly important secondary bond in determining the crystal structure of some polymers is the hydrogen bond. Its strength is greater than that of the other secondary bonds, giving rise to the high cohesive energy of the polyamides.

The average attractive force between portions of polymer molecules

---

* If the monomer is such that the tactic polymer has a regular succession of branches, as is the case with head to tail polypropylene for instance, then it may be classified as isotactic or syndiotactic. If an isotactic polymer is stretched out and viewed along its backbone, all of the branches lie along a single line, whereas in a syndiotactic polymer they alternate between one side of the backbone and the other. In the few cases where both types have been polymerized, we shall, in general, be concerned only with the isotactic polymers; morphological studies have been restricted to them. For a more complete discussion of the nomenclature of sterically ordered polymers, see references 4 and 5.

that are in a crystalline lattice is greater than that between portions which are not in a lattice. The density of a crystalline polymer is as much as 15% greater than that of the supercooled melt at the same temperature. In fact it is these secondary forces that result in the molecules crystallizing, the arrangement that the molecules take in the lattice being that which produces the greatest average attractive force consistent with the maintenance of the primary bonds and a contact force of repulsion. The force of repulsion varies approximately with the inverse twelfth power of the distance between the atoms.

2. STRUCTURE OF POLYMER SOLUTION AND MELT

The structure of polymer melts and solutions is not well defined. In this section we discuss some suggestions for this structure and the associated molecular motions that are being developed on the basis of the recent work on polymer morphology.

As the temperature of any material, polymeric or not, crystalline or amorphous, is raised from $0°K$, the vibration and rotational or torsional oscillation of its constituent atoms or groups of atoms increases. This motion, because of its anharmonic character, results in thermal expansion. It also results in a broadening of Bragg x-ray reflections from crystalline materials. The temperature at which the interatomic restraints are sufficiently reduced to permit a large number of relative, nonelastic displacements of neighboring molecules under an applied stress (thermal, mechanical, electrical, etc.) is known as the glass transition temperature. Its value depends on the method of measurement. When the displacements are due to thermal motions, there are, for instance, rather abrupt changes in heat capacity and the volume expansion coefficient at the glass transition temperature. In fact the glass transition temperature is often defined as the temperature at which there is a change in slope of a plot of volume versus temperature, i.e. it is the temperature at which rearrangements in molecular position occur sufficiently rapidly to permit the formation of "free volume" in experimentally feasible times (6). A more complete discussion of a possible interpretation of the glass transition in terms of molecular motions in crystalline and amorphous polymers is given in Chapter IX.

At temperatures above the glass transition temperature but below the melting point, cooperative motion of groups of atoms, both within a polymer molecule and in neighboring molecules, occurs within the lattice. X-ray diffraction by the lattice is not greatly affected by this motion. X-ray diffraction sees a time average lattice, so that if only a relatively few atoms or groups of atoms are moving at any one instant their effect on the diffraction pattern is negligible. As the relative number of moving groups increases, the Bragg reflections will decrease in intensity, more and more intensity being scattered into a diffuse background. The basic structure of the lattice would be maintained during this motion, the moving groups acting as isolated defects within the lattice. This gradual decrease in intensity of the Bragg reflections as the melting point is approached has often been described in polymer studies as premelting, small crystals melting far below the equilibrium melting point.

The motions believed to occur in long chain polymer crystals would be related to those known to occur in molecular crystals of simpler substances. For instance, large scale free rotation of entire molecules occurs in crystals of a number of low molecular weight substances such as CH_4; oftentimes the lattice is maintained when the rotation starts. Furthermore, it is well known (see, for instance, Daniel (7)) that short chain molecules, such as paraffins, ketones, and alcohols start to twist or rotate in the crystal at temperatures well below the melting point. In pure crystals the transition temperature at which the motion starts is sharply defined but in crystals with only slight admixtures of different length chains the transition occurs over a wide range of temperature (8). Presumably motion occurs most readily in the vicinity of defects in the lattice. The motion in the case of polymers should differ primarily in that only portions of the molecule may be involved at any one time.

Although not greatly affecting x-ray scattering these motions in the lattice would be expected to affect the nuclear magnetic resonance spectrum, NMR line widths being greatly narrowed by rotation of the resonating atoms. For linear polyethylene the broad band attributed to protons in the crystalline regions abruptly narrows at about 120°C, 17° below the melting point (9). A similar change in the NMR spectrum occurs at 20°C in highly crystalline polytetrafluoroethylene (10). In the same temperature range some, but not all, of the x-ray diffraction reflections from polytetrafluoroethylene become diffuse. It has been shown that segments of the molecules in the lattice start to undergo hindered rotation in this temperature range (11). The basic lateral packing in a hexagonal lattice is maintained, but the azimuthal order between neighboring molecules is lost.

An idea of the number of moving groups that can be accommodated within the lattice can be obtained from a realization that low density, branched polyethylene crystallizes in the same lattice, only slightly expanded, as linear polyethylene. However, a typical branched polyethylene has two branches per 100 carbon atoms in the backbone of the molecule. These branches, as will be discussed later, create a much larger defect in the lattice than one of the postulated types of group motion that can result in a screwlike translation of the molecule within the lattice (12). A large number of defects caused by rotating groups can apparently be accommodated within the lattice without greatly disturbing its over-all structure and cohesion.

Rotation within a lattice of a group of atoms in one molecule in many cases requires a cooperative motion of the atoms in adjacent molecules. Although this motion of neighboring molecules need not be large, it does cause restriction of the rotation. The restriction would be expected to be less in the vicinity of lattice defects, permitting the motion at a lower temperature. At the same time as this rotation of isolated groups is occurring, the entire lattice is expanding in a manner analogous to that occurring with nonpolymer type molecules, i.e. all of the atoms vibrate about their equilibrium position to a greater and greater degree as the temperature is increased. At the melting point the molecules are separated sufficiently that the restraining force of the lattice breaks down and there is a catastrophic increase in the number and size of the rotating groups. The sample melts. As in the case of low molecular weight substances there is a loss of long-range order. An interesting feature is the fact that the change in entropy that occurs when paraffins melt is about the same as that which occurs when rotation within the lattice begins (13).

A polymer melt is often pictured as consisting of more or less randomly coiled and intercoiled molecules with little or no restriction on the rotation of the molecular segments due to the proximity of neighboring molecules. Large-scale translation of the molecules normal to their chain axes, however, is restricted by their long-chain nature, requiring the cooperative rotation of a large number of segments of many molecules and giving rise to the high viscosity of polymer melts. Translation parallel to the axis probably occurs readily through the same type of screwlike motion that will be discussed in Chapter V in connection with recrystallization. We do not believe that very much randomizing of the molecules occurs when a polymer melts, although rotation of segments is certainly occurring. Unless the polymer melt is worked (i.e., stirred, extruded, or otherwise disturbed) or the sample was previously in a highly strained state, it is believed that little over-all change in position of the molecules occurs at first when a sample is melted. With increasing time or temperature a somewhat more random state would be approached.

Experimental investigations of ordering in the molten state have not been reported for polymers. Recent work by Rabieski and Kovacs (14) suggests, in the author's opinion, that the general molecular orientation in the solid is maintained for considerable periods of time at temperatures well above the melting point. The crystallization kinetics of samples of polyethylene, for instance, originally molded and thus being oriented at 180°C (40°C above the melting point) do not, during subsequent crystallization experiments, resemble the crystallization kinetics of randomly oriented molten samples unless the sample is heated to a temperature higher than 180°. The density of polymer melts is sufficiently close to that of the crystal that even in randomly oriented samples there must be many localized regions in which molecular segments are parallel, approximating the structure of a crystal but without the crystals relatively rigidly defined lateral and longitudinal order. There is no strong cohesion to these clusters, however, and they therefore do not directly affect the physical properties of the melt. These clusters are maintained in the sample if it is sufficiently rapidly cooled to temperatures near or below the glass transition temperature. The absence of any but vibratory and oscillatory motion at these temperatures prevents rotation of molecular segments to form a perfect lattice.

A polymer solution has also been believed to consist of more or less randomly coiled and intercoiled molecules. The diffusion of molecules is less restricted than in polymer melts because fewer cooperative motions are required. Hydrodynamic forces tending to stir the solutions would allow the molecules to move more readily than in the melt. Depending on the concentration and solvent-polymer interactions, clusters of aligned molecular segments are expected similar to those postulated to exist in the melt.

Although we are suggesting here that little molecular transport necessarily occurs during melting there is also evidence that a considerable amount of molecular transport may occur during crystallization. Molecules move distances up to 1 mm during the growth of hedrites from thin films (Chapter III), a feature which, however, may result from surface tension effects. Even more indicative, perhaps, is the growth of spherulites consisting of well formed crystalline lamellae from previously quenched, glassy solids at temperatures well below the melting point (Chaps. IV and V). Further evidence for the absence of large-scale position changes during melting is also presented in Chapter V. It thus appears that while there are driving forces for molecular transport during crystallization, there are only relatively small forces for large-scale randomization during melting.

When a polymer melt or solution is cooled, crystallization occurs when and if the forces of attraction between neighboring segments in a cluster become great enough to cause many of the rotating segments to lock into and only oscillate about a particular position. Again, because of the restriction on rotation imposed by the lattice and decreasing separation of the molecules, the effect is catastrophic at a particular temperature known as the crystallization temperature. Whereas in nonpolymeric materials the crystallization temperature and the melting point are, under normal conditions, approximately the same, in polymers the crystallization temperature is usually 10° or more below the equilibrium melting point. This, it is suggested, is due to the lamellar nature of the crystals; the energy involved in forming the surfaces of the lamellae is much larger relative to the volume free energy in the case of polymers than in the case of most crystals of low molecular weight material. The equilibrium melting point is the temperature at which the free energy of an infinite crystal (in polymers, consisting of fully extended chains) is equal to that of the liquid. The crystallization point, on the other hand, is the temperature at which the free energy of a stable crystal nucleus, including a surface free energy term, is equal to that of the liquid. (For a more complete discussion, see Section VI-1.)

3. FRINGED MICELLE CONCEPT OF POLYMER MORPHOLOGY

Shortly before it was generally accepted that polymers were long-chain molecules, x-ray studies showed that some polymers, both natural and synthetic, had recognizable crystalline features. The Bragg reflections, however, were considerably broader and more diffuse than those obtained from well developed crystals of nonpolymeric low molecular weight materials. Furthermore, the reflections were observed to rapidly merge with the background away from the main beam. In general, no reflections are resolved at angles greater than 90°.

Diffraction theory in the late 1920's was sufficiently developed to indicate that the above features of the diffraction pattern could be attributed either to small crystal size or defects within an extensive lattice. Although this theory permitted a discrimination between these two possibilities, polymer diffraction patterns were too indistinct at the time to allow it. The effect was attributed to crystallize size. The measured broadening corresponds to crystal dimensions of about 200 A. Initially the molecules were assumed to be this long, the size of the crystal being limited by the length of the molecule (15,16). Following acceptance of the idea that polymer molecules are much longer than 200 A and the idea that molecules in the melt and solution are randomly coiled and intercoiled, crystallization was thought to occur by the precise alignment of neighboring segments of various molecules. This crystallite would grow in the direction of the molecular axes by further alignment of molecules already associated with it as well as growing laterally by the addition of segments from other molecules. It

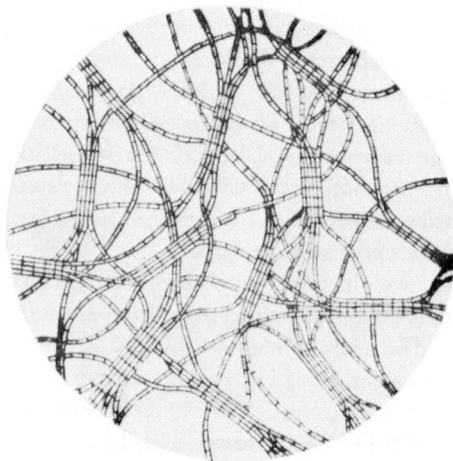

Fig. I-1. Fringed micelle model of a partly crystalline polymer. Molecules extend through a number of crystalline (crystallites) and amorphous regions (Hermann and Gerngross (18)).

was believed that large-scale growth of the crystallite would be impeded by development of entangled and strained regions. Again assuming that all of the x-ray diffraction broadening was due to the size of the crystal, the values of 200 A or less appeared quite reasonable for this picture. The presence of a substantial background of liquidlike scattering was attributed to the entangled and strained portions of the molecules between crystallites, i.e., in amorphous regions. Individual molecules were assumed to extend from one crystallite to another, passing through the amorphous regions and tending to hold the polymer together. The crystallites were considered as physical crosslinks in the polymer solid.

The resulting model (Fig. I-1) came to be known as the "fringed micelle" or "fringed crystallite" model. The fringes are the portions of the molecules in the transition region between the perfect crystal and the amorphous matrix. This concept was first advanced in 1930 by Hermann, Gerngross, and Abitz (17) to explain the structure of gelatin and shortly later was extended to the structure of natural rubber (18). Since that time, with numerous modifications, it has been applied to polymers in general. The fringed micelle concept was a most logical model and has served as a versatile means of correlating a large number of apparently unrelated experimental observations. Furthermore it is essentially statistical in character, permitting theoretical calculation of many features of the thermodynamics and physical properties of polymers. However, the statistical nature of the congealing of the molecules into crystallites and amorphous regions made it difficult to explain the occurrence of microscopic aggregates of crystalline material, known as spherulites (19,20). Light microscopy revealed a radial structure in the spherulites. Attempts to explain the growth and structure of spherulites led to the numerous modifications of the basic fringed micelle model that have been published and described in the last 10 to 15 years. All of these modifications, however, still view a crystalline polymer as containing relatively perfect crystallites on the order of several hundred Angstroms in size, imbedded in an amorphous matrix, with polymer molecules extending from one crystallite to another, passing through amorphous regions and holding the polymer together.

The observations to be described in the following chapters, dating from about 1957, indicate, we believe, that the fringed micelle model is fundamentally incorrect for crystalline polymers. However, it may still be suitable in modified forms for the description of the morphology of crystallizable polymers quenched to a glassy state, and the morphology of highly oriented samples such as drawn fibers. Only a small degree of motion, obtained just above the glass transition temperature, would be sufficient to convert a cluster of aligned molecules, as quenched from the melt or solution, to a fringed micelle type of structure. Once the molecules lock into their preferred position in the lattice they will tend to remain there. A considerable degree of motion must occur before they will recrystallize in the form of a lower energy crystal shape. Experiments indicate, however (Chapters IV and V), that this degree of motion, resulting in the development of lamellae, can be attained well below the melting point.

When a polymer crystallizes from the melt, crystallization starts at various points in the sample and spreads out in a more or less spherical manner. Unless quenched rapidly enough to temperatures below the glass transition temperature, the resulting spherulites will grow until they impinge upon each other, completely filling the sample's volume (see Figure IV-I). Optical and x-ray measurements indicated that the molecules are, on the average, tangentially oriented in the spherulites. One of the most serious objections to the fringed micelle models was the observation that in guttapercha two types of spherulites formed simultaneously from the melt (21). These two types have a different crystal structure, melting point, birefringence, etc. Such an observation, as was pointed out (21), is impossible to explain in terms of fringed micelle crystallites.

Prior to 1958 a number of explanations were put forward to explain the increasing number of observations on spherulites. On the basis of his observations on gutta-percha, Schuur suggested (21) that some form of continuous crystallization must occur within each spherulite, that the crystal structure of each spherulite nucleus is propagated throughout the resulting structure. Although many of the details of his "autoorientation" growth mechanism appear to be incorrect on the basis of subsequent observations, the basic hypothesis of a continuous crystal structure within each spherulite has been confirmed. One of the other more ingenious suggestions was that of Keller (22–24) who proposed a structure in which fibrils originate at the center of a spherulite. Each fibril is composed of a tightly wound crystalline ribbon, the molecules being oriented parallel to the ribbon and therefore nearly normal to the helix axis. The crystalline ribbon was composed of an array of individual crystallites similar, perhaps, to the fringed micelles. With this model Keller was able to explain optical and x-ray properties of spherulites and the orientation processes occurring during drawing and relaxation. It was not anticipated, however, that polymers would crystallize in the manner to be described in the following chapters.

4. X-RAY AND ELECTRON DIFFRACTION APPLIED TO POLYMERS

a. Wide Angle Diffraction

X-ray and electron diffraction permit one to obtain, by Fourier analysis of the diffraction pattern, a time and space average picture of the arrangement of the atoms in a material. Although applicable to gases, liquids, and solids, in the case of polymers its use has been restricted almost entirely to the determination of the arrangement of the atoms in solids and, specifically, the crystalline portion of the solid. By crystalline is meant that material in which there is a regularly ordered arrangement of the atoms in at least one dimension and usually all three dimensions, for distances in excess of 50 A or so; i.e., there is long-range order. For the theory and practice of electron and x-ray diffraction the reader is referred to texts by Pinsker (25), James (26), Klug and Alexander (27), Bunn (28), and others. Following a brief discussion of the principles of diffraction that are necessary for the subsequent discussions we shall consider the application of x-ray diffraction to the determination of the structure of polymers. The discussion of diffraction is based primarily on Pinsker (25) whereas the discussion of the application of x-ray diffraction to the determination of polymer crystal structure is taken from an unpublished report of E. S. Clark (29). In this book the crystallographic symbols defined in the International Tables for X-ray Crystallography (30) will be used to represent the indices of Bragg reflections, and points, planes, and directions of the crystal lattice (Table I-1).

TABLE I-1

Crystallographic Symbols Used in This Book (30)

a, b, c	Length of unit cell edges.
a, b, c	Unit cell vectors.
α, β, γ	Angles between the axes, i.e., $\mathbf{b} \wedge \mathbf{c}$, $\mathbf{c} \wedge \mathbf{a}$, and $\mathbf{a} \wedge \mathbf{b}$, respectively.
u, v, w	Coordinates of any lattice point in terms of a, b, c as units.
$[uvw]$	Indices of a direction in the real space lattice. $[110]$ is the direction of the vector from the origin to the point with coordinates $u = 1, v = 1, w = 0$.
$\langle uvw \rangle$	Indices of a "form" of directions, i.e. all directions which are related by symmetry.
(hkl)	Indices of a crystal face, of a single plane, or of a set of parallel planes.
hkl	Indices of the reflection from a set of parallel planes and the coordinates of a reciprocal lattice point.
$\{hkl\}$	Indices of a form of planes or of reflections from a form of sets of parallel planes, i.e. all those sets of planes or reflections which are related by the symmetry of the unit cell.
	A bar over a particular index indicates a negative value. $[\bar{1}\bar{1}0]$ is the direction of the vector from the origin to the point with coordinates $u = -1, v = -1, w = 0$.

Pinsker (25) points out that diffraction (x-ray or electron) from a three-dimensional lattice can be represented by the superposition of the diffraction from three noncoplanar linear lattices. For the simplest case we let the three linear lattices be orthogonal, one parallel to the incident beam and two at right angles to it and to each other. The diffraction patterns from infinite linear lattices are shown in Figure I-2 as they would be observed on a flat photographic film normal to the incident beam. On a cylindrical film concentric about the linear lattice the hyperbolae shown in Figure I-2a would appear as straight lines. The diffraction pattern from any linear lattice consists of a family of cones centered on the axis of the lattice. Thus, if the lattice is not quite parallel to the incident beam (Fig. I-2b), the center of the diffraction pattern on the film is shifted away from the incident beam; the circles become ellipses. The conical angle (2ψ) depends on the wave length of the incident beam, λ, and the lattice spacing, a, $a(\cos \psi - \cos \phi) = n\lambda$, where ϕ is the angle between the lattice axis and the incident beam.

The superposition of the diffraction patterns from three orthogonal infinite linear lattices, one of which is parallel to the incident beam is shown in Figure I-3. The requirement for diffraction from a three-dimensional lattice is that two of the hyperbolae and a circle intersect. As can be seen, the probability of this occurring for a given crystal and incident beam wave

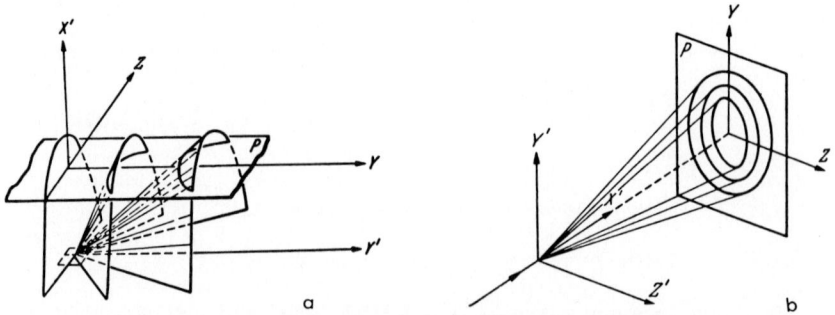

Fig. I-2. Diffraction patterns from infinite linear lattices. (a) Incident beam, X' normal to the row of atoms, Y'. The row of atoms lies parallel to the photographic film, P. A similar pattern rotated by 90° would be produced by a row of atoms parallel to Z. (b) The row of atoms parallel to the incident beam. Scattering is also produced in the backward direction, only a few of the diffraction cones being shown. (Pinsker (25).)

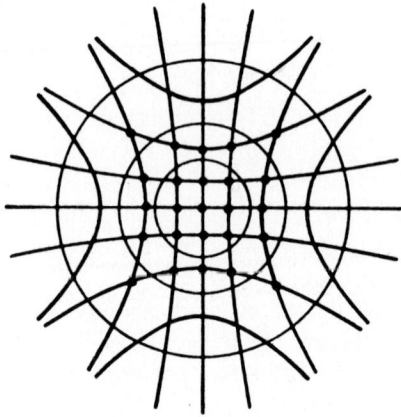

Fig. I-3. Typical superposition of the diffraction patterns from three orthogonal linear lattices. No constructive interference is produced for the lattice spacings and incident beam wavelength represented here. The relative dimensions correspond to x-ray wavelengths and lattice spacings on the order of several Angstroms. (Pinsker (25).)

length is very small. In terms of the reciprocal lattice, which is often used in x-ray diffraction, it is highly improbable that the sphere of reflection will intersect any of the reciprocal lattice points of an infinite space lattice for a given orientation of the crystal and incident beam. However, by tilting the crystal the relative positions of the hyperbolae and circles change and, in certain positions, three of the lines will intersect at a given point resulting in a diffracted beam. At these points Bragg's law is satisfied. In normal x-ray diffraction practice either the crystal is rotated or a powder consisting of numerous crystals randomly arranged is used in order to obtain a diffraction pattern.

The preceding discussion applies to an infinite, perfect lattice and a monochromatic, perfectly collimated beam. Relaxation of any of these conditions results in a relaxation, to some extent, of the requirements for diffraction. A spread in wave length or collimation results in a broadening of the reflection cones, or, in terms of the reciprocal lattice, results in a thickening of the sphere of reflection. In either description the likelihood of intersection of the appropriate lines and surfaces increases as the thickness of the lines and surfaces increase. Of more significance is the effect of a finite and/ or distorted lattice. The total intensity scattered from a finite linear lattice into a given cone is proportional to N, the number of elements in the lattice. In a plot of scattered intensity as a function of θ, the height of the maxima corresponding to the diffraction cones is proportional to N^2. Thus the width of the maxima, the thickness of the line of intersection of the cone and photographic film is inversely proportional to N. Polymer crystals are lamellar; one, and sometimes two dimensions are on the order of 100 A, whereas the other dimensions are often 2 or more orders of magnitude larger. The incident beam, especially in electron diffraction experiments, is often directed normal to the lamella so that the linear lattice parallel to the beam is considerably smaller than the linear lattices normal to the beam. The circles thus broaden whereas the hyperbolae remain relatively sharp. Pinsker refers (25) to the linear lattice parallel to the beam as the third atom row. Figure I-4 shows the diffraction situation for the case of (a) an infinite third atom row, and (b) a finite third atom row. The drawings in Figures I-2 and I-3 are drawn to correspond to x-ray wave lengths. In electron diffraction (as in Fig. I-4), the wave length being much shorter, the conical angle, 2ψ, is much larger, resulting in a closer spacing of the hyperbolae and larger radii for the circles. Distortions of the lattice, due to imperfections of various sorts, also result in a broadening of the diffraction cones. This causes the hyperbolae as well as the circles to become wider. The effect of tilting a thin crystal is shown in Figure I-4c. The center of diffracted intensity is shifted from the direction of the incident beam.

In reciprocal lattice notation the lattice points for a finite crystal are expanded into three-dimensional bodies, any particular dimension of the reciprocal lattice "point" varying inversely with the corresponding real lattice dimension. The reciprocal lattice of a thin crystal or lamella consists of

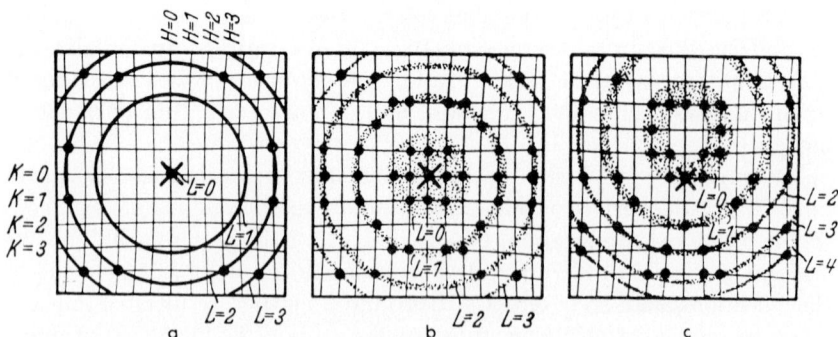

Fig. I-4. Effect on diffraction patterns of finite crystal thickness and tilting of the crystal. The values of H, K, and L correspond to the Miller Indices ($h\ k\ l$) of the reflections, the c axis being parallel to the incident beam. (a) Diffraction pattern from an infinite crystal for which the wave length of the incident beam has been chosen to give rise to some reflections. The diagram is drawn to correspond to electron diffraction. In an electron diffraction camera the reflections that would be observed are those indicated by the black dots, i.e., $\{000\}$ $\{342\}$ $\{543\}$ and $\{612\}$. (b) With a crystal of finite thickness, but infinite lateral dimensions, diffraction zones are formed. All of the reflections shown by the black dots would be visible on the diffraction pattern. On an electron diffraction pattern taken in an electron microscope only $\{hk0\}$ reflections, i.e., those in the central zone, would be observed. Electrons scattered at wider angles are intercepted by the walls and apertures of the microscope. (c) When the thin crystal is tilted the center of the diffraction pattern shifts also. For small degrees of tilt only the reflections in the $L=0$ zone would be present in the electron microscope, the center of intensity of the pattern being shifted from the center of symmetry of the pattern. (Pinsker (25).)

rods instead of points, the rods lying parallel to the normal to the lamella. The sphere of reflection can thus intersect more lattice positions, resulting in more reflections in the diffraction pattern.

Agar, Frank, and Keller have calculated (31) that for a 120-A thick polyethylene crystal, assuming a perfect lattice oriented normal to a highly collimated beam of 80 kv electrons, reflections of appreciable intensity should be obtained corresponding only to spacings greater than 2.5 Å. This includes only two reflections of the polyethylene unit cell. The fact that more than this are observed is evidence for the complex structure of polyethylene crystals that is discussed in Chapter II.

An elaborate nomenclature has been developed to describe the ordered arrangement of atoms in a crystalline material. The basic unit is the unit cell, a parallelepiped containing a certain grouping of atoms which by simple translation in three dimensions defines the crystal structure. Unit cells are classified by shape into seven systems (i.e., cubic, orthorhombic, hexagonal, etc.) whereas the arrangement of atoms within the unit cell is expressed in terms of its symmetry using the 230 space groups. This means of defining crystal structure was devised for lattices of materials composed of atoms or small molecules which behave as discrete units in packing into a lattice. In sharp contrast, polymer molecules are much too long to behave as individual units in packing into a lattice. Furthermore, because of the ability of the molecules to rotate about single carbon-carbon bonds, the molecule may be twisted into peculiar symmetries which cannot be described in terms of the space groups used so successfully for lattices of small molecules. Finally, a feature of many polymer structures is partial crystallization; lattice order may exist in only one or two dimensions. For these reasons it is necessary to redefine crystal structure in terms specifically suited for polymers, maintaining standard concepts and nomenclature whenever they are pertinent.

A simple and highly versatile method for defining polymer crystal structure has been adopted in the Plastics Department of E. I. du Pont de Nemours and Co., Inc. (29). The structure of polymers is described in terms of the following three categories.

1. The *configuration* of a molecule is defined by its chemical structure. The configuration can be changed only by breaking and reforming chemical bonds. Thus, isotactic polypropylene has a different configuration than either syndiotactic or atactic polypropylene.

2. The *conformation* of a molecule refers to the geometrical arrangement of atoms with respect to rotation about single bonds. Isotactic polybutene molecules, which have a helical nature, may have at least two different conformations in the crystalline state. The two conformations correspond to there being either 3 or 4 monomer units per turn of the helix. In the melt or solution, additional conformations are possible. Categories 1 and 2 apply whether or not the polymer is crystalline.

3. The *molecular packing* or *chain packing* refers to the lateral organization of the molecules in the crystal. Included in this category are many types of lattice disorders including such problems as paracrystallinity, point defects, and other imperfections.

Categories 2 and 3 are of primary interest here, the first category, molecular configuration, usually being known. It too, however, can be determined by x-ray diffraction. It is the latter two categories that may be referred to as crystal structure, replacing the concept of the unit cell. The desirability of using these two categories results in part from practical difficulties in determining polymer crystal structure from x-ray data. Polymer crystal structures usually are poorly developed in comparison with crystals of small molecules, resulting in a small number of x-ray diffraction data. A complete three-dimensional analysis and even Fourier projections of the molecules may thus be impossible. The principal limitation of the standard crystallographic techniques is that all details of the structure must be tackled at once. By separating crystal structure into the two categories it is possible, even with meager data, to learn many important details about the conformation and the packing.

Before discussing these two categories in greater detail, Clark describes (29) the x-ray diffraction technique by which most polymer crystal structures have been determined. Although polymer single crystals have been obtained, that being the primary subject of this book, they are too small to use with standard single crystal x-ray diffraction techniques. In the future it is believed that electron diffraction and microbeam x-ray techniques will prove valuable in determining polymer crystal structure. Frank, Keller, and O'Connor (32) have recently published the first polymer structure determination using electron diffraction patterns from single crystals, determining the unit cell of poly-4-methyl-pentene-1. In the past the crystal structure of a polymer has usually been determined from x-ray diffraction patterns of a drawn fiber. Within the fiber the crystalline regions are aligned with the molecular axes parallel to the fiber axes. In most fibers there are numerous crystalline regions in any cross section. The resulting x-ray fiber pattern is essentially identical to a rotation pattern from a single crystal, the rotation usually being about the molecular axes. Fiber patterns from those polymers whose morphology has been most completely determined and which thus serve as the basis for many of the descriptions in the following chapters, are shown in Figure I-5. These patterns were taken on a cylindrical film, the fiber axes being vertical. The reflections occur in rows known as layer lines, normal to the fiber axis. On a flat film the layer lines, as indicated previously, are hyperbolae.

The periodic placement of atoms in a polymer molecule in a crystal is expressed in terms of the "repeat distance" and "repeat unit." The repeat unit is the simplest arrangement of atoms which, by the operation of linear translation (no rotation) and duplication, will generate the structure of the extended molecule. It thus defines the conformation of the molecule in the crystal. The repeat unit usually contains one or more "chemical or configurational repeat units," the chemical repeat unit referring to the smallest unit which by translation, rotation, and duplication can generate the molecule. The axial length of the repeat unit is the repeat distance. The repeat unit and repeat distance are illustrated in Figure I-6 for polyethylene and polytetrafluoroethylene. The repeat unit of polyethylene contains two chemical repeat units (—CH₂— groups) whereas the repeat unit of poly-tetrafluoroethylene has 13 or 15 chemical repeat units (—CF₂— groups) depending on the temperature.

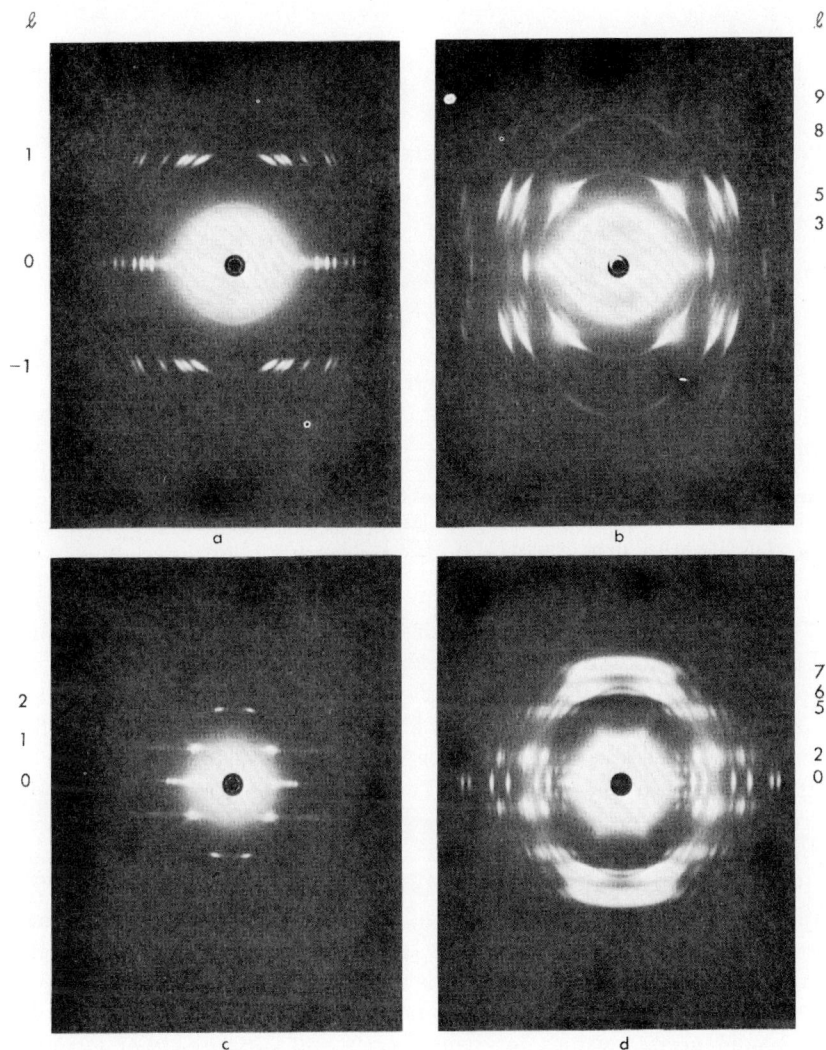

Fig. I-5. X-ray fiber patterns of (a) polyethylene, (b) polyoxymethylene, (c) poly-propylene, and (d) polytetrafluoroethylene (at 0°C). Ni filtered Cu Kα radiation was used. The layer lines are numbered. (Clark (29).)

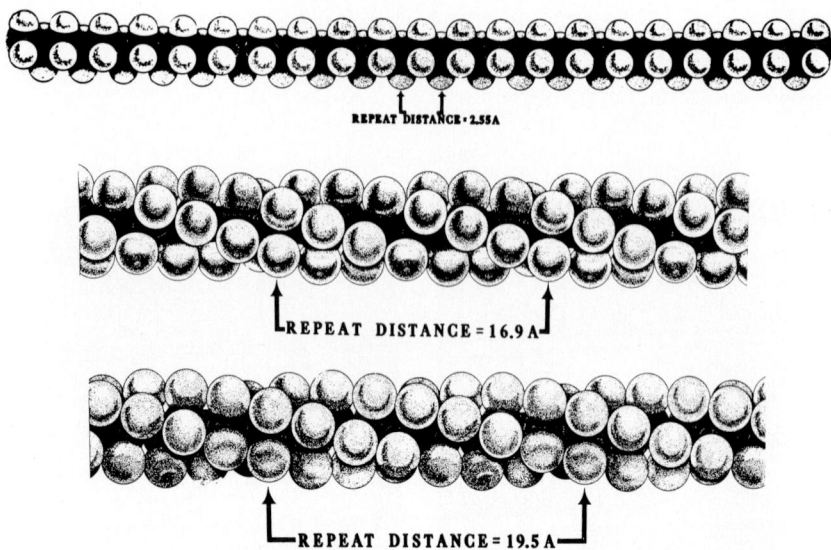

REPEAT DISTANCE = 2.55A

REPEAT DISTANCE = 16.9 A

REPEAT DISTANCE = 19.5 A

Fig. I-6. Molecular model of a portion of a molecule of (a) polyethylene and (b) and (c) polytetrafluoroethylene. The model in (b) refers to the conformation of the molecule below 19°C, there being 13 CF_2 groups per turn of the helix. As the temperature is raised librations occur. Between 19°C and 30°C a 15 CF_2 group per turn helix (c) is stable. Such a helix packs well into the lattice. Above 30°C the librations increase and no specific helix can be defined (11). (Clark (29).)

The determination of the repeat distance is based on the representation of the chain molecule as a one-dimensional lattice of points separated by the repeat distance. This linear lattice gives rise to the layer lines on the fiber pattern. Periodicities normal to the fiber axis result in the discrete reflections within each layer line. The greater the repeat distance, the closer together are the layer lines (Fig. I-5). The repeat distance may be determined directly from the position of the layer lines on a cylindrical film.

The determination of the repeat unit is considerably more difficult. Although the classical x-ray method is most accurate, the observed intensities being compared with those calculated from a proposed unit cell, there usually are insufficient data for this technique in polymer diffraction patterns. More often the conformation is reasoned out from a prior knowledge of the chemical structure (configuration) of the molecule, the repeat distance and normal values for bond angles and distances. For example, knowing that the repeat distance for polyethylene is 2.55 Å, that the chemical repeat unit is a —CH$_2$— group, that the C—C bond length is about

1.54 Å and that the C—C—C bond angle is approximately tetrahedral, the planar zigzag arrangement pictured in Figure I-6 is the only reasonable arrangement.

It is also possible to determine the repeat unit or conformation of the molecule by the "helical structure method" (29). Most polymer conformations can be described in terms of atoms regularly spaced along helices. By defining the radius and pitch of the helix and the separation of adjacent atoms along the axis, the position of all of the atoms on a given helix is defined. This method has been applied by Clark and co-workers (11, 29) to the determination of the conformation of polytetrafluoroethylene and polyoxymethylene molecules. Only the over-all intensity in each layer line, which depends on the order of Bessel function appearing in the equation for layer line diffraction from a helical structure, is needed to determine the conformation. For a helical structure the following selection rule is found to apply

$$\ell = tn + u\,m$$

where

ℓ is the layer line number

t is the number of turns per u atoms on the helix

n is the order of the Bessel function controlling the intensity of the layer and m is an integer, $+$, $-$, or 0.

The selection rule is solved for the proposed helix (u_t) to give the values of n corresponding to each ℓ. The lower the value of n the higher the intensity of the corresponding layer line. The application of this method to polytetrafluoroethylene is given in Table I-2. A comparison of predicted intensity with the observed pattern (Fig I-5) shows good agreement. The most intense layer lines on the observed patterns are those with $\ell = 0, 6,$

TABLE I-2
Selection Rule for 13_6 Helix (Polytetrafluoroethylene)

ℓ	n	ℓ	n
0	0	7	-1
1	-2	8	-3
2	-4	9	-5
3	$-6, +7$	10	$+6, -7$
4	$+5$	11	$+4$
5	$+3$	12	$+2$
6	$+1$	13	0

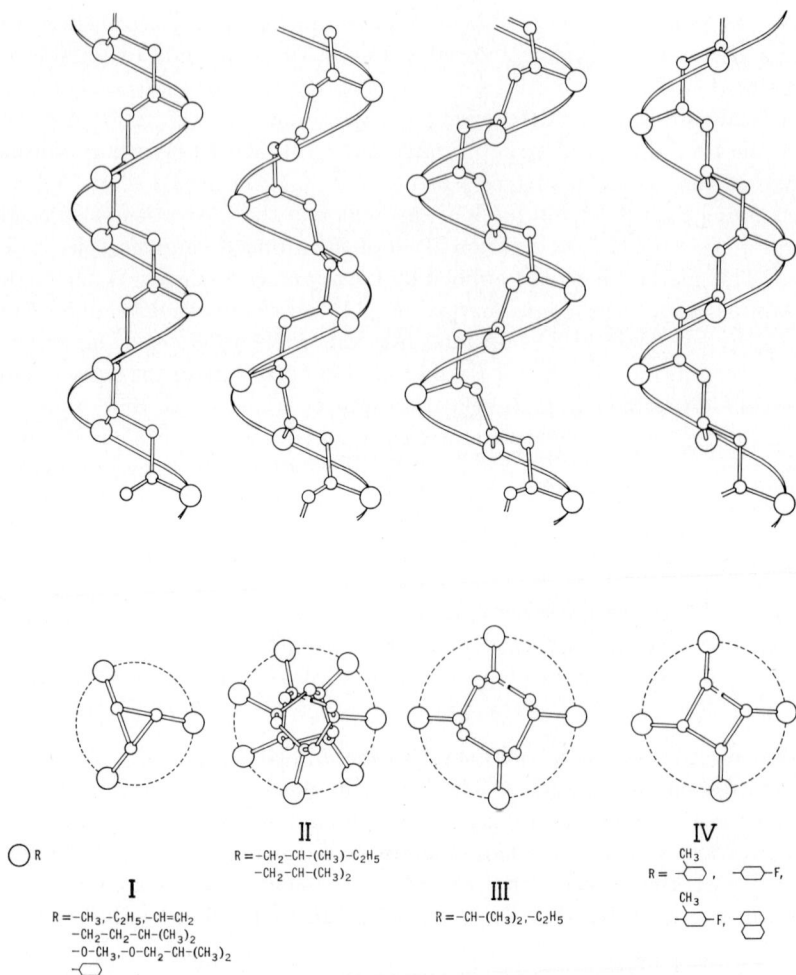

Fig. I-7. The conformation of a number of isotactic polymers can be represented by one of those helicoidal models.   The large balls, representing the branches, can be any of the groups listed below each of the models.   Note that polybutene (branch is C_2H_5) can and does exist in the form of both helices I and III.   (Natta and Corradini (33).)

and 7, intense higher index layer lines being at too large an angle to show on the film.

Figure I-7 from Natta and Corradini (33) shows the conformation for many of the newly polymerized isotactic polymers. The arrangement of the branches on a helix can be seen. In helical structure theory additional helices

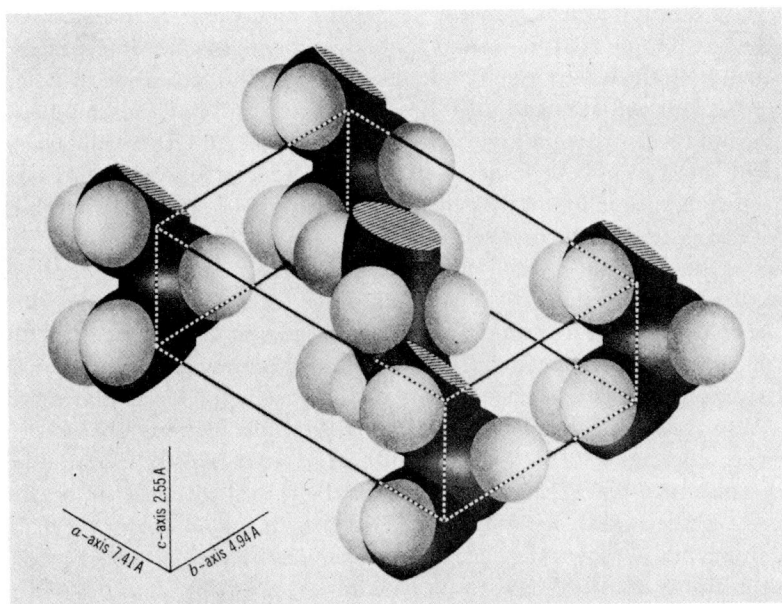

Fig. I-8. Unit cell of polyethylene. The unit cell contains two repeat units, one in the center and $1/4$ at each corner. The line drawing shows the orientation of the planar zigzag backbone of the atom with respect to the **a** and **b** axes. (Clark (29).)

with smaller radii, passing through the main chain atoms, are also considered.

The packing of the molecules is expressed most completely in terms of the unit cell and its contents. The unit cell has the shape of a parallelepiped with axes **a, b,** and **c,** angles α, β, and γ, and contains one or more repeat units. The unit cell of polyethylene is shown in Figure I-8. A complete solution of this type is not always possible. There may either be insufficient data, due to a poorly developed lattice, to permit its determination or else certain aspects of a unit cell may not exist due to special types of disorder. For example, if polyethylene molecules were given random linear translations along their axes, the concept of a unit cell would no longer apply. From x-ray data one could, however, still determine the molecular conformation and obtain certain details of the lateral packing. This is the situation with some polymers.

The dimensions of the unit cell of polymer crystals are determined from the position of Bragg reflection spots on layer lines of the fiber pattern. If the unit cell is orthogonal or hexagonal the procedure is straightforward whereas if the unit cell is skewed, i.e., either monoclinic or triclinic, the procedure is complicated, reflection spots overlapping on the fiber pattern. The reader is referred to a complete discussion of the technique of determining the unit cell axes and angles by Bunn (28). In brief, measurements of each spot on the pattern are converted into reciprocal lattice dimensions and then fitted to a three-dimensional lattice in reciprocal space. When a satisfactory lattice is found it is converted to a real or space lattice which has the dimensions of the unit cell.

Besides defining the crystal structure, one of the most useful quantities obtainable from the unit cell dimensions is the calculated crystalline density. In cases where a determination of all of the unit cell constants is not possible, a partial solution may still be obtained permitting calculation of the crystalline density. Measurements of the spacing of equatorial reflections (zero layer line) combined with the repeat distance may be used to determine a special cell, the "volume indicatrix" (29). This cell, containing one and only one repeat unit, is a right parallelepiped with axes $\mathbf{a}'$, $\mathbf{b}'$, and $\mathbf{c}$ where c is the repeat distance. a', b' and their included angle, γ, can be determined directly from the equatorial spots. The volume occupied by one repeat unit is thus $a'\,b'\,c \sin \gamma$.

The arrangement of repeat units in the unit cell, e.g., the orientation of the planar zigzag of the polyethylene molecule with respect to the $\mathbf{a}$ and $\mathbf{b}$ axes (Fig. I-8), is much more difficult to determine than the unit cell dimensions. Using a knowledge of chain conformation and unit cell dimensions, trial structures can often be deduced by fitting together scale molecular models. Confirmation of a trial structure is based on the agreement of observed and calculated intensities of the Bragg reflections using well-established crystallographic techniques.

The preceding discussion has been in terms of a perfect lattice; real crystals and particularly polymer crystals are not perfect. For simplicity, however, polymer structure is described whenever possible in terms of unit cells in a perfect lattice, disorder being treated as a modification of the idealized crystal structure. Polymer crystal imperfections may result from such factors as terminal groups (end of the molecules), branches and crossovers of the molecules, as well as modifications of the numerous types of defects (34) found in crystals composed of small molecules.

X-ray diffraction can give a great deal of qualitative and, to some extent, quantitative information about the imperfect lattice. Distortion of the lattice results in a broadening of the Bragg reflections; the greater the distortion the more diffuse are the reflections. A recent treatment of the scattering from an imperfect crystal is that of Beeman et al. (35). Unfortunately,

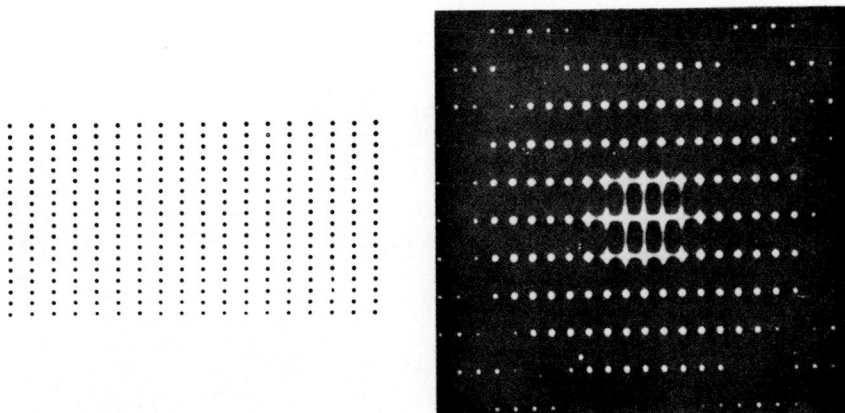

Fig. I-9a. This and the following parts (I-9b to I-9f) of this figure are lattice model and optical diffraction patterns for various types of distorted lattices (Hosemann, Bonart *et al.* (38,39)). Above is a model and diffraction pattern of a two-dimensional ideal crystal. The line width of all reflections, apart from collimation errors, depends only on the length of the edges of the diffracting structure. Their integral intensities are dependent, on the other hand, on the atomic form factor of the spots of the model.

broadening of the Bragg reflections may be due to either distortion of the lattice or a small crystal size. Although the dependence of the Bragg reflection broadening on the Bragg angle differs for the two cases, sufficient data are not yet available for most polymers to differentiate between the two causes. The broadening of the Bragg reflection at an angle θ due to crystal size is proportional to $\lambda/\cos \theta$ whereas that due to lattice distortion is proportional to $\tan \theta$. By using different wave lengths and comparing the reflections at various angles, Fischer has been able to show (36) that the diffuseness of Bragg reflections from polyethylene is due primarily to lattice distortion.

Theoretical quantitative treatments of lattice disorder and the effect of crystal size have been advanced by Hosemann (37–39) and Filipovich (40,41). Hosemann's treatment can be pictured most easily. He defines a paracrystalline lattice which may be considered as a perfect lattice distorted in various characteristic ways. Using statistics based on this model it is possible to draw conclusions about the degree of disorder in a structure from its diffraction pattern. Although numerous optical analogs have been prepared, the paracrystalline theory, as well as the related theory of Filipovich, has been applied in only a very limited manner to actual diffraction patterns. Figure I-9 shows various optical analogs of diffraction patterns

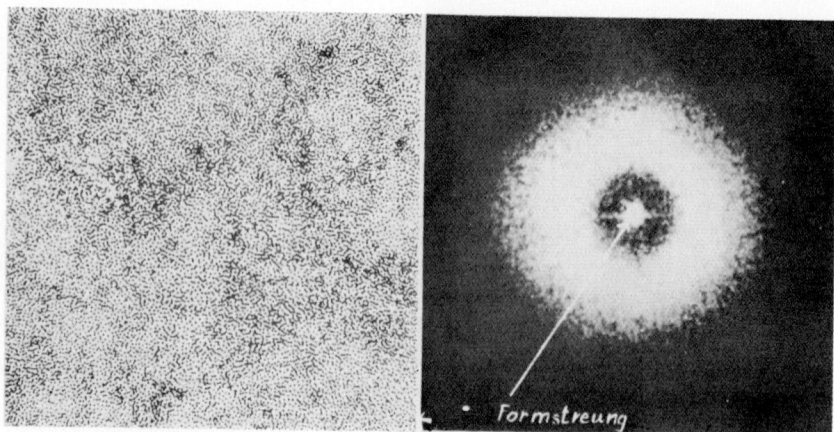

Fig. I-9b. Model and diffraction pattern of a two-dimensional liquid.   The integral breadth of the liquidlike maxima in the diffraction pattern is practically unaffected by the dimensions of the model.

which closely resemble x-ray diffraction patterns from polymers. The para-crystalline theory is applicable to a wide range of distortions, treating perfect lattices, liquids, and all degrees of distortion in between. It is believed that Hosemann's and/or Filipovich's treatments of disorder will play an important part in future descriptions of polymer crystal structure.

Although, as pointed out previously, the availability of polymer single crystals, combined with electron diffraction, should simplify the determination of polymer crystal structure, considerable experimental difficulties are present and must be recognized. Electron diffraction patterns from polymer crystals are most easily obtained in an electron microscope rather than in standard electron diffraction apparatus. The crystal can be located, observed, and photographed in the microscope. Under standard conditions of illumination in the microscope, however, the diffracting power of a polymer crystal lattice is almost instantly destroyed. The electrons in a polymer crystal interact inelastically with an incident electron, whereas they interact elastically with an x-ray photon. Although some polymers may discolor after several days of exposure to a standard Cu K_α x-ray beam, they do not significantly lose their diffracting power.

The destruction of the lattice in the electron beam appears to be primarily the result of radiation damage and crosslinking (42–44). Keller has estimated that every electron striking an atom in the target could be effective in disrupting the crystal lattice (44). The beam damage is proportional to the dose over a large range of intensities. Although one might expect that

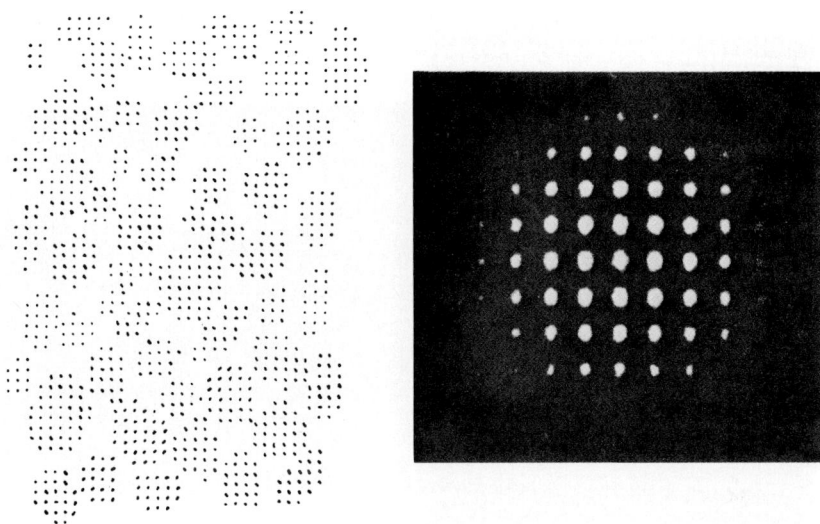

Fig. I-9c. Two dimensional model and diffraction pattern of a network of similarly oriented "crystallites." About 50 crystal reflections can be seen. Their widths are determined mainly by the mean "crystallite" diameter in the model as well as collimation errors. This should be compared with (a).

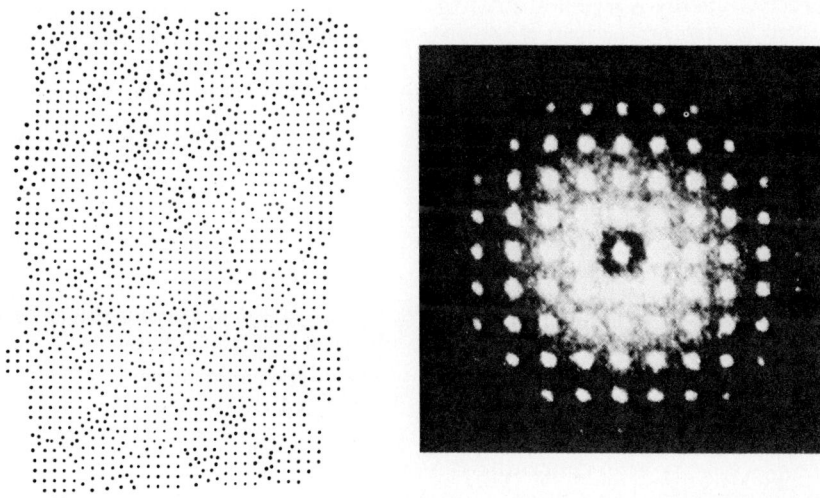

Fig. I-9d. Two-dimensional model and diffraction patterns of a two-phase system of similarly oriented "crystallites" (c) and an "amorphous" structure. The "amorphous" portion was added to the model in (c). As a result, a so-called "amorphous" halo appears in the diffraction pattern. The "crystalline" intensity is the same as in (c), except for a difference in total exposure.

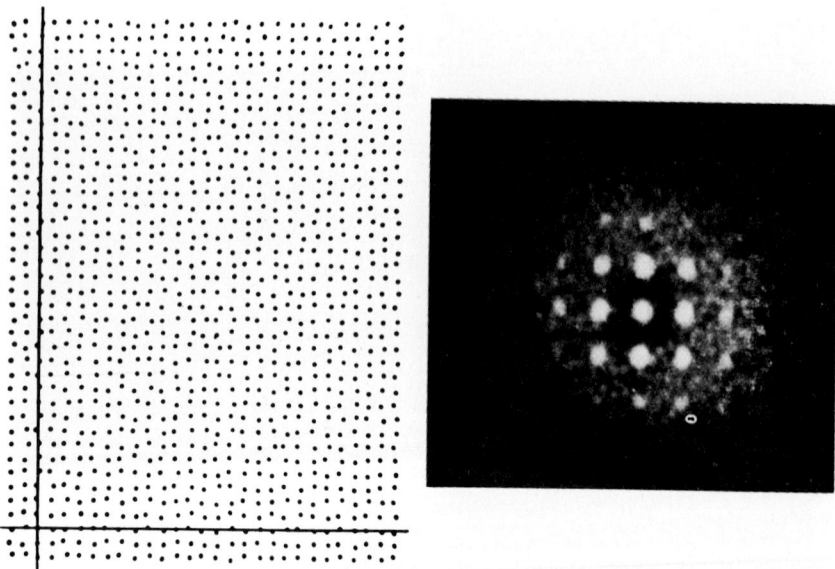

Fig. I-9e. Model and diffraction patterns of a frozen thermal distortion. The lattice lines drawn clearly show that long-range order persists despite substantial distortion. A relatively intense so-called "temperature halo" is seen; this halo is in no way related to an amorphous part of the model's structure, but rather is a measure of the degree of distortion. Because of relatively large distortions in the model (∼ 0.1 of a lattice spacing), only a few reflections appear, their widths being independent of the distortion or the diffraction angle.

heating by the beam could cause melting, he points out (43) that under normal diffraction conditions gutta-percha and paraffins, both of which melt near 50°C, are less affected by the beam than polyethylene which melts near 130°C. Likewise the lattice of polytetrafluoroethylene is more rapidly destroyed than that of polyethylene. The use of cooling stages reduces the lattice damage, but this is probably a result of a lowering of the reactivity and mobility of the radicals produced by the irradiation.

The lattice deterioration effects can be reduced by greatly reducing the incident electron intensity to a level so low that the crystals are barely visible even at low magnifications. As will be discussed in the next section, an electron microscope having a double condenser is especially valuable if it is, as is almost always the case, desirable to photograph the crystal from which the diffraction pattern is obtained. The area illuminated can be kept to the order of one square micron so that the remainder of the sample is not affected during focusing. In order to observe the crystal, even in the

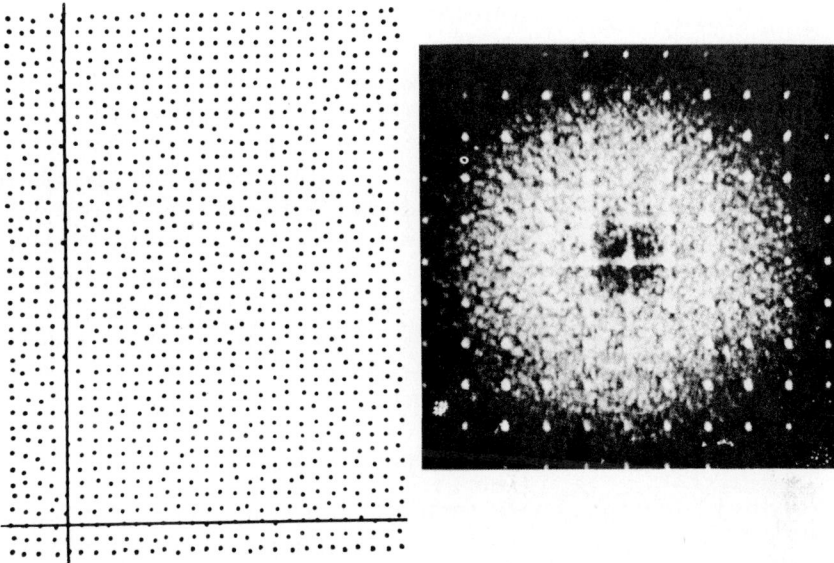

Fig. I-9f. Model and diffraction pattern with statistically, i.e., randomly, distributed defects. The lattice lines drawn show that about 50% of the lattice units are located at the ideal lattice points while the remaining units are somewhat displaced. The width of the reflections is independent of the distortion. The diffuse scattering, however, represents a measure of the number and size of the defects. Here, too, however, it cannot be said that the diffuse scatter is associated with an "amorphous" phase.

best present equipment, the operator requires a considerable period of dark adaptation and, even then, focusing is possible only at low magnifications. Burbank (45), operating under these conditions, has been able to maintain the diffracting power of a polyethylene crystal for periods up to 6 minutes, at an intensity level requiring a 45 sec exposure for a diffraction pattern whereas under normal illumination conditions only a fraction of a second would be required. He indicates that the best results (for diffraction and microscopy) were obtained using 80 kv beam rather than higher or lower values. This agrees with our experience. Higher energies would be expected to cause less deterioration but better contrast is obtained on the micrographs using the lower voltage. Burbank finds (45) that as polyethylene crystals deteriorate in the electron beam the axial ratio (a/b) increases from 1.5 to 1.65. This ratio can thus form an empirical yardstick for the determination of the extent of lattice deterioration. Effects such as this need to be recognized and measured if electron diffraction patterns are to be utilized in unit cell determinations.

A further problem in electron diffraction is the determination of the scattering angle and wave length of the incident beam. This is most easily done by comparing the diffraction pattern with the pattern from a known sample. Frank, Keller, and O'Connor (32), in determining the crystal structure of poly-4-methyl-pentene-1, accomplished this by evaporating a thin layer of a material with a known crystal structure, $TlCl_2$, onto the sample and obtaining both diffraction patterns simultaneously. A further complication arises from the fact that the camera constant, the effective distance from sample to film, varies radially on the film. Although small angles are involved, a measured spacing of $2x$ cm on a film does not correspond to twice the scattering angle of a spacing of x cm. Since the change in the camera constant is smooth, a curve computed from the spacings of the known sample can be used to determine the unknown spacings.

b. Polymer Crystal Structure

On the following pages are given the configuration or chemical structure, the conformation and the unit cell for most of the polymers discussed in the following chapters. In addition, in Appendix I (p. 513) are listed all crystallographic data for polymers which have been reported. This extremely useful compilation of data by Miller and Nielsen is a revised version of that reported in reference 46. In a number of polymers, particularly many of the new isotactic polymers, two or more conformations and unit cells have been observed. As would be expected, the morphology of the polymer differs, depending on which lattice it crystallizes in. Although the molecular packing and conformation can also change after crystallization, a crystalline phase change occurring, the effect of this change on the morphology, if any, has not yet been studied.

(1) Polyacrylic Acid

Configuration (3 chemical repeat units)

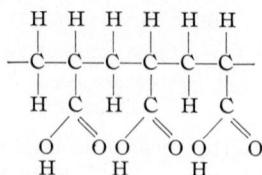

The conformation and molecular packing have not been reported. Various salts of polyacrylic acid can be formed by replacing the hydrogen on the OH group with a different ion.

(2) Polyacrylonitrile

Configuration (3 chemical repeat units)

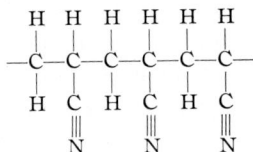

```
    H  H  H  H  H  H
    |  |  |  |  |  |
  —C——C——C——C——C——C—
    |  |  |  |  |  |
    H  C  H  C  H  C
       |||    |||    |||
       N      N      N
```

Conformation and Molecular Packing

The conformation and molecular packing are unknown. In fact, poly-acrylonitrile is generally considered noncrystalline. Even oriented poly-acrylonitrile in the form of fibers yields only a poor diffraction pattern in which only a few equatorial reflections are reasonably well defined.

As shown in Chapter II, it has been possible to grow single crystals of polyacrylonitrile from dilute solution (47). Diffraction patterns from these crystals suggest an orthorhombic type unit cell with **a** and **b** axis repeat distances of 10.6 and 5.8 A. The **c**-axis repeat distance can, of course, not be determined by electron diffraction from the lamellae since the beam is parallel to the **c** axis. It has been reported by Stefani *et al.* (48) from x-ray measurements as being 5.1 A. The values listed above for *a* and *b* are somewhat different than those of Stefani but are believed to be more reliable.

(3) Polyamides (Nylons 66, 610, 6, 7, and 8)

Polyamides crystallize with a number of different types of molecular packing. In many cases the position of the reflections on x-ray fiber patterns varies gradually with the thermal treatment of the fiber. As a result, reports by various authors do not agree. The availability of single crystals may help to clear up some of the present confusion in the crystal structure of the polyamides.

(A) NYLON 66. Nylon 66 crystallizes in at least 3 different forms, α, β, and γ, as well as in mixture of these forms. Furthermore, the lattice spacings vary considerably with the temperature of measurement and even, to some extent, with the original temperature of crystallization. The α and β forms are stable at room temperature whereas the γ, or high temperature form, is not.

Configuration (1 chemical repeat unit)

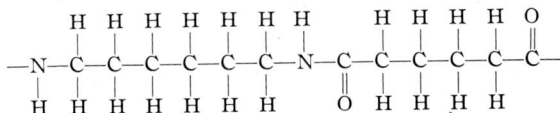

```
     H  H  H  H  H  H  H      H  H  H  H  O
     |  |  |  |  |  |  |      |  |  |  |  ||
  —N——C——C——C——C——C——C——N——C——C——C——C——C—
     |  |  |  |  |  |  |      ||  |  |  |
     H  H  H  H  H  H  H      O  H  H  H  H
```

α and β Forms

Conformation

The molecule has essentially a planar zigzag conformation. The model shown in Figure I-10 depicts the conformation that might be expected in the case of small thermal vibrations and should be compared with that of nylon 6 in Figure I-13.

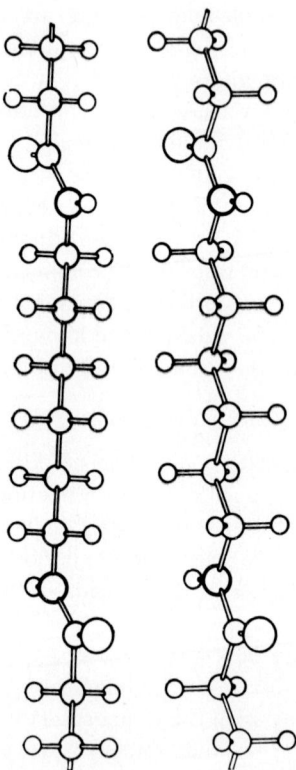

Fig. I-10. Conformation of a nylon 66 molecule distorted slightly by thermal vibration (Natta and Corradini (33)).

Molecular Packing

Hydrogen bonds between the N—H and C=O groups on neighboring chains results in the formation of sheets of molecular segments in which the

intermolecular forces are considerably larger than the forces between segments in neighboring sheets. The triclinic unit cell of the α form is shown
in Figure I-11.

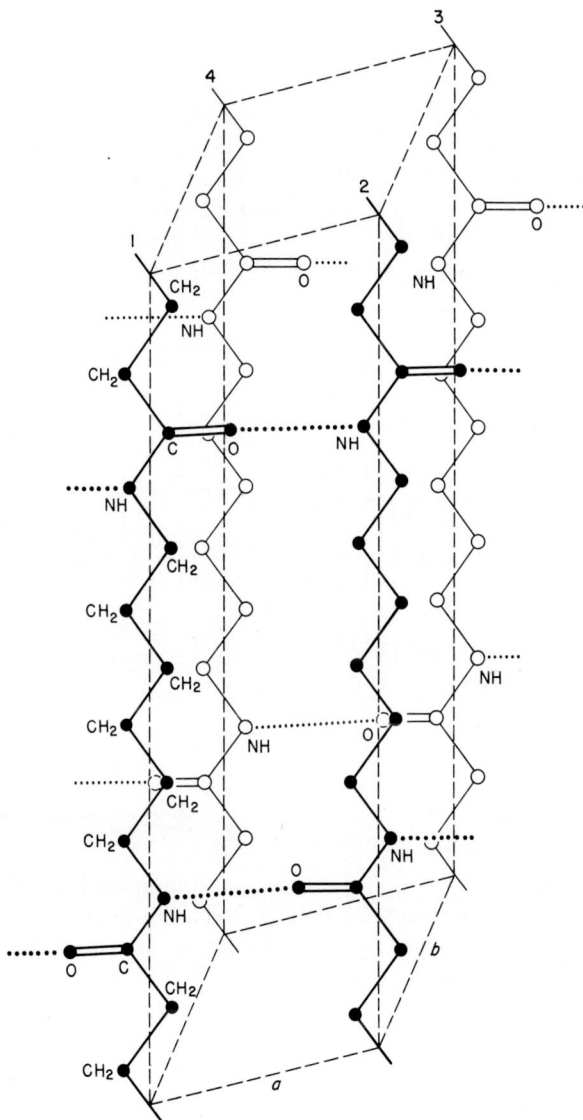

Fig. I-11. Packing of nylon 66 molecules in the triclinic unit cell (Bunn and Garner (49)).

The unit cell constants of the α form are (49)

$$
\begin{aligned}
a &= 4.9 \text{ A} & \alpha &= 48\tfrac{1}{2}° \\
b &= 5.4 \text{ A} & \beta &= 77° \\
c &= 17.2 \text{ A} & \gamma &= 63\tfrac{1}{2}°
\end{aligned}
$$

As can be seen in Figure I-11, not only are the molecular segments shifted parallel to their axis by $1/14$ of the repeat distance (i.e., one zig of the zigzag) in each hydrogen bonded sheet of molecules but each sheet is shifted by $3/14$ of the repeat distance. All segments have the same direction, thus raising a question about how folding and the required reversal of direction can take place. In the α form all of the sheets are shifted in the same direction whereas in the β form the sheets are staggered. As shown in Figure I-12

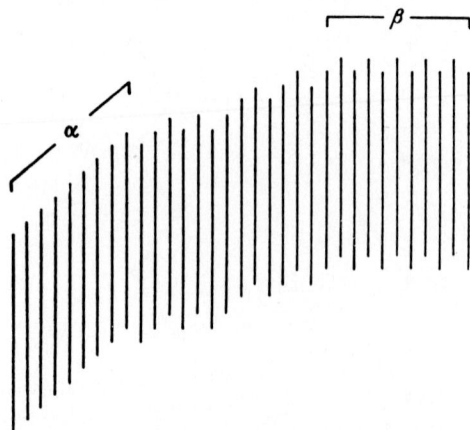

Fig. I-12. α, β, and intermediate forms of nylon 66. The lines represent hydrogen-bonded sheets of molecules seen edgewise. (Bunn and Garner (49).)

intermediate forms are possible and are often found. Keller and Maradudin (50) have suggested that the α-β polymorphism in polyamides may result from a misinterpretation of x-ray peaks really caused by the small size of the crystals.

The cell constants of the triclinic unit cell of the β form are given as (49)

$$
\begin{aligned}
a &= 4.9 \text{ A} & \alpha &= 90° \\
b &= 8.0 \text{ A} & \beta &= 77° \\
c &= 17.2 \text{ A} & \gamma &= 67°
\end{aligned}
$$

The β form unit cell contains two repeat units whereas the α form unit cell has only one.

γ Form

Conformation

The conformation is unknown.

Molecular Packing

It has been suggested that the molecules have a pseudohexagonal packing and that the hydrogen bonds are randomly oriented. This is difficult to reconcile, however, with infrared results which indicate that all hydrogen bonds are made. In fact the γ form may not be a distinct unit cell since the α form spacings gradually approach those of the γ form as a sample is heated.

(B) NYLON 610

Configuration (1 chemical repeat unit)

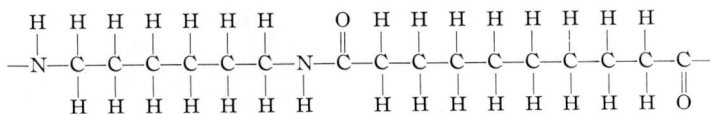

$$
\begin{array}{c}
\text{H} \quad \text{H} \quad \text{H} \quad \text{H} \quad \text{H} \quad \text{H} \quad \text{H} \qquad \text{O} \quad \text{H} \quad \text{H} \quad \text{H} \quad \text{H} \quad \text{H} \quad \text{H} \quad \text{H} \quad \text{H} \\
| \quad\ | \quad\ | \quad\ | \quad\ | \quad\ | \quad\ | \qquad\ \| \quad\ | \quad\ | \quad\ | \quad\ | \quad\ | \quad\ | \quad\ | \quad\ | \\
-\text{N}-\text{C}-\text{C}-\text{C}-\text{C}-\text{C}-\text{C}-\text{N}-\text{C}-\text{C}-\text{C}-\text{C}-\text{C}-\text{C}-\text{C}-\text{C}-\text{C}- \\
| \quad\ | \quad\ | \quad\ | \quad\ | \quad\ | \quad\ | \qquad\ | \quad\ | \quad\ | \quad\ | \quad\ | \quad\ | \quad\ | \quad\ | \quad\ \| \\
\text{H} \quad \text{H} \quad \text{H} \quad \text{H} \quad \text{H} \quad \text{H} \quad \text{H} \qquad \text{H} \quad \text{H} \quad \text{H} \quad \text{H} \quad \text{H} \quad \text{H} \quad \text{H} \quad \text{H} \quad \text{O}
\end{array}
$$

Conformation and Molecular Packing

The conformation and molecular packing of nylon 610 is similar to that of nylon 66. Three forms, α, β, and γ, as well as mixtures are found. Because of the longer repeat unit the unit cells are larger. Cell constants of the triclinic α and β form unit cells are (49)

α Form

$$
\begin{aligned}
a &= 4.95 \text{ A} & \alpha &= 49° \\
b &= 5.4 \text{ A} & \beta &= 76^{1}/_{2}° \\
c &= 22.4 \text{ A} & \gamma &= 63^{1}/_{2}°
\end{aligned}
$$

β Form

$$
\begin{aligned}
a &= 4.9 \text{ A} & \alpha &= 90° \\
b &= 8.0 \text{ A} & \beta &= 77° \\
c &= 22.4 \text{ A} & \gamma &= 67°
\end{aligned}
$$

The γ form has a pseudohexagonal unit cell supposedly similar to that of nylon 66.

(C) NYLON 6. Nylon 6 can crystallize with at least 2, and probably 3 or more, different unit cells. It appears that the description of its molecular packing is even less well defined than is the case for nylon 66. We shall describe below the more significant features of the various forms.

Configuration (2 chemical repeat units)

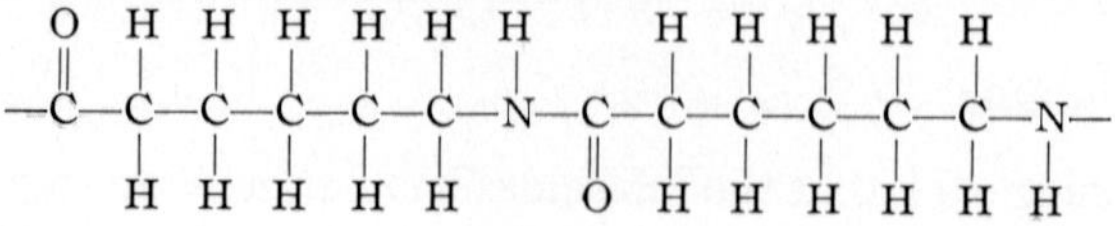

γ and β Forms

Conformation

In the α and β forms the molecule is assumed to have a planar zigzag form. The model shown in Figure I-13, from Natta and Corradini (33),

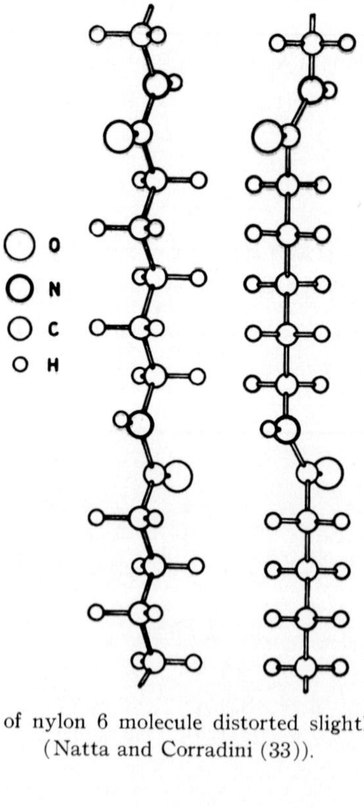

Fig. I-13. Conformation of nylon 6 molecule distorted slightly by thermal vibration (Natta and Corradini (33)).

depicts the conformation that might be expected in the case of thermal vibration. Single chemical repeat units are shifted out of the plane of the
molecule. Comparing this with the similar model for nylon 66 in Figure
I-10, Natta and Corradini suggest that the lower melting point of nylon 6
(215°C) in comparison with nylon 66 (265°C) is a result of the greater
disruption of the lattice during thermal vibrations of given amplitude.

Molecular Packing

The α form has a monoclinic unit cell with the following cell constants
(51)

$$a = \quad 9.56 \text{ A}$$
$$b = 17.24 \text{ A}$$
$$c = \quad 8.01 \text{ A}$$
$$\beta = 67.5°$$

This form is relatively well defined. Note that due to a quirk of crystallographic notation the **b** axis is parallel to the molecular axes. In this form
the adjacent segments in the hydrogen bonded sheets are inverted, permitting all hydrogen bonds to be easily made (Fig. I-14). As in the case of

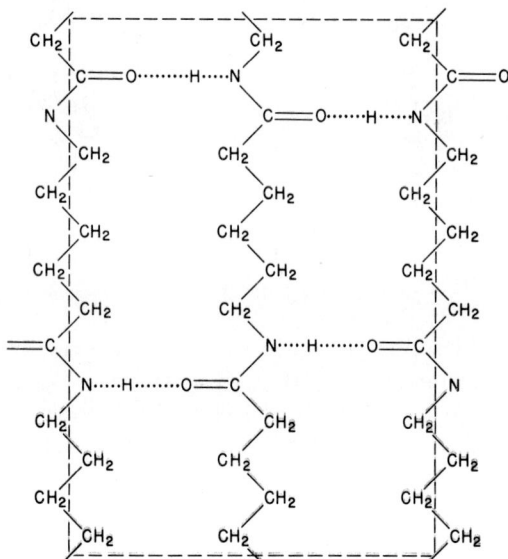

Fig. I-14   Molecular packing of nylon 6 in the hydrogen-bonded sheets of the α form unit
cell (Holmes, Bunn, and Smith (51)).

the β form of nylon 66, neighboring sheets are alternately sheared by $\pm^3/_{14}$ of the repeat distance.

The molecular packing of the β form is poorly defined. Holmes, Bunn, and Smith (51) suggest that the diffraction patterns can best be explained in terms of no shearing of the hydrogen bonded sheets, i.e., all C=O and N-H groups are on the same level in the crystal. This form occurs, sometimes, simultaneously with the α form in drawn filaments. It may be transformed to the α form by immersion in boiling water for 5 or 6 hours.

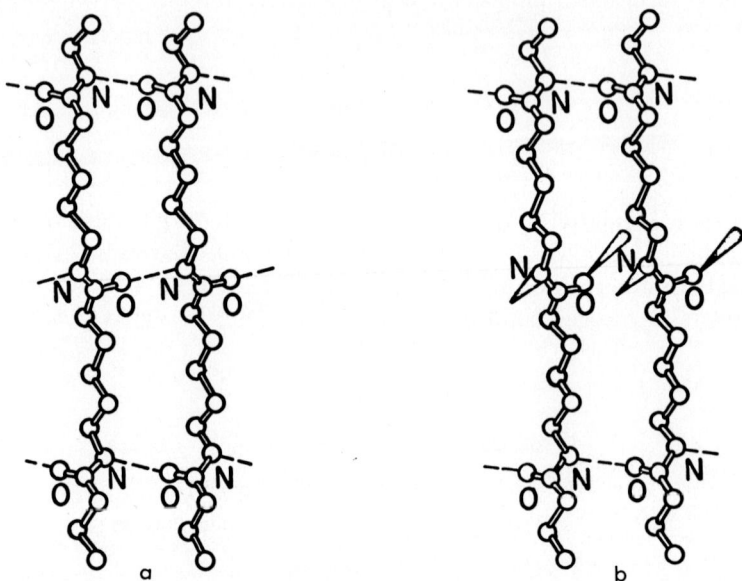

Fig. I-15. Two possible types of hydrogen bonding in nylon 6 which would lead to a pseudohexagonal unit cell. In both types all molecules have the same direction. In *b* the cones represent H bonds out of the plane of the figure. (Vogelsong (52).)

Vogelsong (52) describes a γ form which, he believes, is the same as the β form of Holmes, Bunn, and Smith (51). He suggests (52) that the adjacent molecular segments may all be in the same direction rather than inverted and that as a result the hydrogen bonds are not normal to the chain axis, as in the α and β forms, but are tilted. Of the two possible ways of making the hydrogen bonds, in Figure I-15, Vogelsong prefers that in which the hydrogen bonds along a given chain are made alternately in $\{100\}$ and $\{010\}$ planes. This would appear to require a helical-type molecule. Either of these two forms would have pseudohexagonal symmetry and may be related

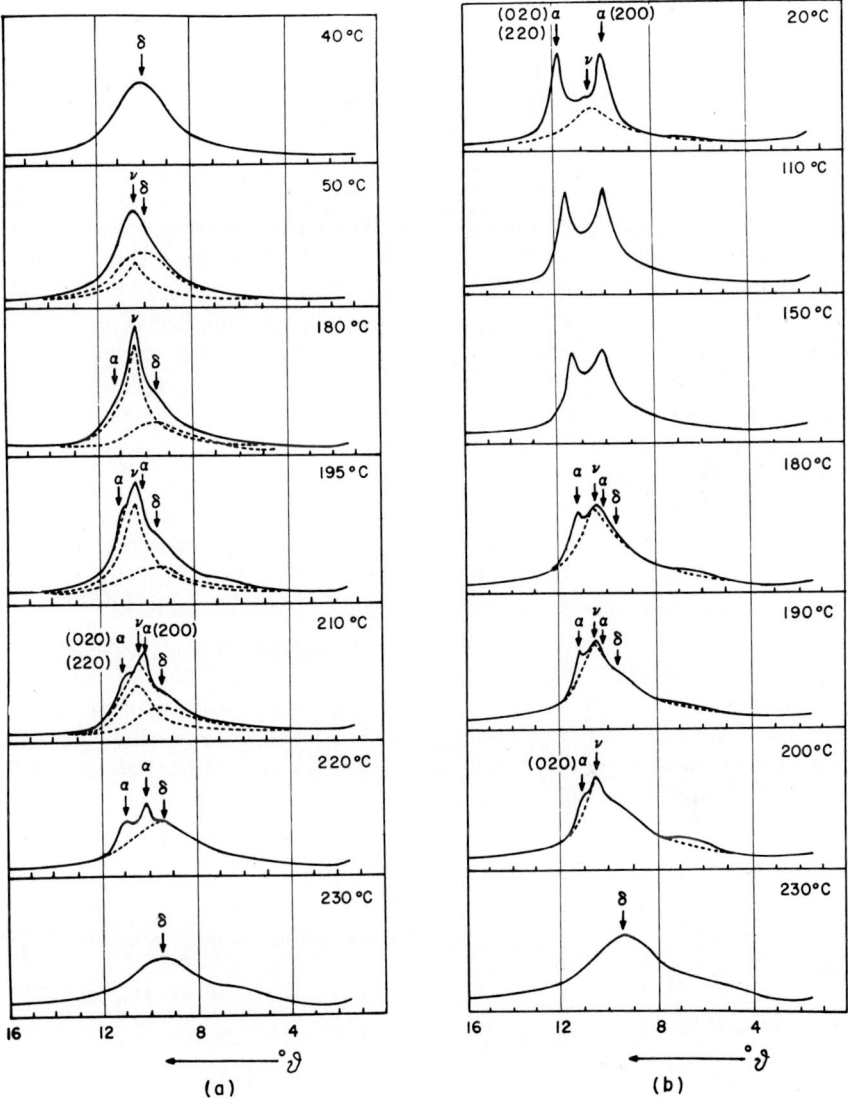

Fig. I-16. X-ray diffraction patterns from nylon 6. (a) Injection molded (quenched) and then annealed patterns measured at the annealing temperature. (b) Sample cooled from the melt, patterns measured at temperature indicated. (Hendus, Schmeider, Schnell, and Wolf (54).)

to what has been termed a high temperature form (53). The dimensions of the rhombohedral unit cell of the γ form are given as (52)

$$a = 4.79 \text{ A}$$
$$c = 16.7 \text{ A}$$
$$\gamma = 60°$$

Hendus *et al.* (54) have followed the changes in the x-ray, infrared and internal friction spectrum in considerable detail. In Figure I-16 are shown the changes in the x-ray diffraction patterns with temperature for a sample quenched from the melt and then annealed and for a sample cooled directly from the melt. The α crystal structure is the same as that described above. In the mesomorphic or liquid crystal γ structure the molecules are in hexagonal close packing and the hydrogen bond statistically distributed. This probably is the same as the β or γ forms above. The δ form is amorphous. In the melt all the hydrogen bonds, at least on a time-average basis are formed, whereas in the quenched specimen only a portion are formed. The reader is directed to the original paper (54) for a rather complete description of the changes in infrared spectra and mechanical properties that accompany the changes in structure observed by x-ray diffraction.

Single crystal diffraction patterns have been obtained (55) which do not appear to be directly interpretable in terms of any of the above forms but instead to a form which is obtained through the treatment of the polymer with an iodine-potassium iodide solution (56,87,88). In this "γ" form the hydrogen bonds are also presumed to be made between parallel rather than antiparallel segments.

(D) NYLON 7

Configuration (2 chemical repeat units)

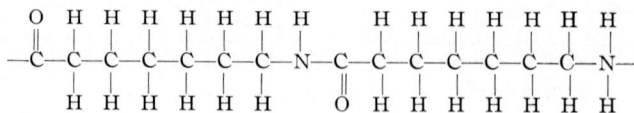

Conformation

The conformation presumably is similar to that of the other nylons, i.e., a planar zigzag.

Molecular Packing

The triclinic unit cell has the following dimensions (56):

$$a = 4.9 \text{ A} \qquad \alpha = 49°$$
$$b = 5.4 \text{ A} \qquad \beta = 77°$$
$$c = 9.85 \text{ A} \qquad \gamma = 63°$$

(E) NYLON 8.   The comments made about nylon 6 apply to this polymer also.

Configuration (1 chemical repeat unit)

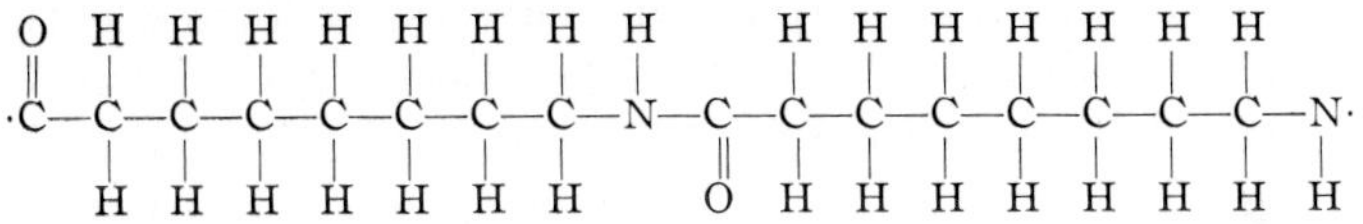

Molecular Packing

Vogelsong (52) lists the following unit cell dimensions:

Monoclinic	Rhombohedral
$a = 9.8$ A	$a = 4.79$ A
$b = 22.4$ A	$c = 21.7$ A
$c = 8.3$ A	$\gamma = 60°$
$\beta = 65°$	

(4) Polybutene

Isotactic polybutene molecules can exist in either of two conformations, a threefold or a fourfold helix (57). The threefold helix is more stable, fourfold helices changing to threefold when the material is drawn or pressed or, slowly, on standing at room temperature. Zannetti *et al.* (58) report that three distinctly different diffraction patterns can be obtained from isotactic polybutene. Form I, corresponding to the threefold helix, is described below. The unit cells of Forms II (corresponding to the fourfold helix) and III have not been determined. Fiber patterns for these forms are not available. Form III is prepared by precipitation from organic solvents, such as benzene, at room temperature through the addition of methanol. Whereas Form II is converted to Form I on standing at room temperature, Form III appears stable. Apparently its infrared spectra, and therefore the type of helix, has not been determined.

Configuration (3 chemical repeat units)

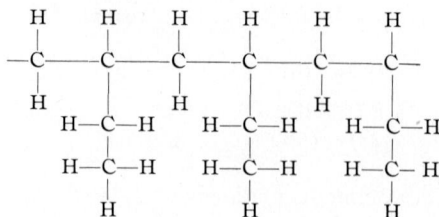

Conformation. The conformation of the threefold helix as seen from the side and end is shown in Figure I-17. The smaller circles represent H atoms.

Packing of the molecules. The unit cell of Form I is rhombohedral, the repeat distance, c, being 6.5 A with the **a**-axis identity period, referred to hexagonal coordinates, 17.7 A long. The projection of the structure on the (001) plane is shown in Figure I-18.

(5) *Polycarbonate (of Bisphenol A)*

Configuration (1 chemical repeat unit)

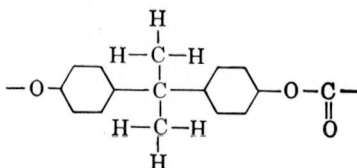

Conformation

The molecule has an extended planar zigzag conformation which can most easily be visualized from the diagram of the unit cell in Figure I-19:

Molecular Packing

Polycarbonate has an orthorhombic unit cell with cell constants (59):

$$a = 11.9 \text{ A}$$
$$b = 10.1 \text{ A}$$
$$c = 21.5 \text{ A}$$

The packing of the molecules into the unit cell is shown in Figure I-19.

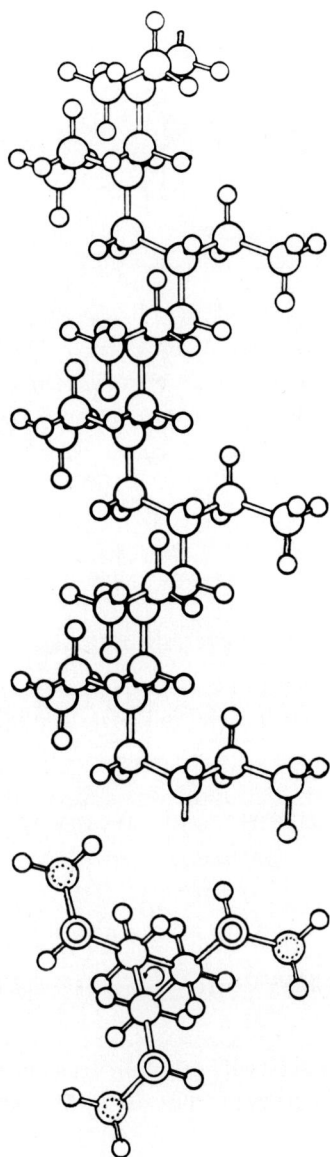

Fig. I-17. Conformation of polybutene in the three-fold helix form. Side and end views. (Natta, Corradini, and Bassi (57).)

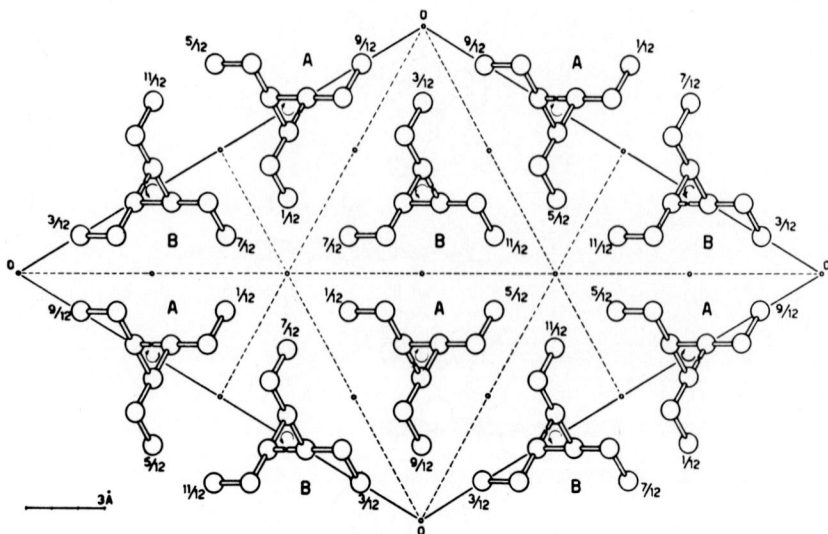

Fig. I-18. Projection of polybutene molecules onto the {001} plane of the unit cell (Natta, Corradini, and Bassi (57)).

(6) Cellulosic Materials

Cellulose can be acetylated to various degrees, its 3 hydroxyl groups being susceptible. Cellulose itself has the following configuration (2 chemical repeat units)

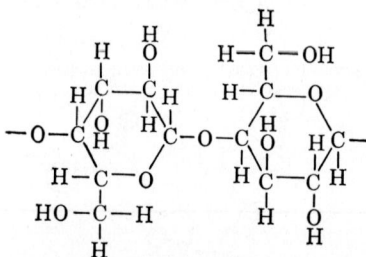

As in the case of cellulose itself, cellulose triacetate (all hydroxyl groups replaced by O—C—CH$_3$ groups) can exist in at least two crystal modifica-
$$\overset{\parallel}{O}$$
tions. The one of interest for later discussions is cellulose triacetate II. Dulmage (60) suggests that the unit cell is pseudoorthorhombic with cell constants:

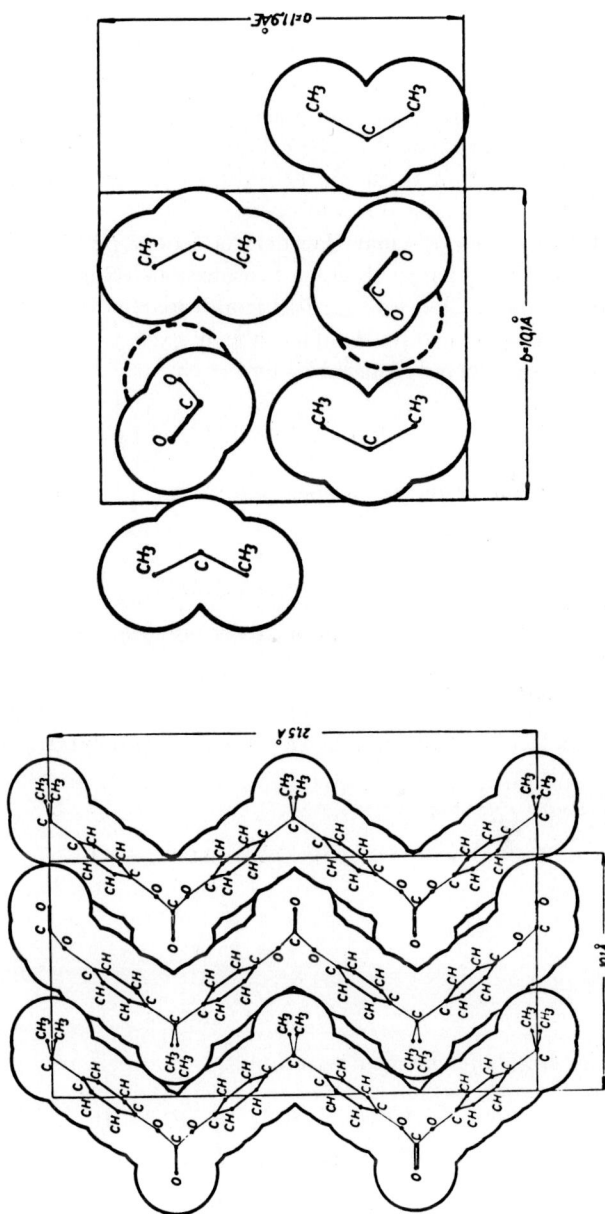

Fig. I-19. Molecular packing in unit cell of a polycarbonate (Prietzschk (59)).

$$a = 24.5 \text{ A}$$
$$b = 11.56$$
$$c = 10.43$$

The molecules are paired in the cell and, having a direction, the pairs are opposite in direction at the center and the corners. The repeat distance corresponds to two chemical repeat units.

Xylans are cellulosic materials derived from a number of different plants. The one of interest for subsequent discussions (Chap. II) is poly-β-D(1–4) anhydroxylose, which was derived from esparto grass. Its configuration is similar to that shown for cellulose above, except that the $CH_2 OH$ groups are not present, being replaced by hydrogen.

(7) Polychlorotrifluoroethylene

Configuration (3 chemical repeat units)

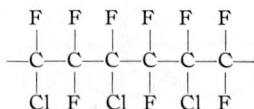

Conformation

The molecule is helical, somewhat resembling polytetrafluoroethylene.

Molecular Packing

The unit cell is hexagonal with cell constants (61)

$$a = 6.5 \text{ A}$$
$$c = 35 \text{ A}$$

(8) Polydioxolane

Configuration (2 chemical repeat units)

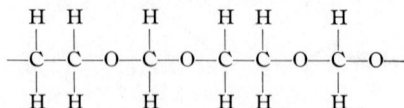

The conformation and unit cell are not known.

(9) Polyesters

The unit cell constants for a large number of polyesters have been reported in the literature and are listed in the compilation of Miller and Nielsen in Appendix I (p. 513). We shall consider only those mentioned later in the text.

(A) POLYETHYLENE ADIPATE

Configuration (1 chemical repeat unit)

```
       H  H       O  H  H  H  H  O
       |  |       ||  |  |  |  |  ||
  —O—C—C—O—C—C—C—C—C—C—
       |  |          |  |  |  |
       H  H          H  H  H  H
```

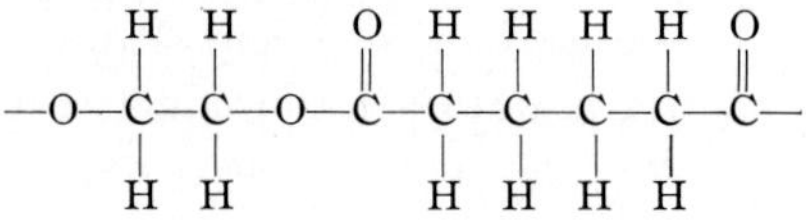

The conformation is indicated as being close to a planar zigzag. The repeat distance is 11.7A (62). The unit cell was described as either orthorhombic,

$$a = 21.8 \text{ A}$$
$$b = 33.3 \text{ A}$$

or monoclinic,

$$a = 25.7 \text{ A}$$
$$b = 30.7 \text{ A}$$
$$\beta = 103.8°$$

A recent paper (89) describes the unit cell as being monoclinic with one chemical repeat unit per repeat distance and two chains per unit cell. The cell constants are given as

$$a = 5.47 \text{ A}$$
$$b = 7.23 \text{ A}$$
$$c = 11.72 \text{ A}$$
$$\beta = 103.5°$$

The authors also found evidence for two other unit cells.

(B) POLYETHYLENE AZELATE

Configuration (1 chemical repeat unit)

```
       H  H       O  H  H  H  H  H  H  O
       |  |       ||  |  |  |  |  |  |  ||
  —O—C—C—O—C—C—C—C—C—C—C—C—
       |  |          |  |  |  |  |  |
       H  H          H  H  H  H  H  H
```

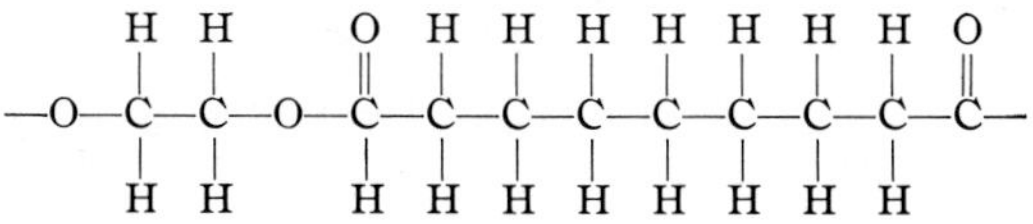

The repeat distance is 31.2 A (62). The other values and remarks made for polyethylene adipate apply to this polymer as well.

(C) POLYETHYLENE TEREPHTHALATE. Polyethylene terephthalate is of considerable interest since uncrystallized portions of a sample can be readily quenched at any time before or during crystallization to a glassy state and maintained in this state at room temperature. The glass transition temperature is 70°C. This permits one to halt crystallization when spherulites are only partially grown. Furthermore, when heated above the glass transition temperature, further crystallization can occur. If the entire sample is in the glassy state, crystallization is initiated and lamellar spherulites develop at temperatures well below the melting point (see Chapters IV and V).

Configuration (2 chemical repeat units)

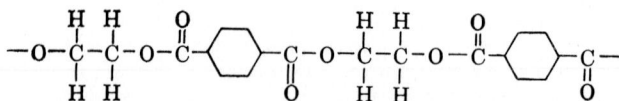

Conformation. The molecule has a nearly planar zigzag conformation with the benzene ring in the plane of the zigzag (Fig. I-20).

Molecular Packing. The unit cell is triclinic with cell constants (63).

$$
\begin{array}{llll}
a & = & 4.56 \text{ A} & \alpha - 98.5° \\
b & = & 5.94 \text{ A} & \beta = 118° \\
c & = & 10.75 \text{ A} & \gamma = 112°
\end{array}
$$

As can be seen in Figure I-20 the benzene rings make a slight angle with the axis of the molecule. The arrangement of the molecules in the crystal is shown in Figure I-21.

(D) POLYHEXAMETHYLENE TEREPHTHALATE

Configuration (1 chemical repeat unit)

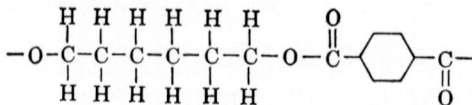

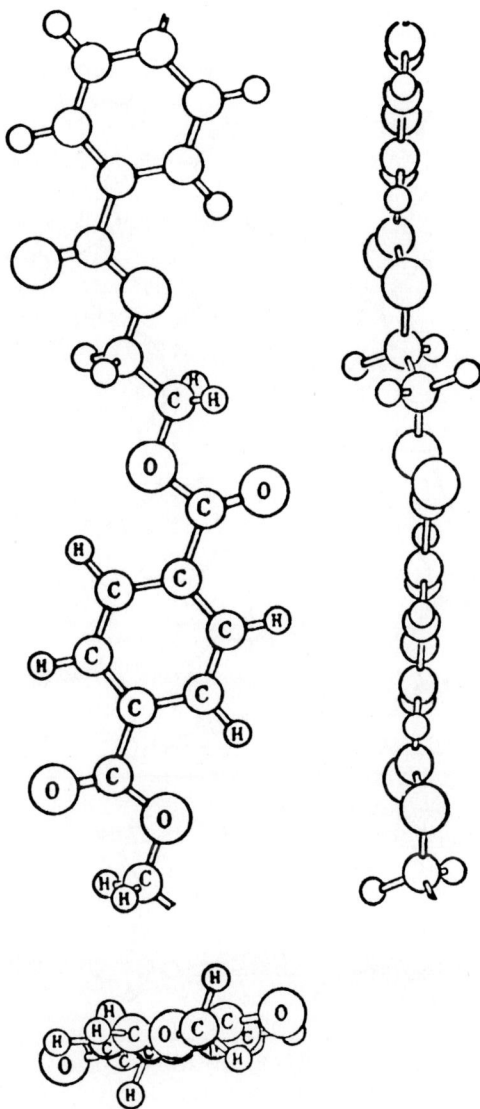

Fig. I-20. Molecular conformation of polyethylene terephthalate (Daubeney, Bunn, and Brown (63)).

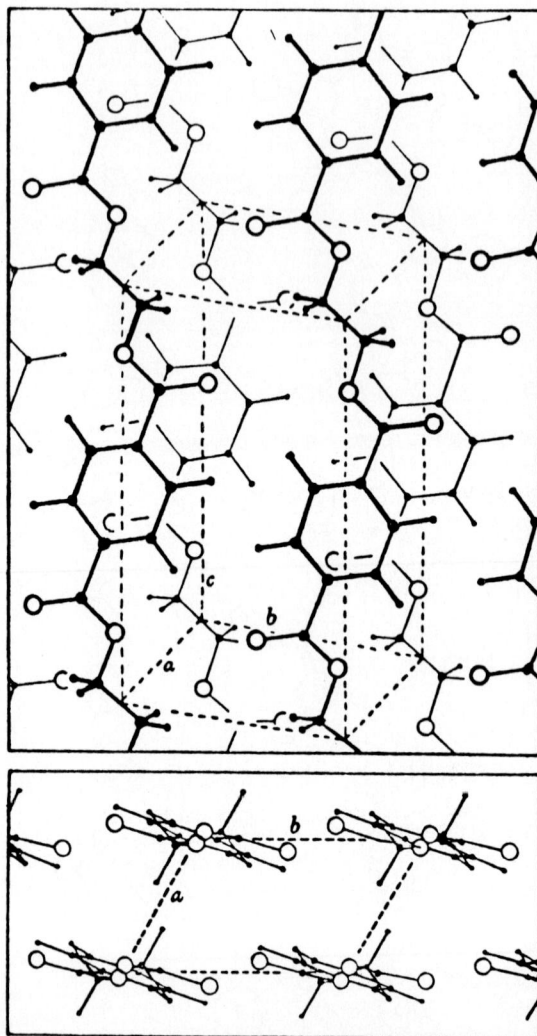

Fig. I-21. Molecular packing of polyethylene terephthalate (Daubeney, Bunn, and Brown (63)).

Molecular Packing

The triclinic unit cell has the following dimensions (64):

$$a = 4.57 \text{ A} \qquad \alpha = 105.5°$$
$$b = 6.10 \text{ A} \qquad \beta = 98.5°$$
$$c = 15.40 \text{ A} \qquad \gamma = 114.5°$$

(E) Polypentaglycol Terephthalate

Configuration (1 chemical repeat unit)

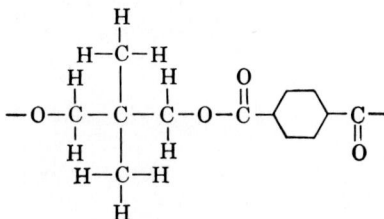

The conformation and molecular packing has not been reported.

(F) Polyesters of 1,10 Decanedial and Various Diacids

Configuration, Conformation, and Molecular Packing

These polymers are formed by the condensation reaction between 1,10 decanedial

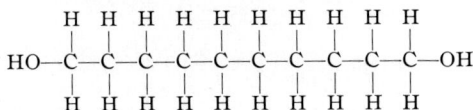

and various hydrocarbon diacids such as azelaic

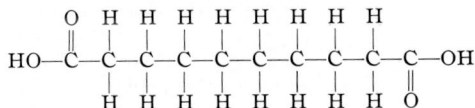

They form linear polymers with

groups as linkages.

The repeat period and unit cell as well as the long period observed in molten samples crystallized from the melt is given in Table I-3. The unit cell parameters are similar to those given by Fuller and Frosch (65). Similar polymers can be formed with other hydrocarbon dials.

TABLE I-3
Long Periods of Melt Crystallized Polyesters (66)

Sample	Length of Chemical Unit, A	Long periods		Subcell parameters. A		
		A	No. of diffraction orders	a	b	c[a]
Polyester of 1.10 decanedial and:						
Oxalic acid	17.5	104	2	6.90	5.24	(?)
Succinic acid	20.0	142	3	7.30	5.04	2.47– 2.51
Glutaric acid	21.2	162	3	7.36	4.96	2.46
Adipic acid	22.5	182	4	7.40	5.01	2.48
Azelaic acid	26.2	208	4	7.45	4.98	2.48
Sebacic acid	27.5	180	4	7.41	4.96	2.47

[a] Distance between C atoms along the backbone.

(10) Polyethylene

Configuration (6 chemical repeat units)

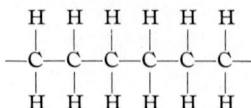

Conformation

The conformation of the molecule, a planar zigzag, is shown in Figure I-6.

Molecular Packing

A drawing of the orthorhombic unit cell is shown in Figure I-8. Unit cell constants, which vary slightly with temperature and the degree of branching, have been determined by Swan (67,68) for a Marlex* 50 polyethylene over a wide temperature range. Fitting his data to a polynomial

$$d = d_0 + d, T + d_2 T^2 + d_3 T^3 + d_4 T^4$$

where $T = °C$, he finds the following results:

	d_0	$d_1 \cdot 10^2$	$d_2 \cdot 10^4$	$d_3 \cdot 10^6$	$d_4 \cdot 10^8$
a	7.3681	0.1427	0.0229	0.230	0.0103
b	4.9350	0.0258	0.0090	0.0065	
c	2.5473				

* Trademark of Phillips Petroleum Co. for its polyethylene resins.

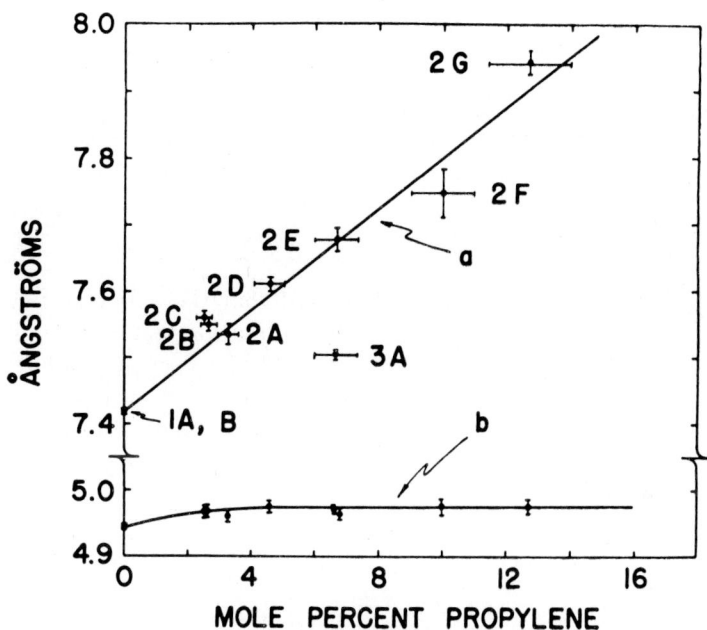

Fig. I-22. Variation of lattice dimensions of ethylene-propylene copolymers as a function of the amount of propylene (CH_3 branches). Sample 3A is believed to be a block copolymer whereas in the other samples the branches are believed to be randomly arranged. (Swan (68).)

Thus, as Swan and numerous other investigators have reported, the a spacing increases rapidly with temperature whereas the b spacing is nearly constant and the c spacing is constant.

At $30°C$, $a = 7.414$ and $b = 4.942$. Bunn, in his initial determination of the polyethylene lattice, reported that the planar zigzag of the molecule's backbone makes an angle of about $41°$ with the b axis (69).

Swan also determined the change in the unit cell as methyl branches are added to the polyethylene molecule (Fig. I-22). The branches result in an expansion of the lattice (the a spacing increases) suggesting that they are incorporated in the lattice. Swan likewise finds that even larger branches can be included in the lattice, also resulting in a slight lattice expansion (Table I-4). The branches are produced by copolymerizing ethylene with other olefins.

X-ray diffraction patterns from polyethylene which has been severely deformed by repeated drawing or rolling are found (90) to have a number of reflections which do not correspond to the orthorhombic unit cell de-

TABLE I-4
Lattice Spacing as a Function of Branch Size (68)

Branch	Comonomer content, mole %	a, A	b, A ·
C_2H_5	3.6	7.521	4.964
C_3H_7	3.1	7.474	4.969
C_4H_9	3.4	7.467	4.965
C_5N_{11}	1.4	7.479	4.956

scribed above. Frank *et al.* have suggested that the new reflections may arise from a regular or irregular sequence of stacking faults (91). Tanaka, Seto, and Hara have recently (92) obtained 15 reflections (not from the orthorhombic unit cell) in a sample which had been drawn, annealed, and then compressed normal to the chain axes. They report a monoclinic unit cell in which the planar zigzags all have the same orientation, normal to the **a** axis. They suggest the following unit cell constants:

$$a = 8.09$$
$$b = 2.53 \text{ (chain axis)}$$
$$c = 4.79$$
$$\beta = 107.9°$$

The calculated density is 0.997g/cc, i.e. slightly less than for the orthorhombic unit cell.

(11) Polyethylene Oxide

Configuration (2 chemical repeat units)

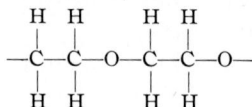

$$\begin{array}{cccc} H & H & H & H \\ | & | & | & | \\ -C{-}C{-}O{-}C{-}C{-}O{-} \\ | & | & | & | \\ H & H & H & H \end{array}$$

Conformation

The conformation has not been reported. The molecule is probably helical.

Molecular Packing

The unit cell dimensions have been reported as (70)

$$a = 9.5 \text{ A}$$
$$b = 19.5 \text{ A}$$
$$c = 12.0 \text{ A}$$
$$\beta = 101°$$

(12) Polyisoprene (Rubber and Gutta-Percha)

Isotactic polyisoprene, having double bonds in the molecular back-bone, can have two different configurations, either *cis* or *trans*. Although the chemical repeat units are arranged in the same order the two resulting polymers, rubber and gutta-percha, differ considerably.

Configurations

Rubber (*cis*-polyisoprene) (2 chemical repeat units)

```
                 H
                 |
           H—C—H  H
             |    |
           H C=C  H  H
            \ /   |  |            H
           —C     C—C             |
            |     | \             C—
            H    H↑H  C=C        / |
                     |   \      H
                H—C—H H
                     |
                     H
```

Gutta-percha (*trans*-polyisoprene) (2 chemical repeat units)

```
                 H
                 |
            —C      H
             |      |
             H  C=C  H  H
              \ /   |  |       H
    H—C—H      C—C  |           |
     |         | \  C=C        H
     H        H↑H  /  \
                  H—C—H    C—
                     |     |
                     H     H
```

The two polymers differ in the relative positions of the methyl groups and the main chain bonds. Since only the main chain carbon atoms on either side of the arrows can rotate about the single bonds, the two polymers cannot be converted into each other.

(A) RUBBER

Conformation

The molecule has a ribbonlike form. The exact conformation differs from one author to another (33,71,72). The form depicted by Natta and Corradini (33) is shown in the model of the unit cell in Figure I-23.

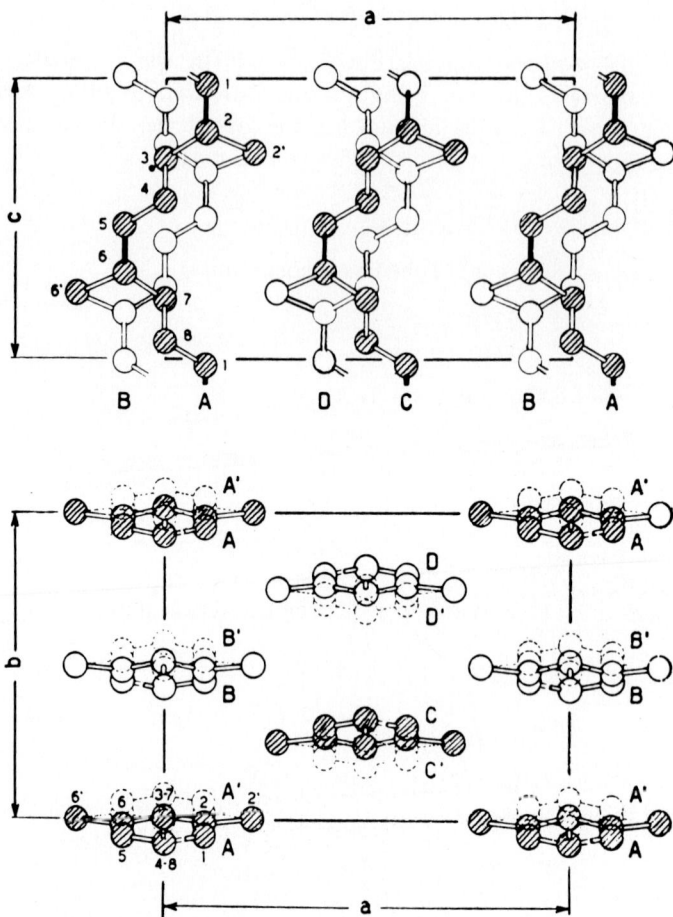

Fig. I-23. Molecular packing of rubber molecules in the unit cell
(Natta and Corradini (33)).

Molecular Packing

The unit cell is orthorhombic with cell constants of (71)

$$a = 12.46 \text{ A}$$
$$b = 8.89 \text{ A}$$
$$c = 8.10 \text{ A}$$

The packing of the molecules into this cell is shown in Figure I-23.
Only the carbon atoms are shown.

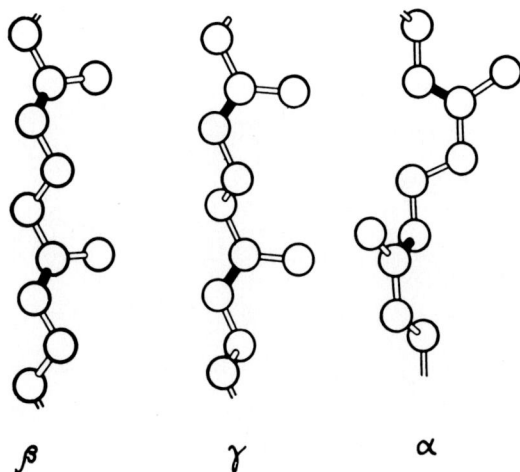

Fig. I-24. Conformation of the gutta-percha molecules in the three different unit cells (Natta and Corradini (33)).

As can be seen, adjacent molecules in the {100} planes are inverted—compare A and B. The dotted molecules indicate another possible position which cannot be distinguished by x-ray techniques from that shown.

(B) GUTTA-PERCHA

Conformation

The gutta-percha molecule can assume three different conformations, all of which are interconvertible by controlling the crystallization conditions. Models of the three conformations are shown below (Fig. I-24). All apparently have a helical structure.

Molecular Packing

The β-form has been studied most closely (71). It has an orthorhombic unit cell with constants

$$a = 7.78 \text{ A}$$
$$b = 11.78 \text{ A}$$
$$c = 4.72 \text{ A}$$

Only the repeat distances, 8.9 and 9.4 A are given for the α form and γ forms, respectively.

The β form is interesting since Bunn indicates (71) that all of the helices within a given crystal have the same hand. A similar feature has been suggested for polyoxymethylene (29).

(13) Poly-4-methyl-pentene-1

Configuration (3 chemical repeat units)

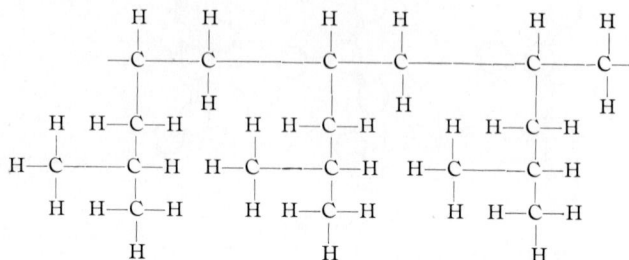

Conformation

The conformation corresponds to type II shown in Figure I-7, seven chemical repeat units occurring in two turns of the helix.

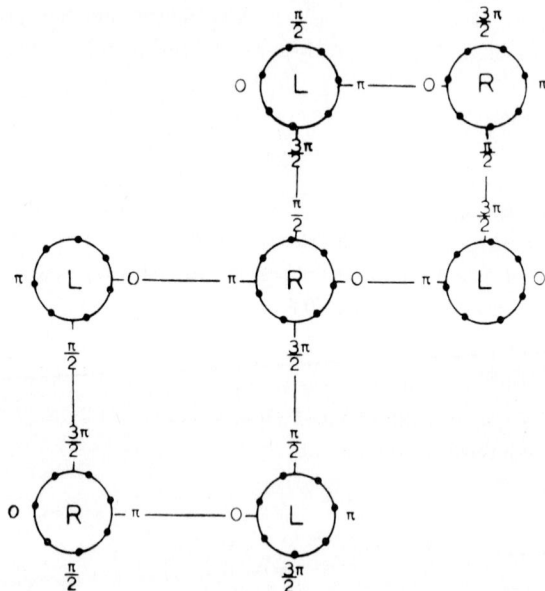

Fig. I-25. Molecular packing of poly-4-methyl pentene-1. The spots on the rings refer to the positions of the backbone C atoms. The sense of helix twist is also indicated. Only a portion of one unit cell is shown. (Frank, Keller, and O'Connor (32).)

Molecular Packing

Using electron diffraction patterns from single crystals, Frank, Keller, and O'Connor (32) have partially defined the unit cell. It is tetragonal with cell constants of

$$a = b = 18.66 \text{ A}$$
$$c = 13.80 \text{ A}$$

Presumably the helices, the side chains producing a helical ridge and groove along the side of the molecule, interlock. The packing of the chains is believed to be that shown in Figure I-25, adjacent helices having opposite hand.

(14) Polyoxymethylene

Configuration (3 chemical repeat units)

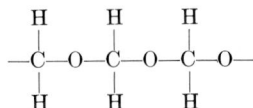

Conformation

The backbone atoms lie on a 9_5 helix as depicted below (Fig. I-26) (73).

Molecular Packing

Unit cell is hexagonal with the following constants (73)

$$a = 4.45 \text{ A}$$
$$c = 17.3$$

As can be observed in the model shown in Figure I-27 the helical molecules interlock tightly. The model and the diffraction patterns require that all of the helical molecules in a given crystal, or large (by x-ray standards) portion thereof, be of a single hand (29).

(15) Polypropylene

Isotactic polypropylene can exist in two crystal modifications, the molecule apparently having the same conformation in both forms. The monoclinic form (75) is the more usual form. An apparently hexagonal form (76) is also seen at times (see Chapters III and IV). If polypropylene is rapidly

Fig. I-26. Conformation of polyoxymethylene (Clark (74)).

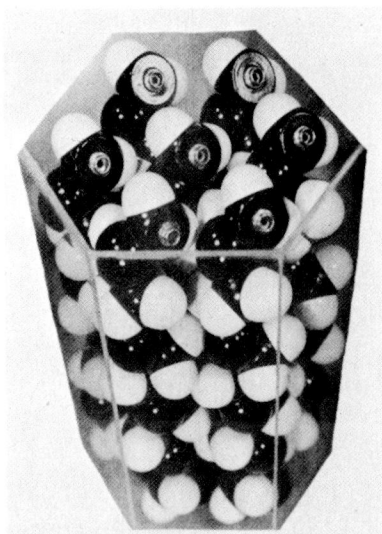

Fig. I-27. Model of packing of molecules in the unit cell of polyoxymethylene (Clark (74)).

quenched still a third type of diffraction pattern is obtained, consisting of 2 very broad peaks. The molecule again has the same conformation but disorder exists in the packing of chains (75). This material reverts to the monoclinic form when heated at temperatures above about 60°C.

Configuration (3 chemical repeat units)

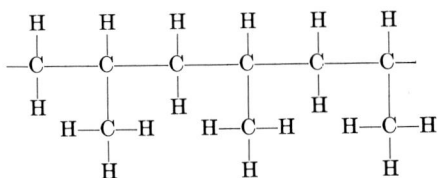

```
     H     H     H     H     H     H
     |     |     |     |     |     |
  —C —— C —— C —— C —— C —— C—
     |     |     |     |     |     |
     H  H—C—H  H  H—C—H  H  H—C—H
            |           |           |
            H           H           H
```

Conformation

The conformation of the molecule viewed from the side and end is shown in Figure I-28. The smaller balls represent hydrogen.

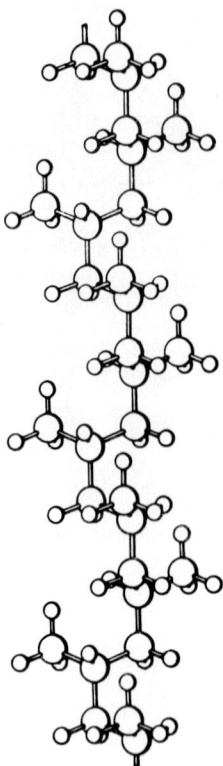

Fig. I-28. Conformation of the polypropylene molecule (Natta and Corradini (75)).

Molecular Packing

Monoclinic Form

The unit cell constants are (75)

$$a = (6.65 \pm 0.05) \text{ A}$$
$$b = (20.96 \pm 0.15) \text{ A}$$
$$c = (6.50 \pm 0.04) \text{ A}$$
$$\beta = 99°20' \pm 1°$$

The chain axes thus make an angle of about 10° with the normal to the base plane.

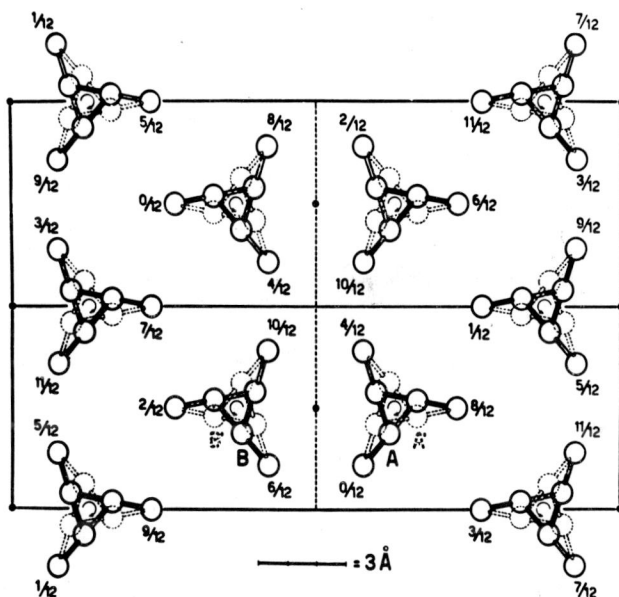

Fig. I-29. Projection of the monoclinic unit cell of polypropylene onto the {001} plane (Natta and Corradini (75)).

The projection of molecular packing onto (001) is shown in Figure I-29. The molecules may have the conformation indicated either by the dotted lines or the full lines. It is not known whether the actual crystal consists of a statistically determined distribution of the conformations or small blocks containing all one form. The numbers refer to the elevation of the C atoms (no H atoms are shown) above the basal plane, the arrows to the direction of rotation of the helix. Since the sense of rotation of a helix is not affected when it is turned upside down, the bridging of two segments by a fold may be restricted to certain planes, a feature which would greatly affect the morphology.

Hexagonal Form

Possible hexagonal unit cells have been suggested by Keith, Padden, Walter, and Wyckoff (76) and by Addink and Beintema (77). No fiber patterns of this modification alone are available.

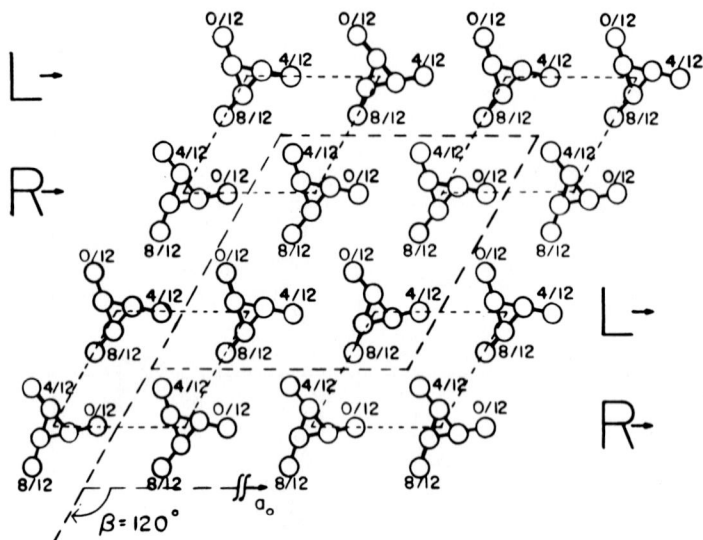

Fig. I-30. Proposed packing of polypropylene molecules in the "hexagonal" unit cell. The hand of the helices in each row is indicated. (Keith and Padden (76).)

Keith *et al.* suggest (76) the unit cell shown in Figure I-30, with cell constants of

$$a = b = 12.74 \text{ A}$$
$$c = 6.35 \text{ A}$$

The helical molecules in alternate rows are of opposite hand. The repeat distance is slightly less than that found in the monoclinic unit cell. Addink and Beintema (77), on the other hand, suggest a different unit cell, having nearly the same spacings, but one in which all of the molecules are of the same hand. The hexagonal form crystallizes from the melt as well formed lamellae (see Chapter III). The presence of folded molecules in the lamellae may be related to the development of this form. The folding within the lamellae does not require a change in the hand of a helical molecule at the folds.

Syndiotactic polypropylene has been characterized by Natta (78).

Configuration (4 chemical repeat units)

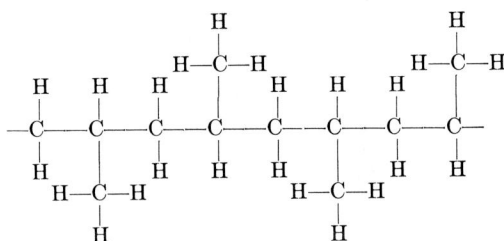

Molecular Packing

Natta reports (78) an orthorhombic unit cell with the following dimensions:

$$a = 14.50 \text{ A}$$
$$b = 5.81 \text{ A}$$
$$c = 7.3 \text{ A}$$

(*16*) *Polystyrene*

Configuration (3 chemical repeat units)

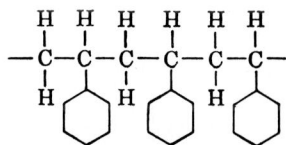

Conformation

The molecule has the type I (Fig. I-7) helical conformation, the plane of the benzene ring being at an angle at about 108° with the axis of the helix (79) (Fig. I-31).

Molecular Packing

As in the case of many of the other helical polymers the packing is complicated by the fact that the side chains can have two opposite inclina-

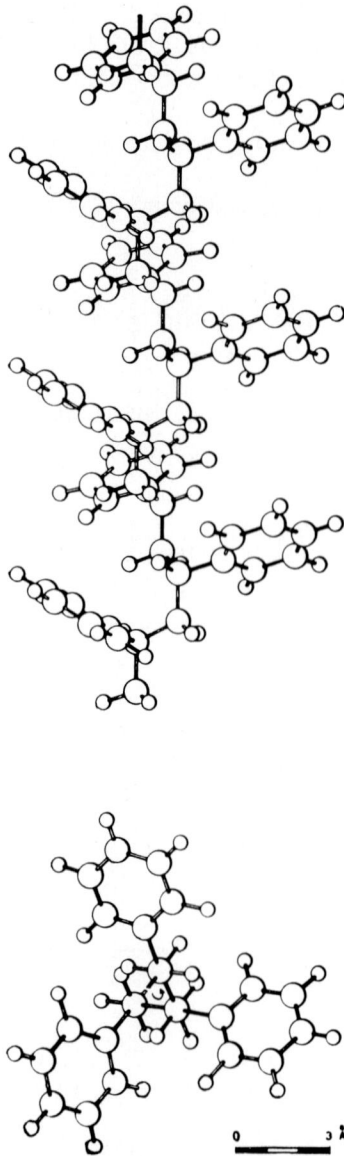

Fig. I-31. Conformation of the polystyrene molecule (Natta, Corradini, and Bassi (79)).

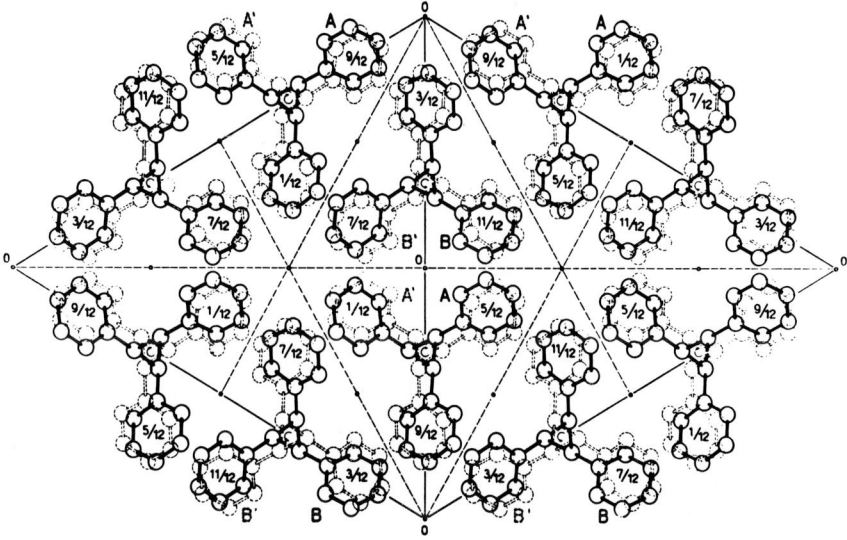

Fig. I-32. Projection of the molecular packing of the polystyrene unit cell onto the {001} plane (Natta, Corradini, and Bassi (79)).

tions to the helix axis. These two forms may occur in the same crystal without significantly disrupting its structure. The dotted atoms in Figure I-32 refer to the alternative form.

The unit cell is rhombohedral with cell constants (79).

$$a = b = 21.9 \text{ A}$$
$$c = 6.65 \text{ A}$$

As can be seen from Figure I-32 the packing is hexagonal in nature.

(17) Polytetrafluoroethylene

Configuration (6 chemical repeat units)

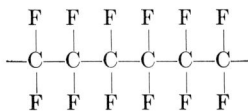

Conformation

The backbone atoms lie on a 13_6 helix below approximately 19°C, a 15_7 helix between this temperature and approximately 30°C (Fig. I-6) and are torsionally oscillating on a helical backbone above this temperature (11).

Molecular Packing (11)

Below 19°C the unit cell is probably triclinic. The molecules pack laterally on a nearly hexagonal lattice but the base plane of the unit cell is skewed with respect to the molecular axis. The repeat distance is 16.88 A while the spacing normal to the chain axis, a', is 5.59 A at 0°C.

Above 19° (the temperatures are approximate and depend on the sample used and its thermal history) the lattice is hexagonal with the dimensions below at 25°C.

$$a = 5.65 \text{ A}$$
$$c = 19.5 \text{ A}$$

(18) Polyurethane

Polyurethanes comprise a family of polymers similar to the polyamides. The 4,6 polyurethane is described below.

Configuration (1 chemical repeat unit)

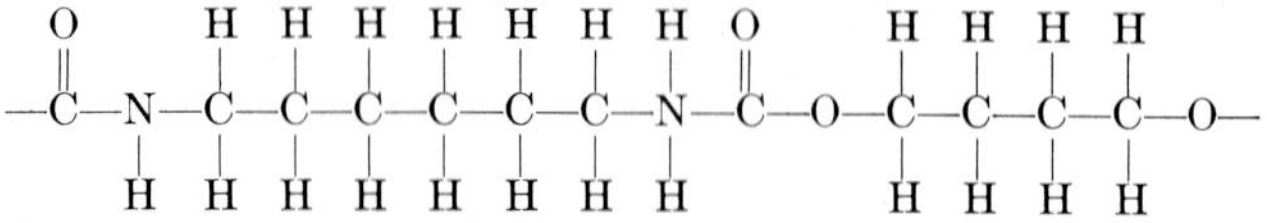

Conformation

The repeat period indicates that the polymer has a planar zigzag conformation.

Molecular Packing

The unit cell is triclinic with cell constants (80)

$$a = 4.95 \text{ A} \qquad \alpha = 60.2°$$
$$b = 19.2 \qquad \beta = 104°$$
$$c = 8.69$$

Hydrogen-bonded sheets are formed in which the molecules are translated parallel to their axis in a fashion similar to that observed for instance in nylon 66.

C. Small Angle X-ray Scattering

Although small angle x-ray scattering ($\sim 0°$–$5°$, 2θ) is of considerable value in determining the size, shape, and surface area of discrete particles such as viruses, proteins, catalysts, etc., both in solution and in solid form, we are primarily interested in its use in measuring periodicities on the order of 50 A to 1000 A. Polymer crystal lamellae are generally in this range of thickness. Thus the simpler theories of Bragg diffraction as discussed for wide angle x-ray diffraction are assumed to apply. The reader is directed to texts by Guinier and Fournet (81) and Beeman et al. (35) for the theory of small angle scattering as applied to discrete particles. Both of these texts also discuss small angle apparatus and the problems involved in interpreting the data.

A major problem in interpreting small angle x-ray data arises from the usual practice of using slits, for intensity reasons, for the collimation of the incident and reflected beams. A pinhole type collimation would be more satisfactory but with present x-ray sources and most diffraction equipment this results in excessively long exposures and/or relatively poor resolution. Photographic film is required to record the pattern. With either type of collimation a cone of diffracted rays is produced from each point of the sample illuminated. With pinhole collimation the resulting pattern on a film consists of rings whose width is related to the diameter of the pinholes. When slits are used, however, the circles from each point of the rectangular area of the sample that is illuminated overlap. The resulting pattern is often detected with a counter having a window also in the form of a slit. As this window, which is parallel to the collimating slit, traverses the pattern in a direction normal to the beam (i.e., where a film is located in a pinhole system) some intensity is recorded at all angles less than the diffraction angle of the scattered beam. The result is so-called "slit-smeared" diffraction scan in which the maxima corresponding to spacings of the order of 100–500 A are shifted by 10–20% toward smaller angles. Unsmearing of this curve is possible under certain conditions (35). The same sort of smearing takes place if photographic recording is used with slit collimation.

5. ELECTRON MICROSCOPY

a. Bright Field

More or less standard techniques of electron microscopy have been developed for the study of, on the one hand, biological specimens and, on the other hand, metallurgical samples (82,83). Polymer microscopists have been able to draw on these techniques in developing their own means of studying polymer structure. The major problem in electron microscopy is that the sample thickness is limited to several hundred Angstroms. The beam in commercially available microscopes will not penetrate thicker material. Further difficulties in polymer studies arise from the heating and degrading effect of the beam and the intermediate (between biological materials and metals) hardness of the samples. Microtoming, to produce a thin enough sample, is in most cases useless because the knife greatly deforms the polymer. The techniques described below, based on shadowing and replication techniques used in the other fields and weak beam illumination, have been developed to permit the study of polymer morphology. We shall describe in some detail those techniques used in our laboratory as examples of those used by others, pointing out at the same time where problems in interpretation of the resulting micrographs arise.

The column of the microscope is under vacuum to permit the transmission of electrons. The lenses of the microscope, performing functions similar to optical lenses, are electromagnetic. Any manipulation of the sample other than simple translation and small amounts of rotation is difficult, both because of the design of the microscopes and the small size of the sample. The total area of sample that can be scanned without removing and changing the sample is limited to an area about 2 mm or less in diameter. The thinness of the samples requires that they be supported on a screen or grid of some type, this screen frequently being covered by a thin, nearly transparent film. Usually we use 200 mesh copper grids, the holes being about 50 microns on a side. The type of film or "substrate" used varies with the sample and is discussed below.

Polymer single crystals, although having a nearly ideal thickness for electron microscopy ($\sim$100 A) suffer in two respects. As described earlier, their crystal lattice is rapidly destroyed in the electron beam. Furthermore, contrast in the image being proportional to the atomic weight of the sample, most polymer crystals have low contrast. To overcome both of these handicaps the crystals are usually shadowed with a thin layer of some heavy metal (we usually use Cr; U, Pt, Pd, Ge and others can also be used). The metal is evaporated in a vacuum from a hot source onto the sample at an angle and essentially stays where it strikes the target. The metal piles up in front of and leaves shadows behind topographical features on the sample. By knowing the angle and measuring the length of the resulting shadows behind steps on the sample, it is possible to calculate the height of the steps.

The substrate found most satisfactory for the single crystal work is a thin, evaporated film of carbon. This film is tough, amorphous, and relatively transparent to electrons. It is produced by evaporating graphite from a carbon arc onto glass slides. Crystal suspensions can then be placed on the slide, the solvent evaporated, and the entire slide shadowed with Cr. The carbon, crystal, metal sandwich can be removed from the slide by scoring it into $1/8''$ squares (to fit the grids), breathing on it and then floating off the film on a water surface. Occasionally the carbon adheres so strongly to the glass that breathing on the slide does not loosen it. This is especially the case if the sample is heated during its preparation. In such cases storing of the slide in a region of high humidity for several hours or more will frequently loosen the film. The sandwich squares floating on the water are picked up directly on the copper grids and, after evaporation of the water, are inserted in the microscope. For electron diffraction the same technique is used although in most cases the sample is not shadowed.

Polymer samples crystallized from the melt, as well as some crystallized from solution, are almost always too thick to insert directly into the beam. A replication technique is used, a thin replica of the surface being prepared. This replica is then inserted into the microscope. The resolution of the resulting observations, limited of course to a study of the samples surface only, is determined by the resolution of the replica. We have found the following technique (84) to produce replicas of polymers of a quality as high as that of any other and to be useable on a wide range of samples.

The sample is shadowed with chromium in a manner and to a thickness equivalent to that used in shadowing the single crystals. It is this layer of chromium, after it has been removed from the sample by a simple but somewhat lengthy process, that is observed in the microscope. The resolution of topographical features, because of the shadowing nature of the process, is comparable to that obtained on the single crystals. Steps on the order of 15 A high can be observed. The metal granulates to some extent so that lateral resolution is limited to about 50 A.

To remove the replicas from the samples small drops of a 3% aqueous solution of polyacrylic acid are placed on the sample and allowed to dry. The polyacrylic acid (PAA) adheres tightly to the chromium and, after drying, the combined Cr-PAA layer can usually be easily removed in one piece. Although the PAA can be dissolved in water, it swells before dis-solving, generally causing the replica to break up. To prevent this the replicas are placed Cr side down on a film of a 3% polystyrene (PS) solution in CCl$_4$, on a glass slide. After drying the PS will then hold the replica to-gether as the PAA is dissolved by placing the PAA-Cr-PS sandwich, PAA side down on the surface of water. About 5 hours is allowed for each of the evaporation and solution steps. The remaining Cr-PS sandwich is then picked up from the water surface on a Formvar* covered grid (for preparation see below) and the PS dissolved by washing the grid on a glass-slide with CCl$_4$. The sample consisting of the grid, film of Formvar, Cr replica can then be placed in the microscope.

The Formvar covered grids are prepared by spreading on water a fresh 0.5% solution in distilled ethylene dichloride. A number of grids are placed on this film. A slide is then inserted into the water through the film so that the grids are held between the slide and the film and then removed. Following evaporation of the water the grids and adhering film can be removed from the slide as needed. Two major defects of these films in comparison with evaporated carbon films are the formation of bubbles if the solutions are old and some motion in the electron beam. Unfortu-nately, carbon films do not adhere tightly enough to the grids to permit their use in the above replicating process.

Extremely rough surfaces can be replicated by the above technique. We have replicated surfaces consisting of fibers 10 to 20 microns high, the fibers remaining standing normal to the surface. It is believed that a layer of the polymer adheres to the chromium as it is removed. The polymer within the fibers apparently also remains, contributing to their rigidity.

Recently we have switched to a Pt replica technique that appears to produce higher resolution replicas. Following Anderson (93), a desired amount of Pt wire is wrapped around the sharpened end of the carbon rod used for producing evaporated carbon substrates and melted at atmospheric pressure. The resulting small bead of Pt adheres strongly to the carbon. The sample is then shadowed under vacuum with the Pt, heating being stopped when all of the Pt has evaporated. This replica, which consists of a thin film of Pt mixed with some C, can then be removed using PAA as above, or in the case of single crystals can be further backed by evaporating the C itself. Following removal of the PAA-Pt sandwich it also can be backed with C and the PAA removed directly by floating on water. If C is applied before the PAA, the PAA does not adhere to the film. C can be

---

* Trade-mark of Shawinigan Resins Co., Inc., for its polyvinyl formamide.

applied directly in those cases in which the sample can be removed by dis-solution. In many cases the PS is not needed as a backing material, nor are Formvar substrates needed. This technique is recommended as a simple, direct way of producing good resolution samples.

For some samples the above technique is not suitable and/or other techniques may be more straightforward. Symons (85) has discussed a technique of replicating polytetrafluoroethylene with evaporated SiO followed by sublimation of the polymer. This can also be used on other polymers. Other workers in the field prefer carbon replicas or plastic replicas, these replicas having to be subsequently shadowed to bring out details.

The only available means of investigating the internal structure of bulk polymers is to fracture the sample brittlely and then replicate the fractured surface. We know of no solvent etching techniques, a technique widely applied in metallurgical studies, which has been shown not to pro-duce an artificial surface on polymers. There is some indication that chemical degradation and bombardment may be useable in etching poly-mers but further work is needed.

b. Dark Field

Dark field or diffraction electron microscopy has been of considerable value in the investigation of polymer single crystals. It permits the de-termination of the crystalline portion of the sample which is reflecting electrons into a particular Bragg reflection. Its utilization suffers, and to a greater extent, from the same experimental problem as electron diffraction, the diffracting power of the sample deteriorating during exposure. Under typical illumination conditions permitting the photographing of a diffrac-tion pattern in about $1/10$ second, a dark field exposure at about $1500\times$ requires a 30 second exposure. Since the polymer lattice deteriorates in about 60 seconds under these conditions, little time is available for focusing or the obtaining of additional exposures using different Bragg reflections. Use of a cold stage considerably reduces the deterioration but can present contamination problems.

Dark field electron microscopy can be easily visualized by considering the formation of images and diffraction patterns in the electron microscope (Fig. I-33). One notes that in both bright field microscopy and electron diffraction, a diffraction pattern of the sample is formed at the plane of the objective aperture. By varying the power or magnification of the lenses either a Gaussian image or the diffraction pattern is projected onto the screen.

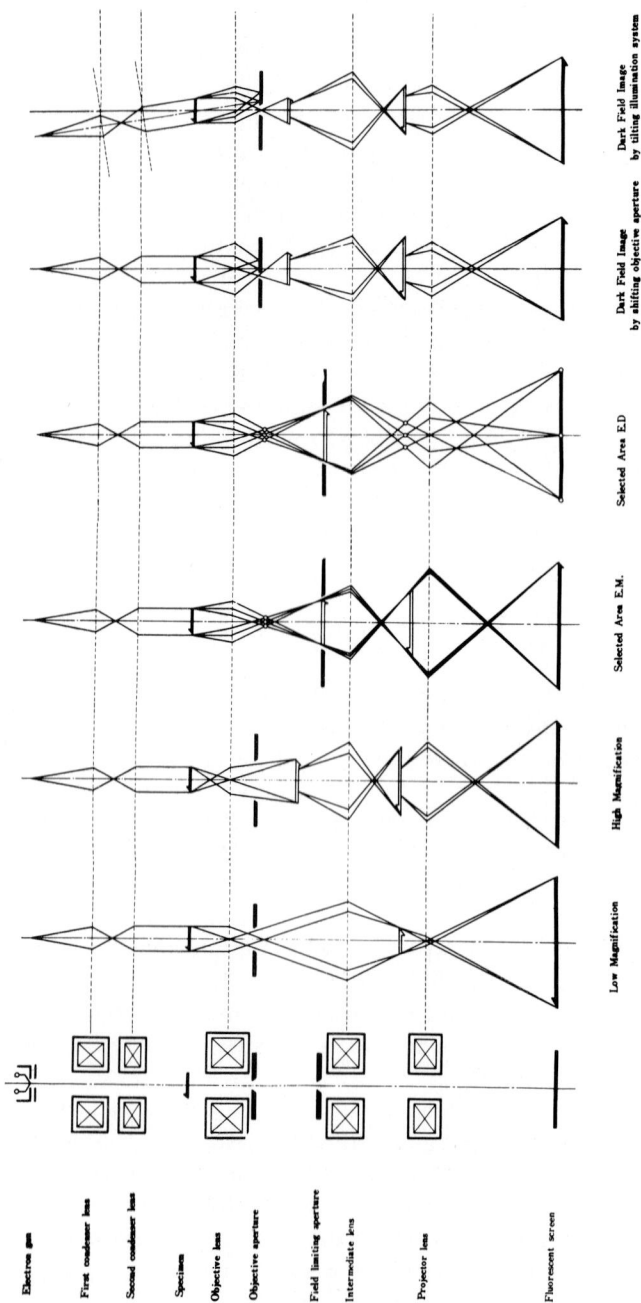

Fig. I-33. Diagram of formation of images in the electron microscope.

Electron gun

First condenser lens

Second condenser lens

Specimen

Objective lens

Objective aperture

Field limiting aperture

Intermediate lens

Projector lens

Fluorescent screen

Low Magnification   High Magnification   Selected Area E.M.   Selected Area E.D.   Dark Field Image by shifting objective aperture   Dark Field Image by tilting illumination system

In bright field microscopy the image seen results primarily from the focusing of electrons which are directly transmitted through the sample. Contrast results from scattering effects, both elastic and inelastic, and is proportional to the thickness of the sample and its electron density (82). Phase contrast has recently also been suggested as being of major importance in forming electron microscope images (83). If the various apertures in the microscope are large enough, electrons diffracted or scattered at small angles will contribute to the image. However, because of aberrations in the lenses and the fact that the diffracted electron make an angle with the axis of the lenses, they will usually not be imaged at the same place as the transmitted electrons. The result is the formation of so-called ghost images (see Figure II-20). With polymer crystals a large proportion of the incident beam is diffracted at small Bragg angles and thus the ghost images can have considerable intensity. If one of the apertures is small enough to prevent all diffracted electrons from reaching the screen, those regions of the crystal which are so oriented as to strongly diffract electrons will be dark but no ghost images will be formed. The dark areas are known as Bragg extinction lines or zones. Agar, Frank, and Keller pointed out the significance of ghost images and Bragg extinction lines in polymer crystal microscopy (31).

In dark field microscopy diffracted electrons alone are used to form the image. A small objective aperture (the smaller the better) is positioned either by translation of the aperture or tilting of the incident beam so as to intercept all of the electrons except those diffracted into a particular portion of the diffraction pattern: namely one Bragg reflection spot if a single crystal is being viewed. The positioning of the aperture can be viewed on the screen when the lenses are set for diffraction. By changing the power of the lenses so as to form an image, leaving the objective aperture in its position, only those electrons scattered into the particular portion of the diffraction pattern outlined by the aperture contribute to the image. In many microscopes the beam can not be tilted on an axis through the sample; therefore, since the electrons contributing to the image make an angle with the lens axis, aberrations limit the magnification to some extent. More significant and present even in microscopes in which the beam can be tilted is the low intensity of the dark field image. The low intensity necessitates long exposure times at high magnifications which are impossible because of the degradation of the polymer's diffracting power with time. A combination of a double condenser system, to permit focusing with sufficient illumination at the high magnifications, a tiltable electron beam, extremely low working illumination and cooling of the sample, should permit considerably higher usable magnification than has been reported to date.

A somewhat related type of dark field image is formed if the diffraction pattern is slightly defocused. A dark field image is formed at the position of each Bragg reflection. Thus on one micrograph the portions of a crystal contributing to all of the reflections can be determined. The magnification possible is even lower than that by the previous procedure and furthermore the technique is usable only with isolated single crystals. With more complicated diffraction patterns, as in the case of twinning, overlapping or rotation of crystals, etc., the dark field images overlap and are difficult to interpret.

6. OPTICAL MICROSCOPY

We shall consider only one aspect of optical microscopy here, the use of an interference microscope to measure the thickness of the lamellae in solution grown single crystals. Wunderlich and Sullivan have described their determinations of the thickness of polyethylene single crystals, using an A.O.-Baker interference microscope (86). This is a polarizing microscope modified into a two beam interferometer. Wunderlich and Sullivan indicate that they are able to measure the thickness of the lamellae to ± 6 A. The value determined depends on the refractive index of the crystals. It was assumed that this was equal to the value determined from a drawn fiber. However, the crystals Wunderlich and Sullivan used were pyramidal in nature and thus the c axis or molecular axis probably was not parallel to the beam. (See Chapter II.) They indicate, using the above determined value for the refractive index, that the optically measured thickness was 134 ± 6 A whereas small angle diffraction from the same sample gave a value of 138 A. This technique should be highly useful, not only for measuring the thickness of isolated single crystals of polyethylene and other polymers, but also for following its changes in thickness during annealing experiments (Chapter V).

REFERENCES

1. N. G. Gaylord and H. F. Mark, "Linear and Stereoregular Addition Polymers: Polymerization with Controlled Propagation" in *Polymer Reviews*, Vol. 2, edited by H. W. Mark, Interscience, New York (1959).
2. F. W. Billmeyer, Jr., *Textbook of Polymer Chemistry*, 2nd ed., Interscience-Wiley, New York (1963).
3. P. J. Flory, *Principles of Polymer Chemistry*, Cornell University Press, Ithaca, N. Y. (1953).
4. G. Natta and P. Corradini, "Conformation of Linear Chains and Their Mode of Packing in the Crystal State," *J. Polymer Sci.*, **39**, 29 (1959).

5. M. L. Huggins, G. Natta, V. Desreux, and H. Mark, "Report on Nomenclature Dealing with Steric Regularity in High Polymers," *J. Polymer Sci.*, **56**, 153 (1962).

6. J. D. Ferry, *Viscoelastic Properties of Polymers*, Wiley, New York (1961).

7. V. Daniel, "The Physics of Long Chain Compounds," *Adv. in Phys.*, **2**, 450 (1953).

8. J. D. Hoffman and C. P. Smyth, "Molecular Rotation in the Solid Form of Some Long Chain Alcohols," *J. Am. Chem. Soc.*, **71**, 431 (1949).

9. D. W. McCall and W. P. Schlichter, "Molecular Motion in Polyethylene," *J. Polymer Sci.*, **26**, 171 (1957).

10. R. K. Eby and K. M. Sinnott, "Transitions and Relaxations in Polytetrafluoroethylene," *J. Appl. Phys.*, **32**, 1765 (1961).

11. E. S. Clark and L. T. Muus, "Partial disordering and crystal transitions in polytetrafluoroethylene," *Z. Krist.*, **117**, 119 (1962).

12. D. N. Reneker, "Point Dislocations in Crystals of High Polymer Molecules," *J. Polymer Sci.*, **59**, 539 (1962).

13. M. Broadhurst (personal communication).

14. M. Gubler, J. Rabieski, and A. J. Kovacs, to be published.

15. K. H. Meyer and H. Mark, "Über den Bau des Kristallisierten Anteils der Cellulose," *Ber. deut. Chem. Ges.*, **61**, 593 (1928).

16. K. H. Meyer and H. Mark, "Über den Kautschuk," *Ber. deut. Chem. Ges.*, **61**, 1939 (1928).

17. K. Hermann, O. Gerngross, and W. Abitz, "Zur Röntgenographischen Struktur.forschung des Gelatinemicells," *Z. physik. Chem.*, **B10**, 371 (1930).

18. K. Hermann, and O. Gerngross, "Die Elastizitat des Kautschuks," *Kautschuk*, **8**, 181 (1932).

19. H. Staudinger and R. Signer, "Über den Kristallbau hochmolekularer Verbindungen. 17. Mitteilung über hochmolekulare Verbindungen," *Z. Krist.*, **70**, 193 (1929).

20. C. W. Bunn and T. C. Alcock, "The Texture of Polythene," *Trans. Faraday Soc.*, **41**, 317 (1945).

21. G. Schuur, "Some Aspects of the Crystallization of High Polymers," Rubber Stichting, Delft, Communication #276 (1955).

22. A. Keller, "The Spherulitic Structure of Crystalline Polymers: Part I. Investigations with the Polarizing Microscope," *J. Polymer Sci.*, **17**, 291 (1955).

23. A. Keller, "The Spherulitic Structure of Crystalline Polymers: Part II. The Problem of Molecular Orientation in Polymer Spherulites," *J. Polymer Sci.*, **17**, 351 (1955).

24. A. Keller and J. R. S. Waring, "The Spherulitic Structure of Crystalline Polymers. Part III. Geometrical Factors in Spherulitic Growth and the Fine Structure," *J. Polymer Sci.*, **17**, 447 (1955).

25. Z. G. Pinsker, *Electron Diffraction*, Butterworths, London (1953).

26. R. W. James, *The Optical Principles of the Diffraction of X-rays*, Bell and Sons, London (1954).

27. H. P. Klug and L. E. Alexander, *X-Ray Diffraction Procedures*, Wiley, New York (1954).

28. C. W. Bunn, *Chemical Crystallography*, Oxford Press, London (1946).

29. E. S. Clark, personal communication.

30. *International Tables for X-ray Crystallography*, Vol. 1, *Symmetry Groups*, N. F. M. Henry and K. I onsdale, eds., Kynoch Press, Birmingham (1952).

31. A. W. Agar, F. C. Frank, and A. Keller, "Crystallinity Effects in the Electron Microscopy of Polyethylene," *Phil. Mag.*, **4**, 32 (1959).

32. F. C. Frank, A. Keller, and A. O'Connor, "Observations on Single Crystals of an Isotactic Polyolefin: Morphology and Chain Packing in Poly-4-Methyl-Pentene-1," *Phil. Mag.*, **4**, 200 (1959).

33. G. Natta and P. Corradini, "General Considerations on the Structure of Crystalline Polyhydrocarbons," *Nuovo Cimento, Suppl. to Vol. 15*, **1**, 9 (1960).

34. C. Kittel, *Introduction to Solid State Physics*, Wiley, New York (1956).

35. W. W. Beeman, P. Kaesburg, J. W. Anderegg, and M. B. Webb, "Size of Particles and Lattice Defects," in *Encyclopedia of Physics* edited by S. Flügge, Springer-Verlag, Berlin, Göttingen and Heidelberg, Vol. 32, p. 321.

36. H. A. Stuart, "Problem of High-Polymer Crystallinity," *Ann. N. Y. Acad. Sci.*, **83**, 3 (1959).

37. R. Hosemann, R. Bonart and G. Schoknecht, "Faltungspolynom und Gitterfactor parakristalliner Gitterwerke," *Z. Physik*, **146**, 588 (1956).

38. R. Bonart, R. Hosemann, F. Motzkus, and N. Ruck, "X-Ray Determination of Crystallinity in High Polymeric Substances," *Norelco Reporter*, **7**, 81 (1960).

39. R. Hosemann and R. Bonart, "Linienbreite und Kristalligrosse bei Hochpolymeren," *Kolloid-Z.*, **152**, 53 (1957).

40. V. N. Filipovich, "Theory of X-Ray Scattering by Distorted Crystals, I. Theory without Atomic Factors," *Soviet Phys.-Tech. Phys.*, **3**, 2486 (1958).

41. V. N. Filipovich, "Theory of X-Ray Scattering by Distorted Crystals, II. Theory with Atomic Factors," *Soviet Phys.-Tech. Phys.*, **3**, 2496 (1958).

42. A. Keller, "A Note on Single Crystals in Polymers: Evidence for a Folded Chain Configuration," *Phil. Mag.*, **2**, 1171 (1957).

43. A. Keller, personal communication.

44. A. Keller, "Electron Microscope-Electron Diffraction Investigation of the Crystalline Texture of Polyamides," *J. Polymer Sci.*, **36**, 361 (1959).

45. R. Burbank, "Molecular Structure in Crystal Aggregates of Linear Polyethylene," *Bell System Tech. J.*, **39**, 1627 (1960).

46. R. L. Miller and L. E. Nielsen, "Crystallographic Data for Various Polymers. II," *J. Polymer Sci.*, **55**, 643 (1961).

47. V. F. Holland, S. B. Mitchell, W. L. Hunter, and P. H. Lindenmeyer, "Crystal Structure and Morphology of Phlyacrylonitrile in Dilute Solution," *J. Polymer Sci.*, **62**, 145 (1962).

48. R. Stefani, M. Chevereton, M. Garnier, and C. Eyraud, "Les structures cristallines du polyacrylonitrile," *Compt. rend.*, **251**, 2174 (1960).

49. C. W. Bunn and E. V. Garner, "The Crystal Structure of Two Polyamides ('Nylon')," *Proc. Roy. Soc. (London)*, **189A**, 39 (1947).

50. A. Keller and A. Maradudin, "Diffraction of X-rays by Fibres Consisting of Small Crystals: Application of Theory to Polyamides," *J. Phys. Chem. Solids*, **2**, 301 (1957).

51. D. R. Holmes, C. W. Bunn, and D. J. Smith, "The Crystal Structure of Polycapro-amide: Nylon 6," *J. Polymer Sci.*, **17**, 159 (1955).

52. D. C. Vogelsong, "Crystal Structure Studies on the Polymorphic Forms of 6 and 8 Nylons and Other Even Nylons," *J. Polymer Sci.*, **1A**, 1055 (1963).

53. I. Sandeman and A. Keller, "Crystallinity Studies of Polyamides by Infrared, Specific Volume and X-Ray Methods," *J. Polymer Sci.*, **19**, 401 (1956).

54. H. Hendus, K. Schneider, G. Schnell, and K. A. Wolf, "Molekülordnung in Poly-caprolactam," *Festschrift Carl Wurster, zum 63. Geburtstag*, Ludwigshafen am Rhein, p. 293 (1960).

55. P. H. Geil, "Nylon Single Crystals," *J. Polymer Sci.*, **44**, 449 (1960).

56. Y. Kinoshita, "An Investigation of the Structures of Polyamide Series," *Makromol. Chem.*, **33**, 1 (1959).

57. G. Natta, P. Corradini, and I. W. Bassi, "Crystal Structure of Isotactic Poly-alpha-Butene," *Nuovo cimento, Suppl. to vol. 15*, **1**, 52 (1960).

58. R. Zannetti, P. Manaresi, and G. C. Buzzoni, "Cristallinita e polimorfismo del poli-alfa-butene," *Chim. e Ind.*, **43**, 735 (1961).

59. A. Prietzschk, "Die Kristallstruktur des Polycarbonat aus 4,4'-Dioxydiphenyl-2,2-propan," *Kolloid-Z.*, **156**, 8 (1958).

60. W. J. Dulmage, "The Molecular and Crystal Structure of Cellulose Triacetate," *J. Polymer Sci.*, **26**, 277 (1957).

61. H. S. Kaufmann, "X-Ray Examination of Polychlorotrifluoroethylene," *J. Am. Chem. Soc.*, **75**, 1477 (1953).

62. C. S. Fuller and C. L. Erickson, "An X-ray Study of Some Linear Polyesters," *J. Am. Chem. Soc.*, **59**, 344 (1937).

63. R. de P. Daubeney, C. W. Bunn, and C. J. Brown, "The crystal structure of poly-ethylene terephthalate," *Proc. Roy. Soc. (London)*, **226A**, 531 (1954).

64. J. Bateman, R. E. Richards, G. Farrow, and I. M. Ward, "Molecular Motion in Polyethylene Terephthalate and other Glycol Terephthalate Polymers," *Polymer*, **1**, 63 (1963).

65. C. S. Fuller and C. J. Frosch, "X-ray Investigation of the Decamethylene Series of Polyesters," *J. Am. Chem. Soc.*, **61**, 2575 (1939).

66. Y. V. Mnyukh, E. M. Belavtseva, and A. I. Kitaigorodski, "Morphology of Molec-ular Packing in Linear Polyesters," *Proc. Acad. Sci. (USSR), Phys. Chem. (Eng. Trans.)*, **133**, 739 (1960).

67. P. R. Swan, "Polyethylene Unit Cell Variations with Temperature," *J. Polymer Sci.*, **56**, 409 (1962).

68. P. R. Swan, "Polyethylene Unit Cell Variations with Branching," *J. Polymer Sci.*, **56**, 439 (1962).

69. C. W. Bunn, "The Crystal Structure of Long Chain Normal Paraffin Hydrocarbons. The 'Shape' of the $\diagup$CH$_2$ Groups," *Trans. Faraday Soc.*, **35**, 482 (1939).

70. C. S. Fuller, "The Investigation of Synthetic Linear Polyesters by X-rays," *Chem. News*, **26**, 143 (1940).

71. C. W. Bunn, "Molecular structure and rubber-like elasticity. I. The crystal struc-ture of β-guttapercha, rubber and polychloroprene," *Proc. Roy. Soc. (London)*, **180A**, 43 (1942).

72. S. C. Nyburg, "A Statistical Structure for Crystalline Rubber," *Acta. Cryst.*, **7**, 385 (1954).

73. E. Sauter, "Ein Modell der Hauptvalenzkette im Makromolekülgitter der Poly-oxymethylene," *Z. physik. Chem.*, **21B**, 186 (1933).

74. E. S. Clark in H. Solow, "Delrin": Du Pont's Challenge to Metals," *Fortune*, **40**, 116 (1959).

75. G. Natta and P. Corradini, "Structure and Properties of Isotactic Polypropylene," *Nuovo cimento, Suppl. to Vol. 15*, **1**, 40 (1960).

76. H. D. Keith, F. J. Padden, Jr., N. M. Walter and H. W. Wyckoff, "Evidence for a Second Crystal Form of Polypropylene," *J. Appl. Phys.*, **30**, 1485 (1959).
77. E. J. Addink and J. Beintema, "Polymorphism of Crystalline Polypropylene," *Polymer*, **2**, 185 (1961).
78. G. Natta, "Progress in Stereospecific Polymerization," *Makromol. Chem.*, **35**, 94 (1960).
79. G. Natta, P. Corradini, and I. W. Bassi, "Crystal Structure of Isotactic Polystyrene," *Nuovo cimento, Suppl. to Vol. 15*, **1**, 69 (1960).
80. H. Zahn and U. Winter, "Über die Langperiodenreflexe im Röntgenogramm von Polyurethanfaden," *Kolloid-Z.*, **128**, 142 (1952).
81. A. Guinier and G. Fournet, *Small-Angle Scattering of X-rays*, Wiley, New York, (1955).
82. C. E. Hall, *Introduction to Electron Microscopy*, McGraw-Hill, New York (1953).
83. M. E. Haine and V. T. Cosslett, *The Electron Microscope*, Interscience, New York (1961).
84. R. G. Scott, "The Structure of Synthetic Fibers," Am. Soc. Testing Mat., Special Tech. Pub. #257, 121 (1959).
85. N. K. J. Symons, "A Method for Replication of Insoluble Particulate Material and its Application to Polytetrafluoroethylene Dispersion Particles" (to be published).
86. B. Wunderlich and P. Sullivan, "Interference Microscopy of Crystalline High Polymers. Determination of the Thickness of Single Crystals," *J. Polymer Sci.*, **56**, 19 (1962).
87. M. Tsurada, H. Aramoto, and M. Ishibashi, "Studies on the Different Crystal Structures of Nylon 6," *Chem. High Polymers (Japan)*, **15**, 619 (1958) (in Japanese).
88. M. Ogawa, T. Ota, O. Yoshizaki, and E. Nagai, "Notes on Nylon Single Crystals," *Polymer Letters*, **1**, 57 (1963).
89. A. Turner-Jones and C. W. Bunn, "The Crystal Structures of Polyethylene Adipate and Polyethylene Suberate," *Acta Cryst.*, **15**, 105 (1962).
90. R. H. H. Pierce, Jr., J. P. Tordella, and W. M. D. Bryant, "A Second Crystal Modification of Polyethene," *J. Am. Chem. Soc.*, **74**, 282 (1952).
91. F. C. Frank, A. Keller, and A. O'Connor, "Deformation Processes in Polyethylene Interpreted in Terms of Crystal Plasticity," *Phil. Mag.*, **3**, 64 (1958).
92. K. Tanaka, T. Seto, and T. Hara, "Crystal Structure of a New Form of High Density Polyethylene Produced by Press," *J. Phys. Soc. Japan*, **17**, 873 (1962).
93. F. R. Anderson, personal communication.

II. SINGLE CRYSTALS FROM SOLUTION

1. INTRODUCTION

As early as 1919, the polymerization of formaldehyde in the form of single crystals was reported (1-3). These crystals of polyoxymethylene are hexagonal prisms (Fig. VIII-3). When crushed or treated with sulfuric acid the crystals split into fibers, whereas in NaOH they develop dark bands along their length (2). The molecules are oriented normal to the hexagonal base. Polymerization occurs by the addition of monomer to the growing molecule, presumably at the end of the crystal. The molecules would thus be fully extended. Recent work (95-97) on the solid state polymerization of trioxane has also resulted in the formation of large single crystals of polyoxymethylene. Further work is required to define the structure of both types of crystals. Although both are single crystals and have the same unit cell as lamellar crystals of polyoxymethylene, their growth and morphology is significantly different than the single crystals to be discussed in this chapter.

The first indication that it was possible to obtain large (greater than 1000 A) single crystals of polymers from solution was published by Storcks (4) in 1938. Prior to this time, and later as well, small crystals of the fringed micelle type were presumed to develop during crystallization from solution as well as from the melt. Storcks cast films of gutta-percha (*trans*-polyisoprene), using special precautions to prevent degradation, from dilute chloroform solutions. Films about 200-A thick were prepared from two different grades of gutta-percha. Diffraction patterns from these films consist of "spotty" rings corresponding to {h k 0} reflections of the β crystal modification. The fact that individual spots could be discerned indicates that the films contain microscopic crystals. Furthermore, the fact that only {h k 0} reflections are observed indicates that the molecular axes are normal to the plane of the film. From tilted films Storcks was able to show that the axes make an angle of less than 4° with the normal to the film. As Storcks points out, "It is surprising that most of the crystallites are oriented with their fiber axis directions normal to the plane of a film the thickness of which is much less than the total length of a macromolecule . . . . It therefore seems necessary that the macromolecular configuration of gutta-percha is not generally linear in these thin films. The

79

gutta-percha macromolecule may possibly fold by a mechanism of rotation around single bonds" (4). These views, as will become evident in this and subsequent chapters, are in agreement with present concepts of polymer crystal structure. Similar experiments were performed on thin films of polyethylene sebacate, a polyester. In these films, however, the crystals are much smaller and the molecular axes are essentially parallel to the film surface.

Fifteen years later, Schlesinger and Leeper (5) prepared single-crystal-like structures of gutta-percha (they used a purified form of the naturally occurring *trans*-polyisoprene, which they termed gutta). These structures, designated crystals, are large enough to be seen in the optical microscope (Fig. III-5). The most perfect structures are elongated hexagonal plates up to 0.35 mm. long and apparently at least several microns thick. They were grown by slow diffusion of the nonsolvent, ethyl acetate, into a benzene solution of the polymer or by slow cooling of the solution. "Crystals" were also grown from a number of other solvents. Samples with molecular weights of approximately 16,000 and 32,000 were used, the "crystals" of the lower molecular weight polymer being better formed.

The "crystals" were observed to be anisotropic between crossed polaroids, with refractive indices of 1.562 parallel to the short axis and 1.526 parallel to the long axis. The authors were unable to obtain x-ray or electron diffraction patterns and, therefore, did not determine the orientation of the molecules. The crystal structure is such that optical anisotropy would be expected regardless of the orientation of the molecules. Melting point experiments indicated that the crystal structure corresponds to the α-crystal modification of gutta percha.

The "crystals" described by Schlesinger and Leeper are presumably related to structures now defined as hedrites. A hedrite, as discussed in Chapter III, is a relatively thick aggregate of single crystal lamellae having a common nucleus and orientation. X-ray diffraction patterns from hedrites resemble single crystal patterns except for a slight amount of arcing of the spots.

The growth of single crystals of a different polymer was required before it became apparent that the early reports were significant. The papers of Storcks (4), and Schlesinger, and Leeper (5) appear to have been overlooked by most polymer structure workers of the time, possibly because of a well-known tendency for gutta-percha to degrade. Furthermore, as recognized by the authors, insufficient work had been done to completely characterize their preparations. Jaccodine, in 1955, reported (6) that he had grown single crystals of polyethylene. Using a linear polyethylene with an average molecular weight of 10,000, he obtained rhombus shaped lamellar growths by crystallizing the polymer from dilute solutions in benzene and xylene. Although the molecular weight of this polymer is lower than that of most commercial linear polyethylenes, it is much larger and the distribution of molecular weight is broader than in the case of paraffins from which similar appearing crystals had previously been grown (7). Jaccodine's observations in the electron microscope indicated that screw dislocations in the lamellae result in spiral growths of additional lamellae. He did not report the orientation of the molecules or describe micrographs of isolated lamellae.

Expanding on this work, three independent investigators, Till (8), Keller (9), and Fischer (10), reported the growth and identification of lamellar single crystals of linear polyethylene in 1957. The recognition and understanding of the form of polymer crystallization is considered to date from these three papers. Since that time, solution grown lamellar single crystals have been identified by electron diffraction and microscopy for a number of other polymers. Micrographs of apparent single crystals have been described for several additional polymers. It is becoming clear that crystallization in the form of a lamella, or in a degenerate form which is ribbonlike, is not only a common form, but probably the typical form, of polymer crystallization from solution.

In the next section, we consider general features of polymer single crystals. Subsequent sections of this chapter contain detailed accounts of various aspects of polymer crystallization from solution in the form of lamellae as well as descriptions of all polymer crystals reported to date.

2. POLYMER SINGLE CRYSTALS

All polymer single crystals grown from dilute solution that have been reported have the same general appearance; they consist of thin platelets or lamellae about 100 A thick. Thickening of the crystals during growth at a constant temperature occurs by spiral growth of additional lamellae from a type of screw dislocation. A typical crystal, of nylon 6 (11), is shown in Figure II-I. Polyethylene crystals are very similar in appearance. Crystals of polyethylene and other polymers are shown in subsequent figures. The size, shape, and regularity of the crystals depend on their growth conditions. Such factors as solvent, temperature, concentration, and rate of growth are important (12–16). The thickness of the lamellae, for instance, depends on the crystallization temperature (12,15,17,18) as well as any further annealing treatments (19). These factors and other distinctive

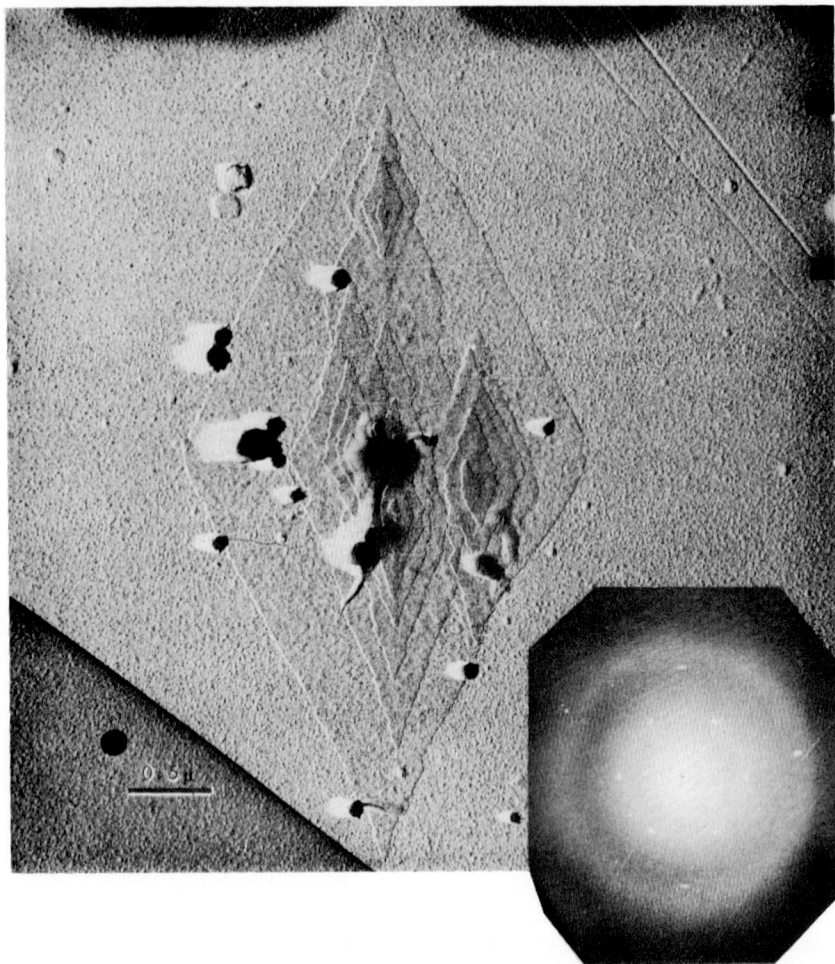

Fig. II-1. Single crystal of nylon 6 precipitated from glycerin solution.   The lamellae are about 60 A thick.   The dark, electron-dense objects are globular particles of nylon. Similar particles also form during the crystallization from solution of other polymers. Their structure is at present unknown.   (Pt-Pd shadowed at $\sin^{-1} 7/11$.)(Geil (11).)

Fig. II-2. Electron diffraction pattern from a single crystal of nylon 6.   The four innermost spots have a spacing of 4.1 A, the other two on the inner ring corresponding to 4.0 A.   Two spots, with a spacing of 2.4 A, are at right angles to the 4.0 reflections. These spacings correspond most closely to the high temperature crystal modification of nylon 6.   (Geil (11).)

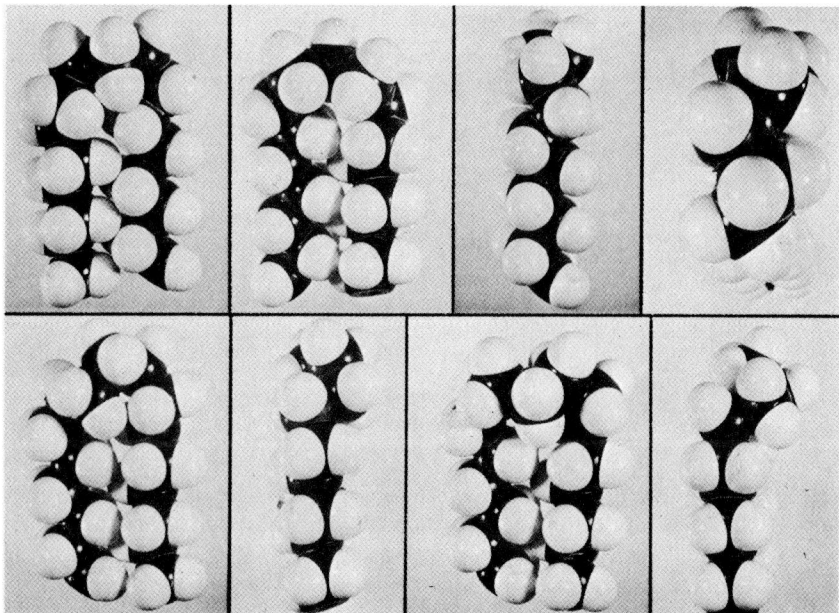

Fig. II-3. Stuart-Briegleb models of folded polyethylene molecules. Approximately five backbone atoms are involved in the fold. Various views of a particular type of fold are shown on the upper line. Two different types of folds are shown on the lower line. The lateral spacing of the straight segments should be slightly larger than shown in these figures. With this slightly larger spacing the configuration of the bonds in the top line can be adjusted so that it consists of three 90° bonds and four slightly twisted 180° bonds. The 90° bonds correspond to the recently calculated positions of the second-lowest potential energy wells for a polyethylene chain (21, 22), while 180° bonds are the normal minimum energy planar zigzag bonds.

features including the formation of "fold domains" and twins during growth, a pyramidal shape and fracture of the crystals, are discussed below.

Electron diffraction experiments reported by Till (8), Keller (9), and Fischer (10) indicated that the molecules are normal to the lamellae. In his paper, Keller pointed out that this, in agreement with Storks' earlier suggestion (4), requires that the molecules be folded back and forth on themselves. With selected area diffraction techniques, he was also able to determine the orientation of the other two axes within the crystal. A diffraction pattern from a nylon 6 crystal (11), typical of diffraction patterns from flat polymer crystals, is shown in Figure II-2. In nylon 6 (11,98) and polyethylene (9) crystals, the **b** axis (hydrogen bond direction in nylon 6)

Fig. II-4. Nylon 610 crystallized from formic acid. The molecules are parallel to the substrate and make an angle of about 70° with axis of the ribbon (Keller (23).)

is parallel to the short diagonal of the crystal. The growth faces are $\{110\}$ planes.

The folded arrangement of the molecules is known to be sterically possible. Cyclic paraffins crystallize in the basic paraffin lattice (ortho-rhombic) with two long, straight segments and a fold at each end (20). This fold, in the cyclic paraffin and the polyethylene crystal, can occur in the direction of closest packing, in the $\{110\}$ planes. Approximately five backbone carbon atoms are involved in the fold itself (Fig. II-3). The carbon-carbon bonds between these atoms are twisted away from the position of minimum energy that they occupy in the straight zigzag conformation in the lattice. In other polymers the position along the chain at which the chain can fold and direction of the fold is limited. In the fibrous or ribbonlike structures of nylon (23) (Fig. II-4) and polyhexamethylene terephthalate (24) (Fig. IV-56 and IV-57) that sometimes form during crystallization from solution, it has been shown that the molecules are at an angle to the fiber axis. These ribbons, in which the molecules are folded, are presumably a degenerate form of lamellae. Unless otherwise noted, the discussion in this and subsequent sections applies to the lamellar type of single crystal, i.e. both lateral dimensions are similar

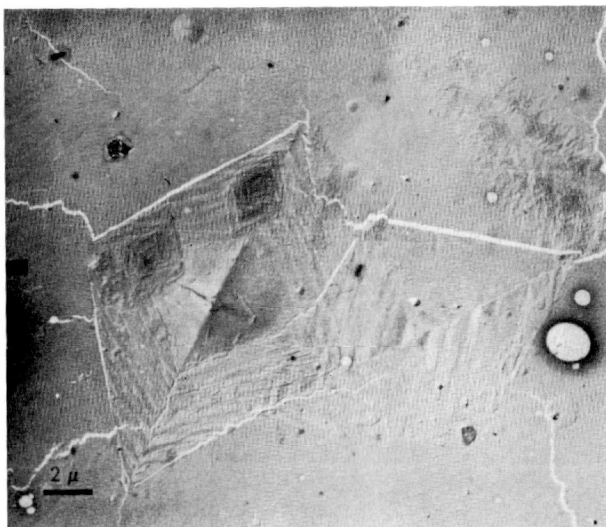

Fig. II-5. Hollow pyramidal crystals of linear polyethylene partially supported to prevent their total collapse. A tetrachloroethylene suspension of crystals was floated on the surface of a polyacrylic acid-water solution. The tetrachloroethylene and then the water was evaporated. If the suspension is deposited directly on a glass slide, evaporation of the solvent results in collapse of the crystal and the formation of pleats. A branch of a dendritic crystal can be seen at the upper right of the figure. (Figure II-22 is of crystals from the same preparation.)

and considerably greater than the thickness. At present little is known of the ribbonlike lamellae.

By means of dark field electron (25–28) and dark ground optical (29) microscopy it has been possible to show that some crystals are not flat lamellae but actually consist of hollow pyramids with 4 or more sectors. Under suitable conditions, the pyramidal structure can be observed directly in the electron microscope (14,30,31)(Fig. II-5). Dark field electron micros-copy is sensitive to the orientation of the molecules. Only those portions of the crystal in which the planes giving rise to the reflection chosen to be imaged are almost parallel to the incident beam (within about 2°) are visible on the dark field micrograph. Only sectors or portions of a sector are seen in those crystals which grew as hollow pyramids. These hollow pyramids should not be confused with the stepped pyramids resulting from a number of lamellae in a spiral growth. As will be shown later, each of the lamellae in a stepped pyramid may also be hollow pyramidal in nature.

TABLE II-1

Polyethylene Crystal Long Periods as Shown on Figure II-6

Symbol	Solvent	Degrees C. T_{xl}	Long period (A)	Per cent conc.	Electron microscope observations	Polymer	Reference
◐	Tetrachloro-ethylene	50	112	0.31		Marlex	36
◑	Tetrachloro-ethylene	60	108	0.25		Marlex	18
O	Tetrachloro-ethylene	68	111	0.25		Ziegler	18
△	p-Xylene	72	110	0.46		Marlex	18
△	p-Xylene	77	116	0.46	Diamond shaped lamellae with spiral growths	Ziegler	18
▲	p-Xylene	78	140	0.58		Marlex	36
▼	n-Butyl acetate	105	147	0.45		Marlex	18
▽	n-Butyl acetate	105	147	0.45		Ziegler	18
▼	n-Butyl acetate	110	162	0.57		Marlex	36
●	Butyl stearate	111	126	0.25	Elongated hexagons	Alathon	35
●	Butyl stearate	111	155	0.25	Large platelets with nearly regular hexagonal spiral growths	Alathon	35
●	Butyl stearate	113.5	166	0.25	Similar to above except fewer spirals	Alathon	35
■	Squalene	115.5	176	0.25	Very large lamellae, a few hexagonal spiral growths	Alathon	35
▼	Glycol dipalmitate	115.4	151	0.25		Alathon	35
▼	Glycol dipalmitate	118	164	0.25	Hexagonal spiral growths, irregular in outline	Alathon	35
▼	Glycol dipalmitate	121	206	0.25	Very large lamellae, nearly hexagonal spiral growths	Alathon	35
☐	Diphenyl ether	120	152	0.37		Ziegler	18
◧	Diphenyl ether	120	202	0.47		Marlex	36
◨	Diphenyl ether	125	173	0.37		Marlex	18
◇	Tripalmitin	121	188	0.25	Thick oval structure with lamellae visible at edges, a few ribbons	Alathon	35

(continued)

TABLE II-1 (*continued*)

Symbol	Solvent	Degrees C. T_{xl}	Long period (A)	Per cent conc.	Electron microscope observations	Polymer	Reference
+	Xylene	50	92.5			Marlex	33, 34
+	Xylene	60	102			Marlex	33, 34
+	Xylene	70	111.5			Marlex	33, 34
+	Xylene	75	118			Marlex	33, 34
+	Xylene	80	120.5			Marlex	33, 34
+	Xylene	85	123			Marlex	33, 34
+	Xylene	90	150			Marlex	33, 34
×	Xylene	75	115	0.1	Complex dendritic crystals	Marlex	15
×	Xylene	80	125	0.1		Marlex	15
×	Xylene	85	133	0.1		Marlex	15
×	Xylene	90	150	0.1	Truncated lamellae (Type D in Section II-4 corresponds to growth from 0.01% solution)	Marlex	15
◆	Melt	120	190	100		Marlex	38
◆	Melt	125	223	100		Marlex	38
◆	Melt	130	355	100		Marlex	38

All the polymers whose crystals have been described and identified to date can be crystallized in the form of single crystals composed of one or more lamellae by cooling a dilute solution from an elevated temperature. With difficult polymers, it is desirable to have crystallization take place at an elevated temperature, as close to the melting point as possible. At these temperatures the molecules have sufficient mobility to readily uncoil, if such is their state in solution, and packed into the lattice. It is thus desirable to use a relatively poor solvent. With the readily crystallizable polymers, such as polyethylene and polyoxymethylene, almost any solvent is suitable. For many polymers even crystallization by evaporation of the solvent leads to films composed of single crystals. (See Fig. II-78.)

The thickness of an individual lamella depends on the crystallization temperature, increasing with increasing temperature (Fig. II-6 and Table II-1). This effect has been shown both by quenching the hot solution to different temperatures (15,32–34) and by crystallizing from different solvents (18,35,36). From the graph (Fig. II-6) it would appear that solvent-

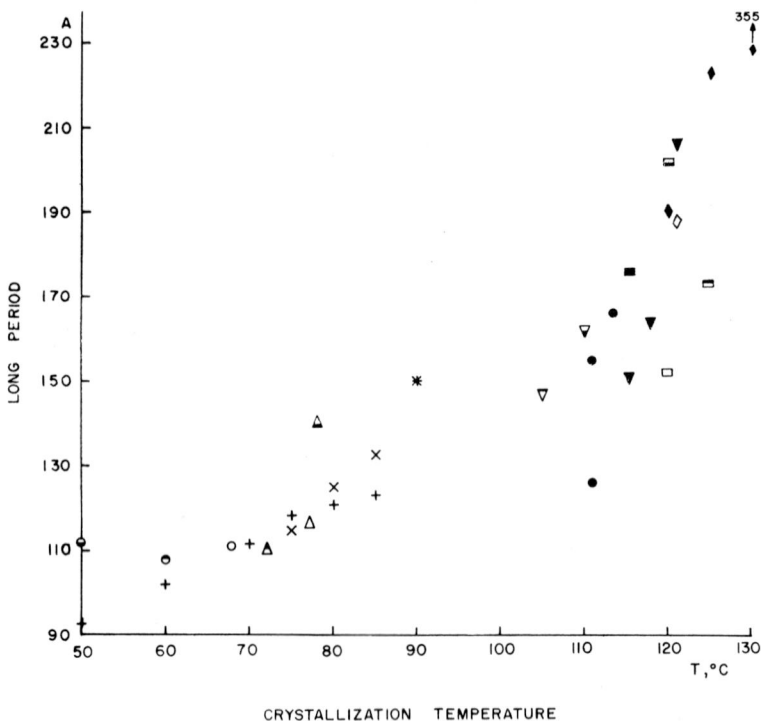

CRYSTALLIZATION   TEMPERATURE

Fig. II-6. Plot of lamella thickness as determined from the long period in small angle x-ray diffraction patterns from sedimented mats of crystals versus the crystallization temperature.   The symbols used for the data are given in Table II-1.

polymer interaction as well as temperature affect the fold period. However, different samples of polyethylene were used in the different experiments and the effect of small and differing amounts of branching on the fold period is not yet known. Molecular weight, which differed among the samples, apparently does not affect the fold period (37) (See Table II-2). Ranby *et al.* (18,36) have reported long periods for a fractionated Ziegler type polyethylene (viscosity average molecular weight 5300) and two Marlex* 50 samples (viscosity average molecular weight of sample in reference 18 was 74,600; the average was not given in reference 36). Some of the differences they observed may be due to differences in branching or to sample preparation as discussed below. The values shown in the graph, Fig. II-6, and listed in Table II-1 are calculated or interpolated from their figures. Ryan (35)

* Registered trademark of Phillips Petroleum Co. for its polyethylene resins.

crystallized an Alathon* linear polyethylene from various solvents. The data of Price (34) was obtained by dropping drops of a xylene solution into a bent metal tube in a thermostated bath, permitting crystallization at low temperatures (rapid cooling), whereas that of Bassett and Keller (15) was obtained by isothermal crystallization of larger amounts of xylene solution and thus is restricted to higher temperatures. Although both sets of experiments were performed with xylene solutions of Marlex 50 on the order of 0.1% concentration and both are fit reasonably well with a straight line, the lines themselves differ considerably. Earlier, low temperature quenching experiments of Keller and O'Connor (32) were known to have resulted in long periods which are too large and, therefore, are not given in Figure II-6. Some of the differences observed among the various sets of data may be due to differences in morphology and in the preparation of the mats used for the small angle x-ray diffraction. The fold period in pyramidal crystals is larger than the thickness of the crystal since the molecules are normal to the pyramid base and not to the sides. When such crystals are flattened, the relationship between the long period and the fold period (and therefore the crystallization temperature) will depend on the details of the flattening process.

The long periods corresponding to the thickness of polyethylene lamellae crystallized isothermally from the melt (38) are also plotted on Figure II-6. These values, as might be expected, appear to fall on the same general curve as that for the long period of polyethylene crystallized as single crystals from solutions in organic molecules of relatively large (compared with the more normal solvents) molecular weights. The values actually plotted are the smaller of the two periods found in melt crystallized samples, this value agreeing with the thickness of the lamellae as observed in the electron microscope (39).

If the temperature is abruptly lowered during crystallization, a small step appears on the surface of the lamellae (15,17,40,41). Of importance to later discussions of crystallization theories is the point that the system appears to adjust to the new crystallization conditions before an observable amount of lateral growth has occurred. The crystals continued to grow with a thickness determined by the new temperature and not the thickness of the substrate; i.e., the part of the lamellar crystal already formed. Like-

---

* Registered trademark of E. I. du Pont de Nemours & Co., Inc. for its polyethylene resins.

wise, a slight increase in temperature during crystallization results in a reverse step in the crystal, the new growth being thicker than the previously crystallized material. Similar results are obtained if a solution is seeded with crystals grown at a slightly lower temperature (15) or if the crystal suspension is heated sufficiently to partially dissolve the crystals (dissolution taking place preferentially at the edges) and then recrystallized at a temperature above that of the original crystallization (15,41). In some cases (Fig. II-7) skeletal crystals are found in which the edge apparently thickened as the suspension was heated, but the interior continued to dissolve (15). Stepwise variation of the temperature during growth can result in several steps on the crystal. In the crystal shown in Figure II-8, at least five changes in temperature apparently occurred during growth. Correlation of the width of the various regions with the length of time at each temperature should permit measurement of the rates of crystallization as a function of temperature. By quenching solutions after specified periods of time, Holland and Lindenmeyer (40) have been able to estimate growth rates. Solutions were brought rapidly to the crystallization temperature by rapidly mixing a small amount of concentrated, hot solution with a large volume of solvent at the desired temperature. Assuming that nucleation and growth began immediately, they found that at a given temperature the growth rate decreases with decreasing molecular weight and decreasing concentration. For a given fraction and crystallization temperature the initial growth rate is linear, with a change in rate of 4 orders of magnitude with a change of about 12°C of supercooling.

For a given molecular weight, concentration and solvent, the morphology of linear polyethylene crystals depends on the temperature of crystallization (Figs. II-9, II-10 and many of the subsequent micrographs). One of the most comprehensive studies of the effect of temperature on morphology has been described by Bassett and Keller (15). They used 0.01% xylene solutions of Marlex 50 which were filtered at the crystallization temperature to prevent further crystal growth during cooling. At 90°C, the highest temperature at which they could grow crystals from this solution, truncated-diamond-shaped crystals formed (Fig. II-11). They indicate that the relative size of the truncated face decreases systematically and apparently continuously as the crystallization temperature is reduced. At 80° true diamond-shaped crystals (Fig. II-9c) are formed which become more and more dendritic as the crystallization temperature is decreased still further (Fig. II-9b). Rapid growth results in the incorporation of imperfections and dendritic growth. However, with very rapid crystallization,

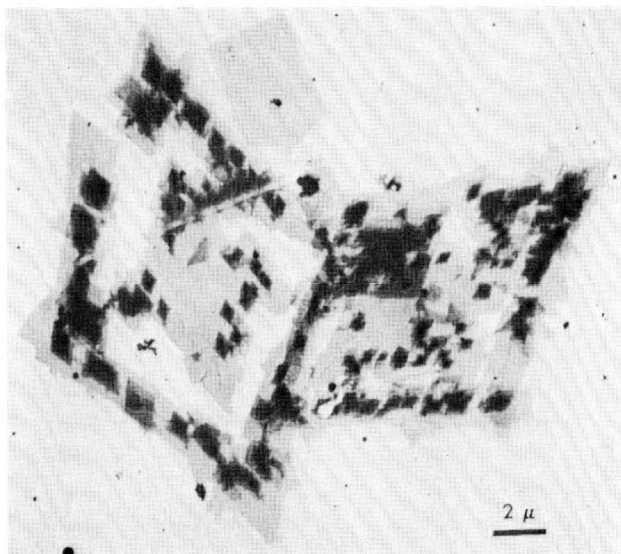

Fig. II-7. Skeletal crystals of linear polyethylene which was heated in solution to a temperature slightly higher than that at which it was originally crystallized (Bassett and Keller (15)).

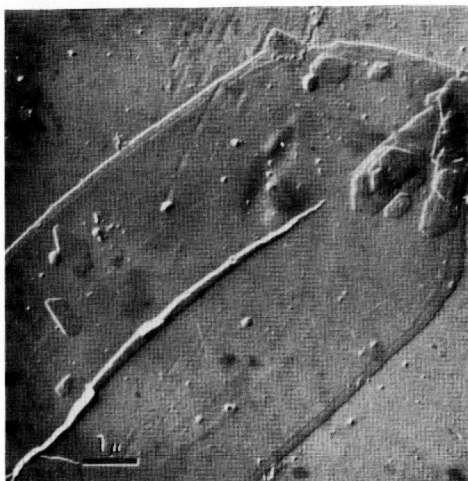

Fig. II-8. Single crystal of linear polyethylene precipitated from an octane solution. The temperature changed several times during growth. A ring of material of smaller fold period developed when the crystal was approximately half grown. Several changes in thickness can also be seen near the crystal's edge. (Cr shadowed at $\tan^{-1} 5/11$.) (Reneker (42).)

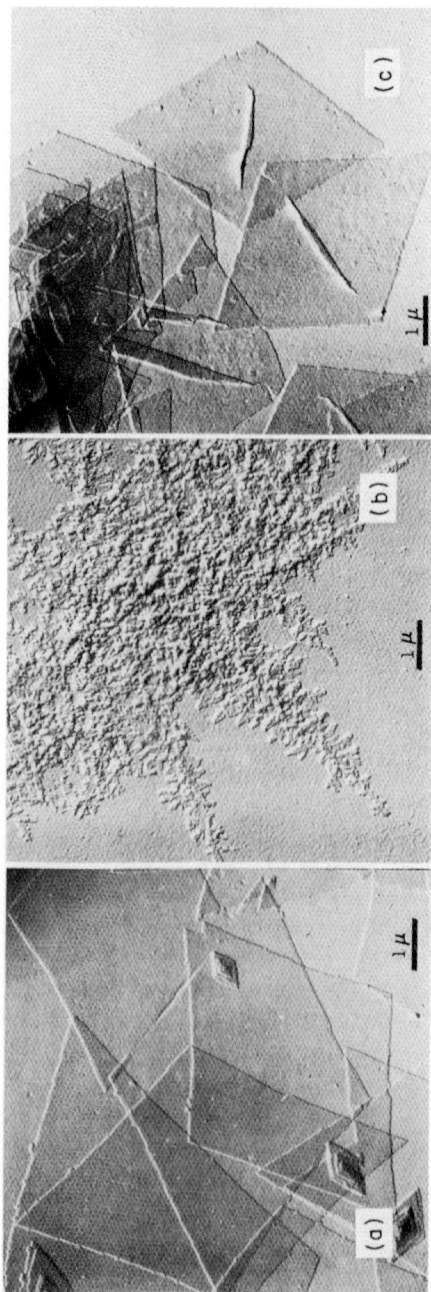

Fig. II-9. Electron micrographs of single crystals of linear polyethylene crystallized from xylene by rapidly dispersing a hot; concentrated solution into the solvent at the desired temperature to produce a 0.1% solution. (a) Sample with molecular weight of about 10,000, crystallized at 40°C. (b) Sample with molecular weight of about 120,000, crystallized at 40°C. (c) Sample with molecular weight of about 120,000, crystallized at 80°C. (Holland and Lindenmeyer (40).)

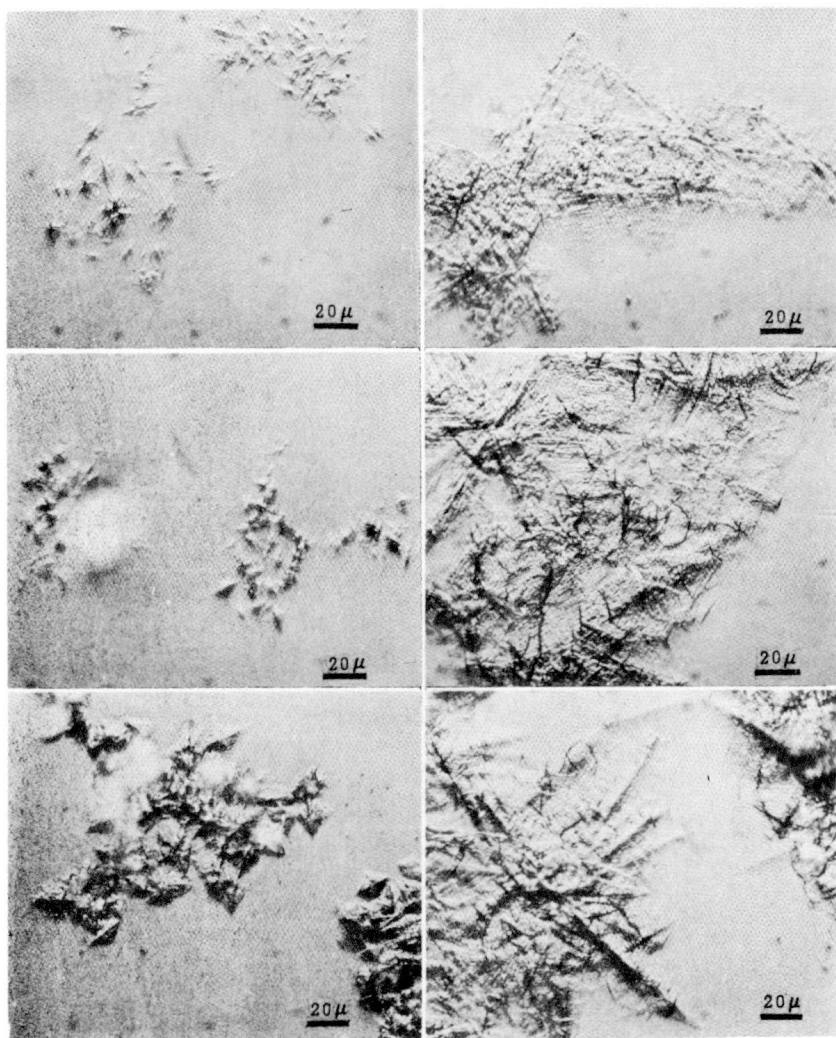

Fig. II-10. Polyethylene crystals grown from tetrachloroethylene solutions of various concentrations, at 58°C. The crystals on the left were grown by spraying the hot solution into a thin-walled glass tube immersed in an oil bath, while those on the right were grown by placing the tube containing the solution into the oil bath. The concentrations from top to bottom are 0.01%, 0.1%, and 0.5%. Similar results are found at higher temperatures also.

as produced for instance by spraying the hot solution into a colder vessel, some small, thin, but relatively simple crystals are formed (14), although with normal crystallization of the same or somewhat higher temperatures dendritic growth is found (Fig. II-10). The cause of this effect is not well defined but it may be related to the numerous nuclei that are probably formed.

For a given solvent and crystallization rate, an increase in concentration leads to an increase in complexity of the resulting crystal in the case of polyethylene (Fig. II-10) and other polymers. The simplest (few or no spiral growths, 4 or 6 sectors) crystals are grown from dilute solution. As the concentration is increased, twins, dendritic crystals and numerous associated spiral growths develop (see subsequent sections). With further increases, hedrites and spherulites are formed.

The type of solvent used, as well as the temperature, affects the shape of the lamellae. In the case of polyethylene, rhombohedral or diamond-shaped crystals can be grown from xylene (6,8–10), toluene (14), tetrachloroethylene (25), trichloroethylene (32), and octane (42)(Figures II-5, II-9, and II-10, as well as many subsequent electron micrographs), diamonds truncated along the long axis to various extents from xylene (26,28)(Fig.

Fig. II-11. Truncated single crystal of polyethylene crystallized and filtered from xylene solution at 90°C. If the solution is allowed to cool to room temperature before removing the crystals, further growth occurs and the final crystals are diamond shaped. (Bassett and Keller (43).)

II-11), amyl acetate (Fig. II-12)(14), and octane (Fig. II-8), and nearly hexagonal lamellae from octane (19) and decane (14)(Fig. III-9). Although crystallization temperature is known to affect the shape (15), further

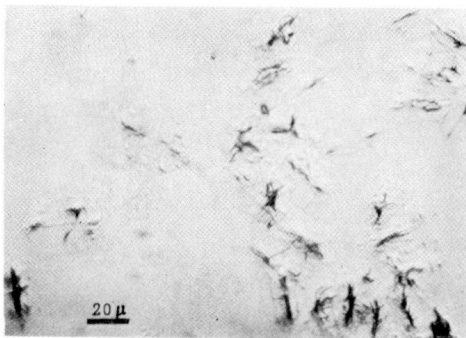

Fig. II-12. Optical micrograph of crystals of polyethylene crystallized from an amyl acetate solution by allowing it to cool to room temperature (incident steep oblique illumination).

work is needed to define the conditions leading to the various types of crystals. For instance, crystals with at least three distinct habits can be grown from octane solutions of the same polymer and of similar concentrations. Crystals from some, perhaps all, of the above solvents are hollow pyramidal in nature.

Several investigators have reported the effect of molecular weight on crystal morphology. Till indicated (8) that, other conditions being equal polyethylene crystals become more complex with increasing average molecular weight of the polymer. Crystals from polymer with a molecular weight of 150,000 ($\bar{M}_w$) were dendritic, whereas simple polygonal crystals were grown under similar conditions from polymers with $\bar{M}_w = 850$. Keller indicated (9) that similar, simple crystals could be grown from polymers of low and high molecular weight. Solution concentration and solvent, as well as crystallization temperature, as indicated, affect the type of crystals that grow. Recent work by Holland and Lindenmeyer (40) and Bassett and Keller (15) tends to confirm both of these reports. Using two fractions with molecular weights on the order of 10,000 (fraction A) and 120,000 (fraction B) Holland and Lindenmeyer found that at 80°C both fractions could be crystallized from 0.01% solutions in xylene as relatively simple crystals with few spiral growths. At 40°C, however, although A yielded similar, simple crystals, B formed complex highly dendritic crystals (Fig. II-9). They report that the maximum temperature at which they could obtain single crystals varied from 84°C for the lowest fraction (A) to 92.2°C for the highest fraction (B). In addition, Bassett and Keller (15) report that the relative size of the truncated face on crystals grown from xylene solution increases with increasing molecular weight. With high

TABLE II-2 (37)

Long Periods of Solution-Crystallized Polyethylene Weight Fractions

1	2	3	4	5a	5b	6a	6b	7	8a	8b
				Low angle X-ray spacing		Molecular weight		Mol. lengths corresp. to values in column 6 (Å)	Calc. by Eq. II-1:	
Fraction No.	Temp. of extraction (°C)	Weight loss (%)	M.p. (°C)	As prepared (Å)	After recryst. at 54°C (Å)	Viscometric	Ebullioscopic		No. of C atoms	Mol. length (Å)
				Solvent: Trichloroethylene						
1	46	1.76	103 ± 1	89.6	—	1,100	900	99 / 81	69	87
						895		81		
						653		59		
						597		54		
2	54	0.77	110 ± 1	95.5	—				87	110
3	62	0.88	115 ± ½	98.0	102				111	139
4	69	0.97	119 ± ½	104.9	105	7,161		710	146	185
5	75	2.98	119 ± 1	102.9	105	7,852		655	146	185
6	81	1.94	122 ± ½	107.9	102	4,864		440	171	216
7	86	4.46	124 ± 1	118.8	104	4,055		377	202	256

—————— Solvent: Xylene ——————

8	84	0.00					
9	92	0.00					
10	100	0.53	130 ± 1	105.9	106		
11	108	26.02	135 ± 2	115	105	443	8
12	115	49.22	Indefinite softening starts at about 135°C	117	107	560	8
13	122	10.61		120	103		
Unfractionated starting material				104		96,000	8,700
						161,000	14,500
						80,000	7,200
						59,000	5,300

The melting point, $T(K)$, of a paraffin with n carbon atoms is given by (48)

$$T = \frac{(0.6035n - 1.75)}{(0.001491n + 0.00404)} \qquad \text{(II-1)}$$

Fractions 1 and 2 crystallized below 54°C. The fractions were obtained by extraction in a Soxhlet extraction apparatus. The various molecular weights listed for a given sample and obtained by viscometry are calculated by various equations; the first two for each sample being based on light scattering measurements as standards and the second two on osmometry.

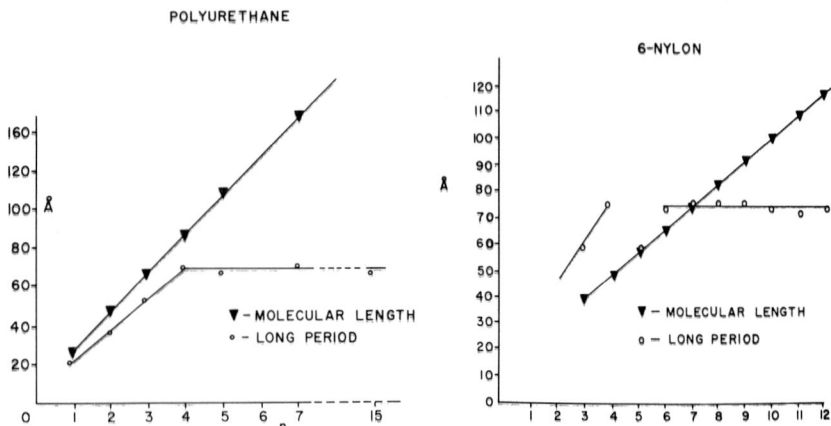

Fig. II-13. Long periods and molecular lengths of oligomers of (a) polyurethane (Kern, Davidovits, Rauterkus, and Schmidt (45)) and (b) nylon-6 (Zahn (46)) as a function of degree of polymerization.

molecular weight polymer they indicate that simple diamond-shaped lamellae cannot be formed at any temperature, dendritic growth beginning before the size of the truncated face is reduced to zero. Somewhat smaller differences in molecular weight had no effect on the morphology of poly-oxymethylene crystals, crystals from polymers with $\bar{M}_w$ equal to 40,000 and 70,000 having the same appearance (17).

The thickness of polymer crystal lamellae is not directly dependent on molecular weight. Keller and O'Connor (37) used polyethylenes of various molecular weights, obtained by fractionating linear polyethylene, to investigate the effect of molecular weight on lamella thickness. Each fraction, of course, had a distribution of molecular weights. It was shown (Table II-2) that the thickness of lamellae from fractions with average molecular weights between approximately 1000 and over 100,000, when crystallized at the same temperature, is the same. The molecular weight will affect the lamella thickness through its effect on the crystallization temperature if samples are allowed to crystallize by gradual cooling. Similar results were obtained by Holland and Lindenmeyer with fractions with a narrower range of molecular weight (40).

Keller and O'Connor (37) were not able to determine the molecular length at which folding begins. The thickness of paraffin crystals with an orthorhombic lattice similar to polyethylene is equal to the length of the molecule. In the longest pure paraffin synthesized so far, $C_{100}H_{202}$, the

molecule is still straight within the crystal (44). Kern, Davidovits, Rauterkus, and Schmidt (45) and Zahn and Pieper (46,47), however, have been
able to investigate the development of folds as a function of molecular
weight. Using oligomers of polyurethane crystallized from solution, Kern
et al. found that the small angle x-ray diffraction long period at first increased with increasing molecular weight and then became constant (Fig.
II-13). Zahn's results, using oligomers of nylon 6 crystallized from solution
are also shown in Figure II-13. The results are interpreted as indicating
that molecules greater than a certain length fold over. Although some
question remains about the fitting of the end groups into the lattice (one
end of the nylon 6 oligomers, for instance, contains a benzene ring) the
folding of the molecules must be such that when one molecule ends within
a lattice row the next one begins. The position of the ends of the molecules
is of interest since similar ends must be accommodated in crystals of high
molecular weight polymers. For the low degrees of polymerization of the
oligomers the molecular length and long period differ because of the unit
cell of the polymer. In the nylon 6 oligomers, for instance, there are two
molecules per unit cell length, these molecules making an angle with the
basal plane of the cell. Zahn and Pieper (47) show that when the samples
are annealed or swollen with a solvent the long period increases to a value
corresponding to the length of the molecules.

The maximum temperature at which crystals will grow from solution
can be increased by applying pressure. Wunderlich, using optical interference microscopy, has found (49) that the thickness of the crystals is
approximately the same for comparable supercoolings, although the crystallization temperature may increase by as much as $90°C$. At any given
temperature the morphology changes from simple single crystallike to
dendritic as the pressure is increased; i.e., as the supercooling is increased.

Branches and other disruptions of the chain affect the structure of and
ease of obtaining crystals. Lamellar crystals of low density polyethylene
containing as many as 2 short branches per 100 carbon atoms have been
grown from solution (50,51). The crystals are small, oval in shape, and
about 100 A thick (Fig. II-14). Thickening of the crystals occurs by the
growth of additional lamellae from apparent double screw dislocations of
opposite sign; i.e., stepped pyramids of complete lamellae, not spiral
growths are seen. Twinning may occur during growth. Electron diffraction
patterns have not been obtained from these crystals of branched polyethylene but it is presumed that their structure is similar to the crystals of linear
polyethylene. Since the branches are randomly located along the chain,

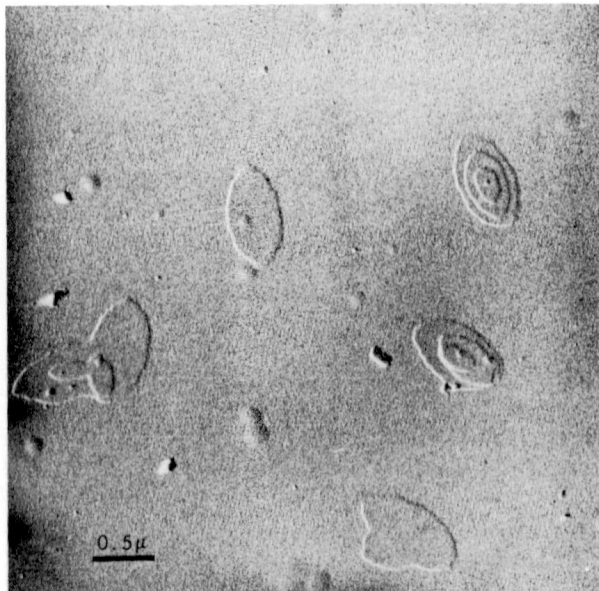

Fig. II-14. Single crystals of a branched polyethylene precipitated from an octane solution. Some fractionation may have occurred during crystallization, much of the sample being covered with a layer of less or noncrystalline material. Terraced pyramids and some indication of twinning can be seen. Spherulites composed of twisted irregular ribbons were observed in the same solution. (Pt-Pd shadowed at $\tan^{-1} 5/6$.)

they must be contained within the lamellae. The lattice can apparently accommodate these branches, the unit cell of branched polyethylene being only 1% larger than that of a linear polyethylene (52). Similar increases in unit cell dimensions, primarily along the **a** axis are observed when propylene or other olefins are randomly copolymerized with ethylene (53) (see Chapter I).

Solvent-polymer interaction can affect the external outline and, apparently, the orientation of the folds in the crystals. The relative degree of development of $\{110\}$ and $\{100\}$ growth faces leads to the various polyethylene crystal outlines described previously. Burbank (54) indicates that he has obtained some polyethylene crystals with $\{530\}$ and $\{540\}$ growth faces, whereas Bassett and Keller report $\{310\}$ faces (15). Variations in shape have also been observed for crystals of polyoxymethylene (17,25) and various nylons (11,24). Further variations of the basic structure are found resulting from various types of dendritic growth and twinning (Sections II-5 and II-6).

Fig. II-15. Linear polyethylene crystallized from a 0.02% xylene solution onto a freshly cleaved NaCl crystal surface. The 100 A thick lamellae standing normal to the surface are aligned parallel to the {110} planes of the salt. (Fischer (57).)

The shape of a crystal is fundamentally related to the molecular packing as evidenced by the structure of the unit cell. Thus, for polymers in which polymorphism is possible, such as the nylons, polypropylene and gutta-percha, it should be possible to grow from the same solvent, crystals with basically different shapes due to a different crystal structure. Some evidence that such occurs, the crystallization occurring at different temperatures from the same solvent, has been reported for nylon 6 (11).

Besides the temperature, rate of cooling, concentration, molecular structure, and solvent, the substrate on which the crystal is grown can affect its morphology. In general, the crystals are nucleated and grown in the solution. However, many crystals are nucleated and grow on the walls of the crystallization vessel containing the solution. The effect of this type of nucleation and growth can be investigated most easily by immersing a substrate, such as a slide, into the vessel during crystallization and washing off any adhering suspension following removal. Allowance for crystals adhering to the slide after growth has started must be made. It has been noticed that glass is a good nucleating surface. Considerably fewer crystals develop on slides which are coated with evaporated carbon (42). One of the most obvious effects of crystallization on a substrate is that spiral growths can develop on only one side of the basal lamella. Furthermore, the crystals nucleated on a glass substrate apparently are constrained to grow on that substrate and, therefore, develop as flat lamellae. This constraint, in the case of crystals which, in suspension, would grow as hollow pyramids, will result in strain within the crystal (25). On crystalline substrates, epitaxial crystallization may occur. It has been shown (55–57) that polyethylene crystallizes on freshly cleaved rock salt in the form of lamellae oriented normal to the surface, with the molecules lying parallel to the $\{110\}$ planes of the salt (Fig. II-15). There is some question as to whether or not this is true epitaxial crystallization since the $\{110\}$ spacing of the salt is considerably larger than the $\{001\}$ spacing of the polyethylene molecules lying on its surface.

3. FOLDED CHAINS

Most polymer molecules can fold easily, at least at some points along their axis. Polyethylene, being a rather simple and flexible polymer, can form a fold anywhere along its length. As pointed out previously, the fold can connect two adjacent segments in a $\{110\}$ plane, the plane of closest packing in the polyethylene unit cell. Models (Fig. II-3) indicate that the fold can be easily made in this distance without affecting either the interatomic distances or valence angles (32). Frank has investigated their conformation in terms of a diamond lattice (30,58,59). However, De Santis *et al.* (21) and McCullough (22) have shown that for polyethylene a lattice should be used in which the low energy bonds are at 90° rather than 120°. The conformation is thus still open to question. The increase in internal energy associated with the fold arises from the fact that the hydrogen atoms on 1st through 5th neighboring carbon atoms are brought closer together than they are in the planar zigzag conformation. It is the interaction with the 3rd, 4th, and 5th neighbors that results in the 90°C—C bond angle having a lower energy than the 120° (gauche) form. The gauche form is a low energy form if only 1st and 2nd neighbors are considered (21,22). Since the molecule can fold back on itself in the $\{110\}$ planes, it can presumably also fold in other planes in which the adjacent molecular segments are more widely separated. Folds in other planes may include more carbon atoms than folds in the $\{110\}$ planes. In the case of some polymers with long repeat periods, for instance polyoxymethylene with a period of 17.3 A, it has been found through the use of models that, because of the conformation of the molecule, folds in a certain direction can be made easily only at certain positions along the molecule. Thus, the fold period should be an integral multiple of this relatively long repeat distance.

Although no direct confirmation of the orientation of the folds has been obtained, there is considerable circumstantial evidence that in diamond-shaped polyethylene crystals most of the folds lie in $\{110\}$ planes. In these crystals, the growth faces are $\{110\}$ planes. Keller and O'Connor (32) pointed out that in all probability the molecules fold against and along this face. The direction of folding and addition of individual segments is thus normal to the over-all direction of growth of the crystal. Growth faces parallel to $\{100\}$ planes are the next most numerous in polyethylene, their extent varying from small in the truncated diamonds (Fig. II-11) to longer than the $\{110\}$ faces in other crystals (Fig. II-8). These $\{100\}$ planes are also the second most closely packed planes in the crystal. The distortion of the molecule in a $\{100\}$ fold is spread out over more carbon atoms. It is not known if the energy involved in creating the fold is larger or smaller than for a $\{110\}$ fold.

The cohesion between adjacent planes containing the folded molecules would be expected to be small. The plane in which a molecule folds, presumably parallel to a growth face, has been defined as a "fold plane" (25). Within a fold plane adjacent segments of the same molecule are connected at the top or bottom of the lamella by a fold and along their length by van der Waals forces. Between fold planes of many polymer crystals, except for an occasional "improper" fold, only van der Waals forces act. An improper fold would result, for instance, if a long molecule started to crystallize with its ends in two adjacent fold planes. These improper folds bridge adjacent fold planes. If few improper folds are present, one would expect that when a crystal is stressed, brittle fracture could occur normal to the fold planes but that stress in other directions would result in unfolding of the molecules and the formation of fibers. Such is found to be the case (13,60,61) (also see Chapter VII). When polyethylene crystals with reentrant growth faces (i.e., dendritic crystals whose formation is described in Section II-5) are scratched mechanically, fracture occurs parallel to the growth faces (Fig. VII-27). Fibers, approximately 50 A in diameter, are formed, however, when the stress is not normal to the growth face. In spiral growths each lamella fractures individually (13). Likewise, when polyethylene crystals are sheared between two slides, sections of the crystal bounded by $\{110\}$ planes are removed (60,61). These sections, which may consist of ribbons parallel to the growth faces of a spiral growth and involving several turns (Fig. II-16), suggest that the fold planes are continuous at the corners on a growth face. In nylon crystals hydrogen bonding, which occurs between segments in adjacent fold planes (11), may result in increased cohesion between fold planes. Fractures which have been observed (Figs. II-1 and II-69) are parallel to the b axis and thus between the hydrogen bonded planes. As is expected, fibers are drawn across the cracks (11,98).

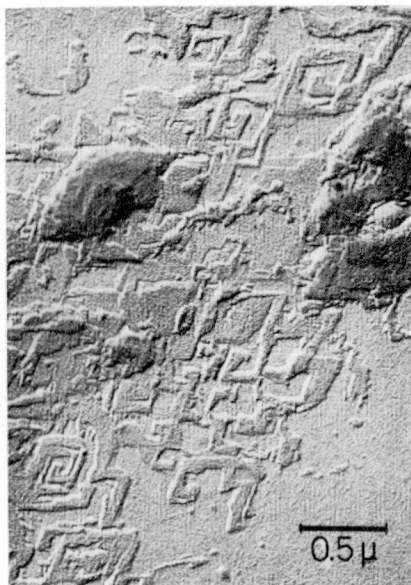

Fig. II-16. Polyethylene crystal destroyed by shearing between glass plates
(Fischer (61)).

The shape of a crystal of a low molecular weight material grown from a pure solution depends on the relative rates of growth of close packed faces, i.e., low index faces. High index faces, if formed, rapidly fill out because of the ease of addition of a molecule to the molecularly rough face. On low index faces molecules are added, it is believed, primarily at kinks in a step resulting from a dislocation. The presence of impurities may greatly alter the relative rates of growth of different faces. However, under any condition, it is those faces with the slowest rates of growth that are found once the crystal is sufficiently larger than the size of the nucleus (62).

In polymer crystals, the length of the molecules, the presence of folds, and the resulting essentially noncrystalline state of two surfaces of the crystal greatly influence the shape of the crystal. The presence of the folds on two surfaces of the crystal effectively prevents surface nucleation on these surfaces. Thickening of the crystal during growth results from the spiral growth of an additional lamella from a type of screw dislocation (for instance, Figures II-1 and II-11). The fold surfaces, having a zero normal growth rate, are thus always visible. The reason for the appearance of the

$\{110\}$ and $\{100\}$ growth faces is apparently more complex. These faces, since they are growth faces, should have a slower rate of growth than any others. This, however, implies that if any face of higher index should develop, molecules will more rapidly fold along this face than along a $\{110\}$ face. The relative amount of energy required to create a different type of fold corresponding to a higher index face is not known. In contrast to the situation in low molecular weight materials, the fold energy may effectively prohibit the formation of certain types of growth faces. However, it must be recognized that in paraffins, which do not contain folded molecules, the same growth faces appear. Furthermore, the thickness of the crystal, it will be shown in Chapter VI, depends on a balance between the energy required to produce a fold and that gained from the addition of the segment to the crystal. Thus, one might expect that the fold periods would be different in those sectors of a crystal in which the fold planes are different. Bassett, Frank, and Keller have shown (28) that sectors containing $\{100\}$ fold planes in a truncated diamond-shaped polyethylene crystal melt at a lower temperature than sectors containing $\{110\}$ fold planes. (See Figure V-1.) In addition, Bassett indicates he has evidence that the sectors containing $\{100\}$ fold planes are thinner than those containing $\{110\}$ fold planes, but that this may have occurred when the pyramidal crystals collapsed (63).

The shape of crystals of nylon and other polymers in which hydrogen bonding occurs, would be expected to be affected by the necessity of forming the bonds. The regular shape of the crystals of nylon 6, as shown in Figure II-1, may be due to the fact that hydrogen bonding in the crystal lattice requires that adjacent molecular segments be inverted. This is readily accomplished by chain folding (51). However, since the hydrogen bonds are parallel to the short diagonal of the crystal (11,98), the various segments of a given molecule are not hydrogen bonded to each other. The thickness of the crystal will also be affected, the molecule being able to fold only at specific points along its length if it is to hydrogen bond to itself or to a neighbor. In nylon 6, the possible thicknesses should differ by factors of 17A (the repeat period) or possibly 8.5 A, depending on the twisting of the molecule in the fold. Keller has noted (23) that in the ribbons of nylon 610 cast from formic acid solution (Fig. II-4) the molecules make an angle of about 76° with the ribbon axis, the **a** axis lying in the direction of the ribbon axis. Nylon 610 cast from formic acid has a triclinic unit cell. The requirement that all hydrogen bonds be made plus a definite fold period would result in the molecules making an angle of this amount with the ribbon axis. The direction of the fold planes, for instance normal or parallel to the ribbon axis, is not known. The factors controlling the thickness of the ribbons are also not known.

The presence of the folds introduces a structural complexity in polymer crystals that is not found in crystals of low molecular weight materials. The orientation of the folds on the surfaces, i.e., the orientation of the fold planes, may vary from one region of a crystal to another even though the lattice in the bulk of the crystal is continuous. These regions, consisting of portions of a crystal in which the fold planes have a common orientation, are termed "fold domains" (25) (see Figs. II-37 and II-38). The orientation of the fold planes changes at each fold domain boundary although the lattice in the bulk of the crystal is continuous across the boundary. In a simple diamond-shaped crystal four fold domains are formed, the fold planes being parallel to the four growth faces. The diametrically opposed quadrants are fold domains in which the fold planes have the same orientation, the growth faces being parallel. It is suggested that fold domains be labeled with the indices of the growing faces, thus a diamond-shaped polyethylene crystal has four $\{110\}$ fold domains, whereas a truncated diamond or hexagonal polyethylene crystal has two additional $\{100\}$ fold domains. In flat crystals, the individual fold domains will be distinguishable only if the orientation of the folds can be measured. They would not be distinguishable, for instance, in an electron diffraction pattern. Although in Section II-6 a somewhat more specific label for a fold domain will be introduced, involving the stacking as well as the orientation of the fold planes, the labels suggested here will be suitable for most discussions. We likewise define a "fold surface" as being the surface of the crystal tangent to the folds and label it with the indices of the crystallographic plane with which it is parallel (i.e., a flat lamella has $\{001\}$ fold surfaces). It will often be convenient, especially when describing pyramidal crystals, to list both the fold surface and growth face following the convention of Bassett, Frank, and Keller (29). Thus, a $\{(112)(110)\}$ crystal is a flat-based, pyramidal, diamond-shaped crystal (see Section II-3) having $\{112\}$ fold surfaces and $\{110\}$ growth faces, the four individual fold domains being $(112)(110)$, $(1\bar{1}2)(1\bar{1}0)$, $(\bar{1}12)(\bar{1}10)$ and $(\bar{1}\bar{1}2)(\bar{1}\bar{1}0)$. Within any fold domain, there will be occasional folds, as previously mentioned, which will bridge fold planes. These "improper" folds form a type of crystal defect not found in crystals of low molecular weight materials. Besides improper folds forming as a result of the ends of a given molecule being incorporated in two different fold planes, they may also result from molecules being added at various points on a fold plane and folding toward each other. When such molecules meet, two improper folds result as the left-over portions of each molecule start a new fold plane. The number of such folds will depend on factors such as the rate of crystallization and the concentration of the solution.

Fig. II-17. Fragmented portion of a "growth twin" crystal of linear polyethylene crystallized from tetrachloroethylene. The composition plane or twin boundary, a $\{110\}$ plane, extends in the direction of the arrow. The presence of the pleats indicates that the crystal was pyramidal in nature. The lath-shaped lamellae lying on the V-shaped pleat at the point of the arrow are portions of another crystal. (Cr shadowed at $\tan^{-1} 5/6$.)

Some polymer crystals have been observed in which the lattice as well as the fold planes change direction at a fold domain boundary (Fig. II-17). In these crystals, twinning has occurred (see Section II-6) during growth and the fold domain boundary should more appropriately be termed a twin boundary or composition plane. In polyethylene, the composition planes are usually $\{110\}$ or $\{310\}$ planes (14,64) although $\{530\}$ and $\{120\}$ composition planes have also been reported (54). Each twin consists of one or more fold domains. Electron diffraction patterns from twinned crystals consist of two (or more if twinning occurs on more than one plane) complete patterns superimposed but rotated by an amount which depends on the composition plane. Care must be taken in all determinations of indices of such features as twin boundaries, etc. to take into consideration the shear and rotation that can take place when pyramidal crystals are deposited on a substrate (30,63).

At this point, it is appropriate to consider how the unit cell of a lamellar crystal is to be defined. There have been suggestions that the unit cell should be considered as being one fold period long and that the cell defined by the repeat distance should be considered a sub-cell of this large unit cell (26,32). In such a case, some of the fold domains, e.g., ($\bar{1}$10) and (110), would be properly termed twins, whereas $\{110\}$ and $\{100\}$ domains would not be related by a twinning operation but rather would involve two different unit cells in the same crystalline structure. Furthermore, such a policy would require an enormous number of different unit cells corresponding to different fold periods, orientations of folds, and, for a given orientation, type of fold. A fold in a $\{110\}$ plane, for instance, can be made in at least five different ways involving only five atoms. Furthermore, as shown later, in crystals composed of many lamellae (as in spiral growths) each lamella grows as an individual lamella. However, a unit cell cannot be so defined that a crystal is only one repeat period thick. Because of this and since the atoms in the folds will not scatter incident radiation (x-ray or electrons) in phase with the atoms in the bulk of the crystal (except for small angle scattering), it is believed desirable to consider the length of the unit cell to be defined as in the past by the repeat period of the molecule. In those publications in which the unit cell is considered to be one fold period long, the indices of lattice points, planes, directions, etc. corresponding to the convention used in this book usually have a subscript s added to the brackets, i.e., $\{110\}$ planes in this book are given as $\{110\}_s$ planes in these publications.

4. HOLLOW PYRAMIDS

Detailed observations have indicated that many polymer crystals do not diffract electrons as a single crystal. In Chapter I, it was pointed out that electron diffraction patterns from a flat lamellar crystal approximately 100 A thick should contain only 1 or 2 orders of reflections. Such patterns are found for crystals of nylon 6, for instance (Fig. II-2). From polyethylene crystals and crystals of some other polymers, however, as many as 7 orders of reflections have been obtained (25) (Fig. II-18). It is thus obvious that some $\{h\,k\,0\}$ planes within the crystal are tilted with respect to the incident beams. This tilting is sufficient to permit their reciprocal lattice points (which are really rods since the crystals are lamellar) to intersect the sphere of reflection.

Diffraction patterns from individual $\{110\}$ fold domains show that different fold domains contribute different reflections to the over-all pattern (28). In many cases the center of intensity of a selected area diffraction

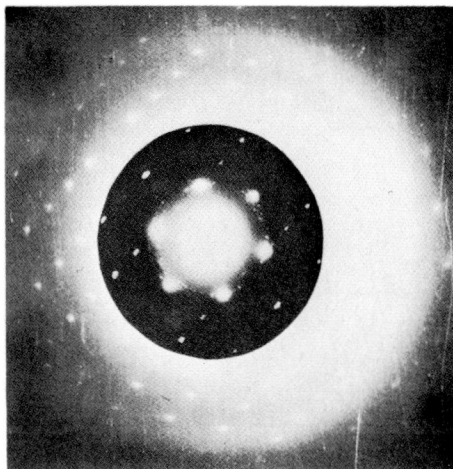

Fig. II-18. Composite electron diffraction pattern from a crystal of polyethylene precipitated from tetrachloroethylene. The central portion was taken with a shorter exposure than the outer portion. On the original, at least 7 orders of reflections as well as the forbidden $\{100\}$, $\{010\}$, and $\{300\}$ reflections can be seen. For dark field micrographs of this crystal, see Figure II-31 (Reneker and Geil (25).)

pattern from a portion of one fold domain is shifted along the non-growth face $\{110\}$ plane toward the center of the crystal (Fig. II-19). The reciprocal lattice is tilted so that more of the reciprocal lattice rods on one side than the other intersect the sphere of reflection, i.e., the lattice is tilted toward the center of this crystal. The diffraction pattern from an adjacent fold domain is similar to that shown except that the center of intensity lies along a line connecting the other two $\{110\}$ reflections. The molecular segments are tilted toward the center of the crystal, essentially parallel to non-growth face $\{110\}$ planes. The diffraction pattern from two adjacent fold domains is symmetrical about a line joining the $\{200\}$ or $\{020\}$ reflections when the fold domain boundary involved is, respectively, a $\{010\}$ or $\{100\}$ plane. i.e., when the boundary is parallel to the **a** or **b** axis respectively.

X-ray diffraction patterns from sedimented mats of truncated crystals grown at 90°C in xylene and taken with the beam parallel to the mats (type D as defined later) show (30,43) that the $\{200\}$ reflection is split azimuthally by 20–30° whereas $\{020\}$ apparently is not split. The $\{110\}$ reflections appear to be slightly split. These results indicate that the $\{200\}$ planes are not normal to the plane of the film whereas the $\{020\}$ are and thus the molecules must be tilted with respect to the normal to the film.

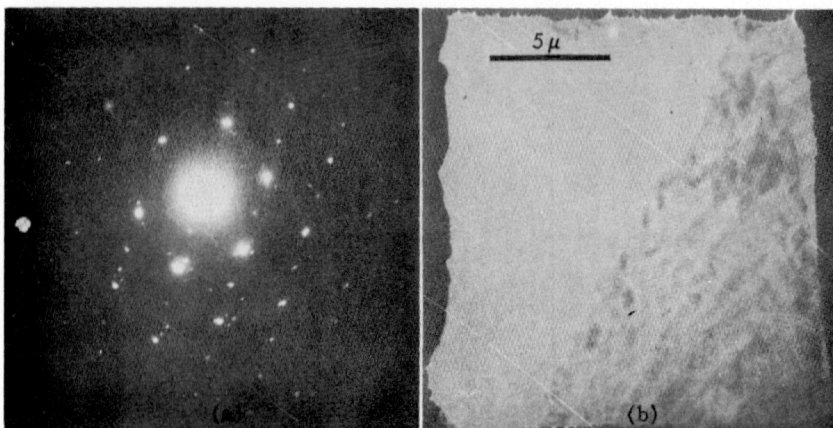

Fig. II-19. Electron diffraction (a) from a portion of a single fold domain (b).   The contrast in (b) is due to Bragg extinction.   The orientation of the diffraction pattern corresponds to that of the crystal shown.

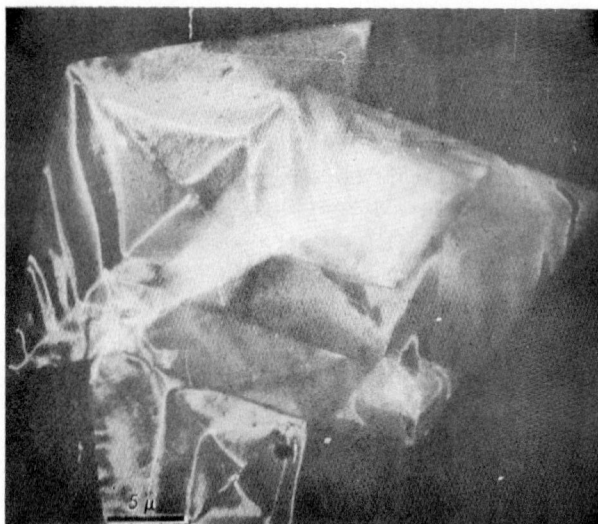

Fig. II-20. Dark field electron micrograph of linear polyethylene single crystals. The reflections imaged were 110 reflections of the upper and lower crystal and a 200 reflection of the central crystal.   The over-all brightness of the {110} fold domains for which the reflection imaged was from nongrowth face {110} planes is greater than that for which the reflection imaged was from growth face {110} planes.   The crystals shown here form pleats when collapsing during solvent removal.

Bassett and Keller point out (43) that this can best be explained by assuming the molecules are tilted within the individual lamellae comprising the crystals, the crystal layers lying flat in the films.

Dark field micrographs also indicate a difference in average orientation of the reflecting planes and thus the molecules in different fold domains, as well as an uneven distortion of the lattice (25–27,30) (Fig. II-20). Bright lines, in which the molecules are oriented so as to reflect electrons into the Bragg reflection imaged, are observed either in two opposite (as in Figure II-20) or all of the fold domains of a diamond shaped crystal. Depending on the reflection imaged two opposite fold domains may also have an over-all brightness considerably greater than the other two. The dependence of the dark field image on the reflection used is discussed in detail later in this section.

All reported electron microscopic observations on polymer crystals have been made on crystals removed from suspension in a solvent by evaporating the solvent. In the case of polyethylene, it was noticed that many dried crystals have a central ridge parallel to the **b** axis (9) (Fig. II-8), others have four sets of parallel corrugations (31,32) (Fig. II-24), one set in each fold domain, whereas still others have a combination of the two effects. The corrugations are apparently not as permanent as the central ridge, in some cases disappearing several hours after removal of the solvent (32), although remaining essentially unchanged for several months in crystals from other preparations (25).

Considerable distortion of the lattice is found in most polyethylene crystals observed directly in the electron microscope. Discrete Bragg extinction lines, essentially parallel to {310} planes, are observed in each domain (32) (Fig. II-21). Moire patterns due to double Bragg diffraction from two superimposed but rotated lattices are seen on some of the spiral growths in this figure. Ghost images of the crystal, the extinction lines and some of the Moire patterns can also be seen in Figures II-19, II-20, II-30, II-31, and II-49. As in the case of the diffraction patterns from individual fold domains, the Bragg extinction patterns indicate that each fold domain is distorted or tilted individually during solvent removal. Although the extinction lines are often continuous across the {010} fold domain boundary (parallel to **a**), they are usually discontinuous across the {100} boundaries. (Fig. II-21). The lines that are bright in a particular dark field image correspond to some of the Bragg extinction lines. The bright lines in dark field images using other Bragg reflections will correspond to the remaining Bragg extinction lines. In truncated crystals a {100} fold domain may extinguish as a unit, extinction lines being visible in the {110} fold domains (28).

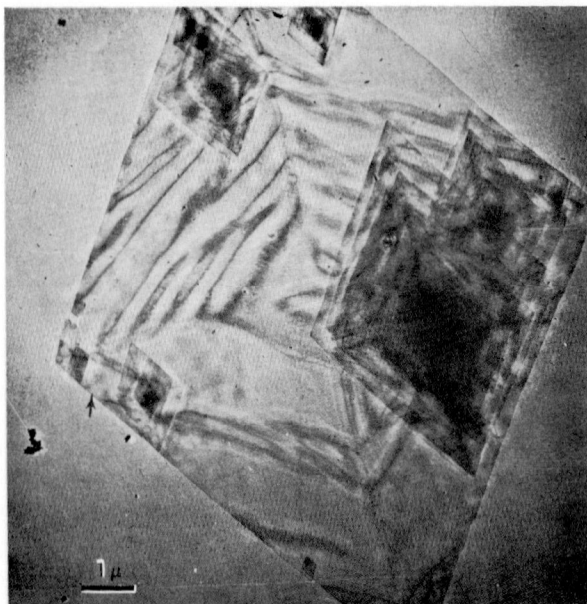

Fig. II-21. Bright field electron micrograph of a single crystal of linear polyethylene on a carbon substrate (no shadowing). The dark, extended, broad bands and lines in each fold domain and along the domain boundaries are Bragg extinction lines. The closely spaced, parallel lines on some of the spiral growths are moire patterns. The white lines, parallel to the Bragg extinction lines, are ghost images. The arrow indicates a ghost moire pattern, this pattern being light in the ghost where it is dark on the spiral growth. A large objective aperture was used for this pattern. With a small aperture the Bragg scattered electrons are intercepted and the ghost images are not found.

On the basis of some of the diffraction, dark field and bright field observations discussed above, Bassett, Frank, and Keller (28) suggested that polyethylene may crystallize in the form of 6- or 4-sided hollow pyramids (depending on whether or not the crystals are truncated diamonds). A similar suggestion, apparently on the basis of similar evidence, was made independently by Niegisch (65). Subsequently, it was pointed out (25) that at least two types of molecular packing (considering the packing of both the straight segments and the folds) as well as intermediate types will lead to hollow pyramidal crystals of polyethylene. This work has been extended still further by Bassett, Frank, and Keller (29,30,41) who have studied two well defined types of polyethylene single crystals. In the following we will consider 4 types of polyethylene crystals in some detail.

Type A (25). Grown from 0.1% tetrachloroethylene solution of an Alathon melt index 3.3 polymer by pouring the hot solution (just below the boiling point) into a separatory funnel and allowing it to stand. The crystals are diamond shaped and fairly simple (Fig. II-22).

Type B (25). A single preparation grown from a very dilute, but poorly defined, tetrachloroethylene solution of the same polymer by slow cooling. These crystals are large, diamond shaped and usually become corrugated during solvent removal (Fig. II-24).

Type C (29,30). Grown from 0.05% xylene solution of a Marlex 50 polymer at 76°C. These crystals are diamond shaped (Fig. II-28).

Type D (29,30). Grown from 0.01% xylene solution of the same Marlex polymer at 90°C and filtered at the crystallization temperature. These crystals are truncated diamonds in which the $\{110\}$ faces are about 3.3 times as long as the $\{100\}$ faces (Fig. II-27).

It has been possible to show (25,28) that the ridges found parallel to the **b** axis of polyethylene crystals, which were originally attributed to a rolling up of the crystal similar to that observed in paraffins (9), actually often consist of two extra layers of material in the form of a triangular pleat (Fig. II-22). These **b** axis pleats are found in type A and D crystals but usually are not found in type B and C. The fractured edges of the pleats in Figure II-22 are complementary indicating that the fracture occurred after growth had ceased and thus presumably during the evaporation of the solvent (25). The pleats thus cannot be associated with the growth of more complex structures as some authors have suggested (26). The width of the **b** axis pleat varies from crystal to crystal, values corresponding to up to about 5% of the length of the crystal being found (25,27,29,30). On some crystals, two or more pleats may be present parallel to the **b** axis. Occasionally, the crystals tear near one or both ends of the **b** axis as well as forming pleats (Fig. II-23). One of the lamellae in the spiral growth on this crystal has also torn, suggesting that the spiral growth was pyramidal also. Fibers, 50–100 A in diameter, are drawn across the tears. Bassett, Frank, and Keller indicate that the lamellae in spiral growths are frequently torn in this fashion (29).

The growth and permanence of corrugated crystals does not seem to be reproducible. Using the same polymer and solvent as for the crystal in Figure II-22, but a more dilute solution and a lower cooling rate, it was possible to grow crystals, Type B, which, following solvent removal, show corrugations only (no **b** axis pleats). These corrugations, as well as the pleats, can be distinguished in the optical microscope using oblique incident illumination (Fig. II-24). In general, **b** axis pleats and corrugations are not

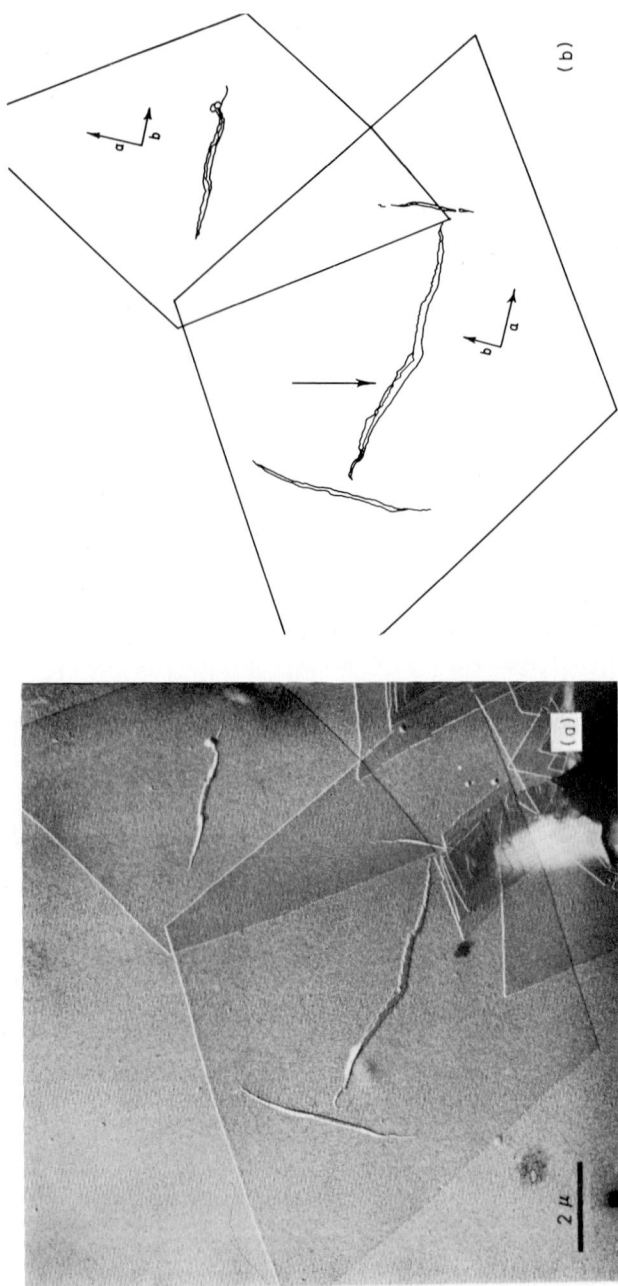

Fig. II-22. Crystals (type A) of linear polyethylene which collapsed to form pleats during evaporation of the tetrachloroethylene solvent. On most crystals the pleats are parallel to the **b** axis, as in the smaller crystal. In some crystals the excess material may remain as a fold normal to the basal lamella, as near the end of the long pleat on the larger crystal. Both edges of this pleat are complimentry, as shown in the tracing (b). On the smaller pleats only one edge has fractured, the excess material collapsing to one side. The growth faces of the crystals are not straight, again indicative of the distortion of the lamellae occuring during collapse of the pyramidal crystal. (Reneker and Geil (25).)

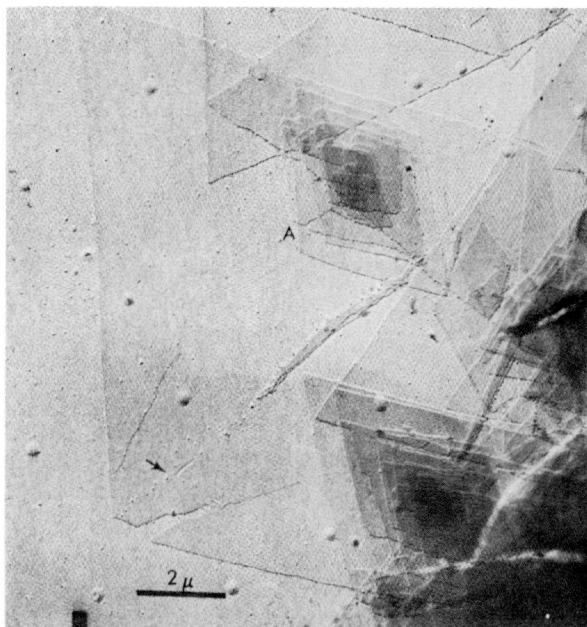

Fig. II-23. Pyramidal crystal of polyethylene which has collapsed with the formation of a pleat and a tear along the **b** axis. Shear between {110} planes and parallel to the fold surface has occurred along the line indicated by the arrow. One of the lamellae in the spiral growth has also been torn (at A) during collapse of the crystal.

observed (14,29) on the same diamond shaped polyethylene crystal. Both features are sometimes seen on complex large dendritic crystals and on truncated crystals. For instance, the truncated, type D crystals have a central pleat and corrugations limited to the {100} fold domains following solvent removal with only occasional corrugations in the {110} fold domains (29). Niegisch and Swan (26) show a micrograph of a truncated diamond shaped crystal containing both a pleat and corrugations in the {110} fold domains. The appearance and permanence of the corrugations, in contrast to the **b** axis pleats, seems to depend strongly on the details of the removal of the solvent. Using the same preparation, type B, corrugated crystals may be absent, present only at the edges of the slide on which the suspension is dried or throughout. The rate of solvent evaporation appears to be important, but all factors are not yet known. Under certain conditions in which the crystal is partially supported during solvent removal, type A

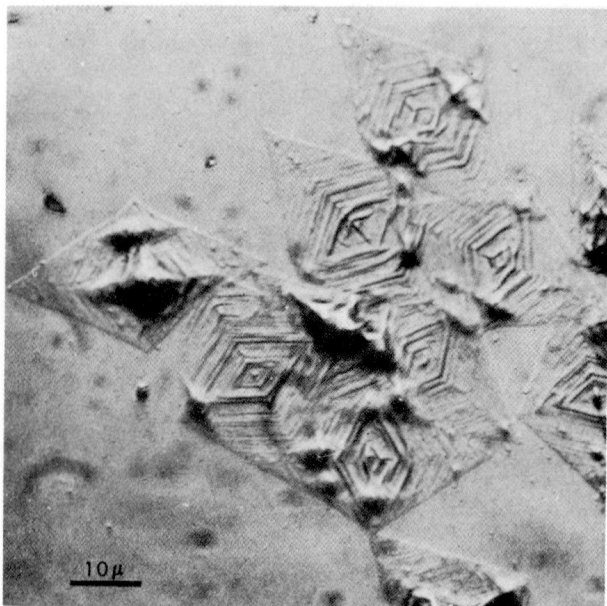

Fig. II-24. Optical micrograph of corrugated crystals of linear polyethylene precipitated from tetrachloroethylene (type B). The corrugations on these crystals remained visually unchanged for at least several months. They are essentially parallel to {310} planes. (Incident steep oblique illumination.) (Reneker and Geil (25).)

(14) and D (30) crystals which normally collapse with the formation of **b** axis pleats may form a type of coarse corrugation instead (see Fig. II-5). Bassett and Keller (29,41,43) have recently indicated that the corrugations can be enhanced by heating a crystal suspension slightly above the crystallization temperature. In the case of type D crystals the enhancement lasts only a few hours, crystals removed from suspension after that time containing **b** axis pleats and {100} corrugations only. Although the corrugations in the crystals in Figure II-24 persisted unchanged for several months, Bassett, Frank and Keller report that the corrugations of type C crystals often disappear a few hours after the solvent is removed (29). Bassett suggests that the permanence of the corrugations in the crystals in Figure II-24 may be a result of their being pleat like (i.e., two extra layers) rather than merely wrinkles in the surface (63). This may occur, depending on the details of the crystal's collapse, if the corrugations are large enough. In such a case the pleats formed are in the ⟨130⟩ direction.

Usually one set of corrugations is present in each fold domain (25). Their

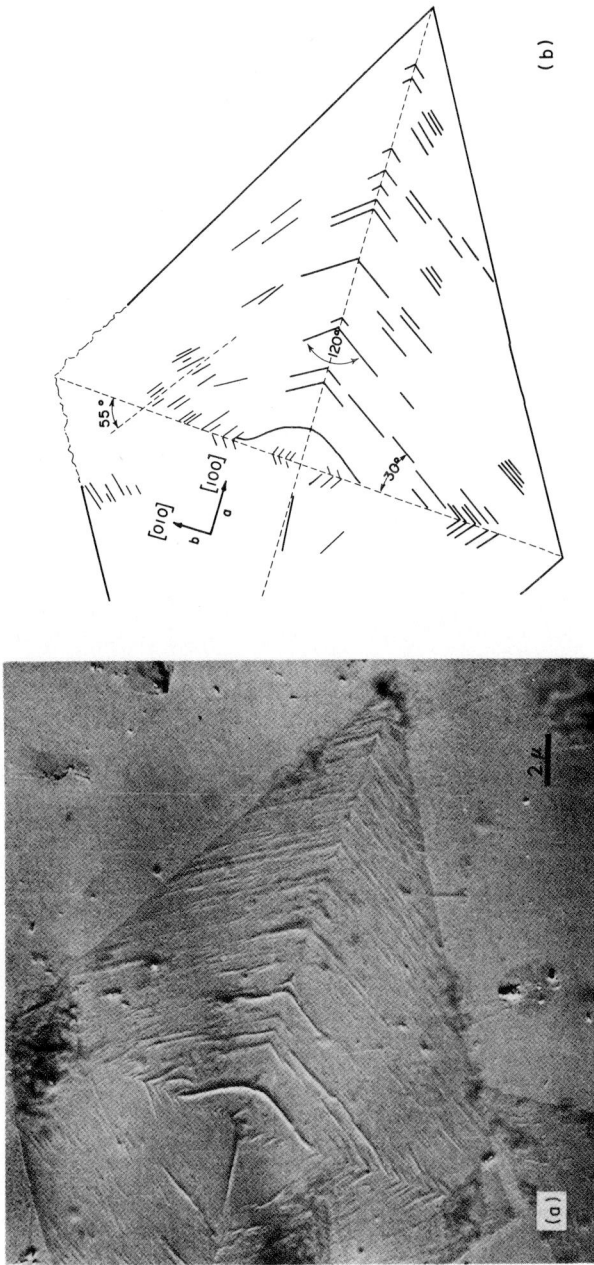

Fig. II-25. (a) Single crystal (type B) of polyethylene precipitated from a dilute tetrachloroethylene solution. (b) Line tracing of the corrugations. The primary set of corrugations are parallel to {310} planes, while the secondary set are parallel to {530} planes.

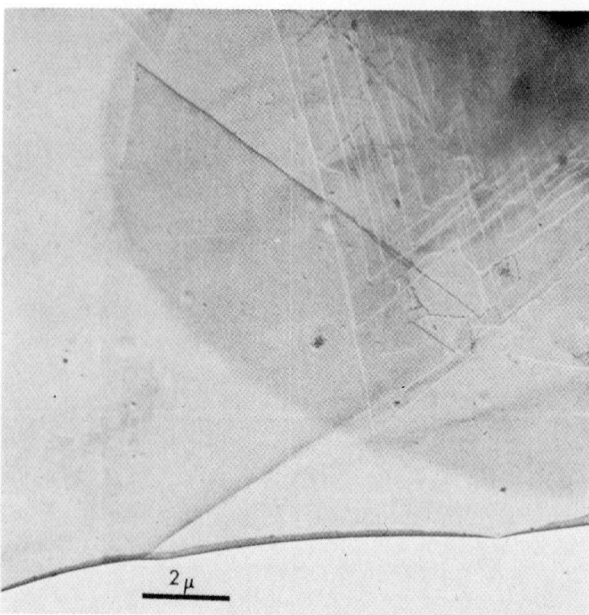

Fig. II-26. A crystal of linear polyethylene folded along its minor diagonal. This suspension of crystals was forcibly projected against the substrate, the fracture of the crystal along the fold domain boundary occurring at this time. On this and other micrographs of collapsed crystals a narrow ridge of material can be observed along the fold domain boundaries. This ridge results from the fact that the crystals are actually three dimensional and represents the material in the acute triangle between the fold domain boundary and a normal to the surface. (Reneker and Geil (25).)

angles of intersection with the **a** and **b** axes (long and short diagonals) of the crystal are slightly larger and slightly smaller, respectively, than the intersections of the {310} planes with the same axes. The angle of intersection with the **a** axis is smaller (closer to the calculated angle for a .{310} plane) at the apex than near the center of the crystal. Occasionally a second set of corrugations is seen on the same crystal. This set appears to be similarly related to the {530} planes of the crystal. (Fig. II-25) (25). Bassett, Frank, and Keller, in their recent papers, describe the primary corrugations as being in the ⟨130⟩, ⟨250⟩ and ⟨120⟩ directions with the secondary set being in the ⟨230⟩ direction (29,43). These directions correspond closely to those listed above.

The formation of the pleats and corrugations is good evidence that in solution some polyethylene crystals are not flat. Further evidence comes

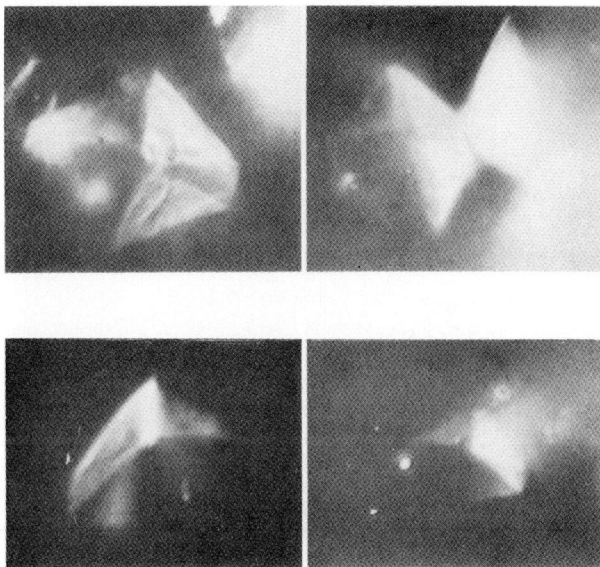

Fig. II-27. Truncated polyethylene single crystals (type D) as observed with dark ground optical microscopy (Bassett, Frank, and Keller (29)).

from forcibly projecting the crystals on to a substrate. Some of the crystals so treated fold back on themselves. In general, the fold is not straight but changes direction slightly where it crosses a fold domain boundary. Brittle fracture occurs along the edge of the fold as the solvent is removed, the two edges being complementary. Although the fold usually occurs with a random orientation, a few crystals have been observed (15,25,26) in which the fold took place along one of the diagonals (Fig. II-26). By measuring the change in direction where the fold crosses the long diagonal, it was possible to show that the crystal in Figure II-26, for instance, must have been a hollow pyramid with a slope of at least 30° measured along a line normal to the {110} growth face.

By evaporating the solvent from the crystals while they are on a liquid substrate and then hardening the substrate, as previously mentioned, it has been possible to observe type A crystals optically and with stereo-electron microscopy while they are still, at least partially, in a pyramidal shape (Fig. II-5). Although coarse corrugations are present on the crystals shown in Figure II-5, the same crystals collapse on a glass slide with a central pleat and no corrugations (Fig. II-22) suggesting that in suspension they are pyramidal and not corrugated.

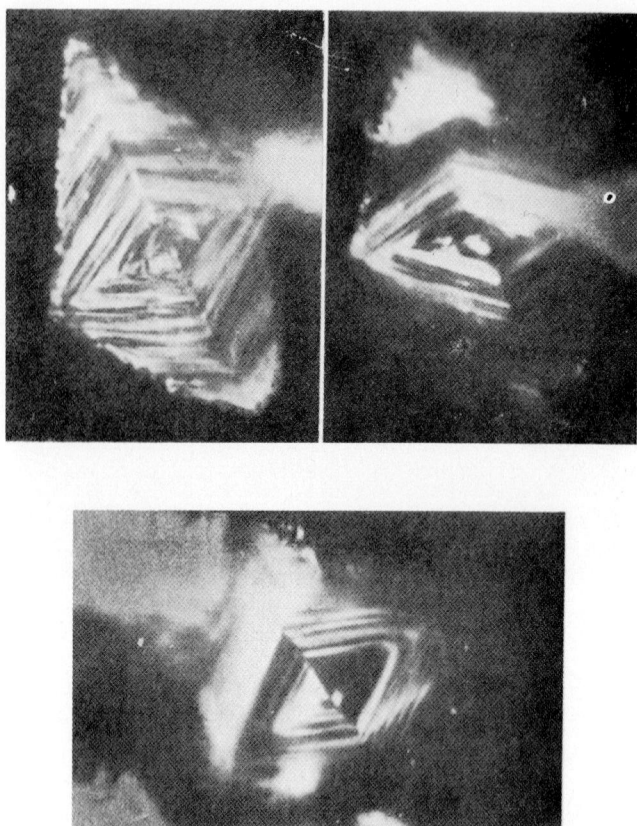

Fig. II-28. Corrugated polyethylene single crystals (type C) as observed with dark ground
optical microscopy (Bassett, Frank, and Keller (29)).

Electron diffraction patterns from the type B crystals with the fine corrugations have more orders of reflections, in the samples we have observed,
than those from the type A crystals which form pleats during collapse. In
both cases forbidden reflections (forbidden by the symmetry of the orthorhombic unit cell) such as $\{100\}$ and $\{010\}$ are present with considerable
intensity (Fig. II-18).The low order reflections, which receive intensity
from portions of all fold domains, are not arced. The narrow width of the
diffraction spots indicates that the fold domains and portions thereof do
not rotate about $\langle 001 \rangle$ with respect to each other to any great extent during
collapse of the pyramid.

Probably the most convincing evidence for the pyramidal nature of

Fig. II-29. Model of a type C, corrugated crystal. In the model the corrugations meet along both the **a** and **b** axes whereas in the crystals they often appear to be out of register along the **b** axis (Fig. II-61). (Bassett, Frank, and Keller (29).)

polyethylene crystals comes from direct observation in the optical microscope. Bassett, Frank, and Keller (29,58) have been able to observe type C and D crystals while still in suspension using dark ground optical microscopy. While the crystal is tumbling in the solution the various fold domains diffract the light in slightly different directions. In a given orientation, certain domains with a particular orientation will be bright, whereas the remaining domains are darker. These observations suggest (29) that the type D crystals are non-flat based hollow pyramids (Fig. II-27), whereas the type C crystals are roughly coplanar but are coarsely corrugated (Fig. II-28). The corrugations, as indicated in the model (Fig. II-29), are in the $\langle 130 \rangle$ directions. It was assumed (25) that type A and B were flat based pyramids. The over-all planarity of the type C crystal and the directions of the corrugations indicate that the fold surfaces are $\{31l\}$ planes (29,30). Type C crystals have been observed while still in suspension in which the central section is pyramidal (similar to the crystal on the left in Figure II-5), but the remainder is corrugated. The edges of the central pyramids in the type C crystals are $\{310\}$ planes, whereas in Figure II-5 they are closer to $\{530\}$ planes. In addition, type C crystals occasionally are found in which only one half is coarsely corrugated, the other two fold

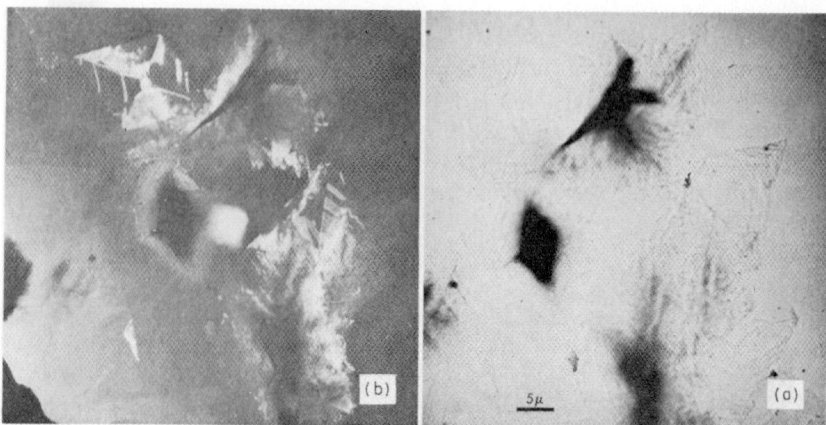

Fig. II-30. Dark and bright field electron micrographs of pyramidal polyethylene crystals which form pleats during collapse. The reflections used for the dark field micrograph (a) were superimposed on the original negative. The selective nature of dark field microscopy can be seen by comparing the number of crystals illuminated in (a) with those present in bright field (b). The reflections imaged on the dark field micrograph correspond to a nongrowth face {110} reflection for the fold domains illuminated in the uppermost crystal.

domains being pyramidal (29,63). These findings present a considerable problem in topography. Bassett indicates (63) that, although diamond shaped pyramids can exist (i.e., noncorrugated in contrast to type C), they were unable to observe and analyze them with dark ground microscopy because of their small size. It is believed that type A crystals are single pyramids but their topology has not been analyzed. Whenever any pyramidal character is present, wholly, at the center or in the form of truncated crystals, a triangular pleat is formed when the crystal collapses during solvent removal (29).

With dark field electron microscopy, it is possible to determine which portions of the crystal and of each fold domain are contributing to a given reflection. With pleated type A crystals, using a {110} reflection, entire fold domains or large portions thereof are bright (Fig. II-30). The remainder of the crystal, except for a few bright lines in ⟨130⟩ directions, is dark. At fold domain boundaries there is a sharp break between light and dark areas whereas within a domain they merge gradually. With a {200} reflection from a pleated crystal, a few broad lines usually parallel to ⟨130⟩ directions may show up as bright in each quadrant. Higher order reflections have not proven suitable for use in dark field microscopy in presently used

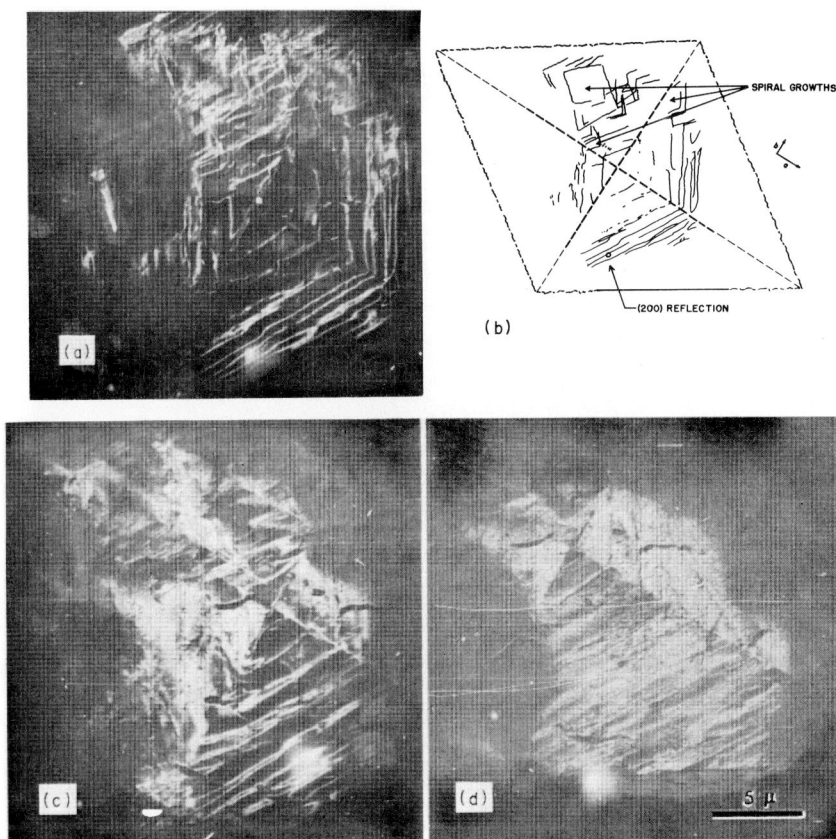

Fig. II-31. Dark field micrograph from a corrugated crystal of polyethylene. The • diffraction pattern from this crystal is shown in Figure II-18. (a) Portions of the corrugated material in all fold domains contribute electrons to the 200 reflection used for this micrograph. (b) Line tracing of material in (a) which is scattering into the 200 reflection imaged. The corrugations in the spiral growths in the (110) and ($\bar{1}$10) fold domains cross those in the fold domains themselves. The approximate outline of the crystal and the arbitrarily chosen axes are shown. (c) Electrons scattered into a $\bar{1}$10 reflection are imaged in this micrograph. The (110 and ($\bar{1}\bar{1}$0) fold domains are bright, whereas only portions of the corrugations are visible in the ($\bar{1}$10) and (1$\bar{1}$0) fold domains. (d) By the time this exposure, using a 110 reflection, was taken, much of the diffracting power of the crystal had been destroyed. The ($\bar{1}$10) and (1$\bar{1}$0) fold domains are still somewhat brighter than the other two, however. The corrugations are not visible. (Reneker and Geil (25).)

equipment. With an {020} reflection, it is expected that all four fold domains would have approximately the same average brightness, the over-all intensity depending on the shape of the original crystal and the details of

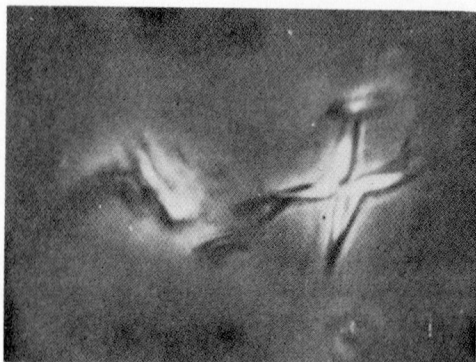

Fig. II-32. A double pyramidal crystal viewed on edge with phase contrast while in suspension (Bassett, Keller, and Mitsuhashi (16)).

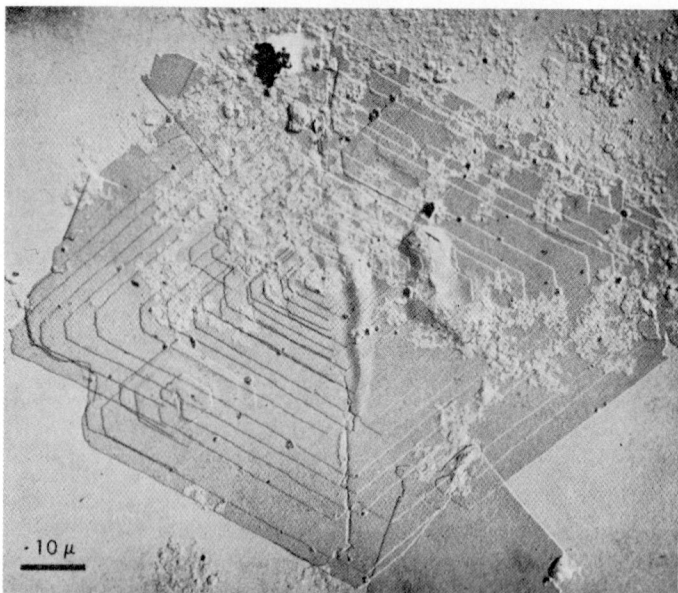

Fig. II-33. Collapsed double pyramid. A pleat has formed in the lamellae which were base down while a crack formed in the lamellae which were base up. (Bassett, Keller, and Mitsuhashi (16).)

its collapse. In bright field, using a small objective aperture, Bragg extinction effects give rise to related effects; i.e., only portions of the crystal are in position to diffract the incident beam. The regions diffracting the electrons at Bragg angles are dark, the electrons being intercepted by the aperture (Fig. II-30). With large objective apertures Bragg extinction lines are also observed at times, aberration in the microscope, as previously discussed, resulting in ghost images (Fig. II-21). In the corrugated type B crystals using a $\{200\}$ reflection, portions of the corrugations in each fold domain are bright, appearing as narrow lines (Fig. II-31). Using a $\{110\}$ reflection, for instance a $(\bar{1}10)$ reflection, the (110) and $(\bar{1}\bar{1}0)$ domains are uniformly bright while portions of the corrugations in the $(\bar{1}10)$ and $(1\bar{1}0)$ domains can be seen. Bassett, Frank, and Keller report similar observations for type C and D crystals (29,30).

In the case of both type A and B crystals, the dark field pattern from spiral growths is independent of that of the basal lamella. In a spiral growth on a $(\bar{1}10)$ fold domain, for instance, corrugations in the (110) and $(\bar{1}\bar{1}0)$ domains of the spiral growth cross the corrugations in the basal lamella (Fig. II-31). It was thus concluded (25) that the spiral growths consist of a separate family of hollow pyramids collapsing independently of the basal lamella. Bassett, Keller, and Mitsuhashi report (16) that occasionally crystals are found in solution which consist of two pyramids joined at their apex (Fig. II-32). When these crystals are deposited on a substrate the upper crystal develops cracks at the **b** axis apices (Fig. II-33).

Bassett, Frank, and Keller have attempted (29,30) to determine the indices of the fold surfaces of type C and D crystals through a number of techniques. As indicated previously, the gross planarity of the type C crystals and the presence of striations in the $\langle 130 \rangle$ directions while the crystals are in suspension (Fig. II-28) requires (29,30) that the fold surfaces be $\{31\ell\}$ planes, these being the only surfaces which will intersect a plane along the $\langle 130 \rangle$ directions. In addition, following solvent removal, the most prominent striations in the type C crystals are in the $\langle 130 \rangle$ directions, a ain suggesting the singularity of the $\{310\}$ planes. In Figure II-5 striations are in the $\langle 350 \rangle$ directions which would thus suggest that these type A crystals have $\{53\ell\}$ fold surfaces (an entirely different suggestion for the singularity of the $\{310\}$ and $\{530\}$ planes is discussed later). The determination of ℓ is more difficult. Through the use of type C crystals in which the center is pyramidal, Bassett, Frank, and Keller (30) have been able to estimate ℓ from direct optical microscope observations. They find ℓ to be 3.9 ± 1, thus indicating that the type C crystals are probably $\{(314)(110)\}$.

It has been possible with dark ground microscopy to see directly that type D crystals are not flat-based (Fig. II-27) (29). The determination of all of the indices has been difficult, however. The evidence presented below is most compatible with the {110} fold domains of the type D crystals being {(312)(110)}. Because of the difficulty of observation and photography only a few crystals have been analyzed directly in dark field optical microscopy.

The observations indicate (30) that h > k, and that the pyramids are steeper than required for 314 but not as steep as required for 311. When type D crystals are deposited on glycerin, a pyramidal center is often retained (as in Fig. II-5). The intersection of the central pyramid and the periphery is approximately ⟨120⟩. The striations that are sometimes observed on these crystals also are fairly close to ⟨120⟩ (30). The transient striations resulting when the crystals are heated slightly appear to be closer to ⟨140⟩. The optical observations thus suggest that the h k indices are between 41 and 21, with values such as 52 and 62 as well as 31 possible.

The acute angle of the triangular pleat that forms on these crystals, assuming that the crystals collapse through tilting of the fold domains toward the center, is in reasonable agreement with $\ell = 2$ (30). This angle is about 5° in type A (25) and type D crystals. Bassett, Frank, and Keller calculate (29) that the angle would be $6^{1}/_{2}$° for $\ell = 2$, and 3° for $\ell = 3$ ($1°10'$ for $\ell = 4$) if the fold surfaces are {31ℓ} planes. In the case of {11ℓ} planes, as suggested by Reneker and Geil (25), the angles would be $8^{1}/_{2}$° for $\ell = 1$ and $2^{1}/_{2}$° for $\ell = 2$. The observed angle would be smaller than the calculated if some of the distortion during collapse took place by molecular shear rather than lamellar tilt (see later discussion).

Electron and x-ray diffraction evidence also suggests that the {110} fold domains in type D crystals are {(312)(110)} (30). Using a tilting stage, Bassett, Frank, and Keller measured the angles of tilt required to bring opposite fold domains to a reflecting position, as determined by dark field observations. The angles obtained agree with those expected for {(312)(110)} fold domains and for {(h 0 ℓ)(100)} fold domains (truncated sectors) compatible with a {312} pyramid. Only regular crystals containing perfect b axis pleats gave consistent results in these experiments. In addition, they noted that the angles of tilt became progressively smaller during the observations as a result of the effect of the beam and thus pyramids somewhat steeper than {312} might be possible. The observed splitting of the {200}, {110}, and {020} x-ray diffraction reflections using sedimented mats of type D crystals is also most consistent with a {(312)(110)} pyramid although the observations might also accommodate {(523)(110)}, {(623)(110)} and perhaps {(423)(110)}. The results, they indicate, are incompatible with the flat based $\{(111)(110)\}$ and $\{112)(110)\}$ pyramids and the $\{(314)(110)\}$ type C crystals. For instance, although the splitting of the $\{200\}$ reflections is compatible with a $\{(111)(110)\}$ crystal, the $\{020\}$ reflection should be split even more than the $\{200\}$ and no split is observable.

In this author's opinion, the present situation regarding the indices of the fold surfaces is not yet completely defined. Polyethylene crystals undoubtedly have the form of hollow pyramids while in suspension. The possibility of flat based pyramids cannot yet be ruled out. Although larger h and k indices may also exist, Bassett, Frank, and Keller's observations (29,30,43) show that at least in some cases the pyramids are nonflat-based (h $\neq$ k) and the evidence suggests that the fold surfaces are $\{31\ell\}$ planes with $\ell = 2$ for a specific type of truncated crystal (type D) and $\ell = 4$ for a form of coarsely corrugated crystal (type C). Pyramidal crystals can be characterized by the direction of the steepest dip of their fold domains as given by the h k indices and the amount of this dip as given by the ℓ index. In the type C and D crystals the direction of steepest dip is identical or nearly so, i.e. $\langle 130 \rangle$ (h k = 31) while the size of the dip is considerably different. Bassett, Frank, and Keller point out (30) that this implies that there is a relationship between the displacement between fold planes and the displacement of the folds within the fold planes. If one increases, so does the other. (See later discussion.) It should be pointed out that all of the diffraction results are based on the assumption that in the crystal as grown, the molecules are parallel to the axis of the pyramid. Agreement of the diffraction results with the direct optical observations tends to confirm this assumption.

The presence of pleats, wrinkles, and uneven shadowing suggests that some crystals of polyoxymethylene may be hollow pyramids (25) (Fig. II-34). Poly-4-methyl-pentene-1 crystals which consist of square lamellae (66) may also form pyramids. Some large complex crystals of this polymer have pleats or folds extending circumferentially or diagonally (Fig. II-35). Furthermore, diffraction patterns from these crystals have as many as six orders of $\{h\,k\,0\}$ reflections (66) indicating that some of the $\{h\,k\,0\}$ planes are tilted. Dark field micrographs were not reported. Nylon 6 crystals from glycerin solution apparently are flat lamellae, as shown by the absence of pleats or corrugations and the presence of only a few orders in the diffraction pattern. Striations that are seen by dark field (11) and occasionally on shadowed samples (11,67) may develop during the heating required to remove the solvent. These striations are exceptionally well developed on a nylon 8 crystal reported by Holland (67) (Fig. II-36). Insufficient evidence has been reported for crystals of other polymers to determine their shape in suspension.

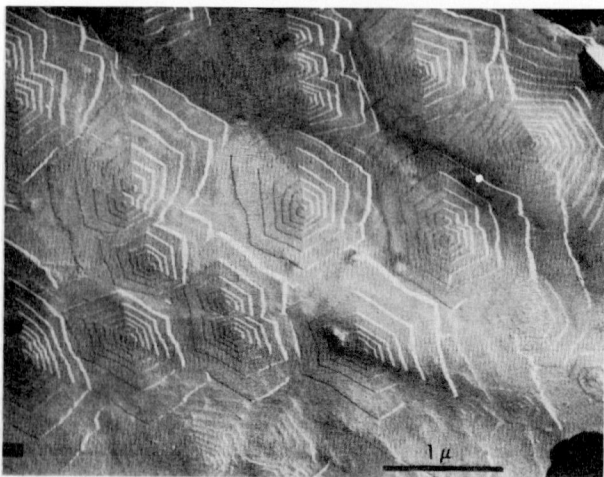

Fig. II-34. Portion of a polyoxymethylene crystal precipitated from a cyclohexanol solution. The uniformity of the shadowing on individual fold domains and its change across fold domain boundaries indicate that each lamella within a spiral growth was sloped, as in a hollow pyramid, when shadowed. Smaller crystals consisting of single lamallae have been observed to have a diagonal pleat along a fold domain boundary.

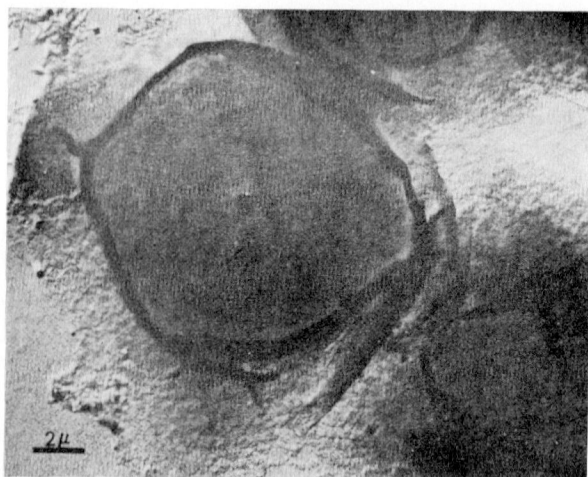

Fig. II-35. Malformed square crystals of poly-4-methyl-pentene-1. The folds or pleats suggest these crystals may have been pyramidal. (Frank, Keller, and O'Connor (66).)

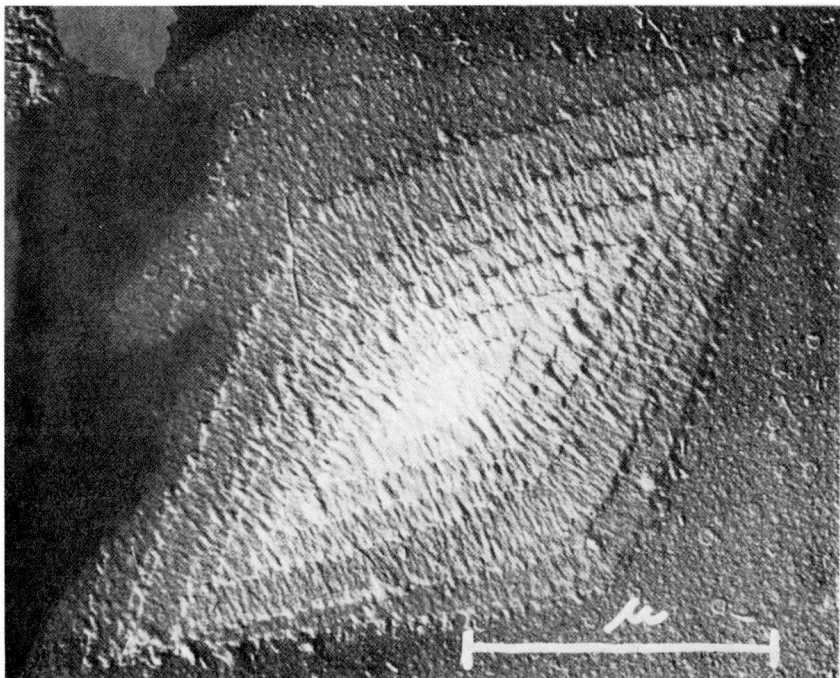

Fig. II-36. Single crystal of nylon 8 precipitated from a dilute glycerine solution.   The glycerine was removed by evaporation in a vacuum oven at ~100°C.   (Holland 67).)

The explanation of the crystallization of some polymers in the form of hollow pyramids and the different modes of collapse has been suggested to lie in the packing of the folds on the two fold surfaces. A detailed description of two modes of packing which lead to flat based hollow pyramidal crystals having different slopes has been described for polyethylene (25). A further variation in the packing will result in a nonflat-based crystal. Similar descriptions apply to other polymers also. If the crystal grows with the pyramidal form (there are suggestions it may develop essentially planar, with fluctuations in fold period, and later rearrange to form a pyramid or corrugated crystal (29)), then it would appear as if the fold period must be uniform over a sizeable number of fold planes. In the following descriptions of the two modes of fold packing, we consider the packing in a (110) fold domain of a flat based pyramid. Similar explanations apply to other {110} fold domains. It is assumed that the conformation of the fold is the same in all folds. Descriptions of the complications introduced by nonflat-based

(a)

REENTRANT FACES

[0l0]
b|
  a ⟶[l00]

FOLD PLANE

(b)

(c)

(00l)

[00l] [0l0]

[l00]

(ll0)

(00l)

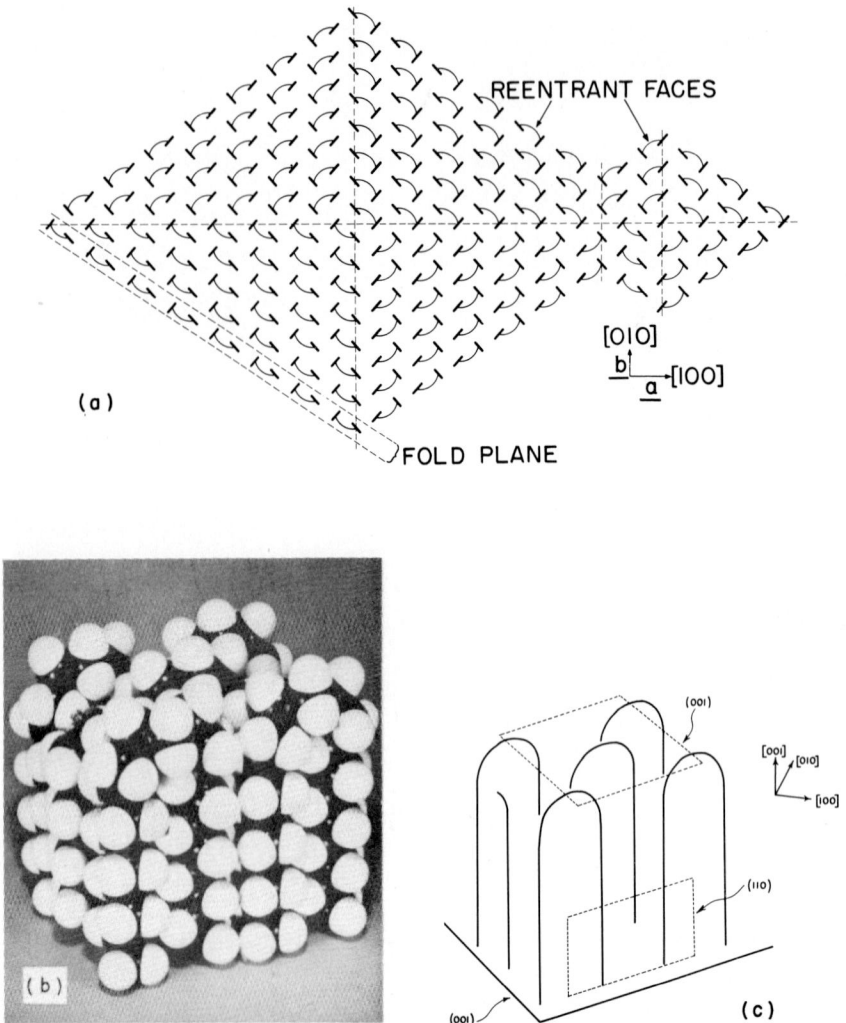

Fig. II-37. Case I fold packing leading to flat or pyramidal polyethylene crystals. (a) Projection of the crystal onto a (001) plane. Planar zigzag of molecular backbone (/). Fold on upper fold surface (⌢), folds on lower fold surface are not shown. The dotted lines are fold domain boundaries. (b) Photograph of a Stuart-Briegleb model showing the orientation of the folds in successive fold planes. The model corresponds to a flat crystal. Each fold plane would be displaced vertically by 1 repeat distance (2 CH_2 units) in a pyramidal crystal. In addition, each fold would be displaced one repeat distance in a nonflat-based crystal. (c) Sketch showing the crystallographic planes and directions for (b). (Reneker and Geil (25).)

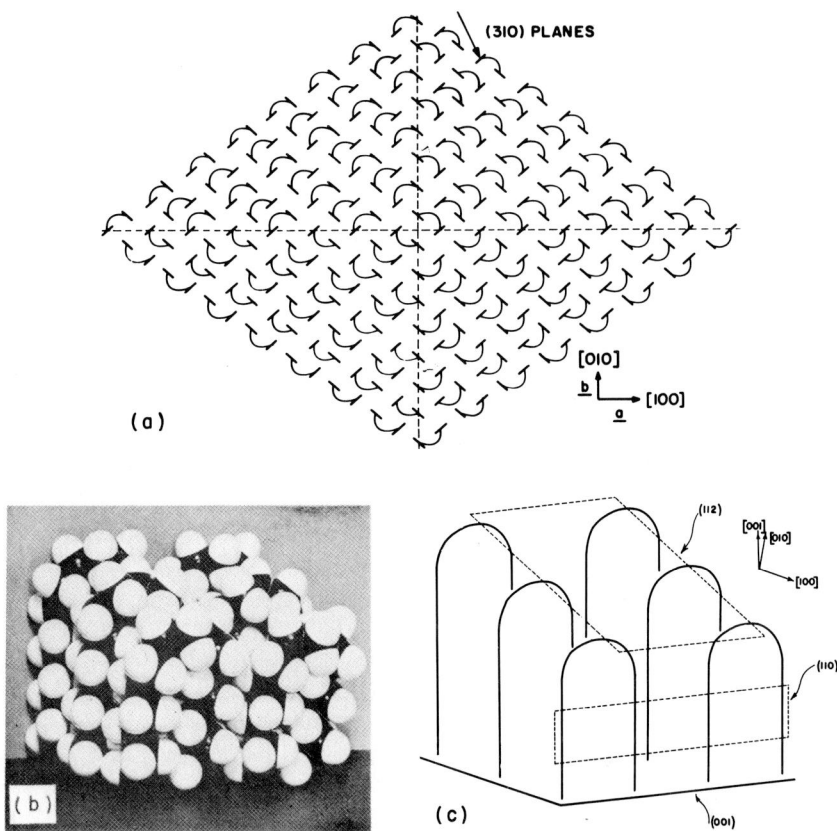

Fig. II-38. Case II fold packing leading to a pyramidal polyethylene crystal. (a) Projection of the crystal onto a (001) plane showing the folds on the upper fold surface only. (b) Successive fold planes are displaced vertically by $c/2$, i.e. one CH_2 group. (c) Sketch showing the crystallographic planes and directions for (b). (Reneker and Geil (25).)

pyramids and the presence of $\{100\}$ fold domains, twinning and dendritic growth are given later in this and other sections.

Case I. In this case, which was described by Niegisch and Swan (26) and Reneker and Geil (25), the folds in successive fold planes are aligned along the [100] and [010] directions (Fig. II-37). The packing of the molecular zigzags in the interior of the lamella requires that the folds either lie in a (001) fold surface or that neighboring folds in successive fold planes be displaced by n repeat distances. For $n = 1$ the folds are displaced by one repeat distance in each successive fold plane and the fold surface is a $(11\bar{1})$ plane. If all of the displacements are in the same direction, a hollow pyramid is formed with a maximum slope of $31.6°$. This agrees well with the value calculated from crystals which have folded along a fold domain boundary (25,26) (as in Figure II-25), and with the width of the larger pleats (as in Figure II-21). Widely spaced changes in the direction of displacement could result in the coarsely corrugated crystals that have been observed on occasion (Fig. II-5). As suggested by Bassett, Frank and Keller (29) the change in observed displacement direction must begin along the **a** axis of the crystal, creating the discontinuities that are observed where the corrugations meet along the **b** axis.

Case II. In this case (25), the folds in successive (110) fold planes are aligned in the [110] and [1$\bar{3}$0] directions (Fig. II-38). Case II differs from Case I in that every other fold plane is rotated by $180°$ about a molecular axis. The packing of the molecular zigzags in this case requires that successive fold planes be displaced by $n + \frac{1}{2}$ repeat distances in the $[00\bar{1}]$ direction. If all the displacements are in the same direction, a hollow pyramid must develop. For $n = 0$ the slope of the pyramid along a line normal to the edge of the crystal is $17.2°$. The folds lie on a $(11\bar{2})$ fold surface. A flat lamella can develop only if the displacements alternate in direction, either regularly or on the average. In both types of hollow pyramids, the molecular segments are not normal to the fold surface. The packing of the segments is similar to some of the "oblique" modifications found in normal paraffins (68,69). When the pyramids collapse, the molecules will make an angle with the substrate if they do not shear within the lamellae.

A nonflat-based pyramid will develop in both cases if the adjacent folds in a fold plane are displaced in a regular fashion as well as the fold planes themselves being displaced (30,43). For instance, a $\{(311)(110)\}$ crystal results if there is a displacement of one repeat distance between each fold plane (Case I) and two repeat distances between adjacent folds in a fold plane, while a $\{(312)(110)\}$ crystal is formed if the displacement between fold planes is one-half of a repeat distance (as in Case II) and adjacent folds in a fold plane are displaced by one repeat distance. A $\{(314)(110)\}$ crystal would require displacements, at least on the average, of one-fourth a repeat distance between fold planes and one-half of a repeat distance between adjacent folds in a fold plane. A type of crystal that one might expect, i.e. Case I with a displacement of one repeat distance between adjacent folds in the fold planes which would be a $\{(211)(110)\}$, has not been reported although, as previously indicated, the corresponding situation for Case II crystals, i.e. $\{(312)(110)\}$ is found. Since it is believed that type C and D

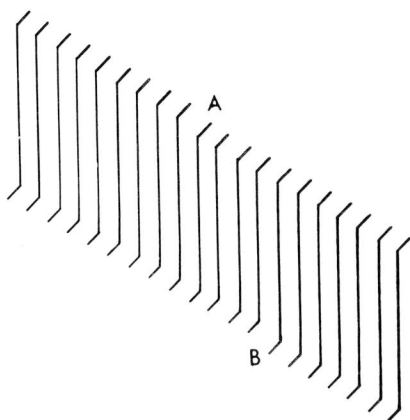

Fig. II-39. Sketch showing staggered packing of fold planes required by the asymmetry of the folds. Fluctuations in fold position due to fluctuations in fold period are more probable in the direction shown for the two fold surfaces than in the opposite direction. At (A) one fold period is slightly shorter than its neighbors to the left of it, whereas at (B) the fold period is longer than the neighbor to the left. The thickness of the lamella, when forming one side of a pyramid, is measured normal to the fold surface and thus is less than the fold period. If the pyramid collapses by the formation of a pleat or corrugations, i.e. the individual fold domains tilting toward the center, this thickness is maintained and the molecular segments make an angle with the substrate. If slip occurs between fold planes when the pyramid collapses, the resulting lamella is as thick as the fold period and the molecular segments are normal to the substrate.

crystals, the only crystals for which appropriate measurements have been reported to date, correspond to $\{(314)(110)\}$ and $\{(312)(110)\}$ crystals respectively (30), it is obvious that it is going to require considerably more study to define and explain the types of fold and fold plane displacement in even polyethylene crystals.

If one attempts to follow the paths of the molecule in the diagrams in Figures II-37a and II-38a, i.e., describe the position of the folds on the lower fold surface as well as the upper, one finds (42,54) that a single spiral winding of the molecules can lead to Case I crystals. This is not possible with Case II crystals, although a double spiral can lead to such a crystal. In this type of treatment, one must consider not only the fact that the molecule is continuous but that the lattice and the lamella in most cases is also. Case II fold packing, for instance, requires that each fold plane be displaced. Thus, in constructing crystals on paper it is necessary to consider the slope of the fold surfaces and the formation of the lattice within the crystal.

Manipulation of Stuart-Briegleb models indicates that the several possible conformations for folds connecting adjacent segments in a $\{110\}$ plane are asymmetric when viewed parallel to the fold plane and fold surface. Several views are shown in Figure II-3. It is not believed that folds significantly more complex than those shown are likely to exist if adjacent segments are connected. The asymmetry of the fold suggests a possible reason for the regular displacement of successive fold planes in the same direction. It is found that the projections of the fold normal to the growth face require a nesting type of arrangement similar to that indicated in Figure II-39. This type of nesting, however, permits (with low energy) only one kind of fluctuation in fold position; i.e., the projections of the folds must, in the diagram shown, be parallel to the fold surface shown or be below it. Whether or not a similar suggestion would apply to the displacements of the folds in a fold plane is not known. The $\{(312)(110)\}$ crystal requires a displacement of two repeat distances (4 CH_2 groups) between adjacent folds.   The models suggest no obvious cause for this large a displacement.

It was suggested (25) that the two types of hollow pyramids corresponding to Case I and Case II fold packing are related to the two types of surface structure observed following solvent removal, i.e., the pleats and corrugations. In both types of fold packing only van der Waals forces act between fold planes. However, in any other set of planes, including nonfold plane $\{110\}$ planes (i.e., $(\bar{1}10)$ in a (110) fold domain), there is the additional bonding between the planes due to folds on the top and bottom surfaces. Certain of these planes (i.e., (100) and (010) planes in a Case I (110) fold domain and (310) and $(\bar{1}10)$ planes in a Case II (110) fold domain) have the peculiar property that the folds which connect segments of molecules in adjacent planes are alternately all on the upper or all on the lower surfaces of the lamella (see Figure II-37 and II-38). A bending moment applied normal to such planes, it was suggested (25), would tend to split them open in pairs, similar to the deformation that occurs if a bellows is bent (Fig. II-40).

Such a bending moment is applied in an essentially radial direction when surface tension pulls down on the peak of a hollow pyramid resting on its base. It was thus suggested that the pleats, which are found primarily parallel to the b axis ($\{100\}$ planes) in some polyethylene crystal preparations, are related to Case I type hollow pyramids and that the corrugations, which are parallel to (310) planes in (110) fold domains, are related to Case II type hollow pyramids. These planes, i.e. the $\{100\}$ and $\{310\}$ along which deformations appear to occur most readily during collapse of a pyramid structure, are the close-packed planes in the two cases which have the particular alternating arrangement of folds noted and are most nearly normal to the bending moment applied to the surface. This suggestion for the singularity of the $\{310\}$ planes is distinct from that arising from the

Fig. II-40. Photograph of a model showing the distortion possibly occurring when a bending mo nent is applied to a family of planes containing segments connected alternately at the top and bottom fold surfaces by folds.   The molecular segments in an actual crystal would, of course, not be rigid but rather might assume more of an S shape.

possibility of growth as $\{(31\ell)(110)\}$ nonflat-based pyramids. Although either or both features may be contributing factors to the distinctness of the $\{310\}$ planes on collapsed crystals, growth as $\{(31\ell)(110)\}$ crystal evidently does occur in those cases in which the corrugations can be identified while the crystals are in suspension (type C). Until the position and packing of the folds can be determined directly in an actual crystal, the relationships suggested above must be considered as tentative only and, in fact, the possibility of collapse with either pleats or corrugations (Fig. II-5 and II-22) indicates the relationships must be at best a simplification of the actual situation.

Adjacent $\{110\}$ fold planes should be very susceptible to relative displacements and separation during collapse of the pyramids. Many crystals are found in type A preparations, for instance, which appear as if they were flat or nearly so, the pleat or corrugations being small or missing. Previously it was noted that the size and number of corrugations, for instance, depend strongly on the details of the solvent removal process. It is suggested that deformation involving the $\{110\}$ planes does occur, but that this deformation results in slip between the fold planes. In Case I crystals an integer repeat distance slip of each successive fold plane would result in a flat crystal with no excess material having to appear as a pleat. The resulting crystal would be thicker normal to the surface after collapse than before (see Figure II-39). The displacement of Case II fold planes cannot be as simple, since a $^1/_2$ integer displacement would alter the lattice. However, small groups of fold planes could slip integer repeat distances as units. The resulting surface would have a furrowed appearance, probably unresolvable in the electron microscope. Niegisch and Swan (26) report being able to observe some fine texture in dark field micrographs of collapsed pyramidal crystals. They suggest that this texture results from localized slip along {100} and {010} planes. Some micrographs of Holland and Lindenmeyer (40) suggest that slip may be easier in crystals of relatively low molecular weight. These crystals appear flat (Fig. 11–9a) whereas those grown from higher molecular weight fractions have pleats. The crystals containing the shorter molecules may have been flat, however, when grown.

Some aspects of the mode of collapse of a hollow pyramidal crystal can be determined from its electron diffraction pattern. If the crystal collapsed by shearing between fold planes, the molecular axes remain normal to the substrate. Electron diffraction patterns from portions of the crystal in which this occurs should be symmetrical about the incident beam and contain only one or two orders of reflections. However, if corrugations or pleats form the (h k 0) planes tilt toward the center of the crystal by varying degrees, permitting higher order reflections to also appear in the pattern. Since the direction of tilt in any one fold domain is more or less uniform, and differs from that in other domains, the diffraction pattern from an individual domain should be asymmetrical (Fig. II-19). The complete diffraction pattern from a collapsed crystal is made up of a superposition of the patterns from each fold domain.

Dark field micrographs should be used in conjunction with the diffraction patterns in determining the mode of collapse. With both types of crystals, corrugated and pleated, in dark field micrographs in which a {110} reflection is imaged, the fold domains are more or less uniformly bright for which the reflection corresponds to a nongrowth face {110} plane. These planes, being essentially radial, are not tilted during collapse of the pyramid. With {200} reflections only those portions of the crystal are bright in which distortion of the crystal has brought the appropriate planes back to a position in which these planes are nearly parallel to the incident beam. This will occur on one side of each corrugation, for instance, if the sides of the corrugation are steep enough. Dark field micrographs of crystals which develop pleats during collapse indicate that some regions of internal distortion are formed within each fold domain. Apparently within these regions, which appear as bright lines or bands in fold domains which should be dark, the (h k 0) planes are tilted sufficiently toward the normal to reflect or else some of the excess material normally appearing in the pleat is accommodated by shear

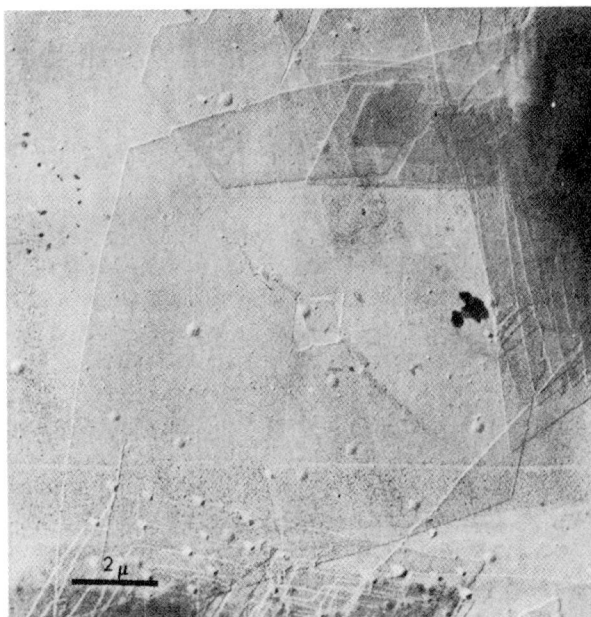

Fig. II-41. Crystal of polyethylene which has collapsed during solvent removal to form a pleat as well as a central diamond-shaped hole. There is a 4.1% deficiency of material at the center of the crystal. The small diamond-shaped piece of material appears to have been rotated by 90° as the crystal collapsed. (Reneker and Geil (25).)

between fold planes. The distortion is also evident from the fact that broad bands or regions may be dark in fold domains which should be bright, indicating that the "radial" planes have also tilted to some degree.

The fold packing that leads to hollow pyramidal crystals may also lead to saddle-shaped crystals (25,28,29). In this case, the angle between the fold domain boundaries is greater than 90° rather than less. A deficiency of material exists at the center. When such a crystal collapses during solvent removal, pleats may form at the apices or a tear may develop in the center. Several crystals have been found with an apparent deficiency of material at the center, a diamond-shaped ring being formed during solvent removal (Fig. II-41). The simultaneous presence of either a pleat near the center, as in Figure II-41, or of tears near the apices, however, leaves some doubt as to whether these crystals were saddle-shaped (25). Niegisch and Swan indicate (26) that they have some evidence that saddle-shaped crystals may exist. In nonflat-based crystals a saddle shape results if the displacement

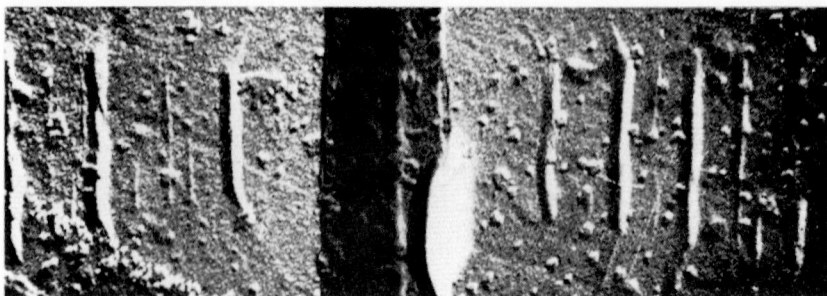

Fig. II-42. Detail of a truncated polyethylene crystal with regular ridges, similar to that in Figure II-11, which has been heated in suspension (Au-Pd shadowed) (Bassett and Keller (43)).

between fold planes is retained but that between the folds in a fold plane is reversed (30).

Although the previous description of the packing of the folds applies to polyethylene crystals having only $\{110\}$ growth faces, truncated diamonds (26,28–30) and hexagonal polyethylene crystals also develop in the form of hollow pyramids. When these crystals, at least some of which are nonflat-based (29), collapse, regularly spaced pleats or wrinkles often develop parallel to the $\{100\}$ planes in the $\{100\}$ fold domains (29)(Fig. II–42). (In reference 19, these wrinkles were assumed to result from thermal expansion during annealing, but subsequent work shows that they were present in the $\{100\}$ fold domains before annealing.) In many of these crystals, for instance, type D, **b** axis pleats are also found as well as occasional corrugations approximately parallel to $\{210\}$ planes (29). The $\{100\}$ fold planes can also be displaced by integer or half-integer repeat distances, depending on the fold packing. Paper experiments, using diagrams similar to those in Figures II-37 and II-38, suggest (42) that complex pyramidal crystals of a flat based type could develop in which the $\{110\}$ fold domains on one side of the long diagonals have a $\{111\}$ fold surface whereas those on the other side have a $\{112\}$ fold surface. The $\{100\}$ fold domains can smoothly join the two sides if the folds are displaced within each fold plane as well as between adjacent fold planes. In dark field micrographs the $\{110\}$ fold domains in truncated crystals appear similar to those seen in diamond shaped crystals. The contribution of the $\{100\}$ fold domains to the diffraction pattern has not been determined. In the dark field tilting stage experiments of Bassett, Frank, and Keller (29), using a 110 reflection, the (100) and ($\bar{1}$00) fold domains are bright at a lower degree of tilt than the

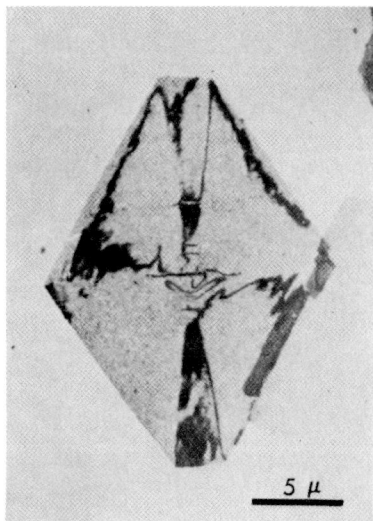

Fig. II-43. Truncated polyethylene crystal in which Bragg extinction occurs primarily in one (100) fold domain (Bassett, Frank, and Keller (28)).

(110) and ($\overline{1}$10) suggesting that the (110) planes in these domains have not tilted to as great a degree as in the {110} domains. In a bright field micrograph of Bassett, Frank, and Keller (28), a (100) fold domain was extinguished whereas only portions of the {110} domains were dark, again indicating the different tilts of the planes in the different fold domains following collapse (Fig. II-43).

Since molecules of polyoxymethylene and poly-4-methylpentene-1 have a helical conformation, the possible types of fold packing in pyramidal crystals of these polymers is not as clear as in the case of polyethylene. Although the unit cell is much longer, the displacement of adjacent fold planes, if it occurs at all, is believed to be about the same as in polyethylene. The helical conformation of the molecule may restrict the directions of the fold at a particular point along the molecule. Experiments with models should help to determine the possible types of fold packing in crystals of these polymers.

5. DENDRITIC GROWTH

In general, polyethylene crystals precipitated from a good solvent by cooling a solution to room temperature are dendritic. An optical micrograph of such a crystal, grown from a 0.1% solution in xylene, is shown in Figure

II-44. These crystals develop, as the solution cools, from initial, relatively simple structures, as can be shown by filtering the solution at an elevated temperature. Apparently, as the solution becomes cooler, growth of the crystals becomes more irregular and the dendrites develop. Part of the irregularity may be due to incorporation of molecules of lower molecular weight and higher degrees of branching; however, as indicated previously, Bassett and Keller report (15) that the tendency toward dendritic growth at a given crystallization temperature increases with molecular weight.

Fig. II-44. Optical micrograph of a dendritic crystal of linear polyethylene.   A double ridge is visible along each of the primary fold domain boundaries and also along some of the secondary fold domain boundaries.   The larger secondary dendrites are seen to overlap near the central portion of the crystal.   (Geil and Reneker (13).)

Unless extremely rapid cooling rates are used (e.g., 50°C/min. or greater), dendritic polyethylene crystals tend to retain the over-all outline of the simpler crystals. Those precipitated from xylene and tetrachloroethylene, for instance, are basically diamond shaped.

In the simplest dendritic crystals, serrated growth faces develop (Fig. II-45). These faces are composed of $\{110\}$ fold planes. As the crystallization temperature is reduced there is evidence that $\{310\}$ faces may also develop (15). Two fold domains are associated with each serration. In the crystal shown, which was complicated but basically pyramidal before it collapsed, the corrugations change direction across each domain boundary (Fig. II-46). Observations of the corrugations (13,15) and dark field micrographs (15) suggest that the domain associated with the reentrant face begins to develop after some lateral growth of the basal lamella occurred.

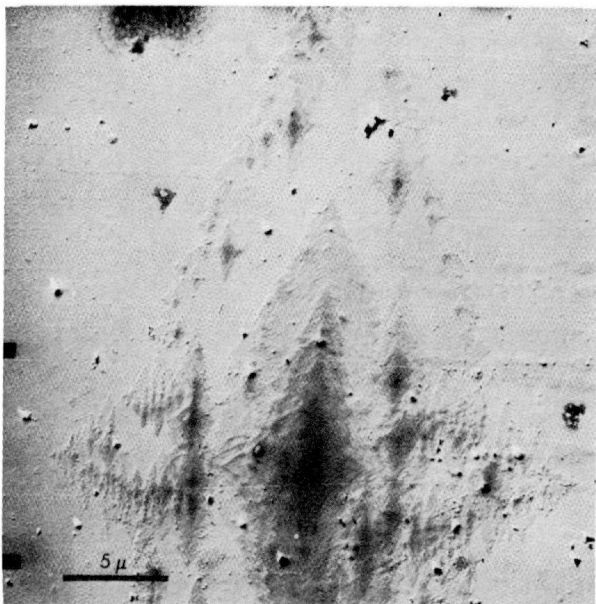

Fig. II-45. Relatively simple, large, dendritic crystals of polyethylene. Secondary dendrites have developed normal to the **a** axis only. Serrated growth faces are present where growth is normal to the **a** axis. The crystal was pyramidal, forming corrugations when it collapsed. (Geil and Reneker (13).)

Geil and Reneker (13) reported a number of observations of dendritic crystals of polyethylene and discussed a possible growth mechanism. Since the linear growth rate in the direction of the apex in dendritic crystals is larger than the growth rate in other directions, it was suggested that new fold planes are nucleated at the apices of the crystal. In the following discussion it should be realized that the crystal may have started growing at an elevated temperature as a simple diamond shaped, probably pyramidal, lamella and that as the solution was cooled the growth habit changed. The growth mechanisms proposed by Bassett and Keller (15) are similar to those given below.

In order to form a reentrant growth face (see Figure II-37a) some of the fold planes must terminate before they have incorporated very many molecules. Small reentrant faces are observed close to the apex suggesting that the terminated fold planes may contain only one or a few molecules. A fold plane may terminate or at least be temporarily pinned when a molecule ends, when impurities or similar imperfections are encountered, or if a mole-

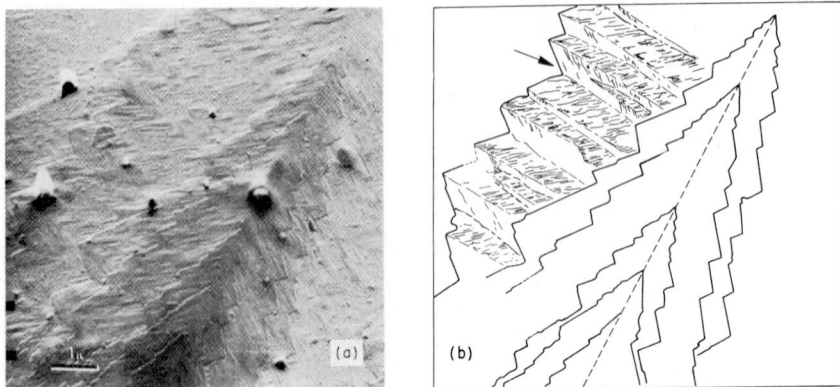

Fig. II-46. Portion of the central spiral growth on the crystal in Figure II-45.   The directions of the corrugations, which are essentially independent in each lamella, change across each fold domain boundary within a lamella.   A tracing of the corrugations is shown in (b).   (Geil and Reneker (13).)

cule turns about and starts folding back toward the apex. As new fold planes are nucleated at the apex they will overtake the "pinned" fold plane and create a jog in the growth face.

Reentrant faces develop if several fold planes are pinned in the same vicinity so that the molecule in the advancing fold plane can turn a corner at the jog and fold along the face of the jog. If only one fold plane is pinned, a defect similar to an edge dislocation is formed and the resulting jog is probably smoothed out with the growth of additional fold planes. When a number of pinned fold planes are involved, and a new growth face develops, the growth of additional fold planes from the original apex preserves the magnitude of the jog (Fig. II-47). The new growth face corresponds to a new fold domain, a new apex also having been formed. The new face can increase in extent, the new fold domain widening, either by termination or reversal of the direction of growth of fold planes from the original apex or by the nucleation of new fold planes at the newly formed apex. Some small re-entrant faces and their associated fold domains that are formed near the apex of these relatively simple dendritic crystals gradually merge into larger faces and domains as growth proceeds. Occasionally, it is observed that even some of the relatively wide fold domains grow out, i.e., a band of corrugations that delineates a fold domain narrows and disappears as it approaches the edge of the crystal.

In discussing the growth of dendritic crystals, Geil and Reneker point out (13) that the presence of reentrant faces on even a flat-based pyramidal crystal introduces complexities beyond those discussed in Section II-4. Figure II-47 suggests two ways in which the reentrant face can develop once it is formed. Either the folds in a fold plane may step up the reentrant face (along line AC) or they may remain on a line in a $\{001\}$ plane (along line AB). In the first alternative, each fold must be displaced from its neighbor in the fold plane by two (Case I) or one (Case II) repeat periods. This kind of fold packing in a fold plane is the same as that giving rise to nonflat based

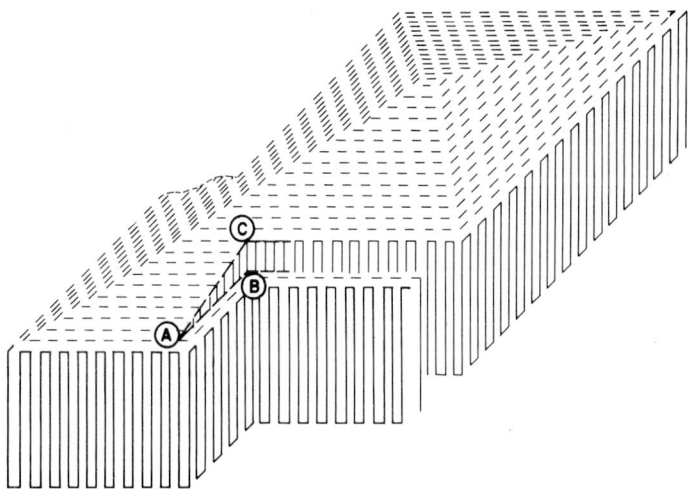

Fig. II-47. Schematic drawing of a pyramidal polyethylene crystal showing the development of a re-entrant face due to the termination or pinning of several fold planes. The growing fold planes have "turned the corner" A and started a new fold plane with a different growth face. The molecules within one of these fold planes may either fold along AB or AC. In the later case, this drawing corresponding to Case I fold packing, adjacent folds within a fold plane must be displaced vertically by 2 repeat distances. (Geil and Reneker (13).)

pyramids in simpler crystals. In this alternative, no aligned molecular segments are exposed. However, if the folds remain along line AB, then as the reentrant face expands, a wedge of exposed aligned segments develops. This wedge can serve as the nucleus for the growth of additional lamellae, i.e., a screw dislocation and its associated spiral growth can develop. The exposed thickness, BC, probably need not be as large as the thickness of the lamellae before the spiral growth begins to develop. Bassett and Keller point out (15) that a wedge could also be formed if one of a pair of reentrant faces in a nonflat-based pyramid (and probably in a flat-based pyramid also)

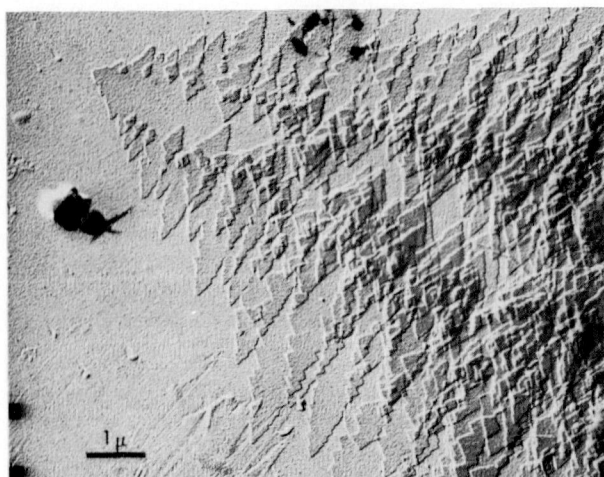

Fig. II-48. Electron micrograph of a portion of a dendritic crystal from the same preparation as the crystal in Figure II-44.   The double ridges along the primary fold domain boundary (here parallel to the **a** axis) in Figure II-44 are seen to result from a double row of spiral growths and overlapping of secondary and tertiary dendrites.   Slip has occurred parallel to the molecules at the edges of the overlapping secondary dendrites. Along the primary fold domain boundary the crystal in many places is only one lamella thick.   (Geil and Reneker (13).)

began to grow more rapidly than the other. The actual situation is undoubtedly complex and all of the above factors may contribute.

Most dendritic crystals are not as simple as that in Figure II-45, instead resembling that in Figure II-44. Figure II-48 shows an area near the tip of a crystal from the same preparation as the crystal shown in Figure II-44. Two rows of spiral growths and overlapping lamellae are seen along each primary fold domain boundary (diagonal of the crystal) corresponding to the double ridges seen in the optical micrograph. Secondary dendrites, growing normal to the primary fold domain boundaries, also have double rows of spiral growths and overlapping lamellae along the secondary fold domain boundaries. Although in this micrograph the secondary fold domain boundaries are parallel to the b axis, similar dendrites develop with domain boundaries parallel to the **a** axis. The presence of pleats and corrugations shows that many of these complex dendritic crystals are also basically pyramidal, the slope changing at each fold domain boundary. The pyramidal nature can be seen on the branch of a dendritic crystal in Figure II-5. Dark field micrographs indicate that the tilt of the molecules, following collapse,

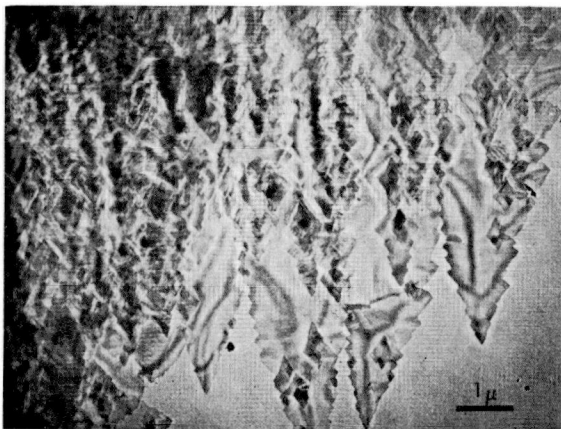

Fig. II-49. Bragg extinction lines in some of the secondary dendrites of a dendritic polyethylene crystal. Moire patterns can also be seen on many of the spiral growths as well as in regions in which the secondary dendrites overlap. · Ghost images of the crystal are visible here as in Figure II-21.

changes across each fold domain boundary. In bright field micrographs Bragg extinction lines also indicate that the dendritic crystals may be pyramidal (Fig. II-49). Some of the spiral growths along the fold domain boundaries probably result from the mechanism suggested above. The origin of others is described below.

When a pyramidal dendritic crystal collapses on the slide during solvent removal, slip occurs parallel to the molecules along one edge of those portions of neighboring secondary dendrites which overlapped during growth. The overlapping regions are left superimposed. This overlapping of adjacent dendrites is believed to be an important feature in the growth of complex dendritic crystals. After the secondary dendrites growing from the apices of the serrated growth faces reach a certain size, the region near their tip resembles the tip of the primary dendrite, i.e., it too is serrated. Tertiary dendrites form. In a pyramidal crystal, neighboring tertiary dendrites on adjacent secondary dendrites will not be at the same level and likely will be growing at different angles above or below the plane of the primary dendrite. As growth continues, the tertiary dendrites on one secondary dendrite will overlap or intersect those growing on the adjacent secondary dendrite. If the crystal collapses, at this stage, the superimposed lamellae observed in Figure II-48 and II-49 are formed. If growth continues, however, the space between the adjacent secondary dendrites fills in and a spiral growth

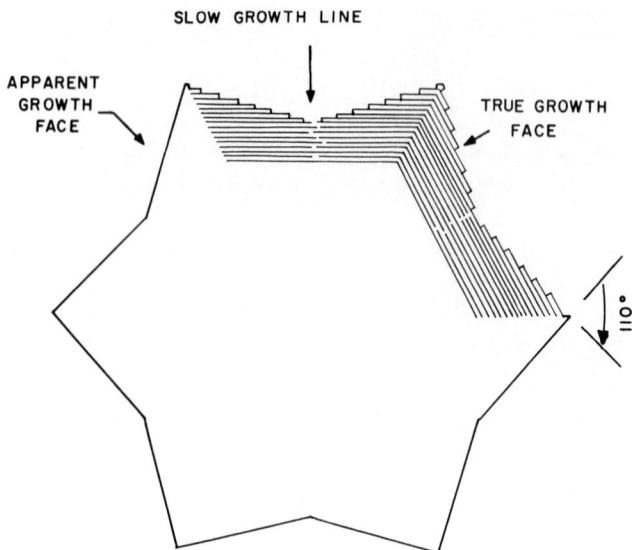

Fig. II-50. Diagram of a star-shaped crystal of polyoxymethylene.   The true growth face is made up of fold planes growing from adjacent apices toward each other.   (Reneker and Geil (25).)

will form where the tertiary dendrites overlap. The intersection and over-lapping of these tertiary dendrites and the structure of pyramidal dendritic crystals leads to so many complications and variations that, beyond the recognition of their possibility, it was not (13) and is not believed fruitful to discuss them further. Similar complicated structures can develop in planar crystals if hydrodynamic forces or impurities cause sufficient distortion to prevent merging of neighboring tertiary dendrites as they are growing.

Polyoxymethylene precipitated from some solvents consists of six pointed starlike crystals (25)(see Figure II-34). Apparently nucleation of new fold planes again occurs primarily at the apices of the basically hexagonal crystals (Fig. II-50). It was suggested that, as the molecules fold back along the growth face, impurities, improper folds, and other defects tend to accumulate at the centers of the resulting reentrant faces. As is discussed in Section II-7, this region serves as the source of numerous spiral growths. The line joining the center of the crystal with the center of the resulting reentrant faces is known as a slow growth line. Serrations are often found in these growth faces, presumably again because of the termination or pinning of some fold planes. Dendritic crystals of nylon 6 appear to develop in a manner analogous to that described for polyethylene crystals. The dendritic nylon 6 crystals observed to date are probably flat.

6. TWINNING

As indicated previously, polymer crystals have been observed in which the basal lamellae and/or the growths from a screw dislocation consist of growth twins. Each twin consists of one or more fold domains, the orientation of the lattice as well as that of the folds being different on opposite sides of the twin boundary or composition plane (Fig. II-51). Various aspects of their appearance have been described by Khoury and Padden (64) and Burbank (54). Burbank has also discussed in detail (54) the types of twins that can develop. His descriptions and diagrams of twins and simpler crystals, in terms of fold packing, orientation, and twinning are reviewed later in this section. With Burbank (54) we define a twin to be a crystal aggregation growing from a single nucleus that gives a diffraction pattern corresponding to two reciprocal lattices with a definite crystallographic relationship between them (Fig. II-52).

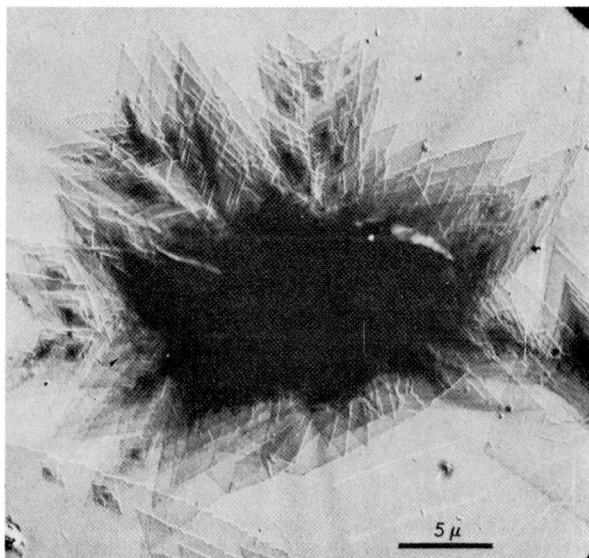

Fig. II-51. Complex "growth twin" crystal of polyethylene from the same preparation as the crystals in Figures II-24 and II-31. The composition planes are {110} planes, radial growth being in the **a** axis direction. A region corresponding to the end of one of the arms is shown in Figure II-17.

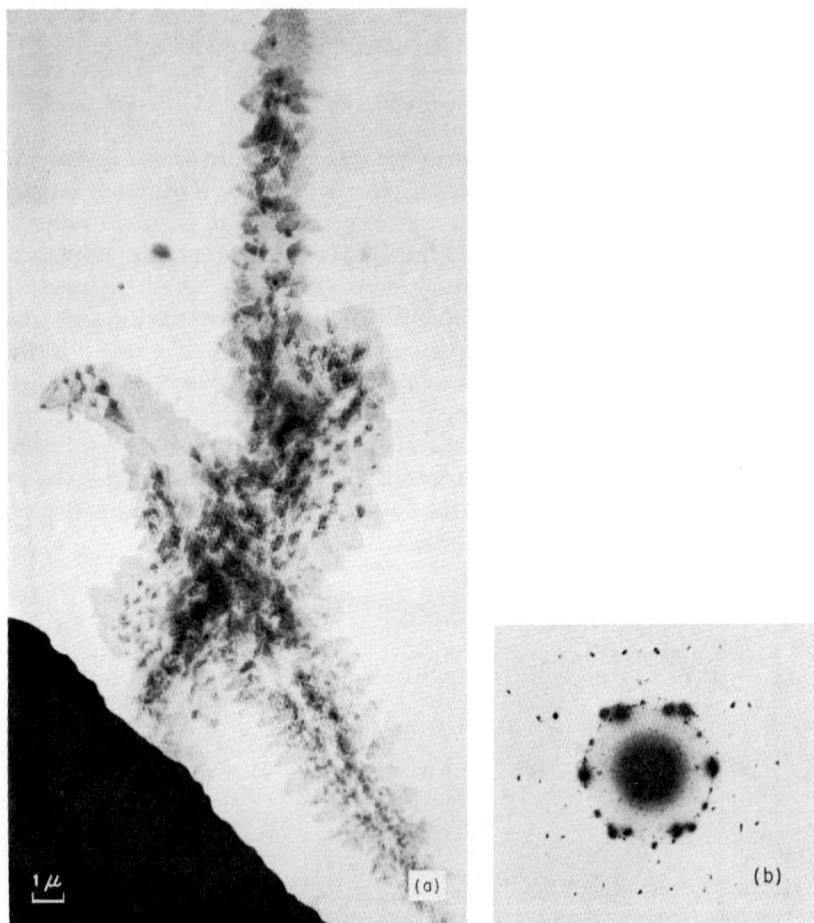

Fig. II-52. Complex twinned polyethylene crystal (a) and its diffraction pattern (b). (110) and (120) growth twins are present, the long arm extending upward being the (120) twin. The cause of the curvature of one of the arms is not known. The Bragg extinction patterns following the serrated growth faces indicate that this crystal was pyramidal in nature. Besides the three superimposed but rotated diffraction patterns from this crystal (b), numerous double Bragg reflections are also visible. These spots, in the ring of {110} reflections, result from the superposition of lamellae which are rotated with respect to each other. Moire patterns result from the same effect. (Burbank (54).)

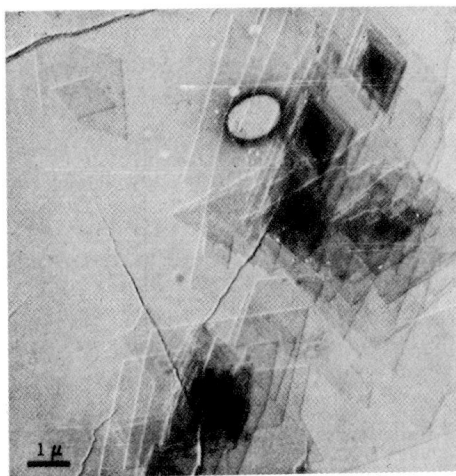

Fig. II-53. Portion of a picture frame-type crystal of polyethylene. Several lath-shaped {110} growth twins have formed, i.e., the composition plane is a {110} plane. Individual diamond-shaped lamellae in {110} twin relationship with each other and the basal lamellae are also present. Most of the spiral growths as well as the development of re-entrant faces took place at the edge of the crystal when it was rapidly cooled following an initial isothermal growth.

Some relatively simple polyethylene crystal growth twins are lath shaped. The composition plane, having some of the characteristics of the slow growth lines in star shaped polyoxymethylene crystals, often is the origin of spiral growths. When spiral growths are formed at the composition plane, the resulting lamellae are diamond shaped. The orientation of the diagonals in the lamellae growing from opposite sides of the composition plane, a spiral developing on each side of the basal lamella, differ by an angle corresponding to the twin relationship (64). In some cases, as in Figure II-53, the lath shaped twins have a definite, sharply defined reentrant face. In other crystals, however, the end of the lath is irregular, the reentrant face being partially filled in (64). The growth rate of the twinned crystals is about the same as that of the normal crystals. Greatly elongated laths of polyethylene, for instance, have not been observed. The ribbons formed during the crystallization of nylon from some solvents (Fig. II-4) are not believed to be due to twinning. Hydrogen bonding affects their shape. Further growth of the lath shaped twins, if in polyethylene crystals dendritic in nature, is believed to lead, through structures like that in Figure II–17, to complex crystals as in Figure II–51. Two or more pairs of arms develop, each half of a pair being related by a twin relationship.

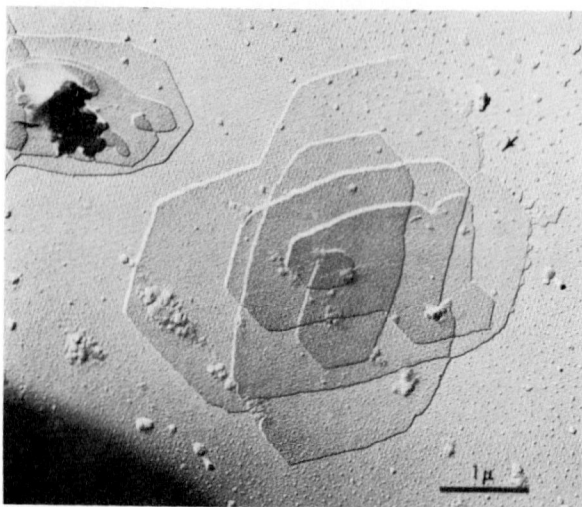

Fig. II-54. Polyethylene (110) growth twin crystallized on a substrate from octane. The composition plane extends in the direction of the arrow. If growth had continued a re-entrant face would have developed at each end of the composition plane. The presence of impurities on the substrate has not greatly affected the crystal's growth except at the lower end of the composition plane. Radial growth is in the **b** axis direction, single crystals in this preparation consisting of truncated diamonds in which the {100} growth faces are larger than the {110} growth faces. (Reneker (42).)

Relatively large numbers of star shaped dendritic crystals, usually six pointed, are found in polyethylene crystal suspensions formed by slowly cooling solutions of greater than about 0.1% concentration, using solvents which normally lead to diamond shaped crystals (see Figure III-2a). The primary fold domain boundary in each arm is parallel to the **a** axis of that arm. It is believed that these star shaped crystals develop from a centrally located spiral growth in which the lamellae in the spiral on each side of the basal lamella are in twin relationship with the basal lamella. The initial stages of the related development of a four pointed star have been observed (Fig. II-54)(42). In this case, the spiral growths appear associated with the actual nucleation of the crystal. In this crystal the **b** axes are radial.

Wunderlich and Sullivan (70,71) report that in the star-shaped six-pointed dendrites (grown from 0.1% or greater toluene solutions cooled with a room temperature gradient) the two side arms are at an angle of 46° to each other, with a 67° angle between the side arm and the primary **a** axis fold domain boundary. The secondary dendrites make an angle of 67° with the primary fold domain boundary on the main arms. On the

side arms the secondary fold boundaries are at an angle of 67° to the primary fold domain boundary of the side arm on on side and 46° on the other
side. A 46° angle could correspond to {120} twinning. At lower concentrations they find that the angle between the side arms decreases (71),
being about 30° at 0.054% and having only single side arms for less than
about 0.01% (as in Fig. II-44). Since all of this work was done with optical
microscopy, they were not able to clearly define the growth axes. From the
relative birefringence they suggest that the side arms also have an **a** axis
fold domain boundary but are two to four times as thick as the longer arm.
Further observations are needed to define more completely the mechanism
of formation of twins, both the star-shaped crystals and the lath-like
lamellae.

In order to adequately label the various types of fold domains involving
different fold planes, fold packing and fold surfaces, Burbank (54) suggests
that a fold domain be labeled with two sets of indices, the first corresponding to the fold plane and the second to the shortest vector displacement on a
fold surface between two equivalent folds in adjacent fold planes. Thus,
Case I crystals would have four {110} ⟨010⟩ fold domains if flat and four
{110} ⟨011⟩ fold domains if pyramidal with a flat base and unit displacement between adjacent fold planes. The simplest Case II crystals would
contain four {110} ⟨$^1/_2$$^1/_2$$^1/_2$⟩ fold domains. This differs from the labeling
introduced by Bassett, Frank, and Keller (30) in which the second portion
of the label refers to the indices of the fold surface itself. Until the entire
situation with respect to the packing of the molecules and the folds in
polymer crystals is clarified, care will be required in clearly defining the type
of labeling that is being used.

Burbank has considered (54) the two dimensional aspects, as viewed on an
(001) plane, of the formation of twins and simpler crystals as it is affected
by chain folding. Although he requires that the molecules be continuous
across the fold domain boundaries, he does not consider the packing of the
molecular zigzags within the lattice and its effect on the relative elevation
of the folds. Burbank considers cases in which not only the orientation of the
folds but also the type of fold packing changes across the fold domain
boundary (i.e., as in truncated crystals) although the continuity of the lattice within the crystal is maintained. He suggests that in this case the term
"phase boundary" be used. The two phases, corresponding to {110} and
{100} fold domains as previously defined, he suggests, and Bassett, Frank,
and Keller report (28), have a different thermal stability.

We shall consider here only some of the types of twinning discussed by
Burbank (54). For further details, remembering that actual crystals are

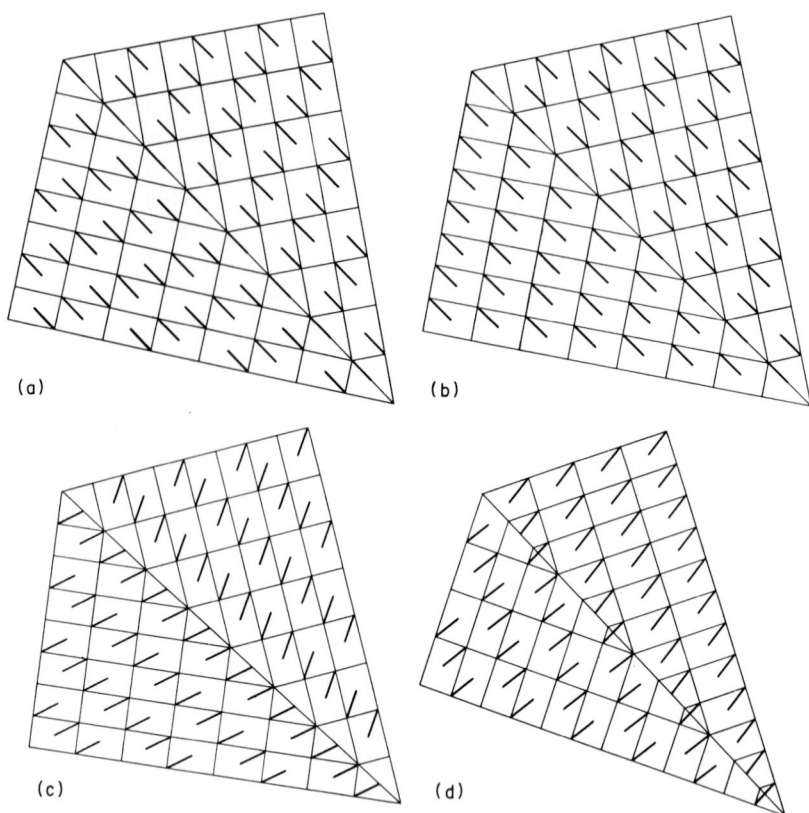

Fig. II-55. Diagrams of various types of twin boundaries. These should be compared with Figures II-37 and II-38 which show fold domain boundaries across which the lattice is not twinned. (a) The orientation of the folds is the same on both sides of the (110) composition plane, only their stacking sequence has changed. This is labeled a (110) $[^1/_2\,^1/_2\,0]$: (110): $[^1/_2\,^1/_2\,0]$ domain boundary in Burbank's terms. (b) Two different types of fold domains are present. This is a (110) [010]: (110): (110) $[^1/_2\,^1/_2\,0]$ phase boundary. (c) Similar to (a) except that the orientation of the fold planes has been changed. This is a $(1\bar{1}0)\,[^1/_2\,^1/_2\,0]$: (110): $(1\bar{1}0)\,[^1/_2\,^1/_2\,0]$ domain boundary in (110) twin. (d) A (310) twin in which some of the folds straddle the composition plane. Furthermore the fold domains are of a different type on opposite sides of the composition plane. This is a (110) [010]: (310): $(1\bar{1}0)\,[^1/_2\,^1/_2\,0]$ phase boundary in (310) twin. (Burbank 54).)

further complicated because of the possibility and, in some cases, requirement that adjacent fold planes be displaced parallel to the chain axis, the reader is referred to the original paper. In describing a twin, Burbank suggests that the indices of both fold domains as well as the composition plane be given. Figure II-55 shows various types of twins in which the composition plane corresponds to a fold domain boundary or a phase boundary. Descriptions of each type are given in the figure caption. These do not correspond directly to observed growth twins since in the latter case the fold planes are presumably parallel to the growth faces. Burbank indicates that there are at least eleven types of twins having a (110) composition plane and nine with a (310) composition plane. Although many of these would not form during growth, they might develop due to deformation of the lattice.

It is suggested that in future detailed descriptions of polymer crystals a notation similar to that of Burbank (54) or Bassett *et al.* (29) be used. In addition it may be necessary to describe the type of fold itself since it is likely that the actual way the molecules fold may differ within the same fold domain. A fold domain should be defined as a portion of a crystal in which all fold planes have the same orientation and, in addition, the stacking sequence of the folds is the same (a (110) [001] fold domain differs from a (110) [$^{1}/_{2}$$^{1}/_{2}$$^{1}/_{2}$] fold domain although the fold plane orientation is the same). The use of the term "phase" to refer to fold domains with different fold plane orientations is not believed necessary since even fold domains with the same fold plane orientation but different fold stacking sequence may differ slightly in stability. All twin boundaries will be fold domain boundaries but many fold domain boundaries are not twin boundaries.

7. SPIRAL GROWTHS

In general, lamellar polymer crystals grown in solution are found to have one or more spiral growths. These growth spirals are similar to those associated with screw dislocation growth in crystals of nonpolymeric substances (see Kittel (72)) in that thickening of the crystal does not require the formation of a new nucleus on a completed growth face. In both the polymeric and nonpolymeric crystals, a permanent step is formed against which the molecules can crystallize. The polymeric and nonpolymeric crystals differ, however, in that, in the case of polymers, little, if any, free energy change results from the crystallizing molecule being in contact with the basal lamella (only the folds are in contact and there is no evidence of epitaxy). In the case of nonpolymeric crystals, however, the crystallizing element is in lattice register with both the face of the step and the basal plane. Although the origin of the spiral growths differs from one polymer crystal to another, most appear to result from the exposure or growth of a wedge shaped portion of a lamella in some fashion or other. This exposed crystalline

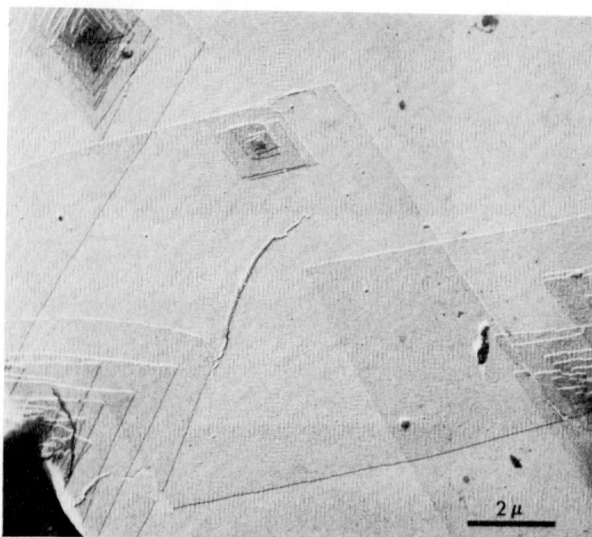

Fig. II-56. Single crystal of linear polyethylene from a dilute tetrachloroethylene solution. The spiral growth is believed to have resulted from a tear in the growth face during growth. See also Figure II-22.

material then serves as the nucleus for the growth of a new lamella which develops into a spiral growth. A number of these sources have been described (13,25).

Tearing or fracture of the crystals at an angle to the growth face by hydrodynamic forces during growth is one of these sources. This source may limit the size to which polymer crystals can be grown as single layers in solution. The relative dimensions of a polymer crystal are such that it is not surprising, despite the stiffening that results from growth in the form of hollow pyramids, that convection currents can cause actual tearing of the crystals. The tearing may involve molecular shear and unfolding, molecules remaining attached to both sides and extending across the tear or actual disruption of the chains. If this occurs, the tear extending inward from a growth face, an exposed face in the form of a wedge would be formed on both sides of the basal lamella. These tears are probably of sufficient length and magnitude that the entire thickness of the lamella is exposed for a portion of its length. Molecules folding against the exposed face prevent the tear from healing and result in the development of a spiral growth on both sides of the basal lamella in a manner similar to that of crystal growth in the form of steps from a screw dislocation. The point of

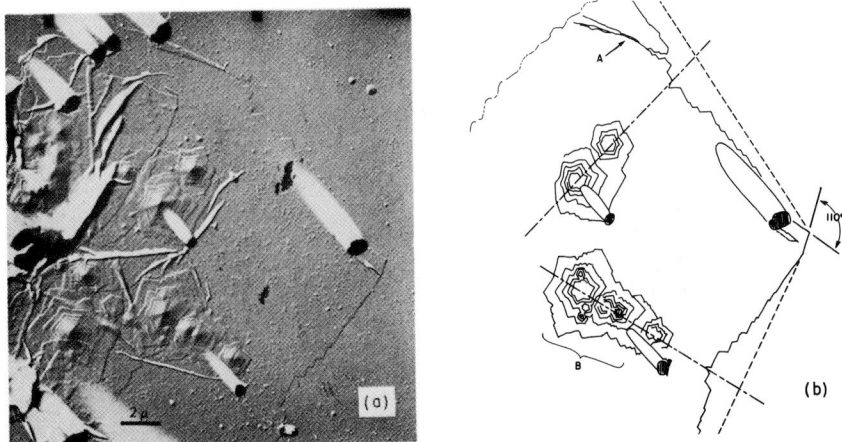

Fig. II-57. Electron micrograph (a) of a region near one apex of a complex star-shaped polyoxymethylene crystal grown in a cyclohexanol solution. The large electron-dense objects are globular polyoxymethylene particles (Cr shadowed at $\tan^{-1} 5/7$). (b) Tracing of (a). The dotted lines show the approximate location of lines connecting adjacent apices. The dash-dot lines indicate the directions of slowest growth along which spiral growths frequently develop. In the region near A the crystal has wrinkled and fractured along a zigzag line. Wrinkles have also developed elsewhere on the crystal. Notches are present where the slow growth line intersects the edge of the crystal. (Reneker and Geil (25).)

the wedge, i.e., the end of the tear, is not the axis of the dislocation. The axis will be at that point along the exposed face at which the face is thick enough to serve as a substrate against which the molecules can fold. As the crystal continues to grow, the split into two lamellae is maintained (Fig. II-56).

Hirai has recently described (99) an additional source of screw dislocations involving the intersection of two originally independent crystals. Subsequently we have found relatively large numbers of such intersecting crystals in which growth has continued after they intersected, in certain preparations. Apparently while growing in solution, either because of the incorporation of the ends of a single molecule into two neighboring crystals or solely because of their proximity, neighboring crystals begin to grow into each other. The line of intersection often serves as the source of several spiral growths.

The source of many of the spiral growths on the star-shaped crystals of polyoxymethylene has been identified (25). On these crystals, numerous spiral growths are formed in pairs of opposite sign, the line connecting their

Fig. II-58. Spiral growths on a polyoxymethylene crystal in the same preparation as the crystal in Figure II-57. The arrow *A* points in the direction of advance of a portion of the growth face near its intersection with a slow growth line. A pair of spiral growths about dislocations of opposite sign are shown. The arrow *B* points to a fault surface which has given rise to a third spiral growth. (Reneker and Geil (25).)

centers lying along the slow growth line (Figs. II-30, II-50, and II-57). As mentioned in Section II-5, impurities and other defects such as improperly oriented folds due to molecules folding from opposite directions, will accumulate and be incorporated in the crystal near this slow growth line. It was suggested that, as a result, the crystal will readily split along the slow growth lines. After further lateral growth of the basal lamella, however, the fault can be bridged and the lamella may grow as a single unit again (Fig. II-58). In effect, an exposed face of aligned segments is left on one or both sides of the basal lamella. This face, being wedge shaped on both ends, serves as the source of two spiral growths of opposite sign on both sides of the basal lamella.

The relative size of the two spirals depends on their separation and the relative time of nucleation of the new lamella and healing of the fault. In general, as in Figure II-58, the new lamella appears to start growth soon after the fault develops. The further observation that the centers in a pair

Fig. II-59. Truncated polyethylene crystal grown at 90°, heated to 94° for a short period of time and then quenched to about 70°C. The thickness of the interior is about 150 A while that of the border is only about 100 A. Fine serrations had developed along {110} growth faces with numerous spirals along the original {100} growth faces. These spirals probably result from an attempt to match the pyramidal topography typical of the two growth temperatures. The two large overgrowths may be real or artifacts deposited when the solvent was evaporated. (Bassett and Keller (15).)

are often separated by only a few hundred angstroms, indicates that the entire thickness of the lamella need not be exposed to serve as a nucleating substrate.

Many crystals of polyoxymethylene, when grown on a substrate, tend to have a spiral growth at their center. It was suggested (25) that this might result if a hollow pyramidal crystal is constrained to grow on a substrate. In these cases, a spiral develops only on the top of the crystal. The crystals in which this was observed, crystallized from phenol, were regular hexagons, i.e., they had no slow growth lines. In contrast to the star-shaped crystals, spiral growths on the regular hexagons are observed only at the boundary or the center. The spiral growths on the boundary developed as the sample was quenched, the slides being removed from the hot phenol solution before crystallization was complete.

Occasionally, either a single lamella or a stepped pyramid of complete lamellae is observed to have grown on a basal lamella (as in Figures II-14 and II-34). The single lamella, it was suggested (13), may result from healing of a fault, the exposed faces slipping back into register after nucleating a new lamella which then continues to grow. Single lamellae may also result

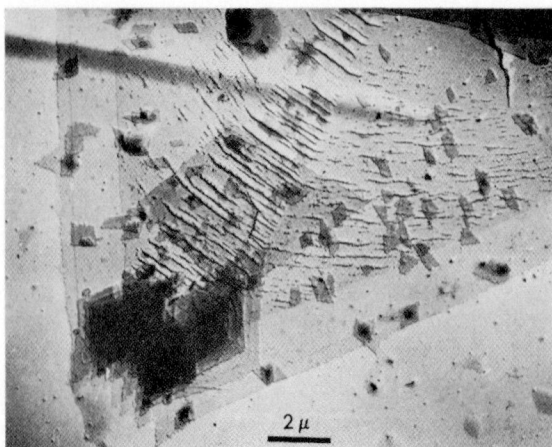

Fig. II-60. Overgrowths on a crystal grown at several temperatures. The over-growths are confined to the original crystal which was heated slightly in solution before the two thinner edges were deposited. (Bassett and Keller (15).)

from nucleation on the surface due, perhaps, to a protruding molecule or group of molecules. This is believed to be a rare occurrence during normal crystallization however. The pyramid of lamellae usually results from having two spiral growths of opposite sign with centers closer than the spacing between steps. A pyramid with a broad top can be formed if the fault giving rise to a spiral and all of the lamellae above it slip by one fold period before growth is completed. These lamellae can continue to grow, with a jog in their boundaries along the line of slip.

Overgrowths consisting of small lamellae or portions thereof are found on truncated polyethylene crystals grown at one temperature, heated slightly while still in suspension, and then cooled quickly (15). When the suspension is cooled a thinner border develops whose morphology is typical of growth at the new crystallization temperature (Fig. II-59). On the {100} growth faces, a {110} type growth leads to the development of a coarsely serrated growth face consisting of a number of diamond-shaped spiral growths. Finer serrations also develop along the large {110} growth faces. A number of very small overgrowths can also be seen on this crystal, restricted to the original fold surface. If the crystals are filtered at the reprecipitation temperature leading to the growth of the boundaries these overgrowths are not found. If cooled quickly to room temperature, however, the size and number of overgrowths can be considerable (Fig. II-60). On this crystal several spiral growths have also developed along the second step (resulting from a

Fig. II-61. Pyramidal crystal of polyethylene with two spiral growths. The distortion in the vicinity of each spiral may have occurred during growth or when the crystal was removed from solution. Note that the Bragg extinction striations are in register along the **a** axis but are not along the **b** axis diagonals. (Bassett (63).)

second change in the reprecipitation temperature). This second step, which is serrated, might be expected to more easily give rise to spiral growths than a smooth edged step (15).

The overgrowths often consist of part of a small crystal, their appearance suggesting that they had originally consisted of a hollow pyramid in which only one fold domain (or perhaps only the apex) was in contact with the basal lamellae (15). During solvent removal the small pyramids may collapse by fracture along the fold domain boundaries, much of the small crystal being washed away with the retreating solvent. In some cases, Bassett and Keller (15) were able to observe small lamellae remaining erect on the surface of the basal lamellae.

The fact that the overgrowths appear only on the original fold surfaces lead Bassett and Keller to suggest (15) these small lamellae may be nucleated by molecules which are not completely incorporated in the basal lamella. These molecular segments may either be occluded during the original crystallization (15,16) or become so during the partial dissolving that takes place when the suspension is heated.

The formation of spiral growths on a fold domain which is a portion of a hollow pyramid presents topographical difficulties in view of the fact that

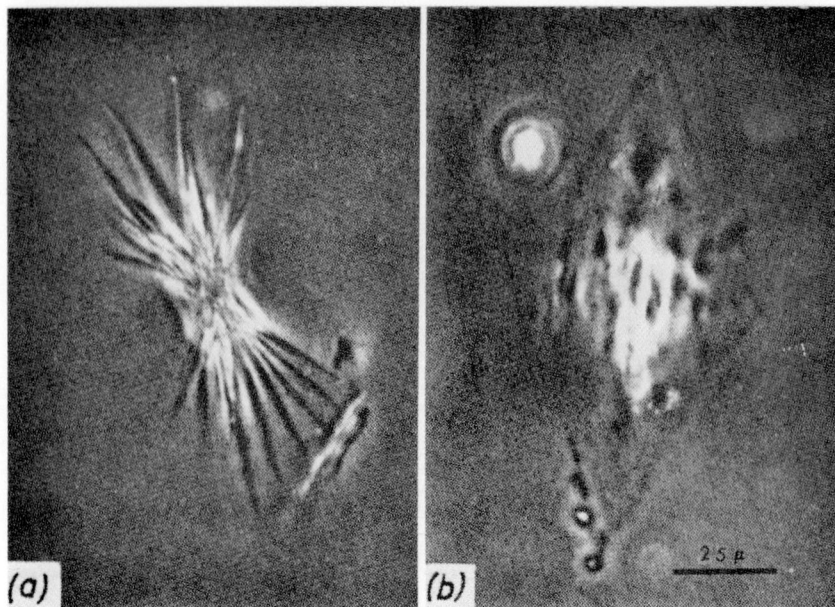

Fig. II-62. Multilayer, dendritic polyethylene crystal observed with phase contrast. The same crystal is shown in (a) edge on and in (b) normal to the lamellae. (Mitsuhashi and Keller (73).)

there is evidence that the spirals (Section II-4) and the overgrowths grow in the form of hollow pyramids. In diamond shaped polyethylene crystals containing only four fold domains either the basal lamellae must be distorted (Fig. II-61) or only one fold domain of the pyramidal spiral can be in contact with the basal lamellae. These correspond to the situations in which a spiral on the outside of a pyramid has either the same or opposite slope as the pyramid itself.

The development of spiral growths on dendritic crystals due to overlapping of adjacent dendrites is discussed in Section II-5. Spiral growths due to these sources develop on both sides of the basal lamella. The individual spiral growths on a pyramidal type dendritic crystal are themselves pyramidal.

The situation in pyramidal type dendritic crystals during growth and collapse is even more complicated than in the case of hollow pyramids consisting of four fold domains. The evidence presented in previous sections shows that each lamella in the spiral collapses individually. The fold domain

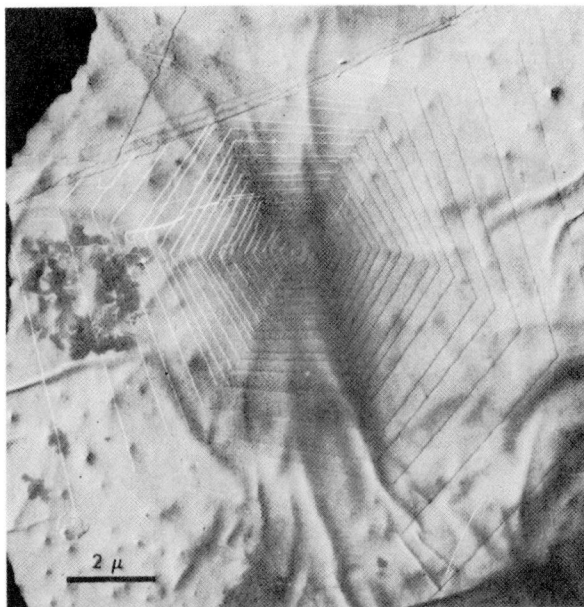

Fig. II-63. Spiral growth on a polyoxymethylene crystal precipitated from a dimethyl phthalate solution. The rotation of the lamellae, although generally in the same direction, is irregular (Cr shadowed at $\tan^{-1} 5/7$). (Reneker and Geil (25).)

boundaries are randomly spaced and there is no indication of an epitaxial alignment of the folds in adjacent lamellae. As a result, the individual lamella in a dendritic spiral must diverge or splay, being in contact only at isolated points and at the center (13,16). The same type of splaying must take place in spiral growths on a coarsely corrugated crystal (type C of Section II-3) (15). The divergence of the lamellae can be observed while the crystals are still in suspension (13,16,71,73) (Fig. II-62). The crystals have a sheaf-like appearance when observed in the appropriate direction.

A further indication of the lack of influence of one lamella on the orientation of an adjacent lamella, other than for their having a common origin, is seen in the fact that in a spiral growth successive lamellae are often rotated by a regular or irregular amount (Fig. II-63). The cause of this rotation is not known at present. It is apparently peculiar to polymer crystals. In a crystal of low molecular weight material, the crystal forces in the underlying plane hold an advancing step in register with the lattice during growth from a screw dislocation. A similar, regular rotation of lamellae in a central spiral growth on polyethylene crystals has been observed by Agar, Frank, and

Fig. II-64. Crystal of polyethylene resulting from a central spiral growth in which successive lamellae have developed with a regular rotation (Bassett, Keller, and Mitsuhashi (58, 74)).

Keller (27) to lead to crystals having a fourfold leaf or bud pattern (Fig. II-64). As they indicate, the leaf pattern arises from successive apices of the lamellae in the spiral being rotated in the same direction on one side of the basal lamella. On the opposite side of the basal lamella the rotation occurs in the opposite direction. The rotation can be of sufficient magnitude that the a axis rotates by 90°, the a axis in the outer lamellae being in the same direction as the b axis in the basal lamella (27,74).

The presence of the rotated layers enables one to study the internal lattice by photographing moire patterns. The superposition and rotation of two equivalent lattices satisfies one of the conditions for the production of moire patterns by double Bragg diffraction. These patterns can be seen in either dark field or bright field illumination. In general, the moire lines are straight and parallel over sizeable regions of the crystal (Figs. II-21 and II-44) indicating continuity and linearity of the lattice planes. Distortion of the lattice shows up as wavy lines (Figs. II-65 and II-66). Although the author has seen no examples of dislocations in the moire patterns from

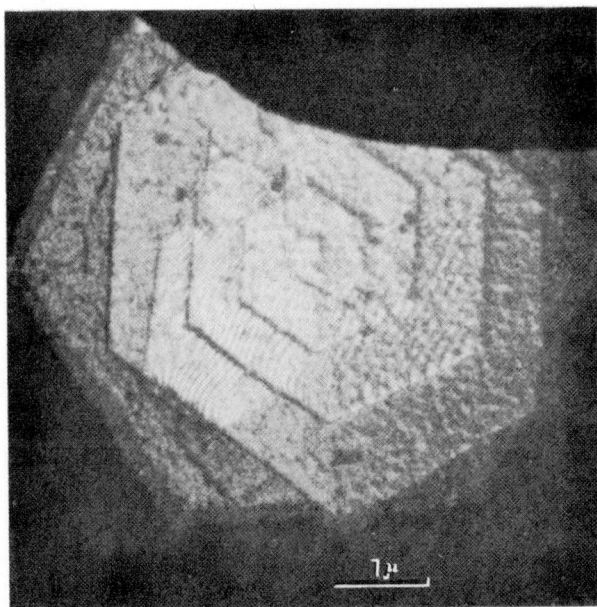

Fig. II-65. Moire pattern on a single crystal of polyoxymethylene precipitated from a phenol solution. Two superimposed but rotated lattices are present. The spacing of the moire patterns depends on the amount of rotation. The wavy nature of the pattern near the apices indicates that one of the lattices is distorted in that region. (Geil, Symons, and Scott (17).)

several hundred crystals, several can be seen in Figure II-66. These dislocations are probably edge-type dislocations involving half-fold planes; however, it must be recognized that other types of dislocations may also give rise to the observed pattern. The use of moire lines, as well as other related techniques, has not yet been exploited to its fullest extent in studying polymer crystal structure. Destruction of the diffracting power of the crystal by the electron beam is the main difficulty to the fuller utilization of such methods.

8. POLYMER CRYSTAL DESCRIPTIONS

In this section, descriptions are given of all polymer single crystals which have been reported to date, the polymers being listed in alphabetical order. In the case of polyethylene and polyoxymethylene, which have served as the basis of most of the preceding discussion, only a few details are given which have not previously been discussed.

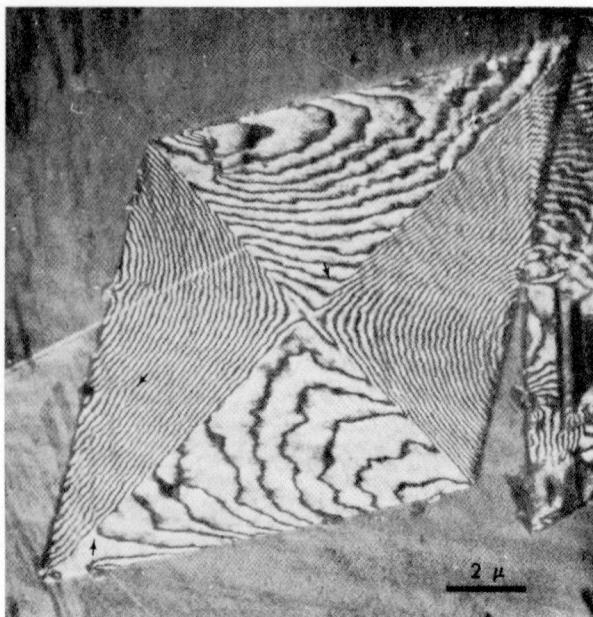

Fig. II-66. Moire pattern on an overgrowth on a polyethylene crystal. Several dislocations (arrows) can be seen. The dark field micrograph was taken using a (110) reflection. (Holland (91).)

a. Polyacrylic Acid

Miller, Botty, and Rauhut (75) have reported the formation of crystals in films of stereoregular polyacrylic acid prepared by successive evaporation

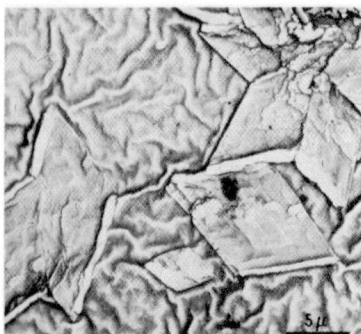

Fig. II-67. Electron micrograph of crystalline polyacrylic acid (Miller, Botty, and Rauhut (75)).

of ten or more layers of a 0.08% (or lower) aqueous solution at 7°C. Storing at this temperature for four to six weeks resulted in the growth of spherulites and a few rhombohedral structures (Fig. II-67). A possible lamellar structure of these "crystals" cannot be determined from the low magnification electron micrograph published. No diffraction evidence was presented.

Subsequently Kargin et al. (76,77) reported the growth of single crystals of both isotactic polyacrylic acid and several of its salts. In some cases helically twisted ribbons were formed. The published micrographs are too poor to permit any resolution of detail. Although Kargin et al. (76) published a single crystal electron diffraction pattern, they did not describe the orientation of the crystal. Presumably, however, the molecules are normal to the crystals. The crystals pictured by both sets of authors appear to be quite thick and might perhaps better be described as hedrites (Chapter III).

b. Polyacrylonitrile (78)

The crystals were grown from 0.1 to 1% solutions in propylene carbonate over a period of several days and at temperatures between 90 and 100°C. Typical ellipsoidal crystals are shown in Figure II-68a. The lamellae are approximately 100 A thick. Twinning was observed in some of the crystals. Electron diffraction patterns (Fig. II-68b) indicate that the molecules are normal to the lamellae and therefore folded.

c. Polyamides

A number of polyamides have been crystallized as single crystals from glycerine solutions (11,67). These crystals grow at an elevated temperature, being formed as a refluxing solution (∼0.01%) is cooled. Spherulites, as well as some single-crystallike structures, can be grown in formic acid solutions at room temperature (23,24,51,79). The spherulites, often composed of ribbonlike lamellae (Fig. II-4) are discussed in Chapter IV. The single crystals of the various polyamides are discussed below.

1. Nylon. 6

Figures II-1 and II-2 show a typical single crystal and diffraction pattern of nylon 6. Geil (11) indicated that spherulites and single crystals are formed in the same solution as it is cooled. Their crystal lattices, however, differ. The diffraction pattern from the single crystals (Fig. II-2) has reflections whose spacings corresponded more closely to the "hexagonal" high temperature modification than to the low temperature form. However

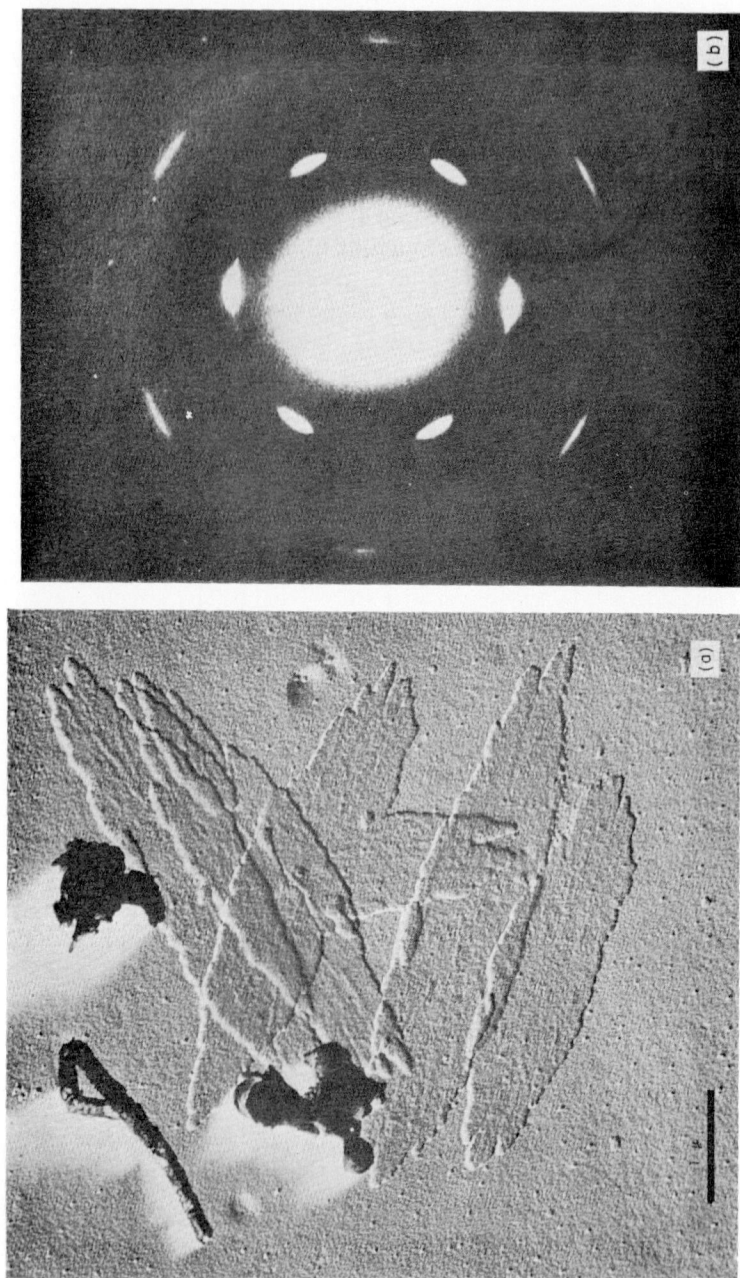

Fig. II-68. (a) Polyacrylonitrile crystallized for 3 days at 95°C from a 0.5% solution in propylene carbonate. (b) Diffraction pattern from a typical polyacrylonitrile crystal. The vertical direction of the pattern corresponds to the long axis of the crystal. Five spacings, with values of 5.3, 5.1, 3.0, 2.9, and 2.6 A, could be measured on the original negative. (Holland, Mitchell, Hunter, and Lindenmeyer (78).)

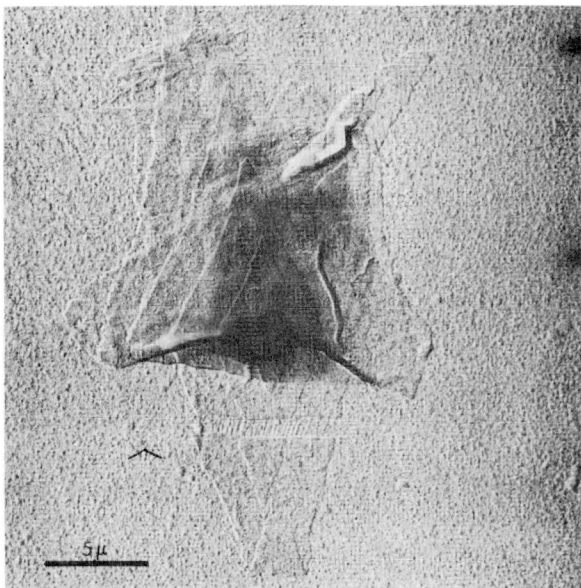

Fig. II-69. Single crystal of nylon 6 precipitated from glycerine. Several of the lamellae have folded back on themselves, whereas the lamella remaining extended has split parallel to the short axis. The splitting is believed to have occurred during evaporation of the glycerine and may be due to thermal contraction effects.

it was clear that the pattern was not hexagonal and that there was a specific hydrogen bond direction. Ogawa *et al.* (98) have recently pointed out that the diffraction pattern corresponds most closely to a γ crystal modification of nylon 6, which is obtained by treating the polymer with an iodine-potassium iodide solution (80,81). They indicate that in this form the hydrogen bonds are made between parallel rather than antiparallel segments as in the more common forms. The diffraction pattern from the spherulites corresponds to the low temperature modification. Globular particles, of unknown structure, are also found in the same suspensions (Fig. II-1).

The conditions required for preferential formation of the various structures are not known. The relative proportion of single crystals, spherulites, and globules varies from one preparation to another. Comparison of selected area single crystal diffraction patterns with the position of the crystal giving rise to the pattern indicates that the molecules are folded within the crystal, the hydrogen-bonded planes of molecular segments being parallel to the short diagonal of the crystal. If the molecules fold parallel to the growth face, as is presumed to occur, then the hydrogen bonds must be made between segments of two different molecules lying in adjacent fold planes.

A number of structural features of the nylon 6 crystals were noted (11). Although most of the crystals appear flat, a few have pleats parallel to the short diagonal. These pleats often extend for an entire fold domain boundary rather than being just in the center of the crystal as in the case of polyethylene crystals. The crystals, composed of lamellae about 60 A thick, thicken by the spiral growth of additional lamellae. In some cases, as in polyethylene and polyoxymethylene (Figs. II-63 and II-64), the crystal axes rotate during the development of the spiral growths. Striations almost parallel to the short diagonal are observed both in bright field and dark field. They have been seen more clearly on nylon 8 (Fig. II-36). These striations may be formed during the heat treatment required to evaporate the glycerine. In some cases, individual lamellae in a spiral growth have folded back on themselves, the remainder of the lamellae remaining extended (Fig. II-69). As pointed out (11), this is consistent with a mechanism involving growth from a screw dislocation. Each lamellae in the spiral grows individually molecules being entirely incorporated into a lamella in a given fold plane before the next lamella growing over the surface of the first covers the fold plane. Many of the crystals are split parallel to the short diagonals (Figs. II-1 and II-69). Fibers approximately 50 A in diameter extend across the breaks. Crystals with a number of reentrant faces, tending toward a den-

Fig. II-70. Single crystal of nylon 7 crystallized from glycerin. A diffraction pattern is shown in the insert. (Holland (67).)

dritic nature, have also been observed. Their development appears related to that described in Section II-5.

2. Nylon 7 (67)

Single crystals of nylon 7 have been grown from glycerine solutions. They are essentially rectangular in shape and consist of a number of lamellae (Fig. II-70).

3. Nylon 8 (67)

Single crystals of nylon 8 precipitated from a glycerin solution are rhombohedral in shape (Fig. II-36). The striations shown on the crystal are similar to and more distinct than those seen on crystals of nylon 6 (11). The molecules are normal to the lamellae.

4. Nylon 66

Single crystals composed of rolled or folded lamellae have been grown from glycerine (11) and methanol-water (50/50)(24) solutions (Fig. II-71). Apparently related, smaller lamellae have also been reported (67). Thickening of the crystal occurs by spiral growths of additional lamellae. Electron diffraction patterns from all of the structures are weak and diffuse. It is believed the structures have the low temperature, triclinic lattice, the long axis of the folded structures lying parallel to the hydrogen bonded planes of molecular segments.

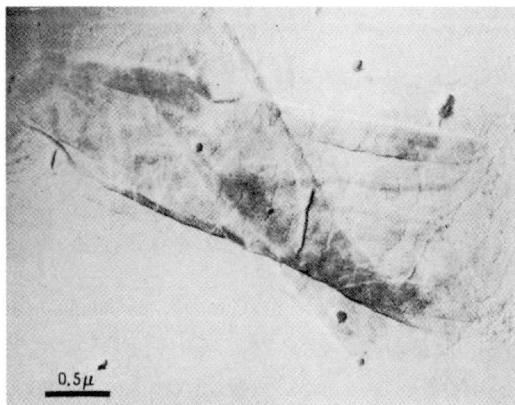

Fig. II-71. Rolled lamellae of nylon 66 precipitated from glycerin solution. Nearly circular spiral growths can be seen. The lamellae are 40 to 60 A thick. (Geil (11).)

Badami and Harris (79) claim to have grown single crystals of nylon 66 from a formic acid-water solution by preferential evaporation of the formic acid. The nylon precipitates as platelets as the water concentration increases. They indicate that the diffraction patterns from areas containing many platelets are consistent with the molecules lying within the platelets, parallel to the substrate. Although the structure of the platelets cannot be determined from the published micrograph, personal observation of the original micrograph confirms the platelike structure. The orientation of the molecules was confirmed by x-ray diffraction from the sedimented platelets

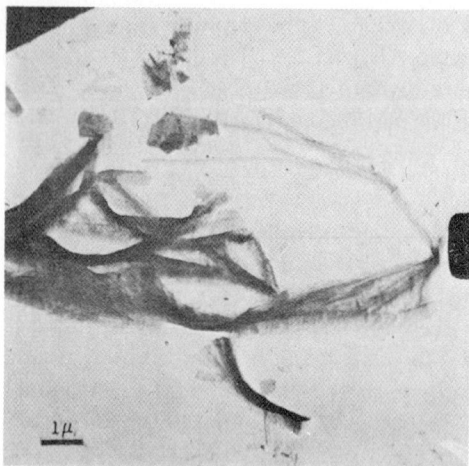

Fig. II-72. Ribbonlike lamellae of nylon 66 precipitated by preferential evaporation of the formic acid in a formic acid-water-nylon solution (Schick (24)).

(82). Using similar conditions the structures shown in Figure II-72 were grown (24). They appear to consist of ribbonlike lamellae, often aggregated into thicker structures. The orientation of the molecules in these structures has not been determined. Further work is required on the lamellae crystallized from both glycerine and formic acid in order to determine their structure and mode of formation, the orientation of the molecules, and whether or not the molecules are folded.

5. Nylon 610

Nylon 610 crystallizes from glycerine in the form of rolled up lamellae (11). Reflections from both the high and low temperature crystal modifications are present in the diffraction patterns suggesting that the lamellae may have crystallized with the high temperature form and then partially transformed to the low temperature form during cooling of the solution.

d. Polybutene

Isotactic polybutene crystallizes from xylene in the form of lamellae (Fig. II-73) and spherulites of irregular structure. It is believed that much of the

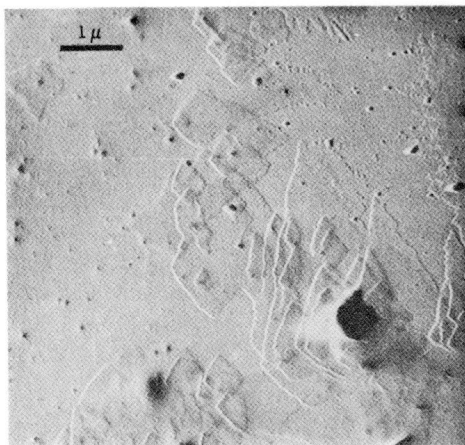

Fig. II-73. Lamellae of isotactic polybutene crystallized from a p-xylene solution (Cr shadowed at $\tan^{-1} 5/7$). (Reneker (42).)

polymer crystallizes during evaporation of the solvent rather than while the solution is cooling. Electron diffraction patterns have not been obtained from the lamellar growths or the spherulites.

e. Cellulose Esters

Ranby and Noe (83) and Manley (84,85) have grown crystals of cellulose acetates from dilute solution. Ranby and Noe report that cellulose triacetate and triethyl cellulose crystallize as diamond-shaped lamellar crystals. Their only published optical micrograph, of a low molecular weight (degree of polymerization approximately 50), partially acetylated cellulose, crystallized from a dilute, buffered aqueous solution at 90°C, shows ellipsoidal or diamond shaped objects. It is reported that spiral growths can be seen. No diffraction data or electron micrographs have been reported.

Manley has reported (85) detailed observations of a fully acetylated polymer with a degree of polymerization of 300. Fractions with molecular weights between 36,000 and 135,000 were used, the crystals being grown from 0.002 to 0.05% nitromethane solution to which an appropriate amount of the nonsolvent, butanol, was added and the solution then cooled. The amount of nonsolvent used depended on the molecular weight. If the non-solvent was not used spherulites and/or thick crystals or hedrites were

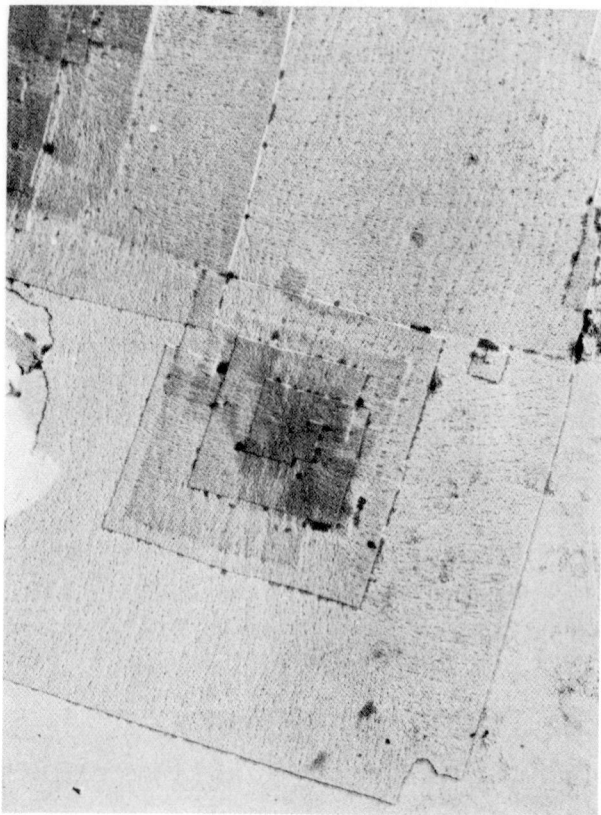

Fig. II-74. Electron micrograph of spiral growths on a single crystal of cellulose tri-acetate (Manley (85)).

formed, whereas under the right conditions large approximately square lamellae are formed (Fig. II-74). With increasing concentration dendritic growth occurs (Fig. II-75). The lamella thickness is 180 to 200 A regardless of the molecular weight fraction used or the temperature of crystallization. Thickening occurs by a spiral growth mechanism.

II. SINGLE CRYSTALS FROM SOLUTION

Fig. II-75. Dendritic crystal of cellulose triacetate (Manley (85)).

Manley indicates (85) that a characteristic feature of the crystals is the system of coarse striations which are normal to the edges and divide the crystal into quadrants (Fig. II-74). Fracture frequently occurs along these striations. The striations occur even in crystals grown from the lowest molecular weight fraction, in which the thickness is about the same as the length of the molecules. In addition pleats are sometimes seen parallel to the striations and extending from the edge toward the center (as opposed to the case in polyethylene where they are located in the center of the crystal). Electron diffraction patterns are extremely difficult to obtain, the crystals rapidly deteriorating in the beam. Those obtained indicate that the molecules are normal to the lamellae. For the various fractions the average molecular length varied from 180 to 2300 A. Except for the lowest fraction, the molecules are longer than the crystals are thick and therefore must be folded, a feature which Manley suggests can occur in about four chemical repeat units despite the supposed stiffness of the chain.

Manley indicates (85) that the a and b axes are parallel to the fold domain boundaries with the growth faces being {210} planes. The packing of the molecules and folds in the crystal is not yet clearly defined. It may be affected by the proposed opposite direction of the molecules in the corners and center of the unit cell (Chapter I). The crystals can be deacetylated and converted to cellulose II without destroying their over-all appearance (84). The crystals shrink laterally by about 30% with the molecular orientation, but not the lateral packing, being maintained. The reaction takes place throughout the crystal and not on the surface only, suggesting that the interior of polymer crystals may be penetrable by small molecules.

f. Polychlorotrifluoroethylene (86)

Kargin et al. (86) have grown two types of single crystals from 0.005 to 0.02% solutions in mesitylene. If crystallized at 160°C, lamellar aggregates of circular shape are formed, whereas at 180°C the crystals consist primarily of rectangular lamellae.

g. Polyesters

Single crystallike lamellae have been grown in intimate association with spherulitic structures, from dilute solutions of polyhexamethylene terephthalate (24). These structures are discussed in Chapter IV. Belavtseva and Mnyukh (87) and Mnyukh, Belavtseva, and Kitaigorodskii (88) have ·reported the growth of single crystals from low molecular weight (2,000–3,000) 1, 10 decanediol (and 1,20 eicosanediol) and a number of hydrocarbon acids. The crystals were grown by cooling of hot solutions in ethanol and amyl acetate. Typical crystals are shown in Figure II-76. The growth faces, as in the case of polyethylene, depend on the solvent. The authors state that the fold period is equivalent to the repeat period (see Chap. I), the crystals from 1,10 decanediol and an acid being 15 to 35 A thick and the crystals from 1,20 eicosanediol and an acid being 40 to 70 A thick. Insufficient information is given to measure the thickness on the micrographs. When crystallized from the melt they report periods on the order of 100 A are obtained, the periods being multiples of the repeat period. Diffraction patterns from the crystals indicate the molecules are normal to the lamellae and therefore folded (Fig. II-77). The authors also indicate (88) that they

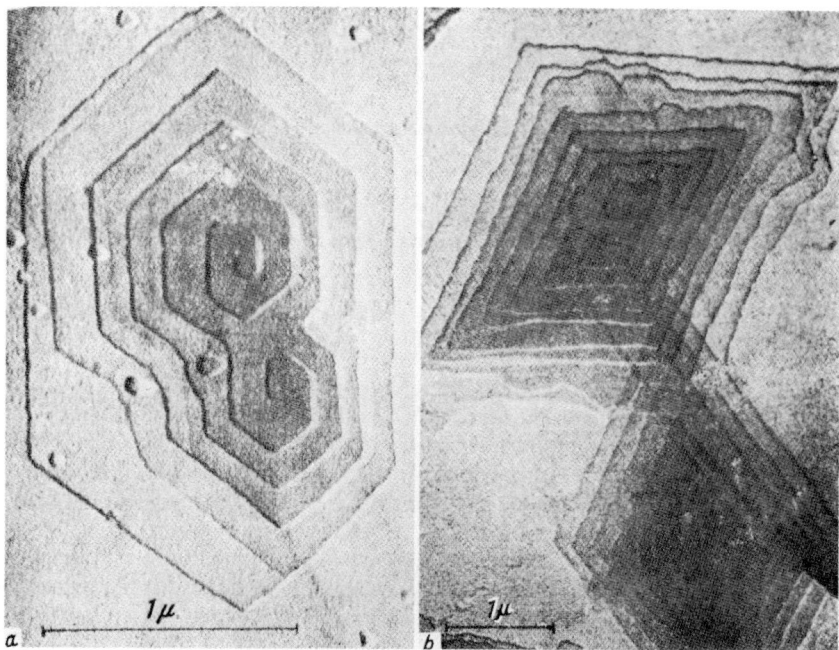

Fig. II-76. (a) Single crystal of polymer of 1,10 decanediol and an unstated hydrocarbon acid, crystallized from ethanol. (b) Similar polymer crystallized from amyl acetate. (Mnyukh, Belavtseva, and Kitaigorodskii (88).)

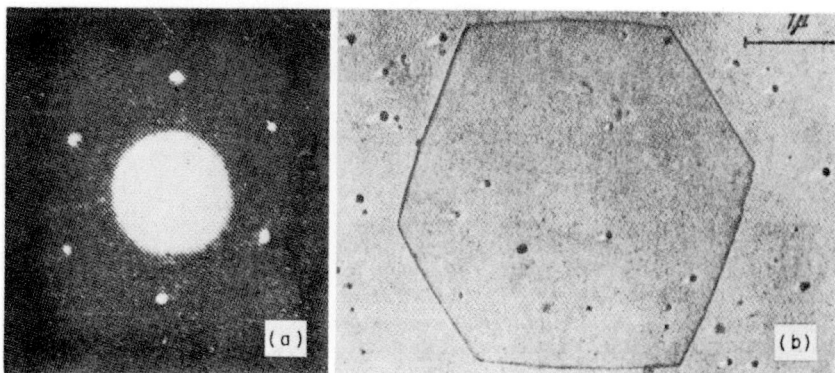

Fig. II-77. Electron diffraction pattern (a) from a single crystal (b) of a polymer of 1,10 decanediol and a hydrocarbon acid crystallized from ethanol (Mnyukh, Belavtseva, and Kitaigorodskii (88)).

could obtain spiral growths of unlimited lamellae, globules developed from spirals, dendrites in the form of monolayers with wrinkles in the center, and fibers from the same solutions as the single crystals.

h. Polyethylene

Crystals of branched and linear polyethylene have been described in detail in the previous sections of this chapter. Several additional features, peculiar to specific preparations, are described in this section. As indicated previously coherent films, prepared by evaporation of the solvent from a

Fig. II-78. Film of polyethylene prepared by evaporation of the solvent at an elevated temperature. Diffraction patterns indicate that each polygonal structure is a single crystal. The striations are parallel to the **b** axis. (Keller (9).)

solution at an elevated temperature, consist of single crystal areas. Figure II-78 is representative of the type of film formed. Keller (9) indicates that a periodicity equal to twice the fold period can be seen on the fibers drawn across the several breaks in the film. He attributes the periodicity to beads of shadowing metal attached to the molecules at the original position of the folds when the film was shadowed.

Wunderlich and Sullivan have found (89) that needles (as observed in the optical microscope) are formed at about 90°C at the interface of 0.1%

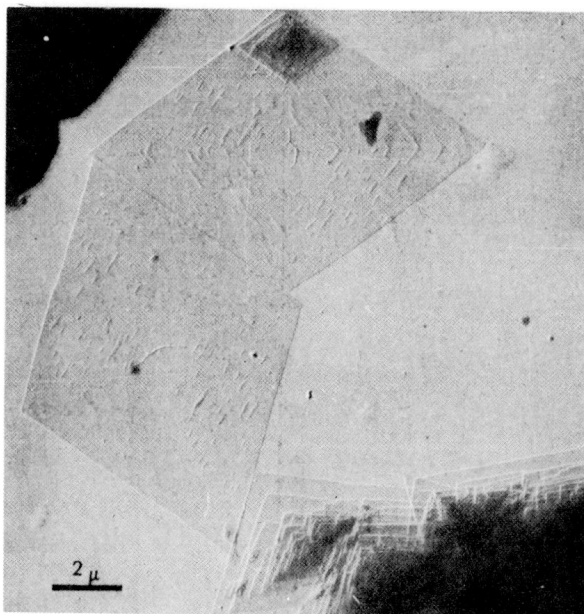

Fig. II-79. Single crystals of polyethylene precipitated rapidly from a 0.1% tetra-chloroethylene solution. Patches of material parallel to {310} planes are observed on some of the fold domains.

solution of linear polyethylene in chlorobenzene over which toluene has been layered. They suggest that the needles may consist of rolled up lamellae. The needles were found in a preparation that also contained large numbers of single crystals.

A peculiar type of pyramid collapse is shown in Figure II-79. These crystals are from the same preparation as those shown in Figure II-22, usually forming pleats. Isolated patches of material are formed, aligned parallel to the {310} planes. The thickness of the patches is about the same as that of the basal lamella. It may be that complete sections of the basal lamella were "popped out" as the crystal collapsed. The apparent localization of the patches to two fold domains in one of the crystals in Figure II-79 is probably a result of the shadowing direction being nearly parallel to the {310} planes in the other two fold domains (63).

i. Polyethylene Oxide

Nearly square crystals of polyethylene oxide have been crystallized from dilute solution (90,91). Spiral growths on a typical crystal, grown from a

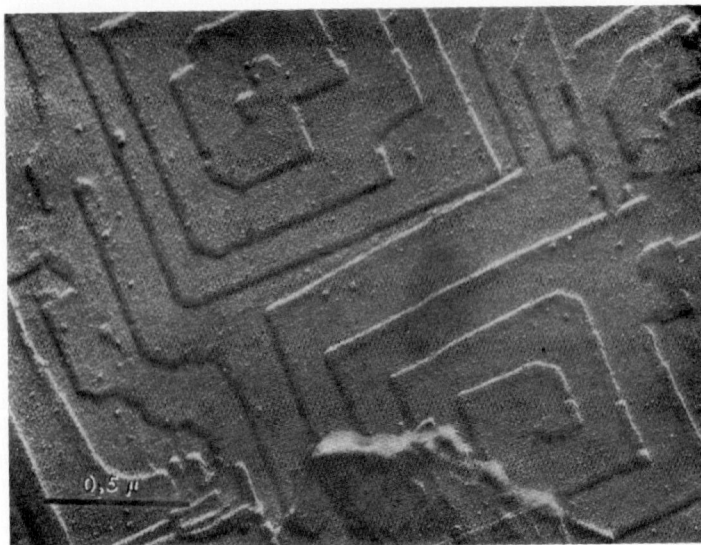

Fig. II-80. Spiral growths on a crystal of polyethylene oxide (Price (90)).

0.05% solution in 20% cyclohexene-80% paraxylene are shown in Figure II-80. The lamellae appear to be less than 100 A thick.

j. Poly-4-methyl-pentene-1 (66)

Single crystals of this isotactic polyolefin containing a sizeable, branched side chain on every other backbone carbon atom and crystallized from dilute (0.01% to 0.1%) xylene solutions, consist of square lamellae (Fig. II-81). Frank, Keller, and O'Connor (66) indicate that the appearance of the crystals is not reproducible, depending in an unpredictable manner on the cooling rate. Thickening of the crystal occurs by spiral growths. The spirals are often particularly common on the edges of the crystals, creating a picture frame effect. As indicated previously, there is a possibility that some of the crystals are pyramidal. Larger, more complex crystals often have thickened regions forming irregular polygons or circles concentric with the crystal boundaries (Fig. II-35). The authors indicate that in some cases the thickened region, which may consist of rolled or folded lamellae, form well defined squares at 45° to the crystal boundaries.

Electron diffraction patterns show that the molecules are normal to the 100 A thick lamellae and therefore folded within them. The unit cell of the polymer was determined from electron and x-ray diffraction patterns from

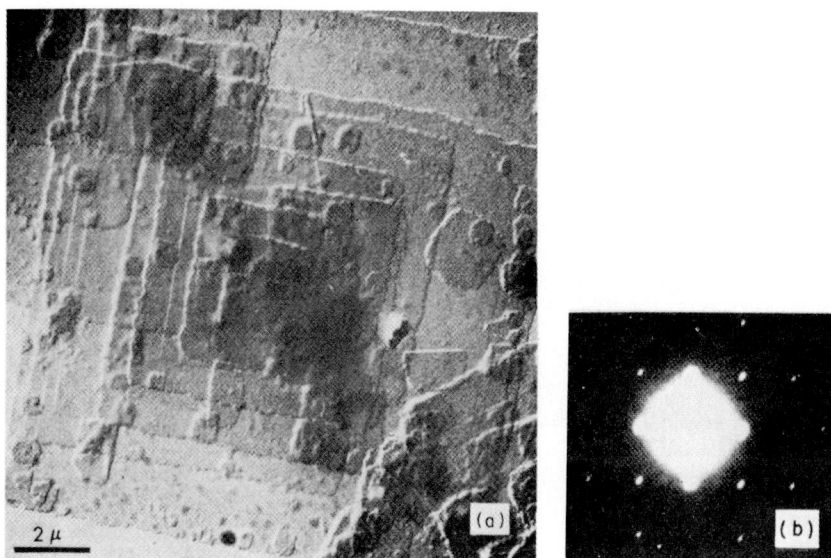

Fig. II-81. (a) Single crystal of poly-4-methyl-pentene-1 precipitated from xylene. Some D shaped growths, of unknown structure, can be seen as well as more typical spiral growths. (b) Electron diffraction pattern from one of the crystals. (Frank, Keller, and O'Connor (66).)

the single crystals. This is the first case in which a polymer crystal structure has been determined using single crystals, most previous determinations being made on the basis of fiber patterns. The small angle x-ray diffraction long period, 108 A, from a sedimented layer of the crystals corresponds to the thickness of the lamellae.

As in the case of polyoxymethylene, the crystals often have a substep near their boundary. In crystals of polyoxymethylene (17) and polyethylene (43) similar substeps have been attributed to a decrease in crystallization temperature, the thinner portion forming during rapid cooling following an initial slow crystallization. Numerous spiral growths and small irregular lamellae are observed to form on this thinner portion in crystals of both polymers. In the case of poly-4-methyl-pentene-1 crystals, we suggest that the thinner boundaries and associated small lamellae may have formed during evaporation of the solvent at room temperature prior to observation of the crystals in the microscope. Dendritic crystals of poly-4-methyl-pentene-1, containing numerous spiral growths, are often observed.

Solutions of the polymer in trichloroethylene produced rows of parallel sheaves when a drop of suspension was placed on a slide and the solvent

Fig. II-82. Poly-4-methyl-pentene-1 precipitated from trichloroethylene at room temperature during evaporation of the solvent (Frank, Keller, and O'Connor (66)).

evaporated (Fig. II-82). The polymer crystallized during evaporation of the solvent. The sheaves are composed of fibrils, electron diffraction patterns showing that the molecules are normal to the fibril axis and therefore folded within the fibrils.

k. Polyoxymethylene (17)

The crystals precipitated from phenol and described in previous sections were grown over a period of several days, the temperature of the solution

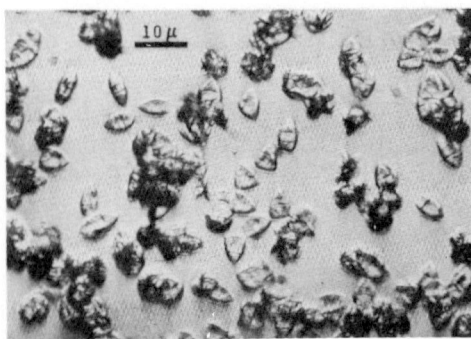

Fig. II-83. Polyoxymethylene precipitated from phenol solution at 86.3°C. It is not known whether the structures shown had grown at this temperature or were formed during the subsequent rapid cooling.

being slowly lowered in steps. Glass slides immersed in the solution were removed at various times.

Crystals similar to that shown in Figure II-65 are obtained if the slides are removed after the solution is cooled to temperatures below 82°C. If removed from the solution at slightly higher temperature, polymer remaining in solution is precipitated rapidly, resulting in a substep followed by a thinner lamella, as well as the formation of numerous spiral growths and incipient lamellae on the boundary. If the slide is removed at a still higher temperature (greater than 85°C) small triangular or nearly oval shaped structures are found (Fig. II-83). It is believed that these structures may consist of hexagonal lamellae partially rolled up. No electron micrographs are available.

1. Polypropylene

Single crystals from dilute solutions of isotactic polypropylene have been reported by Ranby, Morehead, and Walter (18), and by Kargin, Bakeev, Li-shen, and Ochapovskaya (92).

Ranby et al., using a preextracted sample (some atactic and low molecular weight polymers removed) crystallized from p-xylene at 60°C, found a few diamond shaped lamellae with spiral growths (Fig. II-84). Neither the thickness of the lamellae nor diffraction results were presented. Kargin et al., crystallized the polymer from dilute xylene and trichloroethylene solutions. The xylene solutions were cooled from the boiling point to room

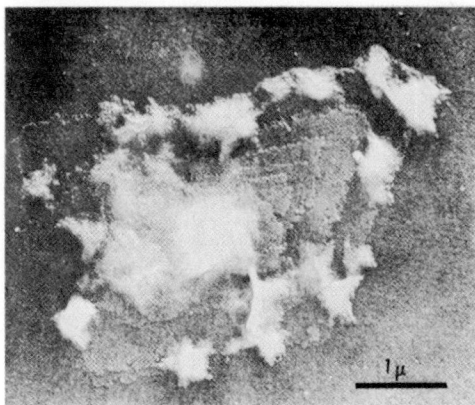

Fig. II-84. Crystal of isotactic polypropylene precipitated from p-xylene. Most of the solution precipitates as irregular structures, sometimes approaching a spherulitic structure. (Ranby, Morehead, and Walter (18).)

temperature over a period of 2 to 3 weeks. Poorly formed, lamellar crystals grew from the xylene solution. Rhombic or triangular crystals, perhaps hedrites, grew from the trichloroethylene solutions. The published micrographs and diffraction patterns are indistinct. The authors indicate that the diffraction pattern shows that the molecules are normal to the lamellae.

m. Polystyrene (92)

Large crystals of a basically rhombic shape have been grown from dilute xylene solutions. Kargin *et al.*, indicate (92) that they observe fibers on the crystals oriented normal to the long diagonal of the diamond and suggest

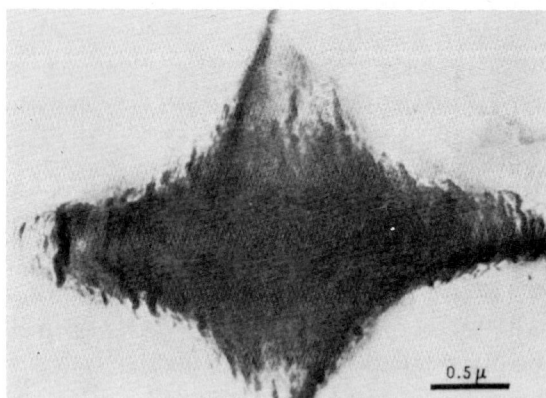

Fig. II-85. Single crystal of polystyrene crystallized from a dilute xylene solution (Kargin, Bakeev, Li-shen, and Ochapovskaya (92).)

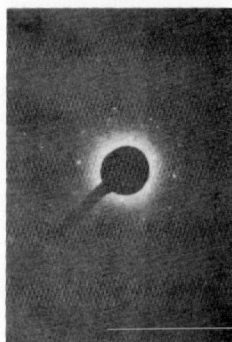

Fig. II-86. Electron diffraction pattern from a single crystal of polystyrene (Kargin, Bakeev, Li-shen, and Ochapovskaya (92)).

that the morphology is not lamellar. The published micrograph (Fig. II-85) strongly resembles those of unshadowed, thick crystals of other polymers which have deteriorated during exposure to the beam. An electron diffraction pattern (Fig. II-86) shows that the molecules are normal to the diamonds.

n. Polytetrafluoroethylene (93)

Palmer and Cobbold (93) have grown single crystals of (Fig. II-87) polytetrafluoroethylene from dilute perfluorinated olefin solutions. The solvent was evaporated in a vacuum oven at 140°C. The crystals consist of hexagonal lamellae 200 to 500 A thick. Spiral growths lead to thickening of the crystals. Electron diffraction patterns indicate that the molecules are normal to the lamellae and therefore folded within them.

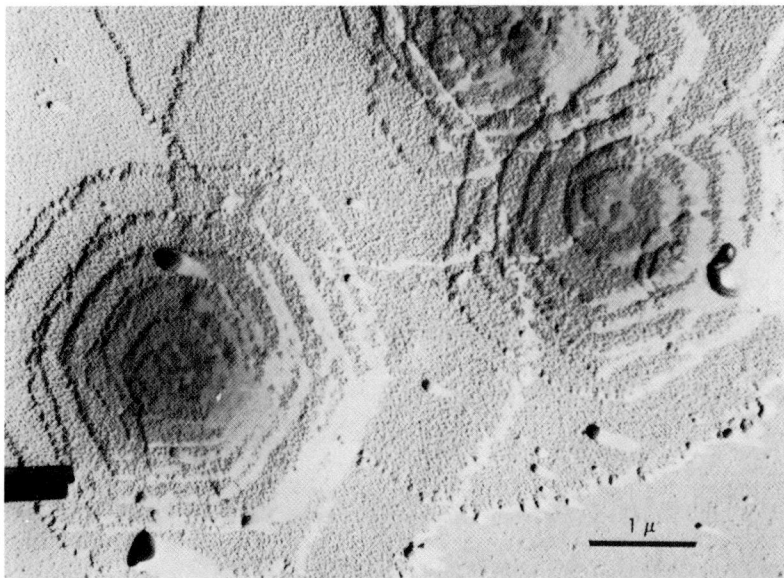

Fig. II-87. Portion of a single crystal of polytetrafluoroethylene (Palmer and Cobbold (93)).

o. Xylan (94)

Single crystals of xylan, a cellulosic type material derived from certain plants, have been grown from 0.025% water solutions. The crystals consist of hexagonal platelets about 50 A thick, which is about one-third the mo-

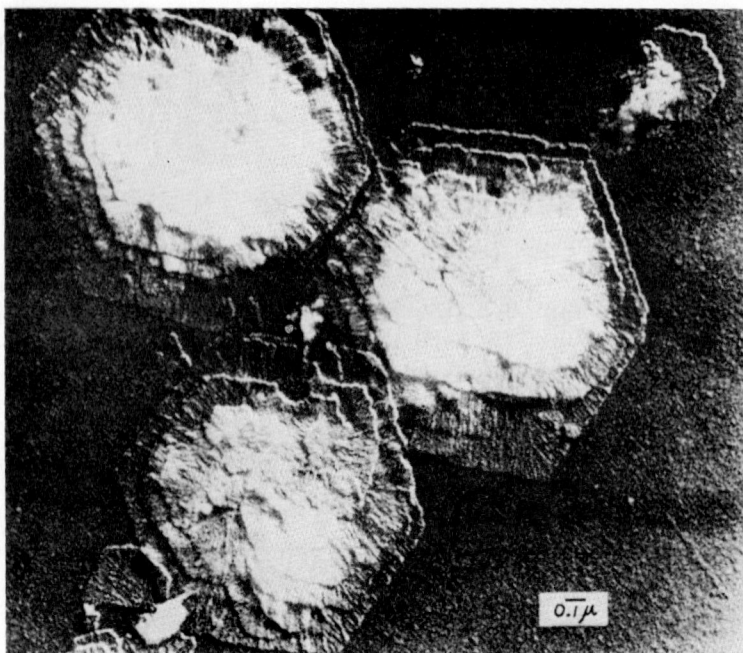

Fig. II-88. Single crystals of xylan (Marchessault, Morehead, Walter, Glaudemans, and Timell (94)).

lecular length of the polymer used (Fig. II-88). Striations are seen resembling those on cellulose triacetate:

REFERENCES

1. O. Schweitzer, *Zeit. Z. Kr.*, **70**, 206 (1919). Original not available, results discussed in H. Staudinger, R. Signer, H. Johner, M. Luthy, W. Kern, D. Russidis, and O. Schweitzer, "Über hochpolymere Verbindungen. 18. Mitteilung Über die Konstitution der Polyoxymethylene." *Ann. Chem.*, **474**, 145 (1929).
2. H. Staudinger and R. Signer, "Über den Kristellbau hochmolekularer Verbindungen. 17. Mitteilung über hochmolekulare Verbindungen," *Z. Krist.*, **70**, 193 (1929).
3. E. Sauter, "Über hochpolymeren Verbindungen. 7. Mittelung "Röntgenometrische Untersuchungen an hochmolekularen Polyoxymethylenen," *Z. physik Chem.*, **B18**, 417 (1932), as well as numerous other papers by Staudinger and his co-workers.
4. K. H. Storcks, "An Electron Diffraction Examination of Some Linear High Polymers," *J. Am. Chem. Soc.*, **60**, 1753 (1938).
5. W. Schlesinger and H. M. Leeper, "Gutta I. Single Crystals of Alpha-Gutta," *J. Polymer Sci.*, **11**, 203 (1953).

6. R. Jaccodine, "Observations of Spiral Growth Steps in Ethylene Polymers," *Nature*, **176**, 305 (1955).

7. I. M. Dawson and V. Vand, "The observation of spiral growth-steps in *n*-paraffin single crystals in the electron microscope," *Proc. Roy. Soc. (London)*, **A206**, 555 (1951), and subsequent papers by Dawson and co-workers.

8. P. H. Till, "The Growth of Single Crystals of Linear Polyethylene," *J. Polymer Sci.*, **24**, 301 (1957).

9. A. Keller, "A Note on Single Crystals in Polymers: Evidence for a Folded Chain Configuration," *Phil Mag.*, **2**, 1171 (1957).

10. E. W. Fischer, "Stufen-und spiralförmiges Kristallwachstum bei Hochpolymeren," *Z. Naturforsch.*, **12a**, 753 (1957).

11. P. H. Geil, "Nylon Single Crystals," *J. Polymer Sci.*, **44**, 449 (1960).

12. A. Keller, "Morphology of Crystalline Polymers" in *Growth and Perfection of Crystals*, edited by R. H. Doremus, B. W. Roberts, and D. Turnbull, Wiley, New York, p. 499 (1958).

13. P. H. Geil and D. H. Reneker, "Morphology of Dendritic Polymer Crystals," *J. Polymer Sci.*, **51**, 569 (1961).

14. P. H. Geil, unpublished data.

15. D. C. Bassett and A. Keller, "On the Habits of Polyethylene Crystals," *Phil. Mag.*, **7**, 1553 (1962).

16. D. C. Bassett, A. Keller, and S. Mitsuhashi, "New Features in Polymer Crystal Growth from Concentrated Solution," *J. Polymer Sci.*, **1A**, 763 (1963).

17. P. H. Geil, N. K. J. Symons, and R. G. Scott, "Solution Grown Crystals of an Acetal Resin," *J. Appl. Phys.*, **30**, 1516 (1959).

18. B. G. Ranby, F. F. Morehead, and N. M. Walter, "Morphology of *n*-Alkanes, Linear Polyethylene, and Isotactic Polypropylene Crystallized from Solution," *J. Polymer Sci.*, **44**, 349 (1960).

19. W. O. Statton and P. H. Geil, "Recrystallization of Polyethylene During Annealing," *J. Appl. Polymer Sci.*, **3**, 357 (1960).

20. A. Müller, "Zur Kenntnis des Kohlenstoffringes XXI. Über einige Röntgenmessungen und hochgliedrigen cyclischen Verbindungen," *Helv. Chim. Acta*, **16**, 155 (1933).

21. P. De Santis, E. Giglio, A. M. Liquori and A. Ripamonti, "Stability of Helical Conformations of Simple Linear Polymers," *J. Polymer Sci.*, **1A**, 1383 (1963).

22. R. McCullough, to be published.

23. A. Keller, "Electron Microscope-Electron Diffraction Investigations of the Crystalline Texture of Polyamides," *J. Polymer Sci.*, **36**, 361 (1959).

24. M. J. Schick, personal communication.

25. D. H. Reneker and P. H. Geil, "Morphology of Polymer Single Crystals," *J. Appl. Phys.*, **31**, 1916 (1960).

26. W. D. Niegisch and P. R. Swan, "Hollow Pyramidal Crystals of Polyethylene and a Mechanism of Growth," *J. Appl. Phys.*, **31**, 1906 (1960).

27. A. W. Ager, F. C. Frank, and A. Keller, "Crystallinity Effects in the Electron Microscopy of Polyethylene," *Phil. Mag.*, **4**, 32 (1959).

28. D. C. Bassett, F. C. Frank and A. Keller, "Evidence for Distinct Sectors in Polymer Single Crystals," *Nature*, **184**, 810 (1959).

29. D. C. Bassett, F. C. Frank, and A. Keller, "Some New Habit Features of Long Chain Compounds, Part III," to be published.

30. D. C. Bassett, F. C. Frank and A. Keller, "Some New Habit Features of Long Chain Compounds, Part IV," to be published.
31. R. Westrik and A. M. Kiel, "Some Electron Microscopic Observations on Single Crystals of Polyethylene," *Nature*, **190**, 162 (1961).
32. A. Keller and A. O'Connor, "Study of Single Crystals and Their Associations in Polymers," *Discussions Faraday Soc.*, **25**, 114 (1958).
33. F. P. Price, "Markoff Chain Model for Growth of Polymer Single Crystals," *J. Chem. Phys.*, **35**, 1884 (1961).
34. F. P. Price, personal communication.
35. J. Ryan, personal communication.
36. B. G. Ranby and H. Brumberger, "Observations on Chain Folding in Crystalline Polyethylene," *Polymer*, **1**, 399 (1960).
37. A. Keller and A. O'Connor, "A Study on the Relation Between Chain Folding and Chain Length in Polyethylene," *Polymer*, **1**, 163 (1960).
38. L. Mandelkern, A. S. Posner, A. F. Diorio, and D. E. Roberts, "Low Angle X-Ray Diffraction of Crystalline Nonoriented Polyethylene and Its Relation to Crystallization Mechanisms," *J. Appl. Phys.*, **32**, 1509 (1961).
39. P. H. Geil, "Recrystallization of Melt Crystallized Polyethylene during Annealing," paper presented at Am. Phys. Soc. meeting, Baltimore, Md., March 1962.
40. V. F. Holland and P. H. Lindenmeyer, "Morphology and Crystal Growth Rate of Polyethylene Crystalline Complexes," *J. Polymer Sci.*, **57**, 589 (1962).
41. A. Keller and D. C. Bassett, "Complementary Light and Electron Microscope Investigations on the Habit and Structure of Crystals with Particular Reference to Long Chain Compounds," *J. Roy. Microscope Soc.*, **79**, 243 (1960).
42. D. H. Reneker, personal communication.
43. D. C. Bassett and A. Keller, "Some New Habit Features in Crystals of Long Chain Compounds Part II. Polymers," *Phil. Mag.*, **6**, 345 (1961).
44. G. Ställberg, S. Ställberg-Stenhagen and E. Stenhagen, "Very Long Hydrocarbon Chains 1. The Synthesis of *n*-Dooctacontane and *n*-Hectane," *Acta Chem. Scand.*, **6**, 313 (1952).
45. W. Kern, J. Davidovits, K. J. Rauterkus, and G. F. Schmidt, "Röntgenographische Untersuchungen an linearen Oligourethanen," *Makromol. Chem.*, **43**, 106 (1961).
46. H. Zahn, "Röntgenstrukter von Linearen Oligomeren," IUPAC-Symposium on Makromolecules, paper IB 8, Weisbadan (1959).
47. H. Zahn and W. Pieper, "Moleküllängenunabhängige Langperioden in Kleinwinkel-röntgenogrammen von Oligomerne," *Kolloid-Z.*, **180**, 97 (1962).
48. W. E. Garner, K. van Bibber, and A. M. King, "The Melting Points and Heats of Crystallization of the Normal Long Chain Hydrocarbons," *J. Chem. Soc.*, **1931**, 1533 (1931).
49. B. Wunderlich, "The Effect of Pressure on the Crystallization of Polyethylene from Dilute Solution," *J. Polymer Sci.*, **1A**, 1245 (1963).
50. P. H. Geil, "Lamellar Crystallization of Low Density Polyethylene," *J. Polymer Sci.*, **51**, S10 (1961).
51. R. Eppe, E. W. Fischer, and H. A. Stuart, "Morphologische Strukturen in Polyathylenen, Polyamiden und anderen kristallisierenden Hochpolymeren," *J. Polymer Sci.*, **34**, 721 (1959).
52. F. C. Wilson, personal communication.

53. P. R. Swan, "Polyethylene Unit Cell Variations with Branching," *J. Polymer Sci.* **56**, 409 (1962).
54. R. D. Burbank, "Molecular Structure in Crystal Aggregates of Linear Polyethylene," *Bell System Tech. J.*, **39**, 1627 (1960).
55. J. Willems, "Oriented Growth in the Field of Organic High Polymers," *Discussions Faraday Soc.*, **25**, 111 (1958).
56. J. Willems and I. Willems, "Über die orientierte Aufwachsung von Polyäthylen auf Steinsalz," *Experientia*, **13**, 465 (1957).
57. E. W. Fischer, "Orientierte Kristallisatron des Polyäthylens auf Steinsalz." *Kolloid-Z.*, **159**, 108 (1958).
58. A. Keller, "Polymer Single Crystals," *Polymer*, **3**, 393 (1962).
59. F. C. Frank, to be published.
60. E. W. Fischer, in H. A. Stuart, "Problems of High Polymer Crystallinity," *Ann. N. Y. Acad. Sci.*, **83**, 1 (1959).
61. E. W. Fischer, personal communication.
62. See for instance N. Cabrera and D. A. Vermilyea, "The Growth of Crystals from Solution, A Review," in *Growth and Perfection of Crystals*, edited by R. H. Doremus, B. W. Roberts, and D. Turnbull, Wiley, New York, p. 393 (1958).
63. D. C. Bassett, personal communication.
64. F. Khoury and F. J. Padden, Jr., "On the Growth Habits of Twinned Crystals of Polyethylene," *J. Polymer Sci.*, **47**, 455 (1960).
65. W. D. Niegisch, "The Nucleation of Polyethylene Spherulites by Single Crystals," *J. Polymer Sci.*, **40**, 263 (1959).
66. F. C. Frank, A. Keller, and A. O'Connor, "Observations on Single Crystals of an Isotactic Polyolefin: Morphology and Chain Packing in Poly-4-Methyl-Pentene-1," *Phil. Mag.*, **4**, 200 (1959).
67. V. Holland, "Electron Microscope Studies of Crystalline Structures Precipitated from Dilute Polyamide Solutions," paper presented at Elec. Micro. Soc. Am. meeting, Milwaukee Aug. 1960.
68. Th. Schoon, "Polymorphe Formen kristalliner Kohlenstoffverbindungen mit langen gestrechen Ketten (Nach Strukturuntersuchungen durch Electröenbeugung)," *Z. physik, Chem.*, **39B**, 385 (1938).
69. A. I. Kitaigorodskii, "The Packing of Chain Molecules II. The Layers of Paraffin Molecules," *Soviet Phys., Cryst.* (Eng. Trans.), **2**, 637 (1959).
70. B. Wunderlich, "Solution Grown Polyethylene Dendrites," paper presented at Am. Chem. Soc. meeting, Chicago, Ill., Sept. 1961.
71. B. Wunderlich and P. Sullivan, "Solution Growth Polyethylene Dendrites," *J. Polymer Sci.*, **61**, 195 (1962).
72. C. Kittel, "Introduction to Solid State Physics," Wiley, New York (1956).
73. A. Keller and S. Mitsuhashi, "The Morphology of Multilayer Polymer Crystals," *Polymer*, **2**, 109 (1961).
74. D. C. Bassett, A. Keller, and S. Mitsuhashi, to be published.
75. M. L. Miller, M. C. Botty, and C. E. Rauhut, "Crystalline Polyacrylic Acid," *J. Colloid Sci.*, **15**, 83 (1960).
76. V. A. Kargin, V. A. Kabanov, S. Ya. Mirlina, and A. V. Vlasov, "Isotactic Polyacrylic Acid and Its Salts," *High Mole. Wt. Comps.*, **3**, 134 (1961) (in Russian).
77. V. A. Kargin, S. Ya. Mirlina, V. A. Kabanov, and G. A. Mikheleva, "Structural Study of Isotactic Polyacrylic Acid and Its Salts," *High Mole. Wt. Comps.*, **3**, 139 (1961) (in Russian).

188   POLYMER SINGLE CRYSTALS

78. V. F. Holland, S. B. Mitchell, W. L. Hunter and P. H. Lindenmeyer, "Crystal Struc-
    ture and Morphology of Polyacrylonitrile in Dilute Solution," *J. Polymer Sci.*, **62**,
    145 (1962).
79. D. V. Badami and P. H. Harris, "Nylon 66 Single Crystals," *J. Polymer Sci.*, **41**, 540
    (1959).
80. M. Tsurada, H. Aramoto, and M. Ishibashi, "Studies on the Different Crystal
    Structures of Nylon 6," *Chem. High Polymers (Japan)*, **15**, 619 (1958) (in Japanese).
81. Y. Kinoshita, "An Investigation of the Structures of Polyamide Series," *Makromol.
    Chem.*, **33**, 1 (1959).
82. P. H. Harris, personal communication.
83. B. G. Ranby and R. W. Noe, "Crystallization of Cellulose and Cellulose Derivatives
    from Dilute Solution. I. Growth of Single Crystals," *J. Polymer Sci.*, **51**, 337 (1961).
84. R. St. J. Manley, "Crystals of Cellulose," *Nature*, **189**, 390 (1961).
85. R. St. J. Manley, "Growth and Morphology of Single Crystals of Cellulose Triace-
    tate, *J. Polymer Sci.*, **1A**, 1875 (1963).
86. V. A. Kargin, N. F. Bakeev, and L. Li-shen, "Investigation of Polytrifluorochloro-
    ethylene Monocrystals" *High Mole. Wt. Comps.*, **3**, 1100 (1961) (in Russian).
87. E. M. Belavtseva and Yu. V. Mnyukh, "Some Forms of Morphological Structures
    in Linear Polyesters," *High Mole. Wt. Comps.*, **3**, 213 (1960) (in Russian).
88. Yu. V. Mnyukh, E. M. Belavtseva, and A. I. Kitaigorodskii, "Morphology of Molecu-
    lar Packing in Linear Polyesters," *Proc. Acad. Sci. (USSR), Phys. Chem. (Eng.
    Transl.)*, **133**, 739 (1960).
89. B. Wunderlich and P. Sullivan, "Interference Microscopy of Solution Grown Poly-
    ethylene Single Crystals," *Polymer*, **3**, 247 (1962).
90. F. P. Price, personal communication.
91. V. F. Holland, personal communication.
92. V. A. Kargin, N. F. Bakeev, Li Li-shen, and T. S. Ochapovskaya, "Electron Micro-
    scopic Study of Crystalline Structures of Polystyrene and Polypropylene," *High
    Mole. Wt. Compounds*, **2**, 1280 (1960) (in Russian).
93. R. P. Palmer and A. J. Cobbold, to be published.
94. R. H. Marchessault, F. F. Morehead, N. M. Walter, C. P. J. Glaudemans, and T. E.
    Timell, "Morphology of Xylan Single Crystals," *J. Polymer Sci.*, **51**, S66 (1961).
95. S. Okamura, K. Hayashi, and M. Nishii, "Polymer Crystals Obtained by Radiation
    Polymerization of Trioxane in Solid State," *J. Polymer Sci.*, **60**, S26 (1962).
96. J. Lando, N. Morosoff, H. Morawetz, and B. Post, "Single Crystal Character of
    Polyoxymethylene Prepared from Single Crystals of Trioxane," *J. Polymer Sci.*,
    60, S24 (1962).
97. S. E. Jamison and H. D. Noether, "Single Polyoxymethylene Crystals by Solid
    State Polymerization of Trioxane," *Polymer Letters*, **1**, 51 (1963).
98. M. Ogawa, T. Ota, O. Yoshizaki, and E. Nagai, "Notes on Nylon Single Crystals,"
    Polymer Letters, **1**, 57 (1963).
99. N. Hirai, "A Source of Screw Dislocations on Polyethylene Single Crystals," *J.
    Polymer Sci.*, **59**, 321 (1962).

III. HEDRITES

Hedrites were originally defined (1) as polyhedral structures crystallized from polymer melts. They were shown to be intermediate in complexity between single crystals and the spherulites usually formed during polymer crystallization from the melt. The definition is extended here to include similar structures crystallized from solution, i.e., hedrites are structures which have a polygonal appearance when viewed from at least one direction. Although they are usually thicker than 1 micron and thus are composed of numerous lamellae, they have a single crystallike structure. In this chapter we shall also consider other structures intermediate in complexity between the single crystals and spherulites.

1. HEDRITES CRYSTALLIZED FROM SOLUTION

The polyhedral structures of gutta-percha described by Schlesinger and Leeper (2) (see the previous chapter) are representative of hedrites crystallized from solution. Optical micrographs of similar structures were published as early as 1929 (3). Subsequently, polyhedral structures were described for polyoxymethylene (Fig. III-1) (4) and polyethylene (5). In general the solution grown hedrites are too small to replicate easily and too thick to shadow and insert directly into the electron microscope. Investigation of their structure is, therefore, difficult.

Slow cooling of linear polyethylene solutions results in the simultaneous development of hedrites and the splaying type of highly dendritic crystals (Fig. III-2). The relative proportion of hedrites to dendrites appears to increase with increasing concentration. Niegisch (5) has attributed the growth of polyethylene hedrites to the use of an extremely linear polymer having a high molecular weight with a narrow distribution and to the use of a relatively poor solvent. We find that similar structures grow from various commercial linear polyethylenes and in numerous solvents (6). Bassett, Keller, and Mitsuhashi (7) have recently reported a detailed study of the structure of solution crystallized polyethylene hedrites (termed axialites in their paper). They report that hedrites crystallize from xylene solutions of Marlex 50 at temperatures above 75°C and concentrations exceeding 0.3%.

189

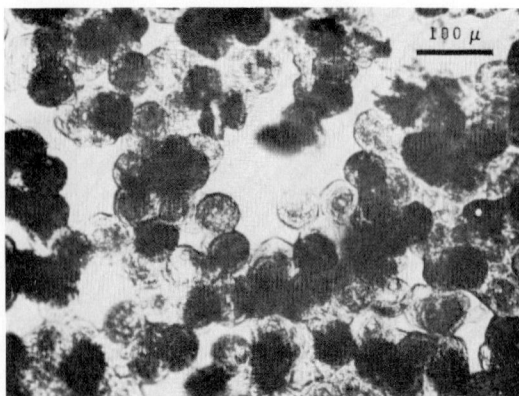

Fig. III-1. Hedrites of polyoxymethylene grown from a 1% solution in phenol.   The bright spot near the center of the micrograph corresponds to the microcamera x-ray beam used to take the pattern shown in Figure III-6.   This film of hedrites formed on the top of the solution as it was cooled stepwise over a period of several days.

There is a possibility that some structures which appear to be hedrites may result from contraction or rolling up of dendritic crystals during solvent removal. When well dispersed systems are observed in the optical microscope following removal of the solvent, the hedrites are usually found on top of the dendritic crystals, the dendrites perhaps remaining extended because of adhesion to the glass slide. In our experience the hedrites are seldom found isolated (6). The dendritic crystals can be seen clearly only with reflected light or phase contrast. In transmitted or polarized light only the hedrites are visible. Crystals are occasionally seen which suggest that only a portion of their material has contracted, the remainder adhering to the slide. Niegisch (8) also has indicated that at least a portion of the structure contracts during solvent removal. Bassett *et al.* (7), however, show micrographs of the compact hedrites still suspended in the solvent (Fig. III-3) and their evidence would suggest that if any contraction or "rolling up" occurs, it occurs while the hedrites are still in suspension. In suspension as well as after removal of the solvent the hedrites they picture are distinctly different in appearance than the splaying dendritic crystals (as in Figure II-62) which may be found in the same preparation.

When large quantities of a suspension of hedrites are dried down to form a film, the resulting film is "fluffy" and low in density, in contrast to the dense films prepared from suspensions of the lamellar crystals. Furthermore the film undergoes considerable lateral shrinkage. The relative pro-

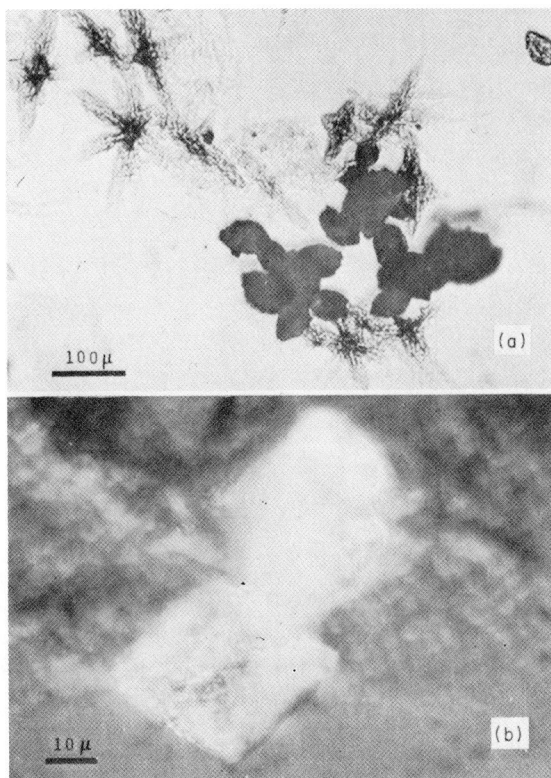

Fig. III-2. Hedrites and dendritic crystals of linear polyethylene crystallized by slow cooling of dilute solutions. (a) Six-sided hedrites (the dark objects) crystallized from o-xylene. (b) Diamond-shaped hedrites crystallized from 0.1% tetrachloroethylene. The hedrites shown are several microns above a background of dendritic crystals. With transmitted light the dendritic crystals are barely visible, whereas the hedrites scatter light strongly. (Incident steep oblique illumination.) (Geil and Reneker (6).)

portion of hedrites to dendritic crystals appears to increase with increasing concentration, a feature which, however, may be related to the resulting larger number of particles per unit volume. When the more concentrated solutions are placed on a slide and the solvent removed, more individual hedrites are left on top of the aggregated dendritic crystal substrate.

Electron micrographs of replicas from surfaces of polyethylene hedrites indicate that they are lamellar and contain numerous spiral growths (6,7) (see later discussion). The thickness of the hedrites corresponds to 100 or more lamellae. Their low density and high light scattering suggest that

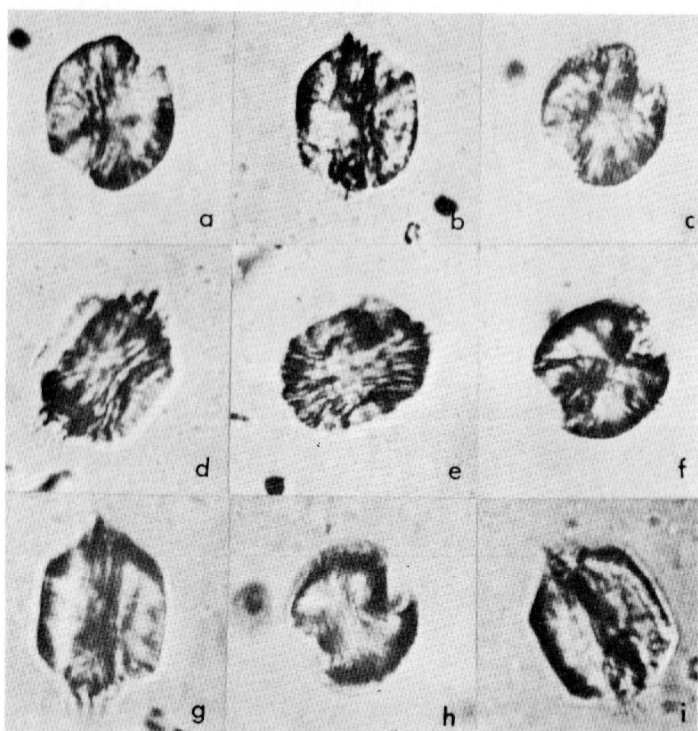

Fig. III-3. Several views of a hedrite in suspension. The sample was mounted on a universal stage and rotated to obtain the various micrographs (see also Figure III-10). (Bassett, Keller, and Mitsuhashi (7).)

they are somewhat porous. Optical micrographs of polyoxymethylene hedrites suggest that the lamellae on the surface are rolled or folded to produce a fibrous appearing structure (Fig. III-4). Although the orientation of this fibrous appearing material is tangential in polyoxymethylene hedrites, in the polyethylene hedrites reported by Niegisch (5) it is oriented parallel to the **b** axis. The structure of the fibrous appearing material observed optically in polyethylene hedrites is even more vague and more poorly defined than that shown in Figure III-4. Similar fibrous appearing material has been observed on crystals of polystyrene (9) (Fig. II-85).

Birefringence measurements indicate that the molecules in hedrites are normal to the basal plane (4,7). Between crossed polaroids, hedrites of polyoxymethylene show no birefringence other than that associated with the apparently wrinkled or rolled up lamellae on the surface. When viewed

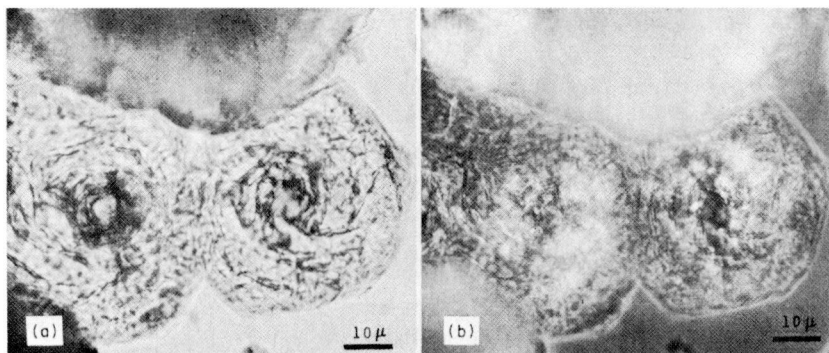

Fig. III-4. Hedrites of polyoxymethylene from the sample shown in Figure III-1. With incident illumination, polyoxymethylene hedrites scatter light strongly. (a) Transmitted light (Geil, Symons, and Scott (4)). (b) Crossed polaroids.

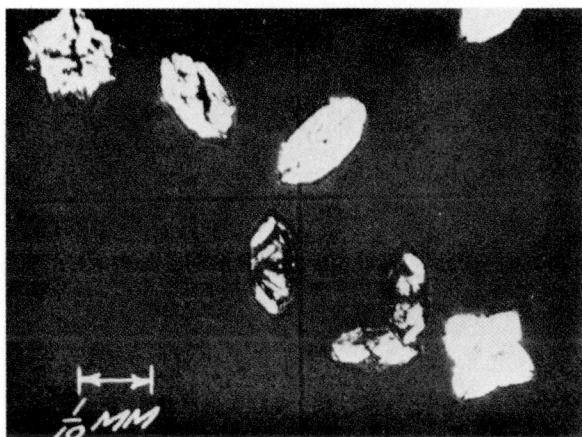

Fig. III-5. "Crystals" of gutta-percha grown by a vapor-diffusion technique in which a nonsolvent (ethyl acetate) diffused into a benzene solution of chicle gutta at the same time that the benzene was diffusing out (crossed polarizers, with the planes of polarization parallel to the cross hairs). (Schlesinger and Leeper (2).)

parallel to the basal plane, they are highly birefringent. Niegisch's micrographs (5) of polyethylene hedrites observed between crossed polaroids can also be interpreted in terms of rolled or folded lamellae on the surface. The compact polyethylene hedrites when viewed parallel to the basal plane, i.e., along their axis, show an extinction pattern resembling the Maltese cross seen in spherulites (Chap. IV) (7). Gutta-percha hedrites are dark when the axes are parallel to the plane of polarization and bright when 45° to this plane (Fig. III-5) (2). This is consistent with the biaxial optical character of the gutta-percha unit cell.

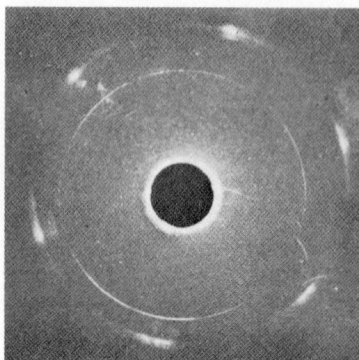

Fig. III-6. Microcamera x-ray diffraction pattern from a single polyoxymethylene hedrite (see Fig. III-1). The reflections shown are {100} reflections of the hexagonal unit cell. The complete ring with a radius slightly less than that of the arcs is an artifact formed during loading of the camera.

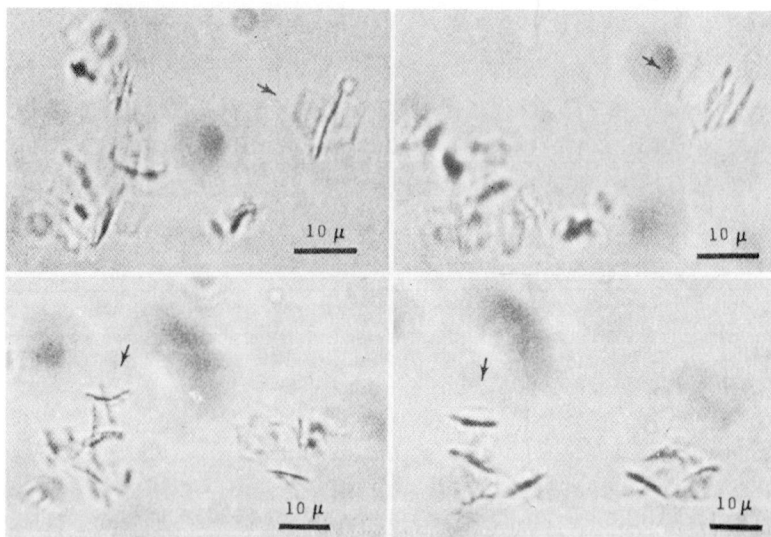

Fig. III-7. Two sets of photographs of polyethylene crystals consisting of two hexagonal layers at right angles to each other. The crystals of interest, suspended in the octane from which they crystallized, are indicated by arrows.

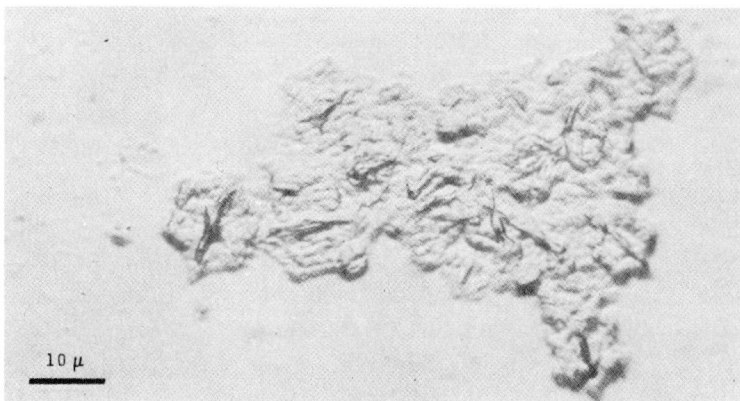

Fig. III-8. Single crystals of polyethylene frozen dried from the sample shown in Figure III-7. In some of the crystals the lamellae remained normal to the slide while the solvent was sublimed. (Incident steep oblique illumination.)

The orientation of the molecules in polyoxymethylene hedrites has been confirmed with microcamera x-ray diffraction patterns from individual hedrites (4). The x-ray patterns (Fig. III-6), which are single crystallike, show that the molecules are oriented normal to the basal plane of the hedrite. The reflections are slightly arced, suggesting that successive lamellae in the hedrite are slightly misaligned. However, as indicated in Chapter II, successive lamellae in a spiral growth are also often rotated. The relative number of molecular segments in the "fibrous" material is apparently small since they do not contribute significantly to the x-ray pattern.

A possible source for the fibrous appearing material, other than the rolling up of surface lamellae, can be seen in some simpler structures than the hedrites. Crystals of polyethylene have been observed optically consisting of two lamellae or sets of lamellae oriented at right angles to each other. Although barely visible they can be observed tumbling about while in suspension (Fig. III-7). In similar but more complex structures containing numerous lamellae, the lamellae appear to be oriented at various angles. The line of intersection is parallel to the b axis. Many, but not all, of the lamellae are found to have a common intersection along the minor diagonal of the crystal. While tumbling in suspension these crystals appear either as sheafs (when viewed normal to the b axis) or as a nearly symmetrical set of radiating lines (when viewed parallel to the b axis). Apparently similar crystals have been described as paddle-wheel-like by Mitsuhashi (10). When the solvent is removed by freeze drying a crystal

Fig. III-9. "Paddle-wheel" crystals of linear polyethylene following solvent removal. (These crystals have been annealed at 120°C, resulting in the irregular edges on some of the crystals. See Figure V-4 for other crystals from this preparation.) The crystal indicated by the arrow *A* is believed to have collapsed while its **b** axis was normal to the slide, forming a rosettelike structure. The crystals indicated by the arrows *B* and *C*, in which the **b** axes are believed to have been parallel to the slide, have collapsed to form sheaflike structures. The others are of intermediate structure. In the crystal at bottom right, most of the lamellae folded to the right. The wrinkles in some of the {100} fold domains are associated with a hollow pyramidal structure.

suspension previously deposited on a substrate, some of the lamellae, or portions thereof, in both the simple (Fig. III-8) and more complex structures remain normal to the substrate. These crystals differ to an extent from the splaying dendritic crystals described by Bassett, Keller, and Mitsuhashi (7) in that the lamellae appear to intersect at discrete angles.

Material appearing fibrous in the optical microscope results from the collapse of these paddle-wheel crystals during solvent removal. When deposited on a slide and the solvent evaporated, the lamellae originally normal to the slide fold over and collapse onto the slide. Depending on the orientation of the **b** axis with respect to the substrate either a rosettelike structure or a sheaflike center form (Fig. III-9). During freeze drying, portions of the lamellae have been observed to remain vertical at least up to the time they are shadowed. When hit by the electron beam in the microscope, however, they tend to curl up and collapse, with only the shadow to show their original position. The growth mechanism of the paddle-wheel crystals is not completely understood. As indicated during the discussion of dendritic crystals, successive lamellae in a spiral growth on a dendritic crystal diverge or splay as they grow. This type of growth, however, would not lead to lamellae oriented at discrete angles to each other, angles which sometimes approach 90°.

It appears possible that on hedrites some of the fibrous appearing material, which has also been called overgrowths (5), may result from a similar collapse of lamellae during solvent removal. The rolling or folding of the lamellae would thus occur at the same time as the lateral dimensions of the hedrite contract. The overgrowths resulting in the observed appearance of polyethylene hedrites (5) evidently occur at various points on the surface of the lamellae, along lines parallel to the b axis.

As discussed in detail by Bassett, Keller, and Mitsuhashi (7), some polyethylene hedrites have a compact appearance while in solution. By mounting the sample on a universal stage they were able to obtain micrographs of the same hedrite from a number of different angles. A stereoscopic presentation of one such set of photographs is shown in Figure III-10. On rotating the sample, which was suspended in a rigid medium consisting of a mixture of ethyl alcohol and polyvinyl alcohol, they found that all appearances between a hexagonal structure and a sheaflike structure could be observed (see for instance the set of micrographs along the vertical diagonal of Figure III-10). Bassett et al. (7) suggest that the hedrites resemble two partly open books, with pages consisting of half of a hexagon, placed spine to spine. By means of electron diffraction they were able to show that the spine of the books, i.e. the axis of the hedrites, corresponds to the b axis of the crystal lamellae. Increasing either the temperature or concentration of the solution tended to increase the relative length of the spine.

Electron microscopy of the compact hedrites, because of their thickness, is both difficult to do and to interpret. It has been possible to retain some of the original structure by heavily shadowing the samples. Freeze drying should also be a help in this respect. Usually it is necessary to remove the polymer before the replicas can be used. The detachment replica technique of Bassett (12) has been especially useful in this respect although it has raised some questions which have not yet been answered.

Figure III-11 shows a detachment replica of a polyethylene hedrite in which one of the "pages" of the book is nearly fully extended on the

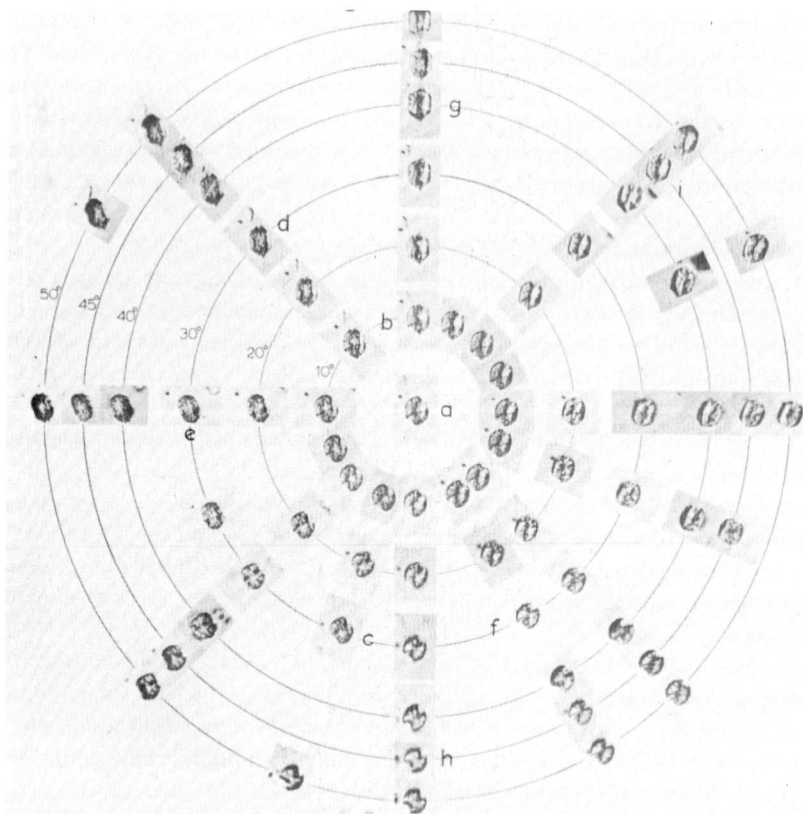

Fig. III-10. Stereographic presentation of a polyethylene hedrite in suspension. The micrographs are positioned according to the setting of the universal stage. Note the variation in outline along the vertical set of micrographs. Figure III-3 shows a set of photographs extracted from this figure, at higher magnification. (Keller (11).)

substrate. Despite the presence of spiral growths in and on the "page" there is a nearly constant thickness of polymer remaining. By comparison of its contrast with that of normal, monolayer single crystals the "page" is found to be about 1,000 A thick although the lamellae within it are of the order of 130 A thick (7). Bassett et al. conclude that there is some form of coherence between the lamellae making up a given "page" although why a uniform slice of constant thickness is extracted by the replica technique is unknown. Only a single lamella is left in contact with the replica in dendritic crystals in the same sample. This coherence between the lamellae is apparently reduced by ultrasonic treatments and is believed

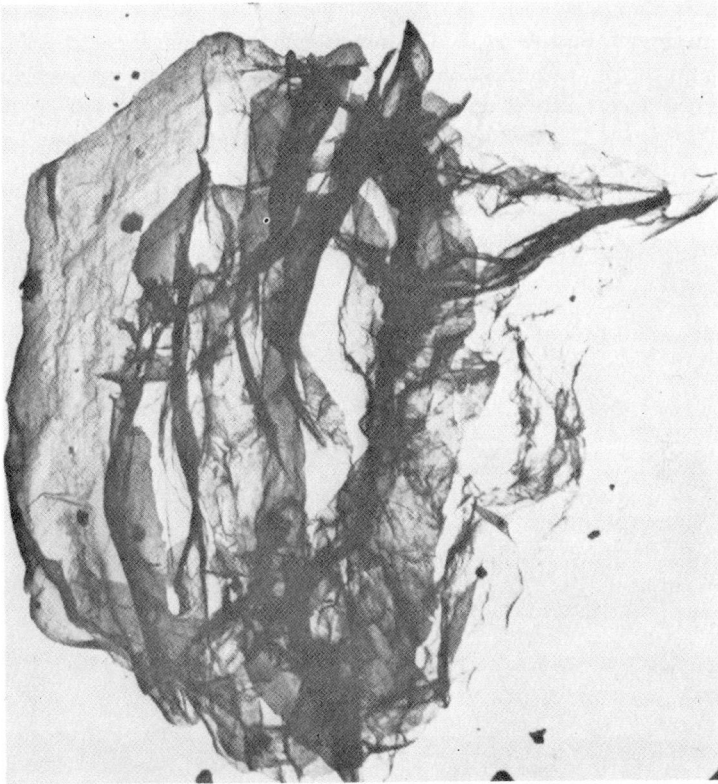

Fig. III-11. Detachment replica (12) of a compact polyethylene hedrite (Bassett, Keller, Mitsuhashi (7)).

to be related to "tie" molecules interconnecting the lamellae (7) (see Chapter IV).

It is possible to obtain somewhat simpler hedrites by interrupting the growth before it is complete (by filtration at the crystallization temperature). Figure III-12 represents a typical hedrite in which several "pages" are still nearly erect. In the optical microscope these hedrites were seen to contain only a few pages. In the detachment replica in Figure III-12 the cohesion between the lamellae within a "page" is present in the upper portion, but nearly absent in the lower portion, of the hedrite. This lower portion appears to consist of lamellae folded back on themselves along the exterior edge. Bassett et al. indicate that in many cases lamellae appear to be peeling off of the pages of the hedrite, the peelings being hinged along the outer boundary (7). This type of peeling, as well as several of the micrographs of Bassett *et al.* (7), again suggests some form of rolling or folding up in the formation of the hedrites, although the contraction can apparently occur during growth or subsequent cooling as well as during solvent removal. Observation at the growth temperature has been impossible.

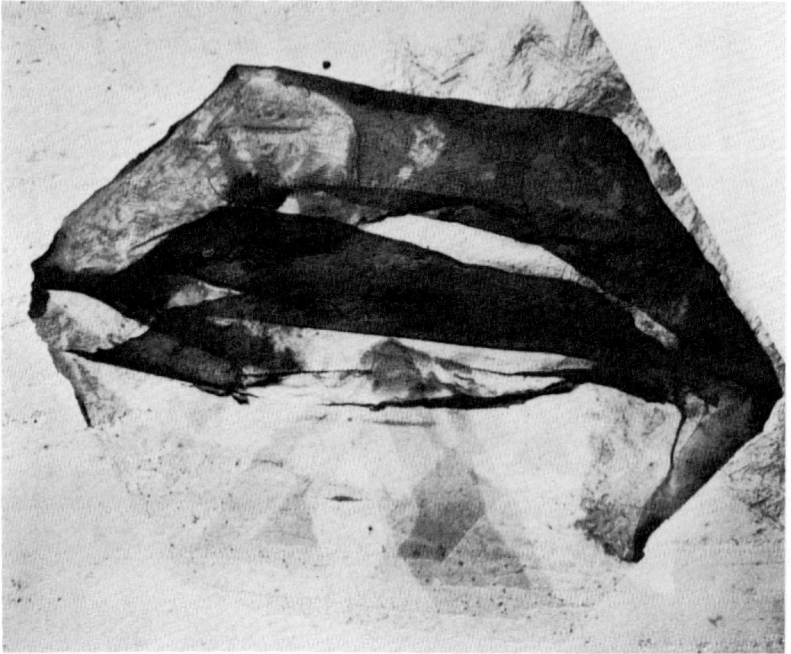

Fig. III-12. Detachment replica (12) of a simple hedrite composed of only a few pages (Bassett, Keller, and Mitsuhashi (7)).

At extremely high concentrations (up to 55%) a gel is produced during crystallization. Bassett *et al.* indicate (7) that this gel still consists of hedritelike structures which become more and more attached to each other as the concentration is increased. At the highest concentrations (40% or more) they become more and more spherulitic in appearance, with concentric extinction rings even becoming visible (Chapter IV). Even so, however, the spherulitic effects do not have a three dimensional symmetry (7).

2. HEDRITES CRYSTALLIZED FROM THE MELT

Polyhedral objects can be crystallized from thin molten films of several polymers, being first reported in 1953 in slowly crystallized thin films of gutta-percha (13,14). So-called "square" spherulites are formed if crystallization occurs at 45°C. Normal spherulites are formed by crystallization at room temperature. These hedrites, which extinguish as uniformly oriented entities between crossed polaroids, melt at 58°C. This is slightly above the melting point, 56°, of spherulites of the α-modification of gutta-percha, but well below the melting point, 65°, of spherulites of the β modification. They are thus probably of the α-modification, the higher melting point being due to the greater perfection and thickness of the crystals resulting from the initial high crystallization temperature.

Additional evidence for the growth of hedrites in gutta-percha as well as related oval shaped structures was reported in 1955 (15). Although crystallization in this case occurred at room temperature, a very thin film was used. Growth of the hedrites took place over a period of 30 minutes. Thicker regions of the same sample crystallized rapidly in the form of spherulites. After a certain amount of growth in the form of a regularly shaped object, the hedrites and ovals tended to split from one edge to the center. Further growth on the split edges occurred rapidly, the structure developing first into a sheaf and finally a spherulite. No reason for the slow initial growth was advanced.

Recognition that melt-grown hedrites are single-crystallike structures, considerably less complex than spherulites, but with a related structure and origin, has followed their discovery in several other polymers (1,16–18). Hedrites of polyoxymethylene (Fig. III-13) have been studied in the greatest detail (1,16) and, unless otherwise stated, the following discussion applies to them and is taken from references 1 and 16. Available descriptions (optical microscope only) of hedrites of several other polymers are discussed in the last section of this chapter.

Polyoxymethylene hedrites are grown from thin films (~30 microns or less) either by slow cooling or by crystallization at constant temperature. Oval structures are usually also formed. Growth of hedrites occurs between 160° and 165°C, the melting point of the polymer being about 180°C. Sufficient residence time in the melt (e.g., 1 hour at 182° or 10 minutes at 190°C) is required to destroy heterogeneous nuclei such as those on foreign particles. Crystallization and growth occur slowly, periods from about one hour up to several days being used. Rapid cooling of films of similar thickness, insufficient time in the molten state, or the use of thicker films results in the growth of spherulites.

Fig. III-13. Hedrites and oval structures of polyoxymethylene crystallized from a molten thin film between 160 and 165°C. (a) Incident steep oblique illumination. (b) Crossed polaroids, (Geil (1).)

Polyoxymethylene hedrites have been grown as large as 1 mm in diameter and may be separated from each other by as much as several millimeters. They usually are thicker than the parent film from which grown and

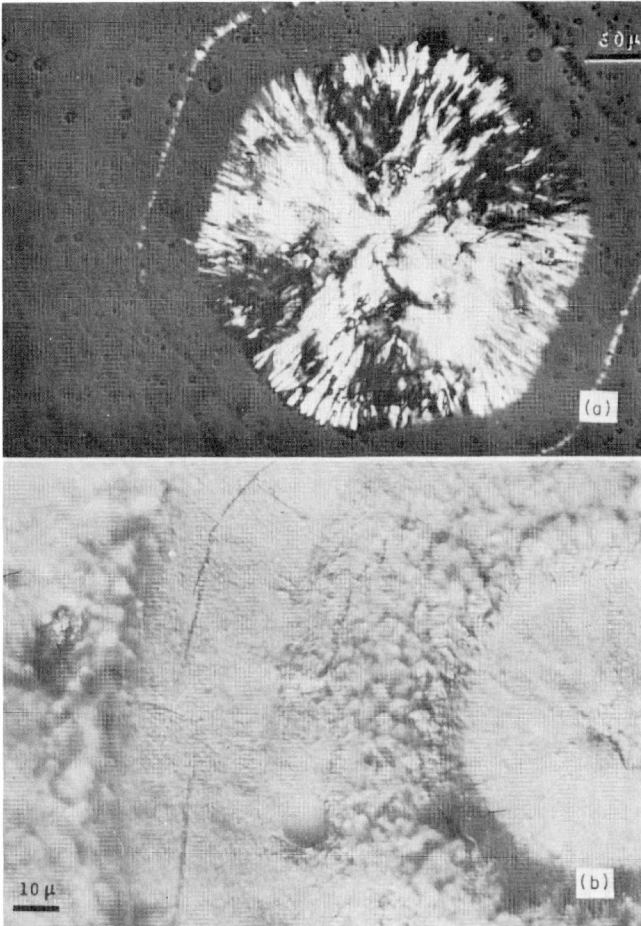

Fig. III-14. Hedrite of polyoxymethylene crystallized at various temperatures between 160 and 168°C on a hot stage.   (a) Viewed between crossed polaroids at 163.5°C.   A ring of material crystallized near 160°C is molten.   (b) Same hedrite as in (a) following a further rise in temperature followed by quenching.   The diameter of the ring, a portion of which is shown here, corresponds to about $^1/_2$ the central portion of the hedrite in (a).   Before quenching, all of the material, except the ring and center which are in focus, was molten.   The coarse structure developed during the quenching.

are considerably thicker than the film remaining between isolated hedrites after their formation. This film, about 1 micron thick, appears to remain molten until the sample is cooled below the growth temperature of the hedrites. The separation of isolated hedrites and the relative thickness of hedrites and parent film indicates that polymer molecules can move distances on the order of millimeters during crystallization.

When grown at a constant temperature, hedrites develop with flat tops and nearly perpendicular sides. Only the diameter increases, and not the thickness, as growth proceeds. This growth continues until, apparently, the molten film remaining between the hedrites is too thin to support further growth at this temperature. If then quenched this film crystallizes and the hedrites retain their perpendicular sides. If slowly cooled, sloping sides develop, a slight increase in diameter occurring.

For a given film the thickness of the hedrite depends on the crystallization temperature, increasing with increasing temperature. If the temperature changes during growth, the thickness of the then crystallizing portion of the hedrite changes correspondingly. If the temperature is raised above the crystallization temperature but not above the equilibrium melting point of the polymer, those portions of the hedrite which crystallized at the lowest temperature melt first (Fig. III-14). They do not recrystallize until the temperature is again lowered, the temperature at which melting occurs being above that for which any crystallization occurs. This is in apparent contrast to results obtained with single crystals precipitated from solution and some spherulitic material (see Chapter V). The dependence on crystallization temperature of the thickness of the ovals has not been determined.

The orientation of the molecules in hedrites and ovals can be determined by optical and x-ray techniques. Well-developed hedrites have no birefringence when viewed normal to their basal plane, whereas the ovals have a high degree of birefringence (Fig. III-13b). This suggested, polyoxymethylene being optically uniaxial, that the molecules are normal to the basal plane of the hedrites and parallel to this plane in the ovals. This orientation was confirmed by x-ray diffraction from individual hedrites. The diffraction pattern from a hedrite consists of discrete arcs, indicating a single-crystallike character, in contrast to uniform rings from spherulites of similar thickness. Some birefringence, as well as a fibrous appearance, is found in less well developed hedrites (as in Fig. III-14a). In these hedrites the molecules are not quite normal to the basal plane.

Hedrites are composed of lamellae. They fracture easily parallel to the basal plane whereas ovals will fracture spontaneously, normal to this

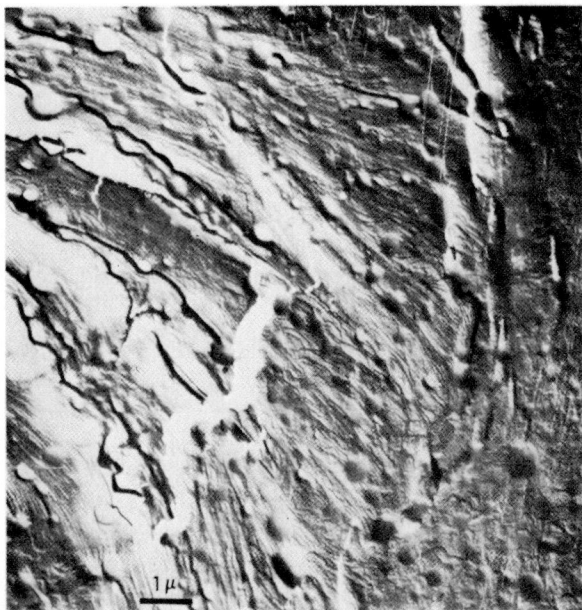

Fig. III-15. Electron micrograph of a portion of the edge of an oval structure of poly-oxymethylene. The lamellae are essentially normal to the substrate. (Cr replica.)

plane and in a radial direction. Optically it is observed that the fracture surface in the hedrites has a layered or steplike structure. Electron micrographs of replicas show that hedrite surfaces consist of lamellae about 150 A thick, lying nearly parallel to the surface. The edges of numerous lamellae are seen around the hedrite circumference. X-ray and birefringence measurements, indicating that the molecules are normal to the lamellae, suggest that the molecules may be folded within the lamellae.

In birefringent hedrites the fibrillar appearance results from the edges of lamellae intersecting the top surface. Apparently in these hedrites the lamellae, due either to the orientation of the nucleus or an imperfect growth process, are not quite parallel to the substrate.

Electron micrographs of replicas of ovals indicate that the sheaflike structure (Fig. III-13) is composed of lamellae oriented normal to the surface (Fig. III-15). The smooth-appearing portion of low birefringence consists of lamellae nearly parallel to the surface. In polyethylene oxide ovals (see Section III-3a) this region fills in after the sheaf has developed. Although the cracks observed in the ovals probably occur between the lamellae, their cause is not known. Since the cracks are normal to the molecular axes, thermal contraction should be primarily parallel to the direction of the cracks.

Pits are seen at the center of hedrites in optical and electron micrographs. Edges of the lamellae encircle the pit. It was suggested that the pit is associated with a screw dislocation. Although these pits may not arise from the same cause, the force field at the axis of a screw dislocation with a large Burgers vector can result in the development of a pit or hollow (19). Thermal contraction may also be involved.

As a result of these observations, it was suggested that melt-grown hedrites (and ovals) develop from a central screw dislocation. Whether or not the pit is directly related to the dislocation, the common orientation of lamellae in the hedrite is indicative of the presence of the dislocation. No evidence of additional dislocations forming in the growing lamellae has been found.

Presumably the initial nucleus consists of a single crystal, primary lamella in which the molecules are folded, in contact with the substrate. Interaction with the substrate results in the molecules being either normal (for a hedrite) or parallel (for an oval) to the plane of the substrate. The lamellae are thus, respectively, parallel or normal to the substrate. A screw dislocation in this lamella results in the growth of additional, parallel, secondary lamellae. Divergence of the lamellae in the ovals, possibly associated with pyramidal crystals or the twisting of lamellae that is observed in spherulites (see Chap. IV), results in the development of the sheaf and the radial cracks. Electron micrographs, to date, have not been clear enough to follow individual lamellae for more than a small fraction of a radius.

The number of lamellae in the central portion of a hedrite is determined, it was suggested, by the original thickness of the molten film and the crystallization temperature. As long as all of the top surface of the hedrite is wetted by the polymer melt, molecules can apparently migrate in the melt to the center and result in the spiral growth of additional lamellae. The final center or nucleus of the hedrite becomes thicker than the original polymer melt. After sufficient lateral growth, the number of lamellae becomes constant and the thickness of the nucleus does not change. At constant crystallization temperature all of the lamellae continue to grow laterally, molecules migrating from the molten film to their growth faces. Growth ceases when the forces causing and opposing the migration are balanced. A thin film of molten polymer remains between the hedrites.

If the temperature is changed while growth is occurring, it was suggested that the thickness of the lamellae, but not their number, changes. As a result the thickness of the growth face changes, increasing or decreasing with the crystallization temperature. No new lamellae need to be nucleated as the thickness increases. Additional observations and measurements are needed to confirm these suggestions.

3. POLYMER HEDRITE DESCRIPTIONS

Hedrites of the few other polymers from which they have been grown are described below. The polymers are listed in alphabetical order. Only optical microscope observations are available. Undoubtedly hedrites can be crystallized from solutions of other polymers than those listed in Section III-1 i.e. gutta-percha, polyethylene, and polyoxymethylene. In fact, as indicated in the last section of Chapter II, some of the larger "crystals" reported in the literature recently might more appropriately be termed hedrites. The hedrite descriptions below are followed by a description of single crystals and other structures, intermediate in complexity between single crystals and spherulites, that can be crystallized from melts of polypropylene by slow cooling. Single crystals of polyamides and polytetrafluoroethylene grown from the melt are described in Chapter IV, Section 5.

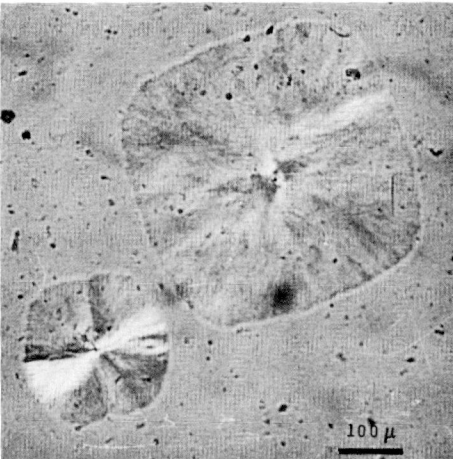

Fig. III-16. Square hedrite of polyethylene oxide growing in a thin molten film as it cooled to room temperature. The smaller structure consists of a birefringent sheaf with nonbirefringent growth occurring normal to the axis of the sheaf. (Crossed polaroids, first order red plate.) (Geil (1).)

a. Polyethylene Oxide

Hedrites of polyethylene oxide are square in outline. In contrast to he-drites of other polymers a fairly rapid growth rate favors their growth. Those shown in Figure III-16 were grown by placing a slide with a thin molten film on the room temperature microscope stage. Slower, controlled crystallization does not prove as effective.

Sheaflike structures, apparently related to ovals of polyoxymethylene, developed simultaneously with the hedrites (Fig. III-17). As indicated previously, regions between the flared ends of the sheafs filled in after some growth of the sheaf had occurred. These two growths are pointed, the apices having an included angle of approximately 70°. Their relation-ship to the hedrites is unknown.

Birefringence measurements suggest that the molecules are oriented normal to the basal plane of the hedrites and the pointed growth areas of ovals and parallel to this plane in the sheaflike portion of the ovals. No x-ray or electron microscope observations are available.

Fig. III-17. Sheaflike structures of polyethylene oxide at various stages of growth.   The sheaves appear to be composed of dendritic, fibrillar structures.

b. Polyisoprene (Gutta-Percha)

A recent Russian paper (20) has confirmed some of the observations discussed at the beginning of Section II. "Bean" or oval shaped structures similar to those in Figure III-4 were described.

c. Poly-4-methyl-pentene-1 (18)

Optical observations on square hedrites of poly-4-methyl-pentene-1 up to 0.2 mm on a side (Fig. III-18), grown by isothermal crystallization between 215° and 230°C of molten thin films, have been reported by Leugering (18). Oval structures are also present. Leugering indicates that similar square hedrites can be obtained from concentrated solutions.

The hedrites have low birefringence. Leugering indicated that no Maltese cross is seen between crossed polaroids during the initial stages of growth. The ovals, on the other hand, apparently have a high degree of birefringence. Spherulites, with a well-defined Maltese cross, are formed if crystallization is rapid. Figure III-18 suggests that the hedrites, as in the case of poly-oxymethylene, are thicker than the parent film from which they are grown. The difference in birefringence between the hedrites and ovals suggests that the molecules are normal to the plane of the hedrites and tangential within the ovals. X-ray and electron microscopic observations were not described.

Fig. III-18. Hedrite oval structure and deformed hedrites of poly-4-methyl-pentene-1 crystallizing in a molten film. Crystallization of the hedrites from high molecular weight polymer occurs in the neighborhood of 230°C, the melting point being about 234°C. (Polaroids crossed at about 85°.) (Leugering (18).)

d. Polystyrene (17)

Low molecular weight isotactic polystyrene can be crystallized from the melt in the form of hedrites and oval structures, as well as spherulites (see

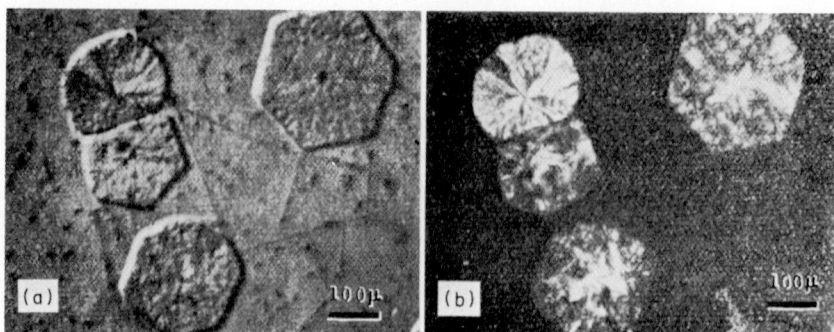

Fig. III-19. Hedrites and an oval structure of isotactic polystyrene. A relatively low molecular weight (less than 20,000) sample was melted at 260°C and crystallized slowly at 190°C. (a) Reflected light. (b) Intense polarized light with crossed polarizers (Danusso and Sabbioni (17)).

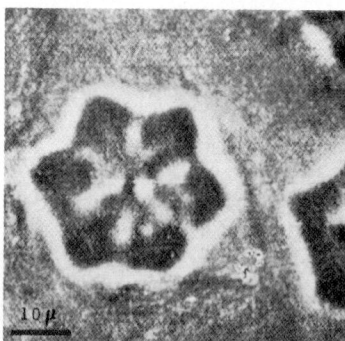

Fig. III-20. Star-shaped dodecagonal hedrite of isotactic polystyrene obtained by melting the sample at 270°C, quenching and then reheating to 180°C at which temperature crystallization took place. (Danusso and Sabbioni (17).)

Chapter IV). The more perfect hedrites consist of flat-topped hexagonal structure with little or no birefringence (Fig. III-19). Oval structures of high birefringence with a sheaflike structure resembling that observed for polyoxymethylene are also seen. As seen in Figure III-19 the hedrites and ovals appear to be thicker than the film from which they grew. Occasionally star-shaped dodecagonal structures are also observed (Fig. III-20). Although the authors suggest that the hedrites are formed only if the length of the chain is of the same order of magnitude as the thickness of the crystal, it is believed, on the basis of the work with polyoxymethylene and other polymers, that hedrites can be grown from higher molecular weight polystyrene and that they consist of lamellae. Although no x-ray or electron microscope data were presented the low birefringence and shape of the hedrites indicates that the molecules are normal to the plane of the hedrites.

Keith has recently discussed (25) the growth of single crystals of isotactic polystyrene from the melt. He indicates that, for all single crystals grown from the melt, both a slow rate of growth (v) and a relatively large self-diffusion coefficient (D) are necessary. In the case of polystyrene he finds that the crystals develop as relatively perfect, thin ($\sim$500 A) hexagons until their sides approach a dimension of $1/\delta = D/v$ (see Section IV-5a). A star-shaped profile then develops, followed by a transformation into polycrystalline aggregates. Keith suggests that a major factor in bringing about this transformation is extensive disorder built into the crystal at those fold domain boundaries that intersect the sides at re-entrant corners (i.e., similar to the slow growth lines in Figure II-50). It is at such regions that impurities and other defects would be expected to collect.

e. Polypropylene

Growth of small oval and rectangular shaped objects has been observed during the slow, controlled crystallization of thin films of polypropylene

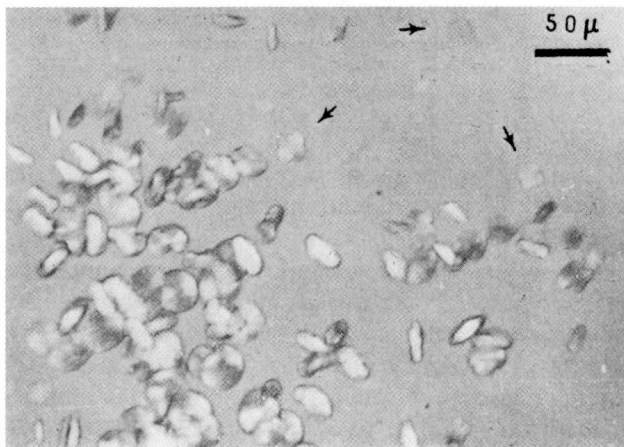

Fig. III-21. Optical micrograph of square hedrites (arrows) and oval structures of polypropylene crystallizing at 139°C from a molten thin film. The birefringence of the oval structures is many times greater than that of the hedrites. It has not been found possible to grow the hedrites much larger than shown here, further nucleation and spherulite growth occurring instead.

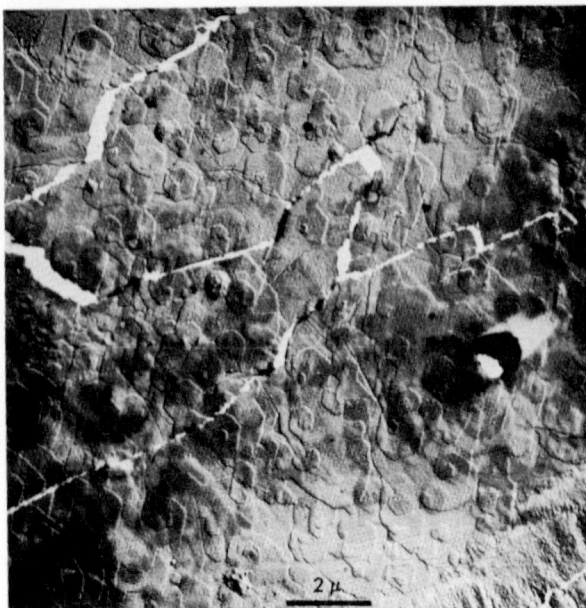

Fig. III-22. Lamellae of hexagonal modification polypropylene crystallized from the melt. This and all following micrographs, except Figure III-23, are from a thin film of polypropylene cooled slowly in a N_2 atmosphere. The slide, covered with a petri disk, was left on a hot plate as the plate cooled to room temperature. The slide and polymer film were subsequently immersed in benzene for several minutes and then replicated with chromium. (Geil (22).)

(Fig. III-21) (21). Migration of the polymer molecules apparently occurs, the objects being thicker than the surrounding film. The crystal lattice and morphology of these objects is not known.

With very slow crystallization some degradation apparently occurs, the isotactic polymer remaining crystallizing from what may be considered a solution in the atactic and degraded polymer. The structures that form have a degree of perfection resembling that of hedrites. These structures, which can be observed by dissolving out the low molecular weight and atactic polymer, are described below (21,22). A considerable number of different morphologies, based on two different crystal lattices, are observed.

That portion crystallizing in the hexagonal crystal modification crystallizes as approximately 150 A thick lamellae (Fig. III-22) (22). Thickening occurs by the spiral growth of additional lamellae. The spirals can be as

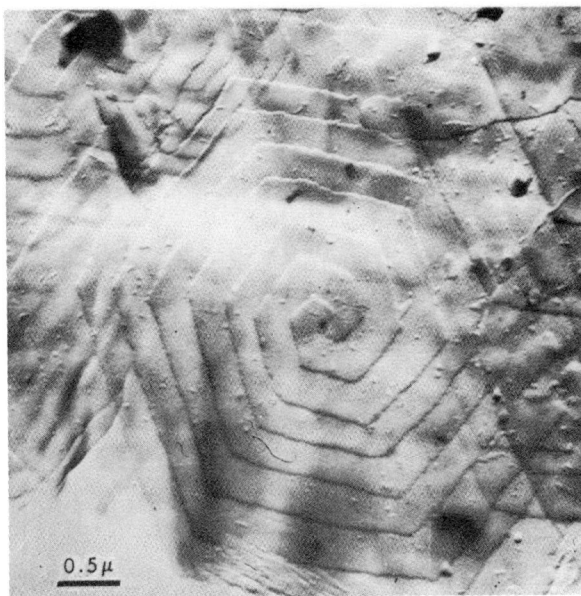

Fig. III-23. Spiral growth in a polypropylene film treated the same as that described in Figure III-22, except that the crystallization was done in air rather than N₂. The lamellae appear to be about 200 A thick. (Geil (22).)

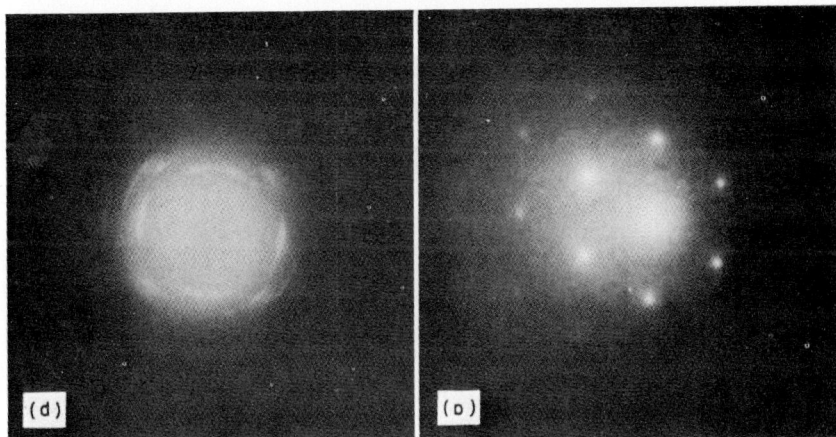

Fig. III-24. Selected area electron diffraction patterns from melt crystallized thin films of polypropylene. (a) Diffraction pattern showing hexagonal packing of the molecules in the lamellar type of crystals. Pattern is from an area similar to that in Figure III-22. (Geil (22).) (b) Diffraction pattern from edge or tip of the broad ribbons (Figs. III-28 and III-29). This material crystallizes in the monoclinic unit cell, the molecules being parallel to the electron beam in the portion of the ribbon which is diffracting.

Fig. III-25. Low magnification electron micrograph of an incipient spherulite of polypropylene.   Several types of different morphological structures can be seen.

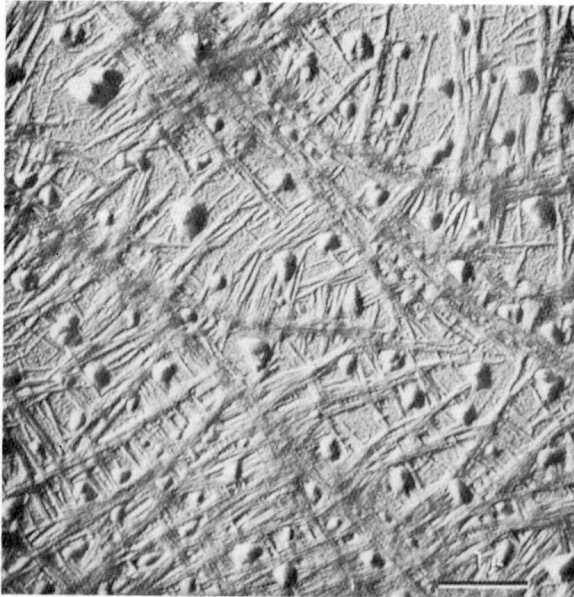

Fig. III-26. "Woven" structure as at the upper left of Figure III-21.   The chromium used for replication was shadowed at approximately $\tan^{-1} 5/10$.

well developed as those seen in solution grown crystals of other polymers
(Fig. III-23). The edges of the lamellae in the spiral growths have a uniform orientation over large areas (as shown in Fig. III-22) that crystallize
in the hexagonal modification. Furthermore, single crystal diffraction patterns are obtained from similar sized areas (Fig. III-24a). Apparently all of
such an area, extending for instance beyond the limits of Figure III-22, is
nucleated at one point and grows as a single crystal.

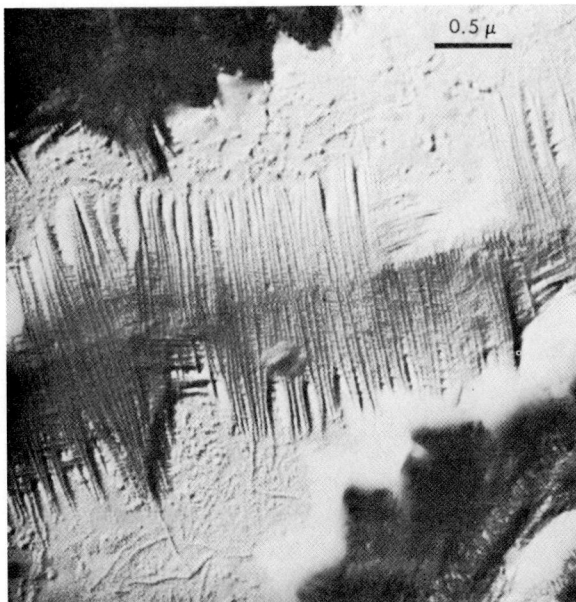

Fig. III-27. Growth from an oriented thread of polypropylene. The film used for this
and the other micrographs of polypropylene was prepared by smearing a small amount of
polymer between two slides. As the slides were separated, fibers drawn from the melt
collapsed back onto the film. The structure shown here is believed to be associated with
one of the fibers.

That portion crystallizing simultaneously with the lamellae but in the
monoclinic modification appears to crystallize with several different morphologies (Fig. III-25) (21). Various aspects of these different morphologies
are shown in subsequent figures. Some areas, as at the upper left, of Figure
III-25 appear to have a woven structure, the relative number of structures
aligned in the two directions varying from almost all in one direction to
equivalent numbers at right angles (Fig. III-26). The individual structures

appear to be thicker (normal to the substrate) than they are wide, suggesting that they may be a lamellar structure oriented normal to the substrate. If lamellae, however, the woven appearance is difficult to explain. At high magnification a 100–500 A periodicity (depending on the width of the structures) is seen along their length. Electron diffraction patterns indicate that the molecules lie in the plane of the substrate. The **a** axis is either parallel or perpendicular to the long dimension of the structures, the orientation being indeterminate because of the "woven" structure.

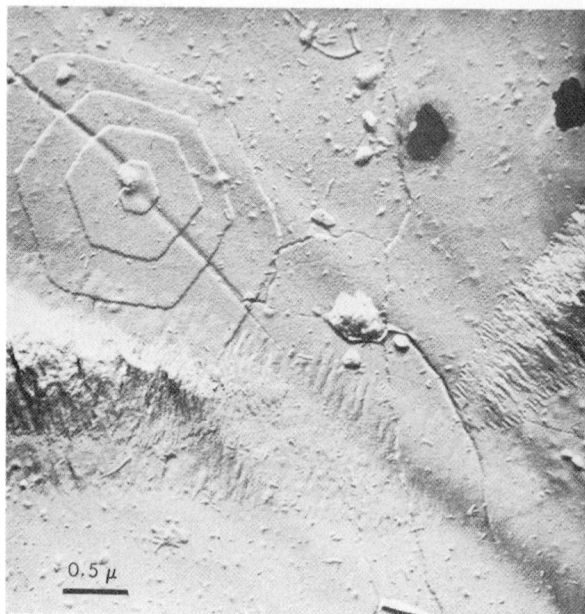

Fig. III-28. The ends of two broad needles, one partially obscured by hexagonal modification lamellae, are shown.   The lamellar nature of the end of the needles at the right is especially evident.

When the melt is oriented before crystallization, long threads of material retain orientation through the slow crystallization process. These threads crystallize as shown in Figure III-27. They appear to be composed of an ordered array of structures similar to those in Figure III-26. A cross striation (in the direction of the thread) can also be seen on some samples. Electron diffraction patterns were not obtained from these structures.

The broad ribbons extending to the lower left of Figure III-25 appear to have a different but, perhaps, related structure. The ends of the ribbons

(Fig. III-28) have a lamellar appearance, the lamellae lying on the substrate. The ribbons often have a dendritic appearance, the edges splitting. As one progresses toward the origin of the ribbon, overgrowths of unknown origin develop in the center of the ribbon. On some micrographs they have the appearance of a second lamella growing on top of the first. However, as one progresses still further toward the origin, the ribbon tends to develop cracks across its width and at an angle to the substrate (Fig. III-29).

Fig. III-29. Thick portion of a broad ribbon. The lamellae edges are especially evident in the lateral crack near the center of the micrograph. As shown by the darkness of the area near the edges of the ribbon, where polymer is still attached, the electron beam cannot penetrate material as thick as these ribbons.

The entire ribbon also has a series of parallel striations. In properly oriented and shadowed cracks one can see that the striations are due to lamellae oriented at an angle to the substrate on the order of 45°. These lamellae are about 100 A thick, the striations being broader because of the angle. A spherulite composed solely of a radiating, dendriticlike array of these needles, apparently crystallized from a partially degraded molten film and described as a crystal of polypropylene, made a striking display when pictured in color in an advertisement (23).

In some areas only the basal lamella of the broad ribbon is found, forming a dendriticlike ribbon (Fig. III-30). The center may either be empty, as in this figure, or consist of a ridge, sometimes alternating from one to the other. along the ribbon's length. The split-edges of the ribbon often have a turned up appearance, as if the lamellae had been trying to grow at an angle to or twist away from the substrate.

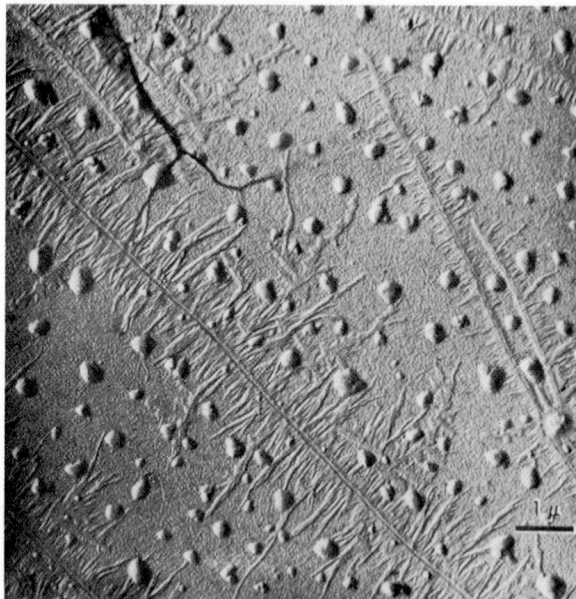

Fig. III-30. Dendritic ribbons composed of single lamellae. The cause of the longitudinal split in the center is not known. In some cases the lateral splitting does not approach the center as closely as here, resulting in a broad flat ribbon with lateral branches corresponding only to the longest side branches shown here.

Electron diffraction patterns from the broad thick ribbons are weak and somewhat diffuse (Fig. III-24b). They indicate that in the portion of the material which is diffracting the beam the a axis is parallel to the ribbon, the molecular axis being normal to the substrate (24). However, it is believed absorption of the electrons takes place in the thicker portions in which the lamellae are oriented at an angle to the substrate. Patterns have not been obtained from the long, single lamella ribbons. Dark field micrographs have not been informative.

The material in the lower portion of Figure III-25 appears to have a structure intermediate between the broad ribbons and the woven type. Its appearance suggests that some of the split-edges of the ribbon, the branches of the dendrite, have grown out into the melt and served as nuclei for further crystallization. A portion of this material, intermediate in structure between the broad needles and the fernlike structure at the bottom of Figure III-25, is shown in Figure III-31.

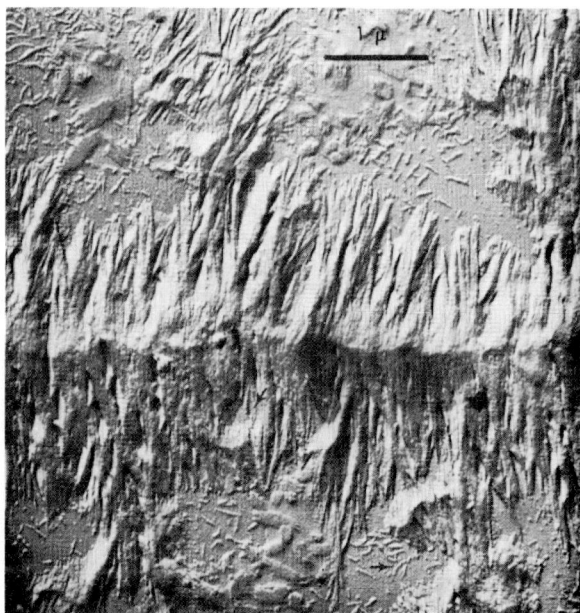

Fig. III-31. Material similar to that at the bottom of Figure III-25. Several of the fernlike growths are extending outward from the broad needle passing through the center of the micrograph. Lateral striations on the needles, as indicated by the arrows, again give the impression of being due to lamellae, about 100 A thick, oriented normal to the axis of the needle.

It is obvious that more work is needed to define the various morphological forms in which the monoclinic modification of polypropylene occurs. The presence of atactic and degraded material must also be taken into account. The structures discussed here also appear in various modified forms in polypropylene spherulites. All of the structures discussed in this chapter need further investigation, the spherulites which form during normal crystallization of polymer melts being more complex entities composed of morphological structures similar to the relatively simple structures composing the hedrites.

REFERENCES

1. P. H. Geil, "Polyhedral Structures in Polymers, Grown from the Melt" in *Growth and Perfection of Crystals*, edited by R. H. Doremus, B. W. Roberts, and D. Turnbull, Wiley, New York, 1958.
2. W. Schlesinger and H. M. Leeper, "Gutta. I. Single Crystals of Alpha-Gutta," *J. Polymer Sci.*, **11**, 203 (1953).
3. F. Kirchof, "Uber die Kristall-Struktur der Tjipetir-Guttapercha," *Kautschuk*, **5**, 175 (1929).
4. P. H. Geil, N. K. J. Symons, and R. G. Scott, "Solution Grown Crystals of an Acetal Resin," *J. Appl. Phys.*, **30**, 1516 (1959).
5. W. D. Niegisch, "Nucleation of Polyethylene Spherulites by Single Crystals," *J. Polymer Sci.*, **40**, 263 (1959).
6. P. H. Geil and D. H. Reneker, unpublished data.
7. D. C. Bassett, A. Keller, and S. Mitsuhashi, "New Features in Polymer Crystal Growth from Concentrated Solutions," *J. Polymer Sci.*, **1A**, 763 (1963).
8. W. D. Niegisch, "The Relation of Single Crystals to Polyethylene Spherulite Formation," presented at Elec. Micro. Soc. Am. meeting, Columbus, Ohio, September (1959).
9. V. A. Kargin, N. F. Bakev, L. Li-shen, and T. S. Ochapovskaya, "Electron Microscope Study of Crystalline Structures of Polystyrene and Polypropylene," *High Mole. Wt. Comps.*, **2**, 1280 (1960) (in Russian).
10. S. Mitsuhashi, personal communication.
11. A. Keller, "Polymer Single Crystals," *Polymer*, **3**, 393 (1962).
12. D. C. Bassett, "Surface Detachment from Polyethylene Crystals," *Phil. Mag.*, **6**, 1053 (1961).
13. G. Schuur, "Mechanism of the Crystallization of High Polymers," *J. Polymer Sci.*, **11**, 385 (1953), and "Spherulicten," *Plastics*, **8**, 312 (1955).
14. G. Schuur, "Some Aspects of the Crystallization of High Polymers," Rubber Stichting Communication No. 276, Delft, 1955.
15. A. Keller and J. R. S. Waring, "The Spherulitic Structure of Crystalline Polymers. Part III. Geometrical Factors in Spherulite Growth and the Fine Structure," *J. Polymer Sci.*, **17**, 447 (1955).
16. P. H. Geil, "Morphology of an Acetal Resin," *J. Polymer Sci.*, **47**, 65 (1960).
17. F. Danusso and F. Sabbioni, "Struttura sferulitiche e poledritiche nel polistrirolo isotattico," *Rend. Inst. Lomb. Sci. e Lettere*, **A92**, 435 (1958).
18. H. J. Leugering, "Uber die Kristallization von Poly-4-methylpenten-1 aus der Schmelze," *Kolloid-Z.*, **172**, 184 (1960).
19. F. C. Frank, "Capillary Equilibria of Dislocated Crystals," *Acta Cryst.*, **4**, 497 (1951).
20. G. E. Novikova and O. N. Trapeznikova, "Concerning a New Type of Crystal Structure of Gutta-Percha," *Soviet Phys. Solid State (Eng. trans.)*, **1**, 1637 (1960).
21. P. H. Geil, unpublished data.
22. P. H. Geil, "Folded Molecules in Lamellae Crystallized from Molten Polymers," *J. Appl. Phys.*, **33**, 642 (1962).

23. Avisun, Inc., Advertisement in *Chem. and Eng. News*, **37,** 67 (1959), and other publications at about the same time.
24. A. Keller arrived at the same conclusion using an apparently similar sample. "Electronenmikroskopie und Rontgenkleinwinkelstreung," *Kolloid-Z.*, **165,** 15 (1959).
25. H. D. Keith, "Morphology of Polymer Crystals Grown from the Melt," paper presented at Am. Phys. Soc. meeting, St. Louis, March 1963.

IV. SPHERULITES

1. INTRODUCTION

The crystallization of low molecular weight substances is reasonably well understood. In general, when a metal or other low molecular weight material is cooled from the melt, crystallization starts at various points, termed nuclei, and spreads out from these points. The result, depending on the number of nuclei, is the development of either a single crystal or a polycrystalline aggregate of a few to many individual grains. Each grain is essentially a single crystal, having developed by continuous, regular growth from a single nucleus. Numerous defects, the number depending on such factors as impurities and rate of growth, are incorporated into the crystal as it grows. Twinning may also occur, resulting in growth of 2 or more grains with different but related orientations from the same nucleus. The relative orientation of grains from different nuclei is not correlated.

Related results, influenced to some extent by the long chain nature of the molecule, occur when a polymer melt is cooled. The hedrites discussed in the previous chapter are the closest approach to a macroscopic single crystal that has been grown from a polymer melt. Although the chain folds introduce planes of discontinuity in the hedrite, the lattice has a uniform orientation throughout the structure. Each hedrite (and oval) grows slowly and relatively perfectly from a single nucleus, the orientation of its lattice being determined by that of the nucleus. Numerous defects such as vacancies and interstitials, related to those occurring in low molecular weight materials, as well as defects such as terminal groups of molecules, branches, and improper folds which are peculiar to polymer crystals, are undoubtedly incorporated into the lattice as the hedrite grows. At faster rates of crystallization the number and magnitude of the defects increases, resulting in the development of less perfect lamellae. They tend to twist, turn, and branch.

Spherulites develop instead of hedrites if crystallization is rapid enough. In contrast to the well defined orientations of the nuclei with respect to the substrate in hedrite growth, the nuclei have a random orientation in spherulitic bulk samples and even in thin films for those polymers in which the substrate has little or no influence. The central portion of the structure which develops from each nucleus, when observed at high enough magnification, often has a sheaflike appearance related to that observed in the

223

Fig. IV-1. Spherulites of a low density polyethylene crystallized from a molten thin film as observed between crossed polaroids.   Besides the radiating fibrous structure and Maltese cross, a ring or band structure is just visible near the center of each spherulite.

ovals. This central portion may be the only part that develops if numerous nuclei start to grow. At sufficiently fast growth rates and with a low enough density of nuclei, the growing structures approach spherical symmetry (or if grown in thin films, circular symmetry). Formation of these nearly symmetric structures, termed spherulites, has come to be recognized as the characteristic mode of crystallization of polymers from the melt. When observed after crystallization is complete, spherulites usually are poly-hedral in shape, having grown radially until the entire volume is filled (Fig. IV-1). Depending on the concentration and temperature, crystallization from solution can result in the development of spherulites as well as the single crystals and hedrites discussed previously. Rapid crystallization from solutions of high concentration or crystallization by evaporation of the solvent (which is actually similar to the first case) often results in the growth of spherulitic structures whereas slow crystallization of the same solutions may result in the growth of hedrites.

Numerous investigations of the structure of spherulites of synthetic polymers, as observed with the optical microscope and x-ray diffraction, have been reported since the observation of spherulites in polyethylene in 1945 (1). Prior to that time, spherulites had been described for polyethylene oxide (molecular weight over 100,000) (2), several natural polymers, and many low molecular weight materials. These and subsequent reports and attempts to relate the observed structure to the fringed micelle model have been reviewed (3–5). We shall consider here those observations useful as background in understanding the structure of polymer spherulites.

A considerable amount of work was done on the structure of spherulites of low molecular weight organic and inorganic materials near the beginning of this century (6–9). The cited authors showed by optical techniques that spherulites consist of radiating arrays of fiberlike crystals. In many cases spherical growth occurs as a result of nearly simultaneous nucleation at numerous points on the surface of a foreign particle followed by dendritic growth in all directions. In other cases, however, the nucleation is essentially homogeneous in character and the spherical growth front develops considerably later during growth. Observations during growth (10,11) suggested that in the latter case the spherulites start growth as sheaflike structures, the ends fanning out as growth proceeds. Morse and Donnay (11) depict numerous variations of the sheaflike structure. With further growth the diverging fibrils from opposite ends of the sheaf meet and a spherically symmetric growth front forms. The central region, however, remains asymmetric. Shubnikov (12) has recently described, in detail, the early stages of growth of spherulites of low molecular weight materials.

It was found possible to crystallize numerous low molecular weight materials in the form of spherulites; spherulitic growth being favored if crystallization occurs in a medium of high viscosity. Morse, Warren, and Donnay (13), for instance, list the properties of spherulites of some 70 different inorganic salts. They indicate that many substances, even some which do not normally crystallize well, precipitate as crystalline spherulites when formed by reaction within a gel. These spherulites were grown by using a gel containing a low concentration of one of the reactants and diffusing in the other reactant in a more concentrated form. Spherulites of organic materials often form spontaneously during crystallization. Their growth is favored by the addition of an impurity to the melt or solution as well as a high viscosity although neither is essential in some cases.

The early authors measured the optical properties of the spherulites. As indicated, after growth had proceeded sufficiently, the spherulites they studied consisted of spherically (circularly if grown in thin films) symmetric aggregates of radiating fibrillar or needlelike crystals. The individual fibers, in general, are optically anisotropic. Thus, when observed between crossed Nicol prisms, a black Maltese cross similar to that in Figure IV-1 is seen, the arms of the cross being parallel to the directions of the prisms. X-ray patterns obtained by Jansen (14) show that the lattice of the crystals in the fibers is the same as that formed during nonspherulitic crystallization. Furthermore he found that the fiber axis corresponds to a particular crystallographic axis and that, within the resolution of his equipment, the other two axes are random, i.e., the sample has rotational symmetry about the fiber axis. Many of the properties of these fibrous crystals are now discussed in terms of whiskers (15).

Concentric bands or spiral structures are also observed in spherulites of some materials. In these cases it was shown that the radiating fibrous units are helically twisted about their axes, the period of the twist corresponding to the spacing of the rings (16). The Maltese cross is due to zero amplitude (17), the vibration directions in the fibers in the arms of the cross being parallel to one or the other of the Nicol prisms. The extinction bands are due to zero birefringence (17) (or very low birefringence (18)), the optic axis of the crystals in the extinguished portions of the bands being paralled to the direction of the light. As Keller has shown (17,19,20), the basic findings of these early authors apply to polymer spherulites also. Keller has also recently confirmed, by optical means, the helical twisting of the fibrous units in spherulites of low molecular weight substances (20).

Because of their probable relationship to polymer spherulites further work is needed with x-ray and electron microscope equipment on the structure of spherulites of low molecular weight substances. The cause of the twisting, for instance, is probably related to the periodic twisting of the lamellae that, as will be shown, occurs in some polymer spherulites. A suggestion by Yoffe (21) needs further consideration. He pointed out that since the lattice parameters on the surfaces of a crystal are different than those in the interior, the resulting stress would cause a thin, lamellar crystal to break up into a mosaic structure and/or twist periodically. A similar suggestion in terms of surface tension was made as early as 1888 by Lehman (7) to explain his observations of helically twisted fibrillar crystals, the period of the twist being smaller the thinner the crystal.

Spherulitic structures in polymeric materials were observed as early as 1905, by Tschirch and Muller (22), during investigations of gutta-percha. Further reports, including photographs, of spherulites in solution cast gutta-percha (23), natural rubber (24), and amylose (25), ($(C_6H_{10}O_5)_x$, a component of starch), as well as rubber crystallized from the melt (26) followed before spherulitic structures were recognized as the characteristic mode of crystallization of a synthetic polymer, polyethylene (1);

Earlier reports of the growth of spherulites in the synthetic polymer, poly-ethylene oxide (2), were apparently overlooked. Since that time they have been reported for a number of other synthetic polymers as well as proteins and synthetic polypeptides. Spherulites of natural polymers appear to be similar in appearance, in the optical microscope, to those of synthetic polymers. Descriptions of spherulites of natural polymers are included in the last section of this chapter. Unfortunately, electron microscopic studies of spherulite structures are not available for many of the polymers, both natural and synthetic.

In the following sections we discuss the optical properties of polymer spherulites, electron microscope observations of spherulites of two polymers, and the structure of spherulites grown from quenched, glassy polymers by annealing at temperatures below the melting point. The last section includes descriptions of most polymer spherulites which have been reported.

2. OPTICAL AND X-RAY DIFFRACTION OBSERVATIONS

The remarks in this section apply to polymer spherulites in general. Most observations which have been reported were made on spherulites grown in thin films, facilitating their observation in the optical microscope. These two dimensional spherulites are usually assumed to have the same structure as diametral sections of three dimensional spherulites. Keith and Padden (24), however, have shown that this is not always correct. In "thin film" polyethylene spherulites nucleation usually occurs on the surface of the film in contact with the substrate. Subsequent radial growth, they showed, occurs at a constant angle (3–10° in various samples) with this surface. Detailed observations, optical measurements and interpretations, thereof, have, as a result, been misleading at times.

Polymer spherulites, as spherulites of low molecular material, have a radiating fibrous appearance when observed in the optical microscope or by transmission in the electron microscope. The "fibers" often appear twisted along the radii. In the optical microscope the fibrous structure can be seen throughout the spherulite by transmitted light. By reflected light some areas appear fibrous and others may appear smooth. In some spherulites, especially those grown slowly, radial cracks may develop between the "fibers."

In spherulites of many polymers it has been possible to determine the orientation of the unit cell by optical means. Between crossed polaroids most polymer spherulites display a Maltese cross whose arms are parallel to the directions of the polarizer and analyzer (Fig. IV-1). For optically uniaxial polymers the orientation of the molecules within the spherulite can

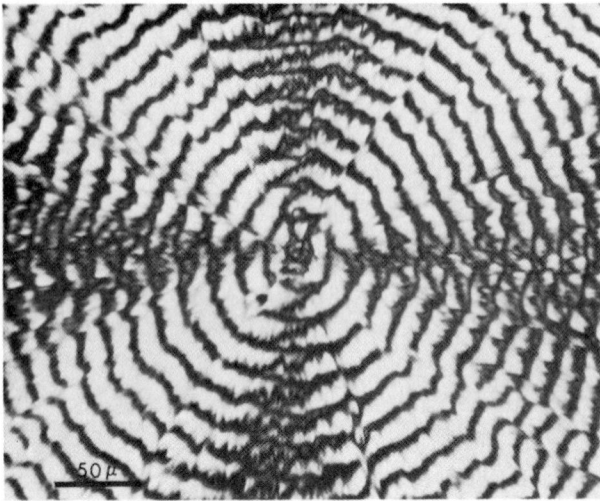

Fig. IV-2. Spherulite of linear polyethylene crystallized from a molten thin film Besides the nearly concentric extinction bands and the zigzag shape of the extinction cross, a number of radial faults are present.   Within each sector all structural units twist with the same hand.   (Crossed polaroids.)   (Keith and Padden (35).)

usually be determined by comparing its optical properties with those of a drawn fiber. Observations indicated that the molecules are, on the average, tangentially oriented. The magnitude of the birefringence is low; in polyethylene, for instance, the measurements of Bryant (28) indicate that the difference between the tangential and radial refractive indices, assuming that the molecules are randomly oriented on tangential planes, is only about $1/10$ that found in drawn fibers. Thed iscrepancy may, in part, be due to the form birefringence of radial cracks.

Optical techniques, however, are not sufficient to determine the molecular orientation in optically biaxial polymers. In these cases, as well as for a few uniaxial polymers, microbeam x-ray techniques have been utilized to determine the average orientation of the unit cells in the spherulites. Work by Herbst (29) and Keller (30) confirmed the tangential orientation of the molecules in spherulites of polyamides, polyethylene, and polyethylene terephthalate. In nylon 610, for instance, two types of spherulites are formed, in both of which the molecules are tangential. In one, however, the hydrogen bonded planes are radial and, in the other, tangential. In polyethylene, which is almost uniaxial, Keller (30) and Point (31,32) found that the **b** axis is radial while the **a** and **c** (molecular) axes are tangential.

As in the case of spherulites of low molecular weight materials, additional features can be observed optically between crossed polaroids. In spherulites of several polymers the Maltese cross assumes a zigzag shape under some growth conditions (Fig. IV-2). In some of these and others, concentric bands or pairs of bands are superimposed on the cross. These results have been shown by optical measurements (17,31–34) to be due to the regular twisting of radial structural units, forming a helical type structure. In unpolarized light the fibrous structures in banded spherulites appear to twist regularly with the same periods as the banding, whereas in spherulites which display only the simple Maltese cross the fibers appear irregularly twisted. Pairs of bands are observed if the polymer crystal is optically biaxial (detailed discussions of the optical properties to be expected in spherulites composed of helically twisted uniaxial and biaxial crystals have recently been published (20,27,35,36) and will not be discussed here). In most polymers, including polyethylene, the period of the twist varies with the crystallization temperature, increasing with increasing temperature (37). Visible light diffraction patterns from these polymers, which often contain a number of orders of discrete maxima, can be related to the spacing of the bands (38–40). The diffraction can, in some cases, be observed visually (33).

The bands are visible by unpolarized light as well as polarized light. Price (18) has shown that the bands in polyethylene spherulites that are bright in polarized are partially opaque in unpolarized light whereas the dark bands are essentially transparent. Correlation of the observed optical effects and the optical properties of the polymer crystal indicate that the molecular axes are tangential and in the plane of the film in the bands which are bright between crossed polaroids. They are normal to the plane of the film in the bands which are dark, giving rise to a zero birefringence effect. Keith and Padden (27) have shown that the extinction bands scatter light feebly with dark field illumination whereas the bright bands scatter light strongly. The cause of all of these observations will be discussed later.

The topography of polymer spherulites can best be observed by crystallizing the polymer as a film with an unrestrained surface. In banded spherulites the bands are visible in reflected light as nearly concentric ridges and valleys. In polyethylene the bands which are bright in polarized light correspond to the ridges. These bands, it has been shown, arise from a double armed spiral (5,37) (Fig. IV-3). Radial faults result in their degeneration into concentric, nearly circular rings. With reflected light the banded structure is seen most clearly when polyethylene spherulites are nucleated within or on the lower surface of the film, in contact with the substrate. When nucleated on the unrestrained surface of a thin film the spherulite develops with a conical shape. Although with polarized light the bands are equivalent to those of bottom surface nucleated spherulites, they are almost indistinguishable in reflected light (Fig. IV-3).

Hendus indicates (41) that when polyethylene is crystallized in a layer of Canada balsam the spherulites that form are conical. The apex angle of

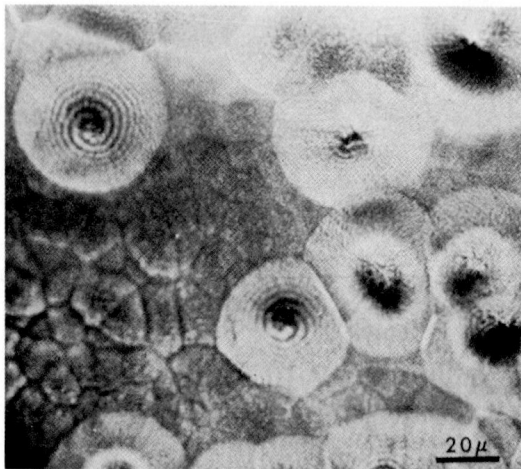

Fig. IV-3. Rapidly cooled (air-quenched) thin film of linear polyethylene as observed with incident steep oblique illumination.   The spherulites with the double-armed spirals were nucleated on the bottom surface of the film and are essentially flat except for ridges corresponding to the spiral.   The conical spherulites were nucleated on the upper unrestrained surface.   Between crossed polaroids both types appear the same, similar to the spherulite in Figure IV-2.   The regions between the large spherulites are "transcrystalline" (see later discussion, page 231).

the cones, centered at the nucleus, is about 160°. Furthermore, although not stated, these spherulites apparently consist of hollow cones; after the balsam has been removed they crack radially when pressed flat. This type of growth may be related to the hollow pyramidal single crystals and/or the findings of Keith and Padden (27) that the "fibrils" in thin film spherulites make an angle of 3–10° with the substrate.

As a result of a number of experiments it was suggested that polymer spherulites grow from a partly crystalline phase and not directly from the melt. For instance, when a polymer film on or between glass crystallizes from the melt during cooling it at first becomes relatively opaque (42–44), a mottled or grey background being observed between crossed polaroids. With further decrease in temperature this is followed by an increase in transparency and the growth of spherulites from randomly spaced nuclei. X-ray studies tended to confirm the suggestion that spherulites formed from this mottled, partly crystalline material. Diffraction patterns from samples which were quenched before the spherulites completely filled the volume were used. Herbst (45) using microbeam techniques, showed that the regions between polyamide spherulites are crystalline. Furthermore, Price (43) showed that partially spherulitic polychlorotrifluoroethylene has essentially the same degree of crystallinity as a fully spherulitic sample.

More recent work, however, indicates that spherulites grow directly from the melt, that they do not result from a secondary rearrangement of previously crystallized material. The glass transition temperature of the polyamide used by Herbst, for instance, is below room temperature and thus, even if quenched to an amorphous state, the polymer will crystallize when warmed to room temperature. Keller (17,30) showed that the non-spherulitic portions of quenched polyethylene terephthalate, whose glass transition is above room temperature, are noncrystalline. Furthermore, Burnett and McDevitt (46) showed that in polyamides (nylon 66) and polychlorotrifluoroethylene, the grey background arose from a "trans-crystalline" layer, i.e., a layer consisting essentially of closely spaced spherulites nucleated on the surface and able to grow only in one direction, inward.

Most of the above observations can be explained in terms of various modifications of the fringed micelle model; the following cannot. In at least two polymers, gutta-percha (3,47) and polypropylene (48,49), two morphologically and crystallographically different spherulites can grow simultaneously at the same temperature (see Fig. IV-35). This differs from the situation in, for instance, some polyamides, in which two types of spherulites develop at different temperatures and apparently have a different morphological structure but the same unit cell. The observations indicate that a continuous crystallization process takes place during spherulite growth, the crystal structure of the nucleus being propagated throughout the entire spherulite (3). This observation of continuous crystallization plus the discovery of single crystals and recent investigations of spherulite structure with the electron microscope have resulted in the refutation of the fringed micelle model for both melt and solution crystallization of polymers.

3. ELECTRON MICROSCOPE AND SMALL ANGLE X-RAY DIFFRACTION OBSERVATIONS

Sufficiently detailed electron microscope observations to permit a reasonably detailed structural interpretation have been reported for only two polymers, polyethylene and polyoxymethylene. Being representative of spherulites with and without bands, they will be considered in this section. Similar but less detailed observations have been made for a number of other polymers and are discussed in Section IV-4.

a. Polyoxymethylene Spherulites (50)

Polyoxymethylene spherulites, which appear to be representative of spherulites without bands, are formed if thin films are cooled rapidly or nucleated with a random orientation, or if films thicker than about 30 microns are crystallized. Controlled slow crystallization of thin films results in the growth of hedrites. Electron microscope observations were

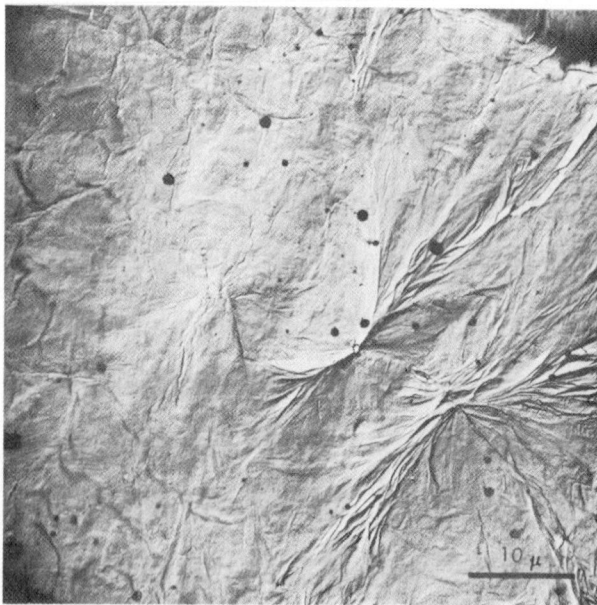

Fig. IV-4. Electron micrograph at low magnification of a Cr replica from a thin film of polyoxymethylene crystallized from the melt. The central portions of the three spherulites as well as the region extending to the lower left of the nucleus of the central spherulite are shown at higher magnification in Figures IV-5 to IV-9.

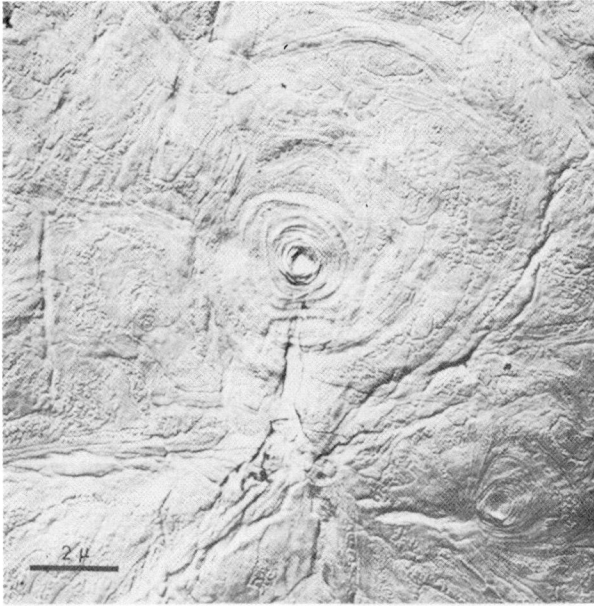

Fig. IV-5. Central portion of the spherulite on the left in Figure IV-4. Actually two spherulites of similar structure were apparently in close proximity, the pit to the lower right being associated with the nucleus of the second spherulite.

made using Cr replicas from the unrestrained surfaces of thin spherulitic films (approximately 100 microns thick) and from surfaces obtained by brittle fracture of $1/8''$ thick specimens at liquid nitrogen temperatures. Optical microscope observations were made using the original films with the unrestrained surfaces.

Several, apparently different, types of nuclei can be observed in the optical microscope and at low magnifications in the electron microscope (Fig. IV-4). The centers of the spherulites often, although not always, have a fibrous, sheaflike appearance. In a few spherulites a central pit is found. If thicker films are used the primary difference is that many of the spherulites are nucleated beneath the surface and therefore, the structure of the nucleus is not observable. The over-all structure, otherwise, appears similar to that seen on thin films. The nuclei of the three spherulities in Figure IV-4 are shown at higher magnification in Figures IV-5, IV-6, and IV-7. The structure of the pitlike nucleus in Figure IV-5 is similar to that observed in the hedrites, while the sheaf in Figure IV-6 resembles the nucleus of an oval. The majority of the nuclei in spherulitic thin films, however,

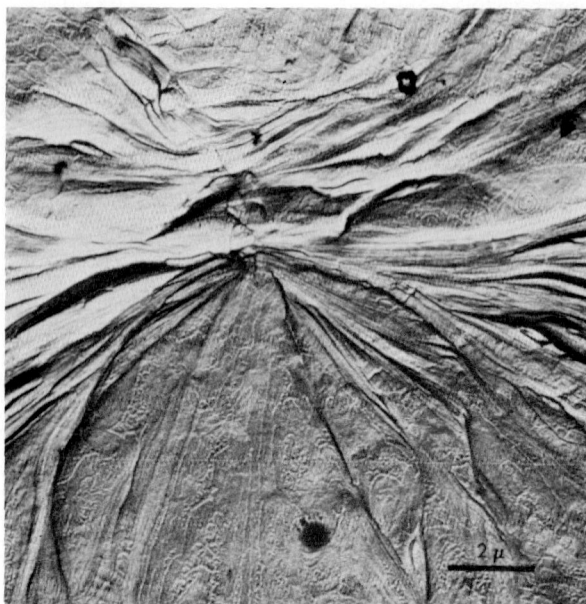

Fig. IV-6. Central portion of the spherulite on the right of Figure IV-4. The nucleus of this spherulite, which is typical of the majority of those observed in thick films, is believed to be beneath the surface.

have a structure similar to that observed in Figure IV-7, i.e., a sheaf with a pit on one side, or occasionally, as here, on both sides of the center.

The nucleus, as well as the remainder of the spherulite, is seen to have a lamellar structure, the lamellae being on the order of 100 A thick. In the nucleus the lamellae are oriented normal to the axis of the pit. When only a pit is present (Fig. IV-5) the lamellae at the center of the spherulite are parallel to the surface. When no pit is visible (Fig. IV-6) the lamellae are normal to the surface and parallel to the long dimension of the sheaf. When a sheaf and one pit are visible (see Fig. 4, Ref. 50), the lamellae intersect the surface at an angle, tilting away from the side of the sheaf having the pit. In the occasional spherulite with two pits, as in Figure IV-7, the lamellae at the center of the sheaf appear to be essentially normal to the surface.

The above observations, in conjunction with those of hedrites and ovals, led to the suggestion that the pit was associated with the axis of a screw dislocation originating in a primary lamella, the nucleus of the spherulite. The secondary lamellae growing from this dislocation are initially normal to the dislocation axis and parallel to each other and the primary lamella.

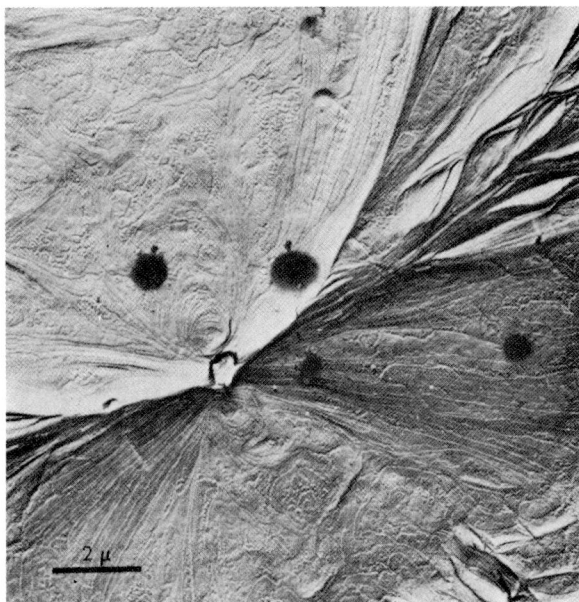

Fig. IV-7. Central portion of the center spherulite in Figure IV-4. The lamellae on both sides of the center of the sheaf are tilted toward the sheaf's center line. On the following two figures it is possible to trace lamellae that can be seen on the lower left of this figure.

Polyoxymethylene spherulites thus start to grow in a manner analogous to the growth of hedrites and ovals. The pitlike nucleus is similar to the nucleus of a hedrite, while the sheaf is similar to that of an oval. Nuclei consisting of a pit and a sheaf correspond to a hedrite whose axis is tilted with respect to the surface. Apparently, in nuclei consisting of two pits and a sheaf, as in Figure IV-7, either the axis of the dislocation is near the surface and possibly curved or two dislocations are involved. In thin films of some other polymers (e.g., polybutene, see Figure IV-48) the presence of two pits or mounds appears to be typical, suggesting that a somewhat different mechanism may occur or that the substrate is affecting the nucleus structure.

In contrast to hedrites, the secondary lamellae in polyoxymethylene spherulites begin to twist, turn, and branch as they grow away from the center of the spherulite (Figs. IV-8 and IV-9). Those areas in which the lamellae intersect the surface at an appreciable angle appear fibrous in reflected light in the optical microscope and at low power in the electron

Fig. IV-8. Region to the lower left of the central spherulite in Figure IV-4.   As the lamellae leave the central sheaf they are seen to twist irregularly.   Spiral growths can be seen on this and the other micrographs of polyoxymethylene spherulites.   The lamellae are about 100 A thick on this surface.

microscope. The lamellae are essentially parallel to the surface in those areas which appear smooth. Radial cracks, especially in slowly crystallized samples, can be seen between lamellae in regions in which the lamellae are nearly normal to the surface. Presumably the fibrillar structure seen throughout the spherulite by transmitted light is due to lamellae within the material oriented parallel to the light beam.

In order to form a spherical object from a radiating array of lamellae, additional lamellae must form as growth proceeds. Many of these tertiary lamellae, needed to fill space in the growing spherulite, apparently result from the development of screw dislocations in the secondary lamellae. Splitting or branching of a lamella, followed by noncoplanar growth on exposed faces during further crystallization, must result in formation of a screw dislocation. When this occurs near the surface the associated spiral growths are visible (Figs. IV-5 to IV-9). In the interior it is likely that only one "additional" lamella is formed for each screw; i.e., the spiral does not have a chance to develop.

Fig. IV-9. In this portion of the spherulite most of the lamellae, which were nearly normal to the surface in the central sheaf, are nearly parallel to the surface. Some of the spiral growths on this (for instance, the two at the upper left) and the other figures are hexagonal in shape suggesting that the hexagonal unit cell and thus the molecules are normal to the lamellae.

Small lamellar aggregates of material, often hexagonal in shape, can be seen on the surface replicas (Figs. IV-6 to IV-9). It is believed that these aggregates form and crystallize near the end of the crystallization process. As a lamella near the surface is growing over the top of one already crystallized it reaches a point at which there are sufficient uncrystallized molecules to permit further growth at a constant thickness. Growth ceases and the remaining material apparently aggregates. The hexagonal faces on the aggregates are not parallel indicating that the aggregates crystallized independently of each other and of the lamella underneath. The smallest aggregates appear to be somewhat thinner than neighboring lamellae, suggesting that they have crystallized at a lower temperature than the bulk of the material. A few lamellae in the interior of the sample may also be nucleated and grow without being associated with a screw dislocation; i.e., a new crystal, as in the case of the aggregates, may be nucleated by a process similar to primary nucleation. Only those whose orientation is similar to that of neighboring secondary lamella will be able to develop to a

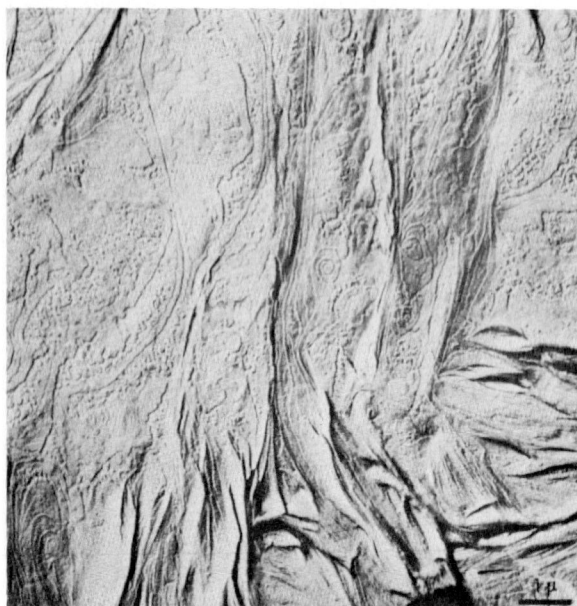

Fig. IV-10. The boundary region (Cr replica) between two spherulites in a thin film of polyoxymethylene. The boundary, parallel to the bottom of the figure, is about 4 microns wide. Within this region lamellae from both spherulites are interleaved. (Geil (50).)

significant extent. The different types of spherulites of polypropylene and gutta-percha suggest that the number of such lamellae, if any, is insignificant in spherulite growth.

The growth front of a spherulite consists of individual secondary lamellae growing outwards. When two spherulites meet during growth, the growing secondary lamellae cross the boundary between the growth fronts (Fig. IV-10). Growth of each lamella, it was suggested, continues as long as there is molten material in front of its growth face. Tertiary lamellae following would fill in the rest of the volume. The boundary between two spherulites thus consists of interleaved lamellae. The amount of interleaving depends on the relative orientations of the lamellae from the two spherulites. The physical bonding between lamellae from different spherulites, at this boundary, should be almost as strong as that between lamellae within a spherulite.

The number of molecules incorporated in two or more lamellae is believed to depend on the crystallization rate. In general a polymer crystal-

Fig. IV-11. Surface replica of a $^1/_8$-inch thick compression molded sample of poly-oxymethylene fractured at liquid nitrogen temperatures. The lamellae to the lower left are oriented at an angle to the fracture surface, whereas those near the middle and right are nearly parallel to the fracture surface. (Geil (50).)

lized slowly is more brittle than if rapidly crystallized. Besides a difference in lamella thickness due to the crystallization temperature, as in the case of solution grown crystals, there is apparently a difference in the bonding between lamellae. If crystallization occurs slowly, the secondary lamellae can incorporate entire molecules before adjoining tertiary lamellae grow over their surfaces. As a result neighboring lamellae are only weakly bound. The physical strength of the polymer may then depend to a great extent upon the mechanical entanglement of twisted lamellae. Spherulite bound-aries would be especially weak. If crystallization occurs rapidly, neigh-boring lamellae may incorporate various portions of the same molecule. These "tie" molecules will result in stronger forces between the lamellae. Rapid crystallization should also increase the number of defects within the lamellae, resulting, perhaps, as in the case of metals, in a tougher sample, as well as increasing the nucleation rate of tertiary lamellae.

Electron micrographs of fractured surfaces demonstrate that the entire sample is composed of lamellae. Slowly cooled, thick samples of polyoxymethylene can be fractured brittlely at liquid nitrogen temperatures. Lamellae nearly parallel to the fracture surface as well as those making an angle with the surface are observed on replicas of these surfaces (Fig. IV-11). Although difficult to determine reliably, the shadowing angle on a fractured surface being unknown, lamellae in the interior appear to be thicker than on the surface of the sample. It was not possible to identify spherulites, their nuclei, or their boundaries in the sample used for these studies.

With the proposed mechanism of spherulite growth one would suspect that small pockets of uncrystallized material would persist behind the growth front. Tertiary lamellae need not be nucleated as soon as two secondary lamellae have diverged by a distance equal to their thickness and even, if nucleation did then occur, some noncrystallized molecules or portions thereof may remain in the thinner wedge between the secondary lamellae. Some of this "trapped" material will crystallize on the face of the spiral growth giving rise to tertiary lamellae, resulting in a "backward" growth toward the primary nucleus. Still other portions may crystallize during the isothermal recrystallization that takes place while the spherulites are still growing. Although this recrystallization is usually associated with annealing subsequent to crystallization (see Chapter V) it will also take place during the initial crystallization. In this way even if small portions of material, for instance in the wedges between diverging lamellae, should crystallize as small crystallites, they likely would melt and their molecules be accommodated in the neighboring, larger lamellae as these lamellae recrystallize.

A somewhat related spherulite growth mechanism has been postulated by Keith and Padden (51,52) on the basis of experiments primarily with polypropylene. This mechanism, which emphasizes the fibrillar or ribbon-like structure observed in spherulites of a few polymers, is discussed in more detail in Section IV-5a. In brief, they suggest that the growth front of a crystal growing out into an impure melt (impurities can be atactic molecules or segments, entanglements or other noncrystallizible entities) will break up into cells. The impurities will be rejected by the growing cells and aggregate along their sides, leading to a fibrillar habit. The spherulite develops through the separation, twisting, and small-angle branching of these fibers.

Long period spacings on small angle x-ray diffraction patterns from polyoxymethylene samples generally are similar to or larger than the thickness of surface lamellae as measured by electron microscopy. In a slowly cooled specimen, similar to that shown in Figures IV-6 to IV-11, the

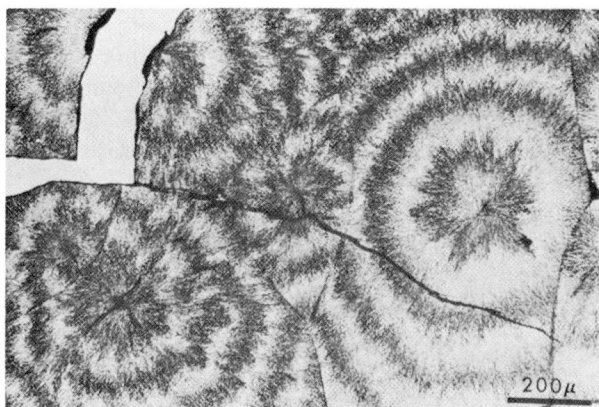

Fig. IV-12. Banded spherulites of polyoxymethylene as viewed in linearly polarized light. Between crossed polaroids a Maltese cross is superimposed on this banding. This sample was cooled slowly and uniformly. The cracks formed as the sample cooled. Note that the bands in the spherulites are somewhat hexagonal in shape.

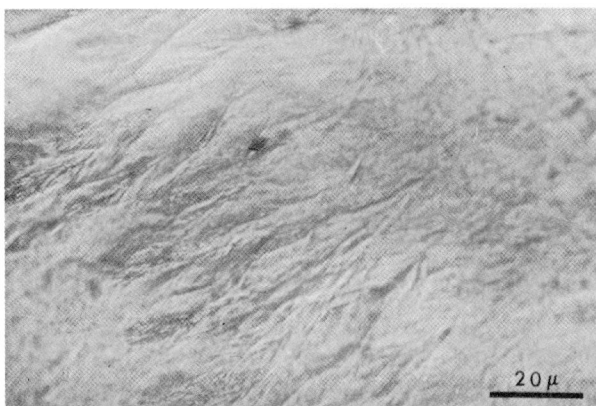

Fig. IV-13. Portion of a light and dark band of the spherulite to the upper right in Figure IV-12. The fibrous-appearing material corresponds to the dark band, the molecules lying in the plane of the film and in, it is believed, lamellae that are normal to the surface (incident steep oblique illumination).

long period spacing was 175 A, whereas the surface lamellae were approximately 120 A thick. In rapidly cooled specimens one might expect a greater difference between long period spacing and surface lamella thickness because of the low heat conductivity of polymers and therefore greater difference in cooling rates between surface and interior. However, in an air-quenched sample the long period spacing was 114 A and the measured thickness of the surface lamellae about 100 A, a difference within the limits of error of the electron microscope measurements. It is not known why the long period is larger than the lamella thickness in the slow cooled specimens. The effects of annealing on lamellar thickness, perfection and small angle long period are discussed in Chapter V.

Occasionally, under conditions approximating those used for the growth of hedrites excepts for the use of slightly thicker films, polyoxymethylene crystallizes in the form of banded spherulites (Fig. IV-12). Although electron micrographs of these banded spherulites are not available, optical

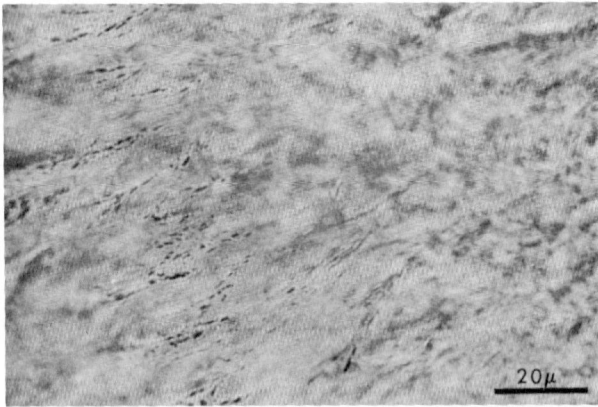

Fig. IV-14. Same region as in Figure IV-13 except viewed in linearly polarized light. The small dark objects are cracks between the lamellae. Similar cracks can be seen in the electron micrographs in Figure IV-4, and in Figures IV-6 to IV-8.

micrographs (Figs. IV-13 and IV-14) suggest that the lamellae are re-gularly twisted with half the periodicity of the banding, i.e., they twist 180° between neighboring bands. In those regions in which the lamellae are normal to the surface, the fibrous appearing bands (Fig. IV-13), cracks between the lamellae (Fig. IV-14) apparently diffract the light out of the objective, creating the dark bands (18). In polarized light the smooth appearing bands extinguish, as is expected, the molecules being normal to to the surface when the lamellae are parallel to the surface. Somewhat similar appearing spherulites of polyoxymethylene have been described in the literature as resulting from repeatedly partially crystallizing and melt-ing a sample (53).

b. Polyethylene Spherulites

The first indication that crystalline polymers had a lamellar structure, prior even to the observation of lamellar single crystals, was presented in an unpublished paper by Peck and Kaye (54). They showed that branched polyethylene, crystallized as a thin film by evaporating a solution and then annealed near the melting point, crystallizes in the form of lamellae (Fig. IV-15). The relationship of these lamellae to the spherulitic structure was not determined. The lamellae were thought to be composed of Keller's

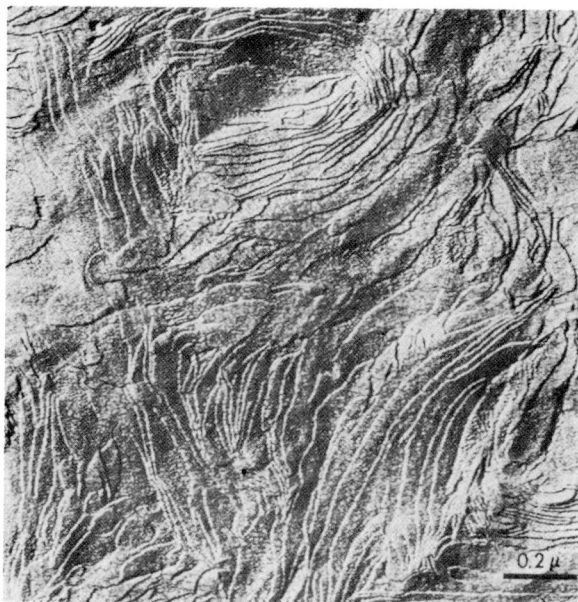

Fig. IV 15. Lamellae on the surface of a low density polyethylene crystallized by casting a film from a xylene solution and then heating for 30 minutes near the melting point (Peck and Kaye (54)).

then postulated helically twisted ribbon structure (17,19,30). The change in structure during the initial stages of drawing, which we shall discuss in Chapter VII, was also described by Peck and Kaye. Subsequently, but still prior to the observation of single crystals from solution, Claver, Buchdahl, and Miller (55), showed that the fibrillar appearance of linear polyethylene spherulites is actually due to the edges of lamellae as they intersect the surface. They attributed the band structure as seen by

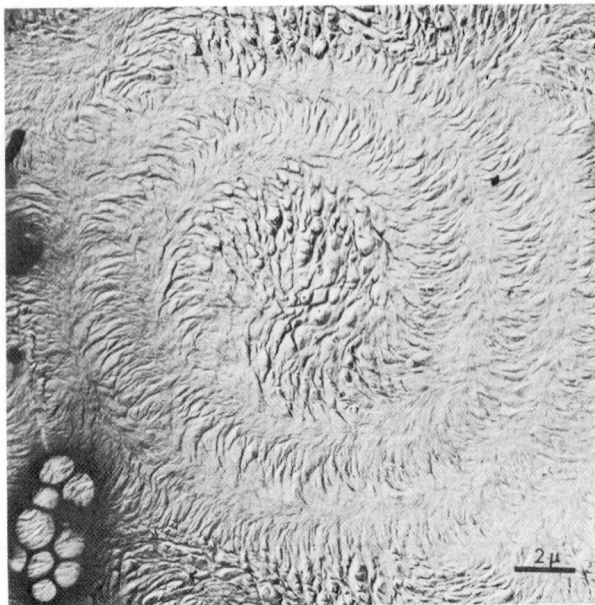

Fig. IV-16. Low magnification electron micrograph (Cr replica) of a spherulite in an air-quenched thin film of linear polyethylene. One radial fault, at 7 o'clock, is present. The double-armed spiral leaves the central complex area at about 6 and 12 o'clock. The bubbly appearing area at the lower left is an artifact in the microscope grid substrate.

transmission in the optical and electron microscope to periodic fluctuations in density or mass of material along the radii.

As indicated by the above and subsequent work, both branched and linear polyethylene typically crystallizes from the melt in the form of banded spherulites composed of lamellae. The concentric band structure, visible by transmitted light and electrons, can also be seen on surface replicas (Figs. IV-16 and IV-17). In linear polyethylene spherulites the bands, which actually result from a double armed spiral when no faults are present (5,37)(Fig. IV-3), remain distinct to the boundary of the spheru-lite. In branched polyethylene spherulites, however, the bands are narrower and usually become diffuse and disappear as growth progresses outward from the nucleus (37) (Fig. IV-1). The periodic effect remains along any radius, but along adjacent radii it becomes out of phase (Fig. IV-17).

Using replicas of thin films, Fischer (56) showed clearly that the banded structure results from the fact that the lamellae are twisted by 180° about a spherulite radius within each band (Fig. IV-18). Subsequently he pub-

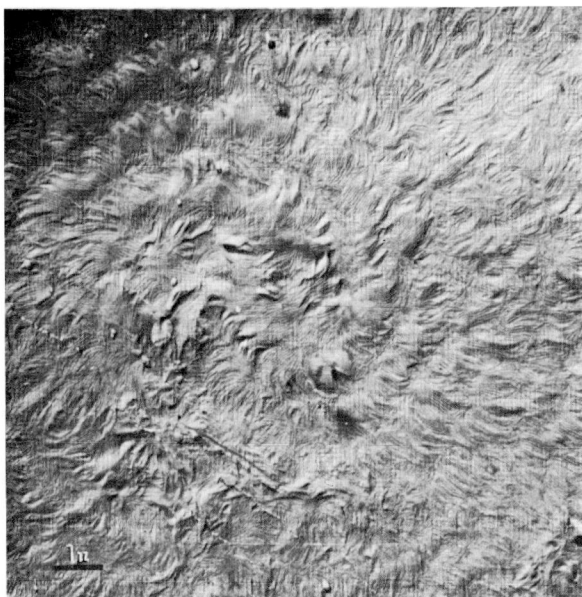

Fig. IV-17. Central portion of a spherulite (Cr replica) of low density polyethylene, nucleated beneath the surface of a slowly cooled thin film. The bands, fairly distinct over most of this micrograph, are seen to result from a double spiral similar to that in Figure IV-16. Near the corners of this micrograph and farther out along all of the radii (spherulite diameter was about 50 microns) the banded structure becomes diffuse. (Geil (37).)

lished a micrograph of a portion of the bands in a spherulite of polyethylene adipate in which twisting of the lamellae within each band also is suggested (40). In both polymers the lamellae are on the order of 100 A thick. Spherulites with clearly defined bands, as in Figs. IV-16 and IV-18, are nucleated on the lower surface of the film. Within these spherulites it is impossible to follow a lamella from one band to the next. Whether this is due to the fact that we are observing only the ends of the lamellae as they approach the surface, as suggested by the work of Keith and Padden (27), is not known. The width of the lamellae normal to the observed edge is not observable. A peculiar and at present unexplained feature is that the lamella edges, when the lamellae are parallel to the surface, are often nearly tangential to the spherulite, curving in the direction of the spiral (Fig. IV-18).

The lamellae do not have a twisted appearance in the relatively few polyethylene spherulites which are nucleated on or near the top, unre-

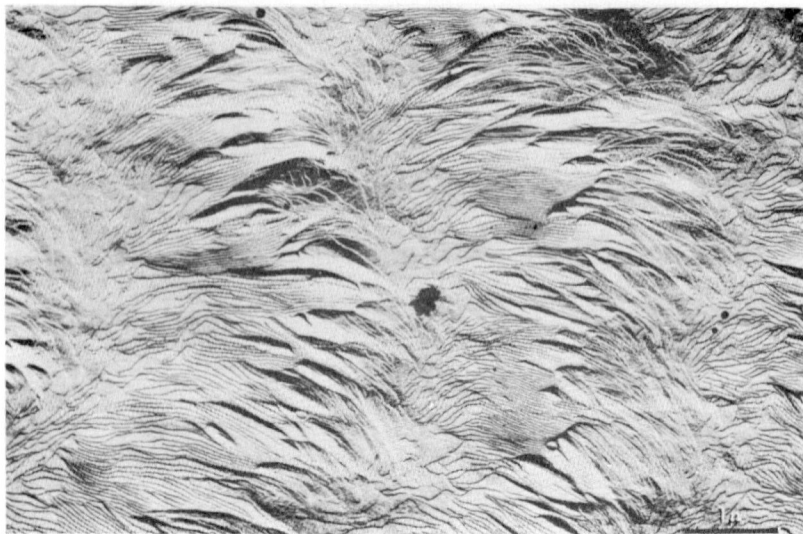

Fig. IV-18. Surface replica of a portion of a linear polyethylene spherulite. By comparison with other micrographs it is believed the spiral was clockwise in this spherulite. The nonextinguishing bands of the spherulite correspond to those regions in which the lamellae are normal to the surface. This spherulite was nucleated beneath the surface of the film. (Fischer (57).)

strained surface of the film. Although the banded structure in linear polyethylene spherulites so nucleated is as well defined between crossed polaroids as in the case of spherulites nucleated on the bottom surface, the bands are less distinct in the optical (Fig. IV-3) and electron microscopes. The surface has the appearance of overlapping shingles (Fig. IV-19). The origin of the secondary and tertiary lamellae in polyethylene spherulites is not known; no spiral growths have been observed on the surface.

In branched polyethylene spherulites nucleated on the upper surface, a lamella can occasionally be followed for several bands, the lamella appearing to twist back and forth within each ring (Fig. IV-20). New lamellae appear on the surface each time the subject lamella is nearly normal to the surface. Further observations, especially of spherulites nucleated on the unrestrained surface, are needed to clarify the type of twisting undergone by polyethylene lamellae.

The nuclei of polyethylene spherulites nucleated near the unrestrained surface usually are sheaflike (Fig. IV-20). The double armed spiral is clearly developed only in spherulites nucleated below the surface. The

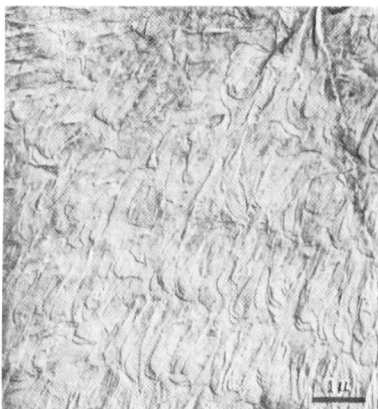

Fig. IV-19. Portion of a spherulite (Cr replica) of linear polyethylene nucleated on the upper, unrestrained surface of an air-quenched, thin film. The lamellae do not appear to twist within the 1.5 micron wide bands.

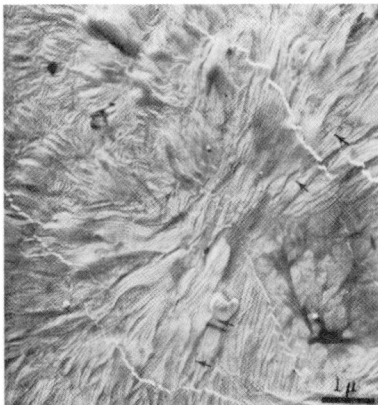

Fig. IV-20. Central portion (Cr replica) of a spherulite of low density polyethylene nucleated on the upper surface of a slow cooled film. New lamellae appear at the surface in the portions of the bands indicated by the arrows. The white lines are cracks in the replica. (Geil (37).)

spiral may twist either to the right or to the left. When the spherulites are nucleated below the unrestrained surface the central portion has a confused appearance (Fig. IV-16). Small clusters of lamellae are randomly oriented. The relationship between the sheaf and the spiral is not known.

Keith and Padden (27) observed that the extinction bands in poly-

ethylene spherulites show a tendency to join continuously across spherulite boundaries. In terms of secondary and tertiary lamellae, as earlier
defined, they suggest that secondary lamellae growing across the interface
between two expanding spherulites can interleave in those areas where
they are parallel. When two spherulites first meet the secondary lamellae
have parallel growth directions (radii of the spherulites), but may not
themselves be parallel; i.e., the growth faces of the lamellae may intersect
at an angle. If they are parallel when they meet, they can interleave, the
twisting of the lamellae on the two sides then becoming locked together.
In thin films, during further growth, the lamellae could interleave at the
boundary within those portions of each band in which they are parallel
when they first intersect. Keith and Padden's observations (27) indicate
that the twist tends to change sufficiently, near the boundary, to permit
interleaving of the lamellae. They indicate that in spherulites of polyethylene adipate and polyethylene azelate the bands are not continuous
across the boundaries, suggesting that the "fibrils" are rounded. In polyethylene spherulites the lamallae growing outwards behind the secondary
lamellae would have the same rate of twist, their thickness being the same
as the secondary lamellae (see Chapter VI). The sense of the twist must
also be the same as that of neighboring secondary lamellae.

The double armed spiral structure observed in well-formed polyethylene
spherulites suggests that the sense of twist of the secondary lamellae
develops as they start to spread out from the nucleus. This sense is then
maintained throughout the growth of the spherulite. In less perfect spherulites radial faults are often found (see Figs. IV-2, IV-3, and IV-16). Although Keith and Padden suggest (27) that the sense of the twist changes
across the fault, this would appear not to occur when only a single fault is
present (Fig. IV-16) and has not been reported as being observed in electron
micrographs. Those faults across which the twist is reversed may arise
from a single diverging secondary lamella starting to twist in the opposite
direction to that of its neighbors or to several closely located primary
nuclei. A single radial fault near the nucleus results in the double armed
spiral degenerating into a single armed spiral. As shown in Figure IV-3
(upper left spherulite) the fault may also develop after the spiral has wound
through one or more turns. Two or more faults result in the development
of concentric, nearly circular rings.

The appearance of the bands between crossed polaroids and in unpolarized light and dark field illumination is consistent with a twisted
lamellar structure. The molecules, folded within the lamellae, are normal
to the surface in the bands which are dark between crossed polaroids and

parallel to the surface in the bright bands. The lamellae are normal to the surface in the bands which scatter light strongly and are opaque in unpolarized light. Cracks between the lamellae may contribute to the scattering (see Figs. IV-13 and IV-14). As Keller suggested (20), total reflection may occur from these cracks.

The twisting of the molecules within each band has been confirmed by wide angle x-ray diffraction from large, thin polyethylene spherulites.

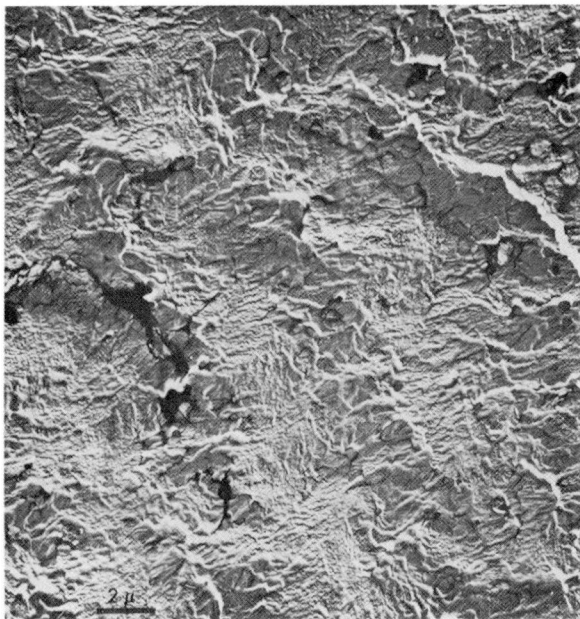

Fig. IV-21. Surface (Cr replica) of a 0.060-inch thick compression-molded, rapidly cooled film of linear polyethylene fractured at liquid nitrogen temperatures. The lamellae are believed to be nearly normal to the surface in the bands which appear granular.

Fujiwara (58), using a beam whose diameter was $^1/_4$ of the band width, showed that the c and a axes rotate about the b axis by 180° within each band. The c-axis, as expected, is normal to the plane of the film in the portion of the band which extinguishes between crossed polaroids, i.e., the portion in which the lamellae are parallel to the surface. Small angle diffraction patterns from the same sample showed sharp orientation of the long period spacing normal to the spherulite radius, an observation con-

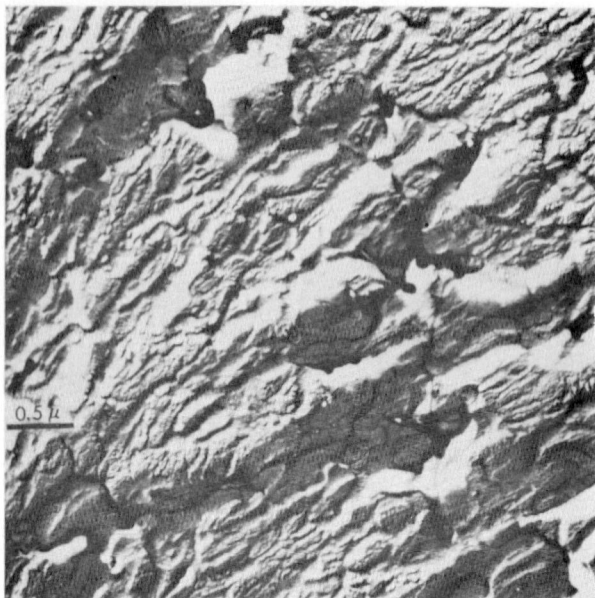

Fig. IV-22. Portion of a surface similar to that in Figure IV-21 but at higher magnification. Considerable disruption of the lamellae occurred during the fracture. In fact, the fibrous areas in which drawing occurred appear, at first glance, to be more lamellar than the areas shown here. Lamellar and fibrillar structures can only be differentiated when the lamellae can be seen nearly parallel to the surface.

sistent with lamellae being parallel to the radius. Small angle patterns were not obtained from portions of a band.

Linear polyethylene samples with molecular weight distributions typical of commercial polymers and in which the banded structure is well developed, do not readily break by brittle fracture. The polyethylene spherulites described previously were grown at intermediate cooling rates, obtained, for instance, by air quenching a molten film on a slide or placing the slide on a table top. During fracture in liquid nitrogen of thicker samples, crystallized at equivalent rates, brittle fracture occurs in some areas (Fig. IV-21) whereas localized drawing occurs in other parts of the fracture. The drawing occurs despite the fact that the initial temperature is below the glass transition temperature. In some of these drawn areas individual fibers can be seen standing normal to the surface. As suggested previously, during rapid crystallization a number of molecules may be incorporated in in more than one lamella. During fracture, sufficient energy must be

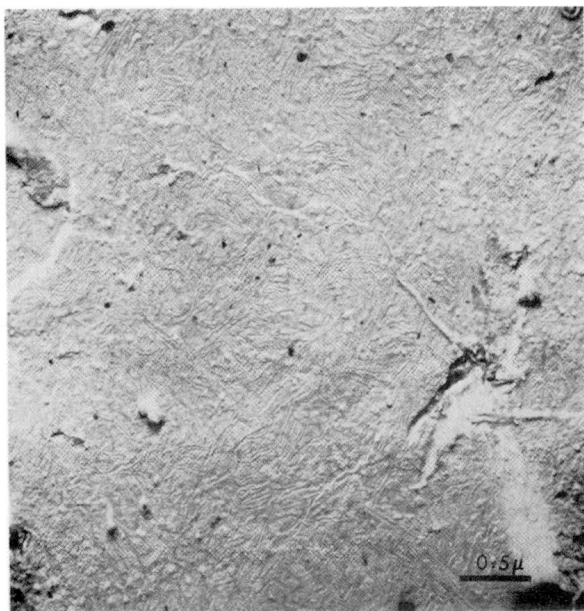

Fig. IV-23. Lamellar, nonspherulitic surface of a dry ice-acetone quenched film of linear
polyethylene (Cr replica).

supplied to break the covalent bonds in the backbone of the molecule or
to tear the "tie" molecule out of lamellae.

The band structure of the spherulites is readily visible on the surface
exposed by brittle fracture (Fig. IV-21). The band spacing is essentially the
same as that on the unrestrained surface of the sample. In those portions of
the bands in which, from the general appearance, one would expect the
lamellae to be parallel to the fractured surface, close examination of the
micrographs (Fig. IV-22) suggests that they are present. However,
considerable disruption of their structure has occurred. They appear to be
somewhat thicker than on the unrestrained surface of the sample.

If molten thin films (250 microns thick or less) are quenched in dry
ice-acetone or equivalent mixtures, spherulites do not develop (Fig. IV-23).
Small aggregates of lamellae do form however. Thicker films are spherulitic in the center even if quenched because the low heat conductivity
gives rise to slower cooling. The lamellae on the surface, 60–80 A thick,
are measurably thinner than in samples of the thin films with intermediate
rates of cooling. Their lateral extent, i.e., the distance the edge can be

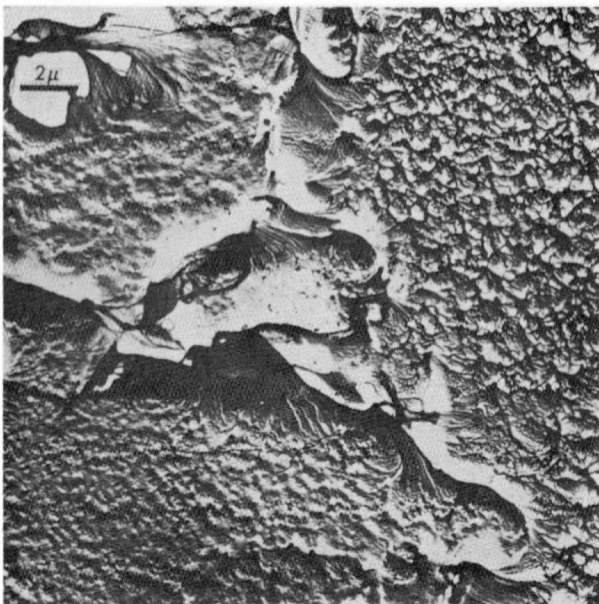

Fig. IV-24. Fractured (at liquid nitrogen temperatures) surface of a 0.060-inch thick dry ice-acetone quenched sample of linear polyethylene. Three different stages of drawing can be seen: large fibrous sheets near the center, smaller fibers with somewhat rounded ends at the right, and small rounded pits and fibers on the left. The rounded ends are believed due to melting during fracture, having developed before the sample was replicated. (Cr replica.)

observed, is also less. Between crossed polaroids the films are grey, being essentially transparent by transmitted light.

The exposed surfaces of quenched, 250 micron thick samples, fractured at liquid nitrogen temperatures, do not appear lamellar (Fig. IV-24). In some areas small pits and mounds, on the order of 400 A in diameter, are seen on the surface. In other areas an irregular surface is seen. We again attribute this appearance to localized drawing, in this case accompanied by melting and retraction of the fibers in some areas to form the mounds. The surface structure seen is thus not typical of the sample's interior.

When samples of linear polyethylene are slowly cooled, so that crystallization takes place at a relatively elevated temperature (e.g. cooling rates slower than about 1°C/min., in the vicinity of 125°C or higher), spherulites are formed which do not have bands. Oftentimes crystallization takes place slowly and perfectly enough and with so many nuclei present that

Fig. IV-25. Surface replica of a portion of a spherulite in a slowly cooled film of linear polyethylene. The cause of the beads on the edges of the lamellae is unknown. The regions of varying density are due to the fact that portions of the lamellae remained attached to the replica. The center of this sheaflike spherulite is to the left of this micrograph, the spherulite being about 80 microns long by 40 microns wide. A diffraction pattern from the attached lamellae in a similar area is also shown.

that only a sheaflike structure develops. Spherical (or circular) symmetry would be approached if the structures were able to grow considerably larger before meeting each other. The lamellae on the surface of these samples are on the order of 150 A thick (Fig. IV-25). If sufficiently slowly crystallized, polyethylene can be fractured readily at liquid nitrogen temperatures.

Well-formed lamellae are visible on portions of the fracture surface (59,60) (Fig. IV-26). Anderson reports (60) that with equivalent crystallization conditions, polyethylene with a narrow molecular weight distribution, obtained for instance by fractionating normal commercial polymer, will fracture more readily than the original polymer. Very low molecular weight samples are extremely brittle, but even the highest molecular weight fractions produced fractures showing lamellae on some portions of the surface. Ribbonlike structures, discussed below, are seen on other

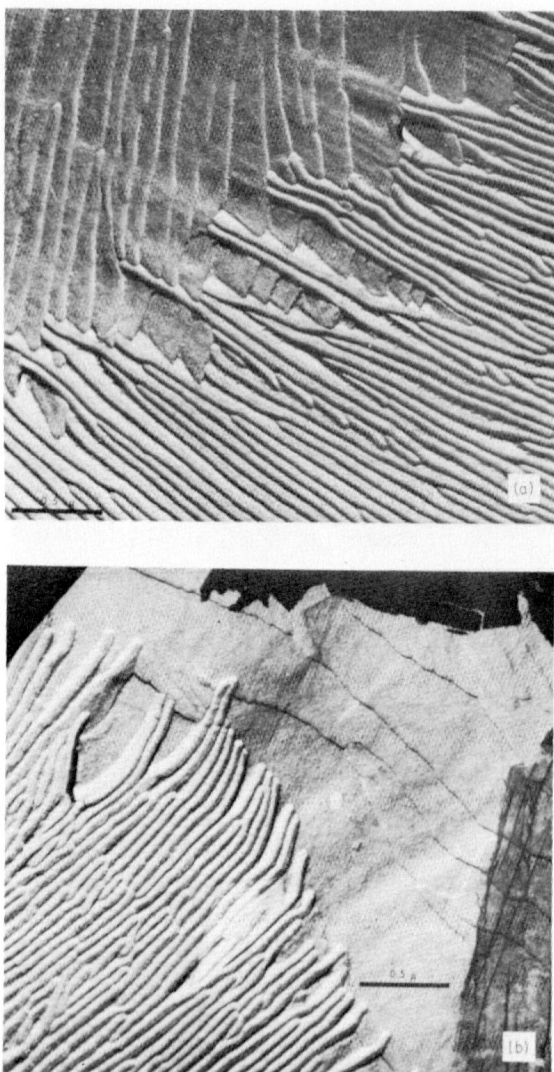

Fig. IV-26. Fractured surface of a polyethylene sample crystallized at 130°C for four days. The lamellae and the ribbons are about 250 A thick while the ribbons are 600 to 800 A wide. In (a) the ribbons can be seen under a single broad lamella, whereas in (b) they extend over the broad lamellae, thus indicating that they are not edges of lamellae oriented at an angle to the fracture surface. (Anderson (60).)

portions of the surface in Figure IV-26. There are apparently few or no molecules extending between the lamellae in these slowly crystallized samples. In fact, in Figure IV-25, portions of the lamellae adhered to the replica, cohesion between the evaporated chromium and the polyethylene being greater than that between the lamellae. Diffraction patterns from these lamellae (Fig. IV-25) show conclusively that the molecules are essentially normal to the lamellae and therefore folded. In addition, Anderson reports (115) that he is able to also find "extended chain" lamellae on fracture surfaces of both whole Marlex 50 polymer and molecular fractions thereof with molecular weights below about 12,000 (chain length = 1100 A). Their appearance is similar to that of the bands on fracture surfaces of Teflon (Fig. IV-68). Anderson assumes that the molecules are fully extended in this form of lamellae, the lamella thus being essentially a paraffin crystal.

Mandlekern *et al.* (61) have reported values for the long period of melt-crystallized polyethylene as a function of the crystallization temperature (Fig. IV-27). Extremely large long periods have been reported (850 A for crystallization at $131°$ (61), 1100 A for crystallization at $131.3°$ (62)). As Mandlekern pointed out (61), many of the molecules in the samples they used are near this length, suggesting that numerous molecules are incorporated into the lamellae with few or no folds if the lamella thickness is the same as the long period. That this is possible, lamellae forming in which all molecules fold only once and in which, therefore, the ends must meet in juxtaposition, has previously been shown by the work on oligomers of nylon 6 and polyurethane (Chapter II). As indicated in Figure IV-27, it is impossible to crystallize thick samples isothermally at temperatures below about 125°C. Crystallization takes place so rapidly that the temperature of the sample does not drop to that of the bath before crystallization is complete. This may be the reason the lamellae in Figure IV-22 appear thicker than surface lamellae on the same thick sample, the polymer in the interior crystallizing at a higher temperature than on the exterior.

TABLE IV-1 (63)
X-Ray Long Periods (A) of Various Polyethylenes

Degree of branching	After slow cooling		After quenching	
$CH_3/1000$ C	First period	Second period	First period	Second period
1.5	344	165	222	86
5	310	140	204	78
15	280	85	184	60
25	218	—	154	—

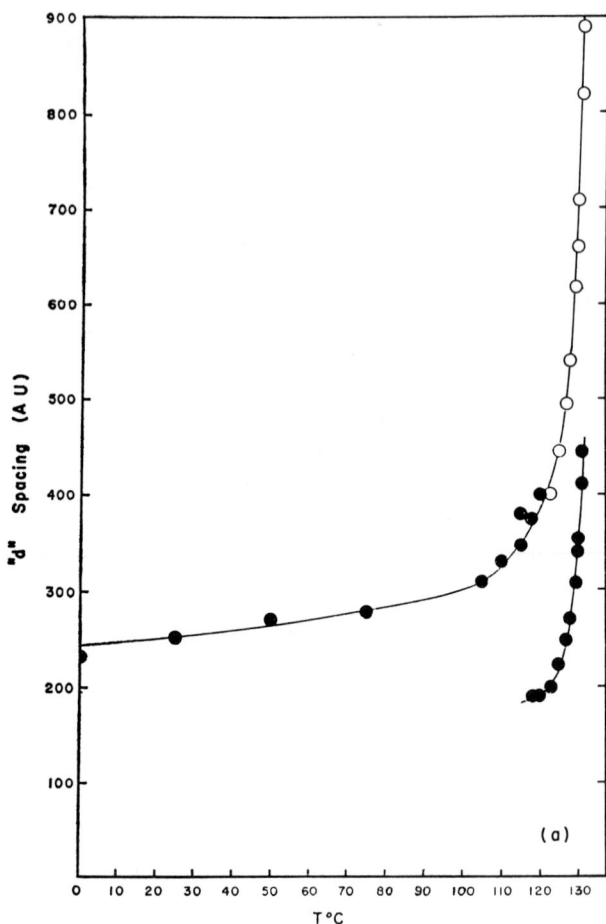

Fig. IV-27. (a) Plot of low angle spacings against crystallization temperature for samples of unfractionated Marlex 50 linear polyethylene. The open circles are calculated values based on the assumption that the measured values are second order spacings. (b) Plot of "first order" long period spacing against $T_m/\Delta T$. The upper scale corresponds to the crystallization temperature. (Mandlekern, Posner, Diorio, and Roberts (61).)

Hendus (63) has shown that the small angle x-ray diffraction long period depends on the rate of crystallization and the degree of branching (which determines the crystallization temperature for a given rate of cooling) (Table IV-1). Somewhat similar results have been report by Sella (64). Hendus points out (63) that the smaller of the two periods for each sample

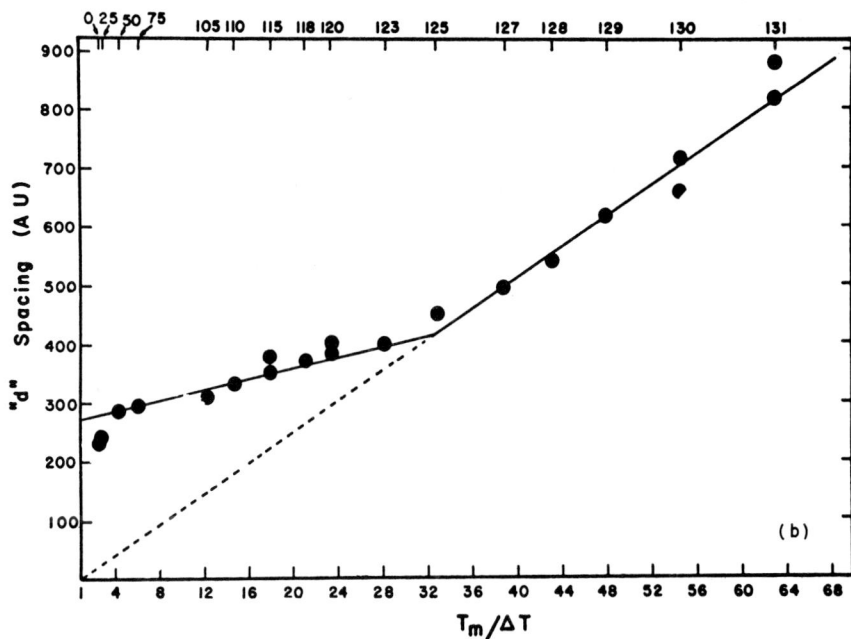

Fig. IV-27 (b).

(Table IV-1), which is of low intensity, in general is not a second order of the larger long period. The smaller of the two appears to correspond to the thickness of the lamellae observed on an unrestrained surface of a sample given a similar thermal treatment. He indicates that the larger is in relatively good agreement with the corrected width of the $\{110\}$ reflection. For a perfect crystal this width is determined by the size of the crystals normal to the direction of the chains, i.e., the lateral dimensions within the lamellae. The long period spacing were obtained on photographic film and measured visually and with a densitometer (65).

The long period spacings of the films cooled at rates equivalent to those used for the samples in Figure IV-21 to IV-25 are listed in Table IV-2. They also indicate the effects of crystallization rate on the long period. The values listed, measured with a small angle x-ray diffractometer, have not been corrected for the effect of the slits. In work with biological macromolecules it has been shown that correcting for the smearing effect of the slits can shift the peaks toward larger angles (smaller spacings) by amounts corresponding to up to 10% of the long period, the effect being greatest for the lowest order reflections (66). In the case the quenched sample at least, however, the smaller of the two spacings is not a second order of the larger, more intense reflection.

The source of the two apparently different reflections in polyethylene is not yet completely resolved. A somewhat similar effect is seen in poly-chlorotifluoroethylene crystallized at a series of temperatures. Only one long period is observed for each sample. It is several hundred Angstroms larger than the thickness of the lamellae observed on exterior and fractured surfaces from the same sample (see Section IV-5-f). Hendus has shown (65), by using polyethylene samples of various thicknesses, that it cannot be due to different lamella thicknesses on the surface and in the interior, as might result from either the low heat conductivity and therefore slower cooling of the interior or differences in free energy between the surface and interior. The relative intensity of the reflections did not change with sample thickness. Fujiwara has shown (58) that a sharp small angle diffraction pattern is obtained normal to the radius of a polyethylene

TABLE IV-2

X-Ray Long Periods as Function at Crystallization Rate

Crystallization rate	Thickness (inches)	1st Period (A)	2nd Period (A)
Quenched	~0.001	208	65
	0.010	216	95
	0.060	210	85
Rapidly cooled	~0.001	252	—
Slowly cooled	~0.001	294	130
	0.010	325	147
	0.060	330	—

spherulite, suggesting that it is due to radially oriented lamellae. However he was able to observe only one reflection and was not, as in the case of his wide angle work, able to limit the beam size to a portion of one of the bands. Furthermore the resolution of his equipment may be too low to permit observation of the larger long period in his slowly cooled samples.

A possible explanation is found in Anderson's micrographs of fractured surfaces of fractionated polyethylene (Fig. IV-26) (60,115). The ribbonlike structures have a width that is similar to the first order in small angle x-ray diffraction while their thickness and that of the lamellae is similar to the smaller of the long periods. Whether or not there are sufficient ribbon-like lamellae to give rise to the observed x-ray scattering intensity is not known. Also the origin of the ribbons and their relationship to the lamellae and the spherulites is not known. Possibly these are related to the cells in the spherulite growth mechanism postulated by Keith and Padden (51,52) (Section IV-5a). The breadth of the {110} reflections would then be directly related to the width of the ribbons. The "extended chain" lamellae may also give rise to a long period spacing.

As indicated earlier, it is believed that the twisting of the lamellae may be due to a surface strain related to their nearly two dimensional character (67,68). In a crystal of low molecular weight material the lattice spacings on the surface are larger than in the interior because of the nonsymmetry of the lattice forces. Although the effect is most noticeable normal to the surface, there is a strain developed parallel to the surface. In a lamellar crystal of a low molecular weight material this strain, tending to expand the surface with respect to the interior, is accommodated by the formation of a mosaic structure and/or twisting of the lamellae (21). The size of the mosaics and the amount of twist will vary with the thickness of the crystal, larger mosaics and smaller twists occurring in thicker crystals.

In polymer crystals an additional surface strain may be introduced by the folds (67). Stuart-Brigleb models suggest that the folds are slightly bulkier in projection parallel to the molecule than the molecule itself. The thickness of polymer crystals depends on the crystallization temperature. Thus, as is found, the spacing of the bands in polyethylene and similar spherulites should decrease with decreasing crystallization temperature. Theoretical aspects of the twisting are discussed in Chapter VI.

Although the previous discussion has been based on polymer spherulites crystallized from the melt, spherulites crystallized from solution are similar in appearance. Thin film polyethylene spherulites produced by evaportion of the solvent, for instance, have a band structure whose period depends on the crystallization temperature. Although transmission optical and electron micrographs suggest that they are composed of twisted fibrils, surface replication of the same films indicates that they are lamellar (54,55). The periodically varying density in the electron micrographs which corresponds to the bands seen visually was attributed to periodic variations in density (i.e., crystallinity) or mass (55). Similar suggestions have been made on the basis of optical observations (69–71). Recently it has been suggested that the extinction bands in the electron microscope result, instead, from "electron scattering at the surfaces of the fibrils" (27) or from the fact that in a portion of the band the lamellae are parallel to the beam (20). In thin films this region would be expected to be thicker. The first of these explanations is not completely satisfactory since it is known the Bragg scattering from polymers is rapidly destroyed in the electron beam. With respect to the latter suggestion, it is not known if the difference in thickness is sufficient to produce the observed intensity differences. Some phase effect may be present.

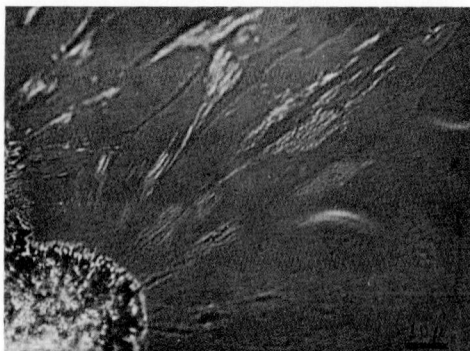

Fig. IV-28. Linear polyethylene spherulite crystallized near the miniscus of an octane solution. Individual lamellae, on which additional lamellar growth in two bands can be seen, extend outward radially. Measurement of the apex angles of the lamellae on this and similar micrographs indicate that the lamellae are growing in the direction of the **b** axis. (Incident steep oblique illumination.) (Reneker (72).)

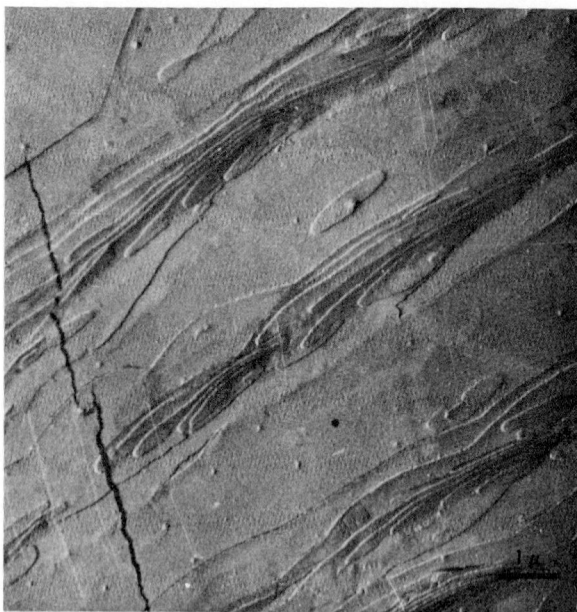

Fig. IV-29. Electron micrograph of a portion of the lamellar material in a spherulite similar to that in Figure IV-28. The area shown corresponds to one of the bands, the radius of the spherulite extending from upper right to lower left (Cr shadowed at $\tan^{-1} 4/7$).

From the x-ray measurements which indicate that the **b** axis is radial in polyethylene spherulites (30,31,58) it is believed that the lamellae grow in the direction of the **b** axis. Some observations by Reneker (72) on spherulites crystallized near the meniscus of a polyethylene solution also appear to indicate that growth is primarily in the <010> direction (Fig. IV-28). The central portions of these spherulites resemble spherulites slowly crystallized from the melt. Individual lamellae can be seen extending radially from the periphery. Although most of the lamellae have a rounded growth face, a few have an angular apex corresponding to the ($\bar{1}$10) and (110) planes, i.e., growth in the <010> direction. The rounded growth faces which often appear during crystallization from concentrated solution or the melt, can also be seen in Figure IV-25.

The micrographs of Reneker (72) suggest that the banding is due to a periodic tertiary nucleation. Isolated lamellae which extend for distances corresponding to more than one band are seen to have periodically spaced patches of lamellae (Fig. IV-28). Electron micrographs (Fig. IV-29) suggest that the tertiary lamellae do not result from screw dislocations as in the dilute-solution grown single crystals. Double dislocations of opposite sign, separated by some distance may be involved, but it is believed that the situation may be even more complex. This type of crystallization should be investigated more thoroughly since it would appear to relate single crystal and spherulitic growth.

4. CRYSTALLIZATION FROM THE GLASSY STATE

Maximum crystallization rates of some polymers including polycarbonate, polyurethane, and polyethylene terephthalate, are low enough that it is possible to quench them to temperatures below their glass transition temperature, T_g, before crystallization can occur. They will remain in a "glassy" state (probably retaining the structure of the melt) as long as they are held below T_g. However, if their temperature is raised above T_g crystallization and, in some cases, the growth of spherulites occurs. Spherulite growth in polyethylene terephthalate, for instance, occurs when the temperature is raised to about 110°C (73,74). The melting point, however, is about 275°C. The spherulites resulting from such treatment are small but are well formed (Fig. IV-30). No electron micrographs are available showing either the original structure of the glassy material or the structure of the spherulites. Spherulites in samples cooled slowly from the melt appear to consist of radiating lamellae.

Extremely interesting results have been obtained by Kampf (75) who has followed the development of crystallinity from the glassy state in

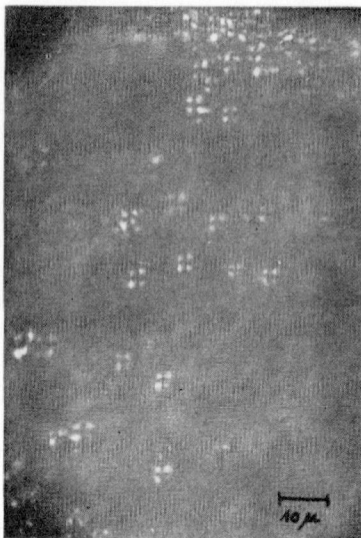

Fig. IV-30. Spherulites of polyethylene terephthalate crystallized by heating a thin, previously quenched and structureless film to 110°C for 35 minutes. The spherulites form as this temperature is reached. (Zachmann (73).)

polycarbonate. T_g for this polymer is 149°C. Films approximately 100 microns thick were prepared either by extrusion from the melt of precipitation from methylene chloride solution. Under normal conditions a structureless, noncrystalline sample results. Kampf showed, however, that spherulites can be formed directly from the melt, presumably by very slow crystallization, and from highly dilute solution. When the 100-micron thick samples were annealed at 130°C (19° below T_g) for 16 hours in air, all surface structure resulting from the method of preparation disappeared, probably as a result of surface tension. As Kampf points out (75), this indicates that the molecules near the surface, at least, have some mobility even below T_g.

Well-developed spherulites form when the noncrystalline sample is annealed above T_g, at 190°, for 8 days under dry nitrogen (Fig. IV-31). Some indication of spherulitic structure is observed after 8 days under nitrogen at 180°. The spherulites formed at 190° have a sheaflike nucleus with a "bud" shaped mound on either side (Fig. IV-32). When observed at higher magnification there is no question that the material is lamellar. The lamellae are about 60 A thick. Similar lamellae are also formed when the sample is slowly cooled from the melt.

Fig. IV-31. Polycarbonate film annealed 8 days under nitrogen at 190°C (41°C above T_g). Some of the polymer has remained attached to the carbon replica, resulting in the high contrast. (Kampf (75).)

Fig. IV-32. Spherulite crystallized from a "glassy" polycarbonate film by annealing for 8 days at 190°C (Kampf (75)).

Fig. IV-33. Lamellae formed in nonspherulitic regions of samples of polycarbonate film annealed for 8 days at 190°C (Kampf (75)).

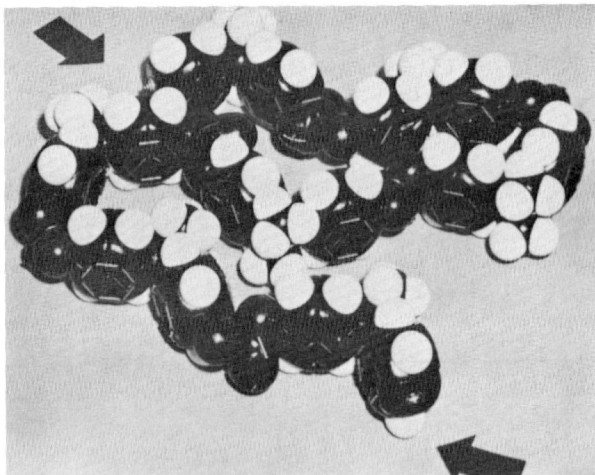

Fig. IV-34. Photograph of a model of a portion of a folded polycarbonate molecule (Kampf (75)).

In other regions of the same samples, following the 8-day annealing treatment at 190°, lamellar structures resembling solution cast single crystals in other polymers, having regular boundaries and forming pyramids as a result of spiral growths, are found (Fig. IV-33). Kampf indicates

(75) that electron diffraction patterns (none published) suggest that the molecules are normal to the 60-A thick lamellae and, therefore, although usually considered quite stiff and bulky, the molecules are folded within the lamellae. Although he shows a photograph of a model of the folded chain (Fig. IV-34) we believe the molecules pack more closely than in the illustration. The fold period of the model shown corresponds to approximately 2 repeat distances whereas the measured fold period corresponds to 3 periods.

Eppe, Fischer, and Stuart (73,76) have reported the development of spherulitic structure from "glassy" polyurethane during annealing. Electron micrographs show a number of interlacing sheafs of undetermined morphology. As these structures form, a small angle x-ray diffraction long period also develops. The dependence of the long period on the annealing temperature is discussed in Chapter V. Structure also develops on glassy films of polychorotrifluroethylene when heated at temperatures above the glass transition temperature (73) (185° for 10 hours, $T_g = 52°$, $T_m = 216°$). Because of the information available concerning the motion of molecules both above and below T_g, considerably more study of the effects discussed in this section is needed.

5. FURTHER OBSERVATIONS OF POLYMER SPHERULITE STRUCTURE

In the remaining pages of this chapter we summarize presently available morphological studies of spherulites of natural and synthetic polymers. In all cases considerably less is known than is described in Section IV-2 for polyoxymethylene and polyethylene. Spherulites of polypropylene, which are discussed first, have what is probably the next best defined morphological structure. The mechanism of spherulite growth proposed by Keith and Padden (51,52) is discussed in this section. The remaining synthetic polymers for which information is available are listed alphabetically, followed by several natural polymers. Although many more crystalline polymers have been synthesized than are discussed in the following pages, little or nothing has been reported of their spherulitic structure. For many of the polymers that are discussed only optical observations have been reported. Although no attempt is made here to refer to all optical observations, representative reports are listed. In all of the following descriptions the statement, much more work is needed to understand the morphology, applies.

a. Polypropylene

This polymer is of especial interest since at least four distinctly different (in the optical microscope) types of spherulites can be grown from the melt, at least two of them at the same temperature (48) (Fig. IV-35). Two different crystal lattices are associated with their growth (48,49); two of the spherulites, termed types I and II, consist of the monoclinic crystal modification whereas the other two, termed types III and IV, consist of the hexagonal modification (see Chapter I).

Fig. IV-35. Types I and III spherulites of polypropylene formed by air quenching a molten thin film. The brighter spherulites are type III and have a hexagonal unit cell. In some preparations the Maltese cross in the type I spherulites is more clearly defined than here, these spherulites having a somewhat "mixed" character; i.e., some of the radii correspond to type II birefringence (crossed polaroids).

Type I and II spherulites, as observed in the optical microscope, differ primarily in the sign of their birefringence, type I being weakly positive and type II weakly negative (48). Type II grows at temperatures above 138°C and type I below about 134°C. Between these temperatures a mixed type of spherulite is formed. Both type I and type II spherulites melt at 168°C, the melting point of the monoclinic crystal modification. The difference in birefringence may be primarily due to form birefringence, the basic morphology appearing to be similar. Distinct radial and boundary cracks are observed in slowly crystallized samples, the boundary cracks developing to such an extent that a slowly cooled sample may fall apart into individual spherulites (48). X-ray diffraction patterns indicate (49) that the molecules make an angle of 65°–70° with the spherulite radii.

As type I spherulites are heated the birefringence changes from positive to negative (i.e., becomes similar to type II) before the sample melts. Likewise the mixed type becomes negative and a well formed Maltese cross develops. The negative birefringence is maintained if the sample is cooled without being melted (48).

The growth of type III spherulites was reported to be enhanced by relatively rapid cooling rates (48). In steam quenched films we find that up

Fig. IV-36. (a) Micrograph of the spherulite in Figure IV-35 as observed between crossed polaroids on a hot stage at 151°C. The type III spherulites have melted. A "granular" over-growth can be seen (arrows) on the type I spherulites. (b) With further time (several minutes) at this temperature or higher, the polymer in the type III spherulites recrystallizes. This micrograph was taken at 158°C. The appearance shown here is maintained between room temperature and the melting point of the mono-clinic modification.

to 50% of the sample may crystallize in the form of type III spherulites, the remainder being type I and mixed. Even with very slow cooling rates, however (see Chapter III), a portion of the material crystallizes with a morphology and crystal structure typical of type III spherulites. Type III spherulites have a negative birefringence much greater than type II spherulites. X-ray diffraction patterns from type III spherulites in thin films indicate that the chains are tangential (74). A hexagonal pattern is observed when the beam is either normal to the plane of the film or parallel to it, indicating that all orientations of the molecule about the radius are found.

In the temperature range between 128°C and 132°C, Padden and Keith (48) indicate that a banded type spherulite may form sporadically. This type IV spherulite also crystallizes in the hexagonal crystal modification. The band spacing is temperature independent. Their optical microscope observations indicate that optically uniaxial crystals are oriented with the optic axis (which should correspond to the molecular axis in a hexagonal unit cell) tangential and that this orientation twists cooperatively and uniformly about the radii.

The melting point of type III and IV spherulites is about 150° (Fig. IV-36). If type III spherulites are heated between this temperature and the melting point of the monoclinic crystal, 168°C, a spherulitic type structure of low birefringence develops and is maintained down to room temperature. X-ray diffraction patterns show that an unoriented monoclinic lattice is present (74). These spherulites melt at 168°C. Apparently the hexagonal modification melts and the molecules recrystallize in the monoclinic modification. In type IV spherulites, similarly treated, the cooperative twisting is maintained; a double set of bands indicating a twisted biaxial crystal develops (48). As in the case of type III spherulites the annealed type IV spherulites yield an essentially unoriented monoclinic modification type x-ray pattern (49). The optical results, however, suggest that the molecules recrystallize in the new lattice in situ; that they do not melt in the sense that a random orientation is assumed. (See also Chapter V.) It is expected that smaller diameter x-ray micro-beams would show a preferred orientation within a portion of a band as is seen with linear polyethylene spherulites (58).

These observations, and related ones for gutta percha (3), require that some form of continuous crystallization occurs during spherulite growth. The unit cell of the nucleus is propagated throughout the spherulite. Secondary nucleation of crystallites, as envisaged in various modifications of the fringed micelle model, cannot explain this type of growth. The

Fig. IV-37. Portion of a type III spherulite (Cr replica). Besides the lamellae, occasional spiral growths can be observed on the surface. This sample was washed in room-temperature benzene to remove some of the atatic material, which usually obscures the surface, before it was replicated with chromium.

secondary nucleation involved in spherulite growth must be on the surface of the crystal itself, i.e., the nucleation of new fold planes on a completed fold plane. These results would tend to indicate, as stated previously, that all new, tertiary lamellae result from branching in already growing lamellae or ribbons.

Padden and Keith (48) describe a granular overgrowth that develops at temperatures above 140–145°C and persists to the melting point (Fig. IV-36a). Our observations suggest that this material is an exuded liquid, probably low molecular weight polymer, which at temperatures above 150°C may spread over the entire surface and become essentially invisible. During subsequent cooling it may crystallize. Unless removed by etching with a selective solvent such as benzene, such material covers over much of the surface detail during replication for electron microscopy.

Type III spherulites is composed of radiating lamellae (Fig. IV-37). As on the case of slowly cooled, partially degraded polypropylene (Chapter III) the morphology of the polymer crystallizing in the hexagonal modi-

Fig. IV-38. Several of the different morphologies associated with Types I and II spherulites. The small differences in birefringence seen in the optical microscope probably result from differences in form birefringence. (a) Broad ribbons, sometimes ribbed or corrugated as well as (b) slightly twisted, ropelike structures can be found. (Cr replicas.)

fication is most clearly defined. Spiral growths, although less numerous than on polyoxymethylene spherulites, are observable on the surfaces of the lamellae. The spiral growths tend to be hexagonal in shape. The general features appear similar to polyoxymethylene spherulites. Presumably in type IV spherulites the lamellae are regularly and cooperatively twisted. The molecules are probably folded within the lamellae.

The morphology of type I and II spherulites appears to be related to the broad needles observed during slow cooling (see Chapter III). At times a ropelike structure is also visible (Fig. IV-38). The fact that the molecules make an angle of about $70°$ with the radius is consistent with the possibility that the needles are composed of lamellae oriented at an angle to the needle's axis. Further observations on samples whose surfaces are not obscured by exuded material are needed.

When polypropylene is quenched rapidly, as in dry ice-acetone, a nonspherulitic structure is formed (Fig. IV-39) (74). Small clusters, on the order of 200–500 A in diameter are formed. This material is apparently paracrystalline, x-ray diffraction patterns consisting of two broad reflections only (77). Upon heating at temperatures above about $60°C$ the x-ray pattern changes to that of the monoclinic modification. The peak widths, however, remain broader than that associated with slowly cooled monoclinic modification material, suggesting that the crystal perfection increases but that

Fig. IV-39. Quenched, paracrystalline polypropylene crystallized from a molten film with an unrestrained surface. This structure remains as the sample is heated until the melting point is reached. Depending on the quenching rate and, to some extent, the polymer sample used, all variations in surface topography from a nearly smooth surface to spherulites can be obtained (Cr replica).

the crystallites remain small. In contrast to polycarbonate quenched to a true glassy state, little or no change is observed in the morphology of the material unless it is heated above its melting point, 168°C. In some samples of polypropylene the micrographs suggest that the clusters are composed of a few small lamellae.

As indicated in Chapter III (Section 3-e, p. 212), slow cooling of thin films of polypropylene results in the formation of individual structures with a number of different morphologies, including several which are ribbon-like. Using slightly thicker films, the fibers in the sheaflike structures in Figure IV-40 were found to grow out into the melt at a relatively high temperature with the material in the background only crystallizing at a considerably lower temperature. Using radioactive tracer techniques, Padden and Keith (78) have shown that atactic and other poorly crystal-lizable molecules such as low molecular weight waxes are "pushed" aside by the growing fibrils and aggregate in the intervening spaces. These

Fig. IV-40. Sheaflike spherulites of polypropylene crystallized by slowly cooling a molten film. The regions between the sheaves crystallized at a considerably lower temperature than the fibers within the sheaves (crossed polaroids).

"impurities" crystallize, if sterically possible, at lower temperatures. When structures such as those in Figure IV-40 are deformed, the initial deformation is almost entirely restricted to the material between the crystalline fibers (79).

Based on the results described above, the known relationships between the presence of impurities and spherulite growth in low molecular weight materials (Section IV-1) and analogy with the effect of impurities on crystal growth in metals, Keith and Padden (51,52) have postulated a cellulation mechanism for the growth of polymer spherulites. They suggest that low molecular weight polymer, branches, atactic or other stereoirregular molecules and entanglements or crosslinks play the role of impurities during polymer crystallization.

In metals it has been shown (see, for instance, reference 80) that in an impurity containing melt a planar growth face on a growing crystal is unstable. The growth face will become broken up into a number of columnar cells or protusions in which nearly pure metal crystallizes while the rejected impurities aggregate along the cell boundaries. The diameters of the cells are given by $\delta = D/v$ where D is the diffusion constant of the impurity and v is the rate of growth of front of the cells. For metals, Keith and Padden indicate (51,52) that typical values are $D = 10^{-5}$ cm^2/sec, $v = 10^{-3}$ cm/sec, and $\delta = 10^{-2}$ cm.

Keith and Padden suggest that crystallization in polymers differs in two ways from that in metals, both effects tending to emphasize the cellular habit of growth. In the first place the impurities in polymers will probably crystallize at a much slower rate, if at all, than impurities in a metal. Thus, although the growth front of a metal crystal is slightly roughened by cellulation, the cells in a polymer may be quite long, remaining separated by amorphous or poorly crystalline material. The cells would have a fibrillar habit as is observed in polypropylene under some conditions (Fig. IV-40). In the second place, they suggest that D will be much lower for polymers than for metals and, therefore, δ will be on the order of microns or smaller. This is the observed order of magnitude of the fibers in polymers.

The individual cells or fibers in a polymer which develops from an original crystal need not continue to maintain a parallel arrangement but may twist and turn. In addition, Keith and Padden suggest (51,52) that small angle branching of individual fibers may occur. They point out that if the mosaic blocks in the substructure of a growing fiber (i.e., subgrains of slightly different orientation, known to be present in metals and probably present in polymers also) are of the same order of size as the fiber, then there is a reasonable probability that some of them near the tip of the fiber will be able to serve as nuclei for branches. These branches will have the same crystal structure as the parent fiber but will differ slightly in growth direction and could give rise to the new fibers needed to fill space in an expanding spherulite.

In the description of their mechanism, Keith and Padden (51,52) do not consider the detailed morphology of the resulting fibers, although they recognize that in a number of polymers the fibers are lamellar in habit. In such cases, individual fibers may be composed of more than one lamellae, perhaps the clusters of lamellae which, in polyethylene spherulites, appear to twist in unison. (In Figure IV-18, clusters of 4-5 lamellae separated by voids can be seen in regions in which the lamellae are normal to the surface.) In such cases, however, the growth mechanism described in Section IV-2-a (p. 234) is believed more applicable, the thickness of the lamellae being determined by the crystallization temperature and branching resulting from formation of a screw dislocation in the growth front. In actuality, however, the two mechanisms described in this chapter are fairly closely related, both allowing for the segregation of impurities. They differ primarily in their emphasis on the lamellar or fibrillar habit of polymer crystal growth and either or both may apply to individual polymers. In polyethylene, for instance, the ribbons observed by Anderson on fracture surface (60) (Fig. IV-26) may be a result of cellulation of a growing lamella.

b. Polyacrylonitrile (81)

Holland (81) has reported that polyacrylonitrile can be crystallized from dimethyl formamide and dimethyl acetamide in the form of nonbanded spherulites with negative birefringence. He indicates that some spherulites are formed with zero birefringence, being totally dark between crossed polaroids. His electron micrographs suggest that banded spherulites can form under some conditions. In relatively thick films the spherulites are lamellar, the lamellae being about 150 A thick (Fig. IV-41). In very thin films, fibrillar sheaf are formed. It is not known if the different morphologies are related to the different solvents.

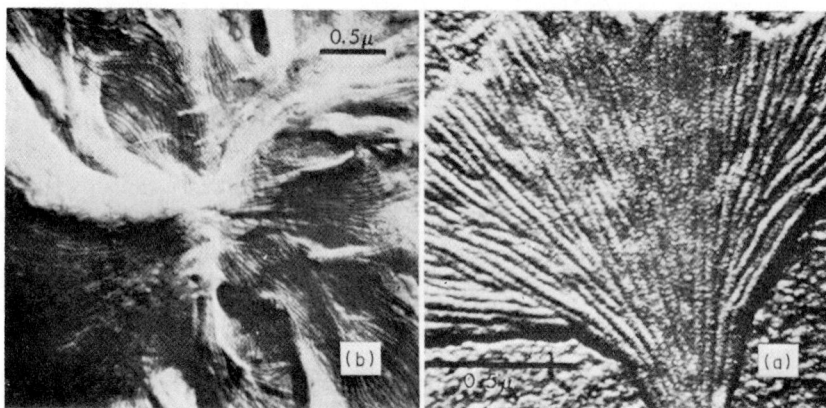

Fig. IV-41. Replicas of solvent cast films of polyacrylonitrile. (a) Lamellae in a portion of a spherulite crystallized in a relatively thick film. (b) Half of a fibrillar sheaf. The fibrils are about 350 A wide. (Holland (81).)

c. Polyamides

Electron micrographs (Fig. IV-42) of melt-crystallized polyamides (nylon 6, 66, 610, and copolymers) are generally indistinct. Less well developed spherulites are often sheaflike. Steps can be seen (76,82) but these may be due to either a fibrillar or lamellar structure. On fracture surfaces the spherulites can be distinguished, the structure elements appearing fibrillar. Kassenbeck describes (82) a high magnification electron micrograph of a small portion of a nylon (type unspecified) spherulite (C replica) in which a periodic structure on the 100 A scale is seen. Although a replica is used, the micrograph has the appearance of a transmission-type micrograph, suggesting that some polymer, as we often find, may have remained attached to the replica. The periodicity is visible in two directions, each

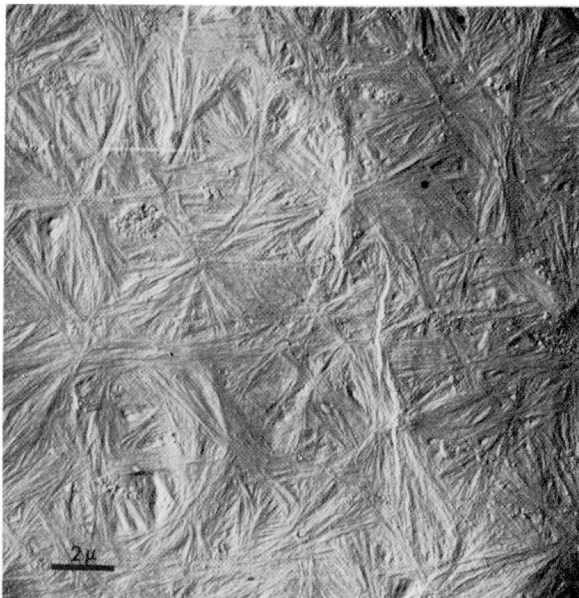

Fig. IV-42. Spherulites of nylon 6 crystallized from a molten film with an unrestrained surface. Fibrillar-appearing sheafs are present. Similar micrographs have been obtained from films of nylon 6, 610, and copolymers. The white lines are cracks in the Cr replica.

making an angle of about 20° with the spherulite radii. On the basis of this micrograph, Kassenbeck concluded that the molecules make an angle of about 70° with the spherulite radii.

In many of the polyamides different types of spherulites are formed at different crystallization temperatures. Morphological studies of the different types have not been reported. X-ray and optical microscope studies in considerable detail have been reported by a number of authors (17,19, 30,83–90).

The morphology of solution cast polyamides is variable, depending on the polymer, temperature, and solvent. Globular particles of unknown structure are often formed either with or instead of spherulites and single crystals. The single crystals grown from glycerine solution at elevated temperatures are described in Chapter II. Present results are described below for spherulites of the individual polymers. Also included are some preliminary studies of polyamide single crystals grown from the melt.

(1) Nylon 6

Spherulites as well as single crystals develop when glycerine solutions are cooled (91). The spherulitic material, which is in the low temperature (α or β) crystalline modification, appears to consist of irregular ribbonlike lamellae, twisted together to form a sheaflike center (Fig. IV-43). Somewhat similar spherulites have been obtained by crystallization during cooling of a 50/50 methanol, water solution in a sealed tube (92).

Crystallization at room temperature in formic acid solutions results in the development of compact, fibrillar appearing, sheaflike spherulites (76). Solvent cast continuous films resemble the melt cast films shown in Figure IV-42.

Spit has been able to stain solution grown nylon 6 spherulites (114). Using 2% phosphotungstic acid solutions as a staining agent, he found that the fibrils in the spherulites formed in the continuous films appear, on

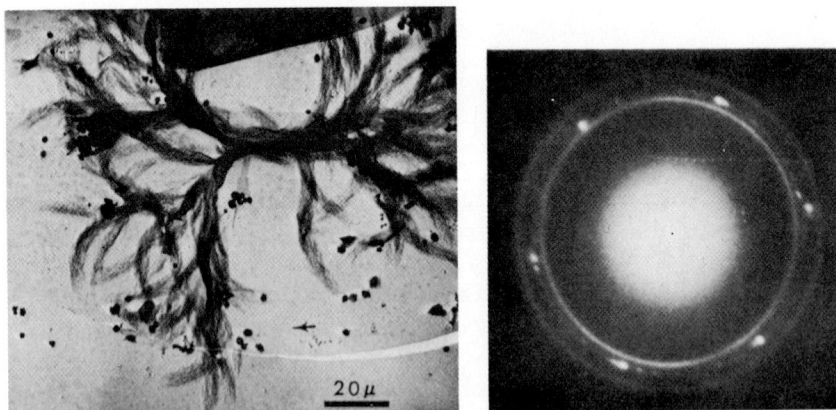

Fig. IV-43. Spherulites of nylon 6 crystallized from glycerin solution. A single crystal (arrow) as well as a number of globular particles of nylon are visible. A diffraction pattern from an area containing spherulites and single crystals is also shown. The spherulites give rise to the diffraction rings (Cr shadowed at $\tan^{-1} 5/7$). (Geil (91).)

micrograph positives, as white bands bordered on both sides by black, electron dense bands. Each black band is about one half the width of the white bands. The results suggest that the heavy metal staining agent is being absorbed on the surfaces of the fibrils. Spit reports (114) only poor or negative results with other staining agents and also for other polyamides. Although nylon 6 single crystals absorb the stain in selective areas on their surface, the observations have not yet been interpreted in terms of structure.

(2) Nylon 66

As previously described, nylon 66 crystallized from glycerine, consists of rolled or folded lamellae (91). Further development of this structure leads to spherulites. Similar material has been crystallized from 0.1% solutions in 50/50 methanol/water (92). Eppe, Fischer, and Stuart (76) indicate that lamellae are formed in solution-cast (probably formic acid) films. Keller (93) and Scott (94), on the other hand, find that films of nylon 66 cast from formic acid consist of fibrillar spherulites (Fig. IV-44). In these thin films the hydrogen bonded planes, and thus the molecules, lie in the plane of the film. Keller (93) suggest that thicker films develop by a rolling up of thin films which are themselves composed of fibrils. The morphology of these structures is obviously in need of clarification.

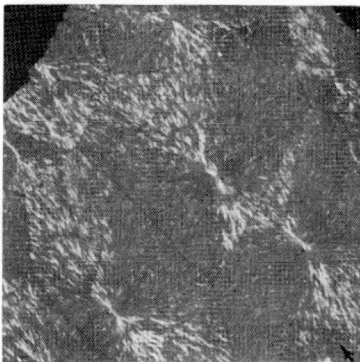

Fig. IV-44. Dark field electron micrograph of a solvent cast film of nylon 66. The bright spots are crystalline regions scattering electrons into the aperture which has been displaced in the direction of the arrow from the main beam. Scott suggests that the light gray regions may be due either to regions in which the molecules are radial rather than tangential or to the inclusion of a second, less intense Bragg reflection within the aperture. (Scott (94).)

(3) Nylon 610

Spherulites composed of rolled or folded lamellae are formed during crystallization from glycerine (91). Spherulites composed of ribbonlike lamellae are formed during crystallization by the addition of benzyl alcohol to a dilute *m*-cresol solution (93) (Figs. II-4 and IV-45). The ribbons, about 150 A wide, appear twisted together in the thicker portions, forming a sheaflike center. From electron diffraction patterns, Keller concluded (93) that the molecules, lying in the plane of the ribbon, make an angle of 65–75° with the axis of the ribbon and thus are folded.

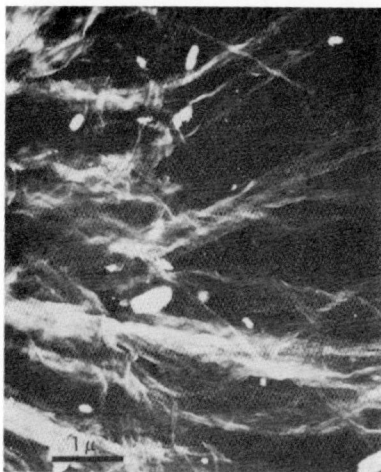

Fig. IV-45. Nylon 610 crystallized from a dilute *m*-cresol solution at room temperature by the addition of benzyl alcohol. The individual ribbonlike lamellae are shown in Figure II-4. At lower magnification the spherulites resemble those of nylon 6. (Keller (93).)

(4) Nylon 11

Eppe, Fischer, and Stuart (76) indicate that solution cast films consist of lamellae.

(5) Melt Grown Nylon Single Crystals (95,113)

Magill and Harris have reported growing large single crystals in thin films of a number of polyamides. The films of nylon 57 and nylon 8, shown in Figure IV-46, were crystallized isothermally at temperatures near the melting point for several hours. Electron diffraction patterns

Fig. IV-46. Spherulites composed of single crystal platelet sectors crystallized from the melt. (a) Nylon 57 (Magill and Harris (95)). (b) Nylon 8 (Magill and Harris (113)). The individual sectors extinguish as a unit and give rise to the diffraction patterns shown in Figure IV-47.

(Fig. IV-47) show that the molecules are oriented normal to the surface of
of the platelets, and Magill and Harris presume that they are folded.

d. Polybutene

Polybutene spherulites are lamellar, resembling the type III spherulites
of polypropylene as to irregularity of the lamella edges, scarcity of spiral
growths, and poor resolution due to noncrystalline material. Most polybutene spherulites have an appearance similar to that described by Popoff
(10) for spherulites of low molecular weight materials, i.e., a double leaf

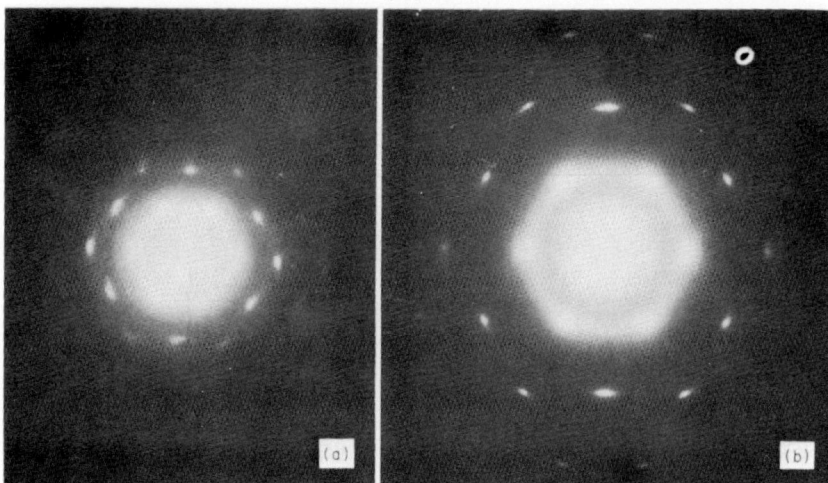

Fig. IV-47. X-ray diffraction patterns corresponding to individual sectors of the thin
films shown in Figure IV-46.   (a) Nylon 57 (Magill and Harris (95)).   (b) Nylon 8
(Magill and Harris (113)).

structure at the center (Fig. IV-48). The ends of the central sheaf spread out
and meet some distance from the center. In contrast to the descriptions
of Popoff and those for many polymer spherulites as well, the two leaf
structures are mounds, not pits. The sheaf takes the form of a valley or
saddle between the two mounds. These mounds, which appear structureless, may be composed of noncrystalline material segregated during the
growth of the spherulite. At higher magnifications the fibrillar structure
on Figure IV-48 is seen to result from lamella edges.

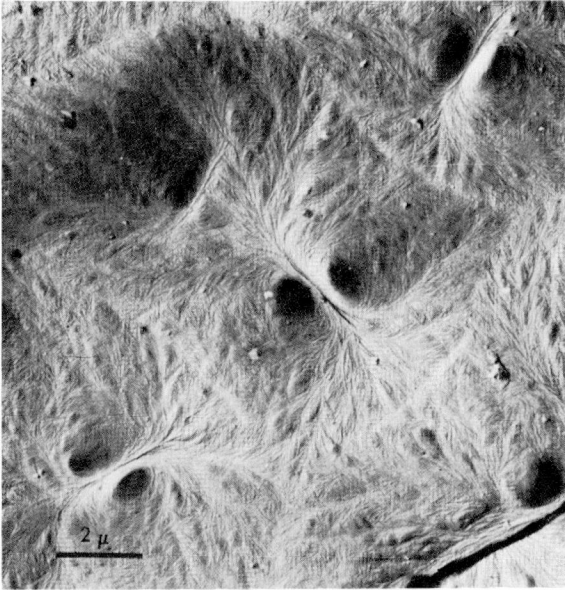

Fig. IV-48. Spherulites of polybutene (Cr replica) crystallized from a thin, molten film with an unrestrained surface.

e. Polycarbonate

Films cast from methylene chloride solutions are composed of fibrous appearing, nonbanded spherulites. The structure of spherulites crystallized from the glassy state and the melt is discussed in Section IV-3.

f. Polychlorotrifluoroethylene

Polychlorotrifluoroethylene crystallizes from the melt in the form of spherulites in which the Maltese cross seen between crossed polaroids has a zigzag structure superimposed on it (see Fig. IV-64). Although it can also be quenched to a glassy state, the morphology of the glass and its change during annealing has not been studied. When a thin film with an unrestrained surface is crystallized, for use in electron microscope studies, primarily only sheaflike bundles of lamellae are observed on the surface (Fig. IV-49). These lamellae are usually nearly normal to the surface. Spherulitic structures develop within the film and in a few areas on the exterior surface (Figs. IV-50 and IV-51). In air-cooled samples the lamellae are about 250 A thick. Spiral growths can be observed in the few regions in

Fig. IV-49. Sheaflike clusters of lamellae observed on exterior surfaces of melt-crystallized polychlorotrifluoroethylene.   A fracture surface from this sample (No. 7 of Table IV-3) is shown in Figure IV-52.

which they are nearly parallel to the surface. Lamellae are also observed on the surfaces of fractured samples (Fig. IV-52).

The structure of the nucleus of the spherulite on the right of Figure IV-50 is shown at higher magnification in Figure IV-51. It has not been possible to completely determine its structure. The shadowing, by comparison with the shadows of dirt particles on the surface, indicates that the nucleus is a mound, not a pit.

As in the case of polyethylene and polyoxymethylene, there is some question concerning the interpretation of the small angle x-ray diffraction long period in terms of the thickness of the lamellae. Using 150 micron thick samples prepared by Hoffman and Weeks (96), the values shown in Table IV-3 were obtained. Sample #7, which was about 3 mm thick, could be fractured (Fig. IV-52). The lamellae on the interior surfaces were also about 450 A thick. It is not known why the lamella thickness for this sample is less than that for sample #3 which was crystallized at the same temperature. There is also no explanation for the observed thickness of

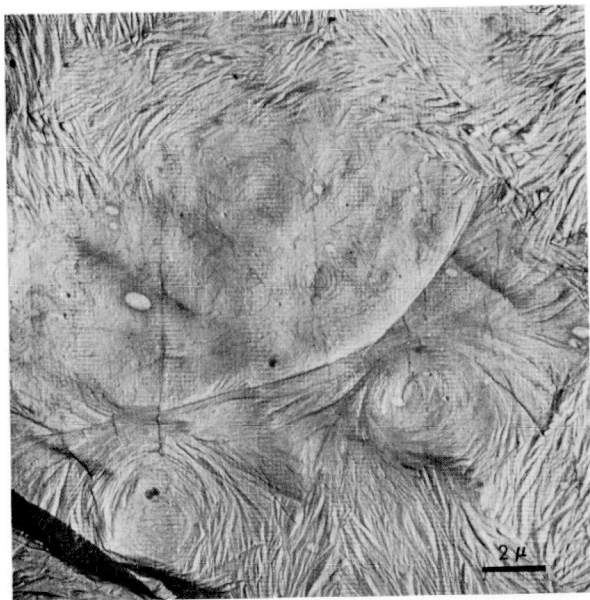

Fig. IV-50. Polychlorotrifluoroethylene spherulites (Cr replica) crystallized with an unrestrained surface from a molten film. Most of the surface resembles that shown around the edges of this figure.

the lamellae in sample #6. The value listed for the E.M. observed thickness is the minimum measured from groups of five or more lamellae oriented normal to the surface and thus would represent a maximum value for the

TABLE IV-3

Long Period and Observed Lamallae Thickness in Samples of Polychlorotrifluoroethylene

Sample[a]	Tc (°C)	Time (min.)	Small angle long period (A)	Electron microscope observed thickness (A)
1	181.3	36	500	310
2	186.1	129	570,280	330
3	191.2	125	710	555
4	190.9	925	670	555
5	196.2	1273	960	600
6	201.7	233[b]	1250	555
7	191.2	125		450

[a] Samples were melted at 305°C, held in an oil bath at the crystallization temperature shown for the specified time, then removed (97).
[b] Hours.

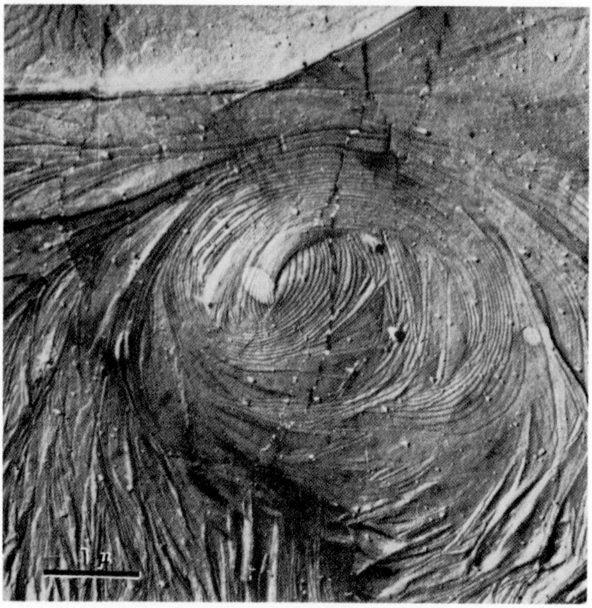

Fig. IV-51. Nucleus structure of one of the spherulites in Figure IV-49.   Some sort of
double spiral seems to be present.

actual thickness (lamellae oriented at an angle would appear thicker than
their actual thickness). Thus it appears that the long periods are on the
order of 100 A or more larger than the thickness of the lamellae.

Hoffman and Weeks (96) have found evidence that the fold period in
polychlorotrifluoroethylene increases while the sample is held at the crystallization temperature. In an excellent study of the melting process and
the rate of crystallization (see Sections V-5 and VI-2) they suggest, on
the basis of their melting point determinations, that the fold period one
observes may be as much as 3.5 times that of the original fold period of the
crystallizing lamellae. No corroborative small angle x-ray data are available.

g. Polydioxolane

The process of crystallization and spherulite growth in polydioxolane is
unusual. Crystallization occurs slowly at room temperature. As a thin
molten film crystallizes or as a water solution evaporates, the spherulites
that begin growing have essentially no birefringence. They can be seen

Fig. IV-52. Fracture surface of sample No. 7 (Table IV-3) of polychlorotrifluoroethylene crystallized from the melt.

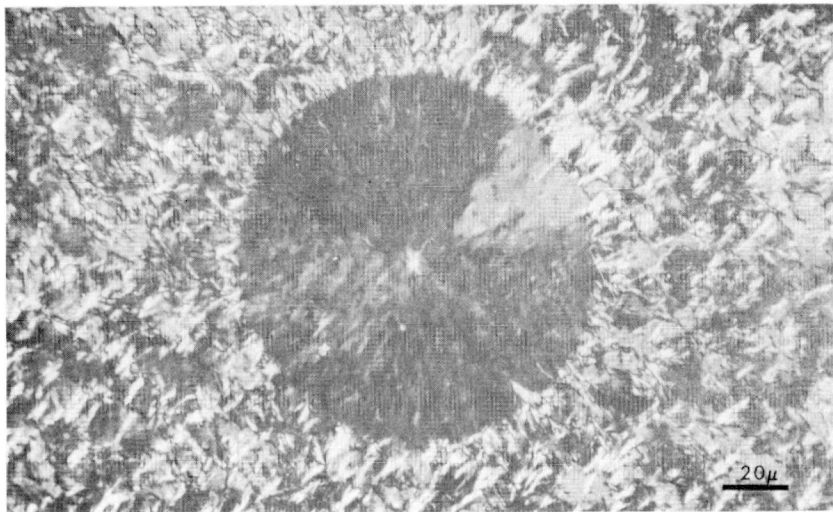

Fig. IV-53. Central portion of a spherulite of polydioxolane crystallized as a thin film by evaporation of a water solution (crossed polaroids).   Similar spherulites are observed during the crystallization of thin molten films.

only poorly with transmitted light. At some stage in their growth a crystalline phase change apparently occurs, usually beginning at some point on the perimeter of the spherulite. The transformation spreads outward from this point across the entire spherulite that has grown up to that time. If the spherulites are growing in a thin film, a uniform extinction can often be observed between crossed polaroids (Fig. IV-53). In other spherulites in the same film various sectors with sharply defined radial boundaries extinguish as a unit. Extinction of the entire spherulite or the sectors occurs every 90° as the sample is rotated. If two spherulites are in contact at the time of the transformation the phase change usually propagates immediately from one to the other although occasionally a delay of several

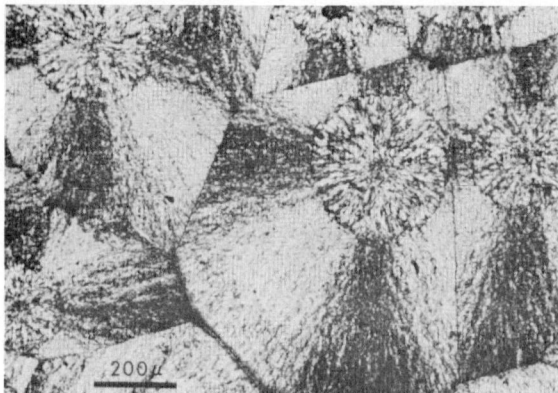

Fig. IV-54. Spherulites of polydioxolane crystallized from a molten film by air quenching. The central, coarsely fibrillar region corresponds to the uniformly dark center of the spherulite in Figure IV-53 (crossed polaroids).

seconds occurs. Following the transformation, further growth is fibrillar-appearing. In somewhat thicker films the central portion has a high, more or less uniform birefringence both before and after the transformation, which is hardly visible, and the fibrillar material displays a Maltese cross (Fig. IV-54).

Electron micrographs indicate that the spherulites are composed of lamellae (Fig. IV-55). In the region of uniform birefringence the lamellae in the thin film spherulites are essentially parallel to the surface. This orientation suggests that the central portion may resemble a hedrite in growth. The portion appearing fibrillar in the optical microscope is due to irregularly twisted lamellae. Cracks occur between the lamellae.

X-ray diffraction patterns takens from a large area of a thick film as it was crystallizing tend to confirm the suggestion that an actual phase change is taking place. A reflection at 22.1° 2θ at first increases in intensity and then decreases as crystallization takes place. While the 22.1° peak is

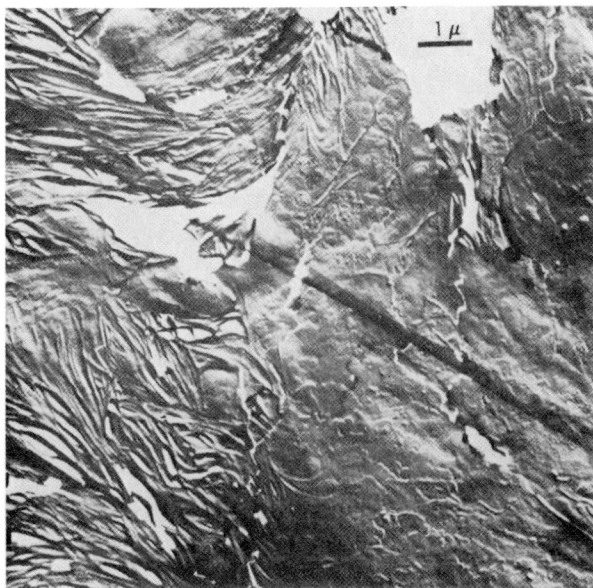

Fig. IV-55. Portion of a spherulite of polydioxolane (Cr replica) crystallized from solution. The portion to the right corresponds to the central area of the spherulite, while that to the left has a fibrillar nature. Several spiral growths are evident.

decreasing, a peak at 22.7° 2θ increases. Following complete crystallization, a trace of the 22.1° peak remains, suggesting that the phase change is not really complete.

h. Polyesters

Many of the polyesters can crystallize in more than one type of spherulite. Usually one of these types is banded, the bands often having a doubled appearance (Fig. IV-56). Those polyesters which form double-banded spherulites are optically biaxial, the double bands resulting from the rotation of the biaxial crystals about the radii.

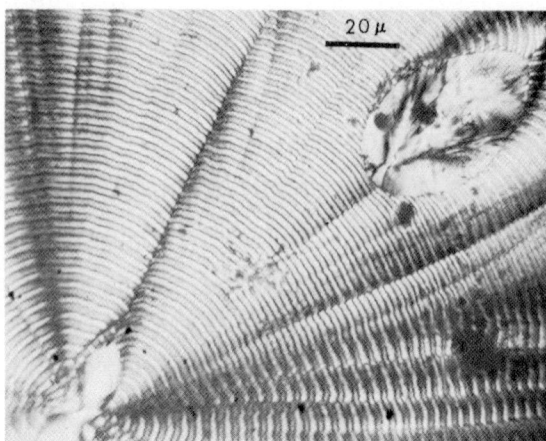

Fig. IV-56. Portion of a double-banded spherulite of polyethylene adipate.   As this spherulite was growing, a nonbanded spherulite was nucleated and started to grow also. (Point (34).)

1. Polyethylene Adipate

Dauscher, Fischer, and Stuart (40) indicate that the band structure in polyethylene adipate (Fig. IV-56) results from a twisting of lamellae about the radii similar to that observed in polyethylene. Detailed optical microscopic observations of the growth and structure of spherulites of polyethylene adipate have been reported (30,33,98,99). Radial cracks, presumably between the lamellae, develop spontaneously in slowly cooled samples (98). Point (33) has shown, with microbeam x-ray patterns, that the **b** axis is radial, while the **a** and **c** (molecular) axes are randomly (on the scale of the x-ray beam) rotated about the **b** axis.

2. Polyethylene Azelate

Polyethylene azelate also crystallizes, near the melting point, in the form of double banded spherulites which are optically similar to those of polyethylene adipate (27). On the basis of the fact that the bands in neighboring spherulites do not become continuous across the boundary, as occurs in polyethylene, Keith and Padden (35) suggest that the structural elements in both polyethylene azelate and polyethylene adipate may be twisted fibrils rathen twisted lamellae.

3. Polyethylene Terephthalate

Detailed optical microscopic observations have been reported for polyethylene terephthalate (17,19,30,100–102). Spherulites of this polymer are also double banded, the double bands developing from a zigzag Maltese cross as the crystallization temperature is raised. By varying the moisture content during fusion the over-all sign of the birefringence can be either positive or negative, probably due to degradation resulting in a lowering of the molecular weight when water is present (17). The double bands result from the fact that as the biaxial crystal rotates about the spherulite radius, two positions of maximum birefringence, of different magnitude and in some cases opposite sign, separated by extinction positions, occur every 360° of rotation. In a uniaxial crystal, on the other hand, the two positions have the same birefringence and only one type of band is formed, as in polyethylene. This polymer is of especial interest since it is commercially available and can be easily quenched to a glasslike state. It is thus possible to produce isolated spherulites imbedded in an amorphous matrix by quenching after growth is partially completed. Furthermore, as indicated previously, spherulites can be grown by annealing the glassy material at temperatures between the glass transition temperature and the melting point. No electron microscope studies of the morphology of these spherulites have been reported.

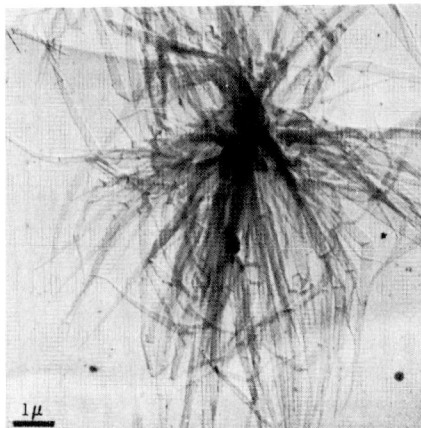

Fig. IV-57. Spherulite of polyhexamethylene terephthalate crystallized from a 0.005% solution in dimethyl formamide. It is stated that the individual tapered ribbons, 50 to 100 A thick, apparently are not single crystals, but may be some sort of helical aggregate of crystals. (Schick (92).)

4. Polyhexamethylene Terephthalate (92)

Two different types of spherulitic structures have been crystallized from 0.005% solution. From dimethylformamide and dimethylacetamide aggregates of long, tapered ribbons, 50–100 A thick, are obtained (Fig. IV-57). Electron diffraction results suggest that the ribbons are aggregates of single crystals, perhaps helically arranged around the ribbon's axis. From tetrachloroethylene, spherulites are formed consisting of a dense

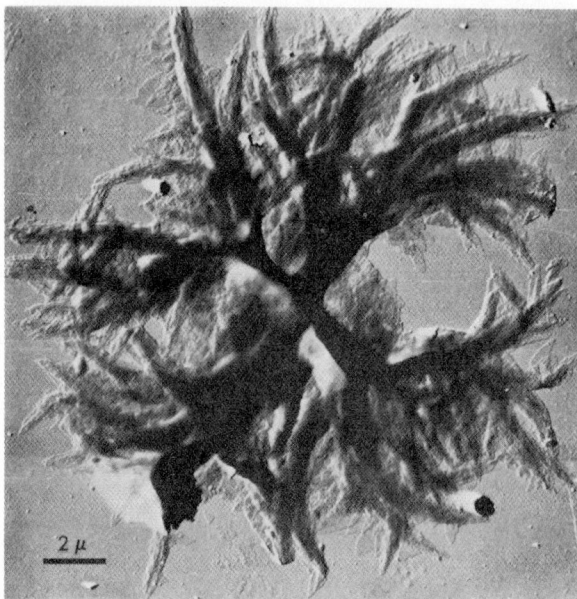

Fig. IV-58. Spherulite of polyhexamethylene terephthalate crystallized from tetra-chloroethylene (sample supplied by M. J. Schick). Single lamellae are seen growing out normal to the arms. Most of the spherulites in this sample had only one central sheaf rather than the two crossed sheafs shown here (Cr shadowed at $\tan^{-1} 5/7$).

central sheaflike structure˒ from which arms composed of "splintered" ribbons extend (Figs. IV-58 and IV-59). Diamond-shaped lamellae, which which have been identified as single crystals consisting of folded chains, are found associated with the central sheaf (Fig. IV-60). Schick indicates that isothermal crystallization at 77–80° leads to larger lamellae with few spiral growths, whereas crystallization at lower temperatures or by cooling results in numerous spiral growths. The relationship between the lamellae and the splintered ribbons is not known.

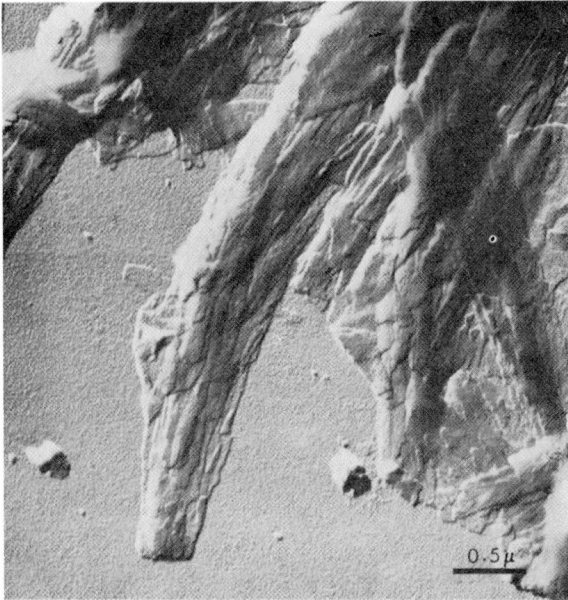

Fig. IV-59. "Splintered" ribbons from the arms of a spherulite of polyhexamethylene terephthalate as in Figure IV-58. The lamellae in the ribbon and the diamond-shaped platelets (Fig. IV-60) are about 275 A thick.

5. Polypentaglycol Terephthalate (92)

Spherulites of polypentaglycol terephthalate crystallized from dimethyl formamide resemble those of polyhexamethylene terephthalate crystallized from perchloroethylene and shown in Figs. IV-58 to IV-60. From cyclohexa-none-acetone mixtures, cigar-shaped structures are formed, composed of irregular ribbonlike lamellae.

i. Polyethylene Oxide

Polyethylene oxide crystallizes from the melt as very large, lamellar spherulites. As in the case of spherulites of polyoxymethylene, lamellae can be followed for considerable distances on the surfaces of the spherulites. Although the surface is often obscured by low molecular weight or non-crystalline material, well-formed spiral growths can be observed on some (Fig. IV-61). This polymer should be well suited for investigations of the effect of molecular weight since a wide range of molecular weights is· available commercially.

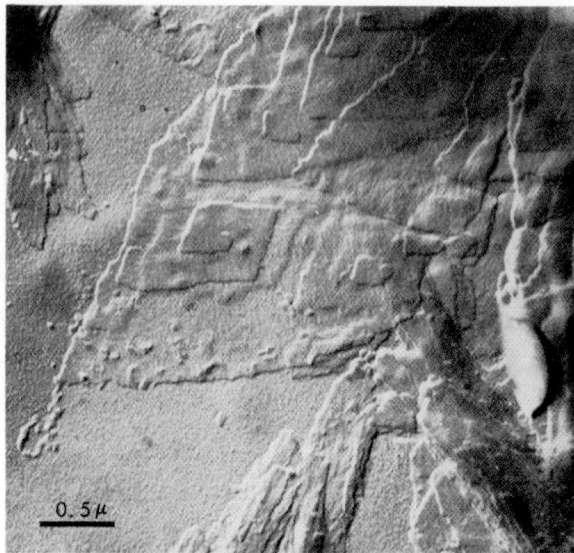

Fig. IV-60. Single crystal lamellae associated with the central sheaf of the polyhexa-methylene terephthalate spherulites.   The lamellae that grow from the arms are usually less well formed.

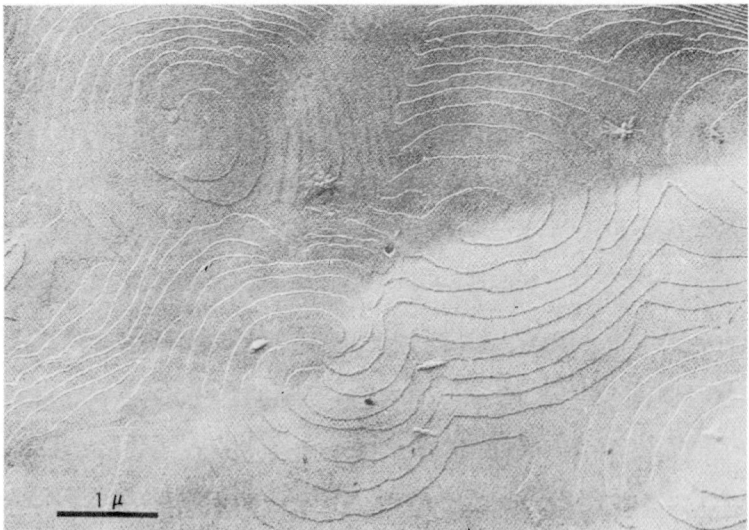

Fig. IV-61. Portion of a surface of a film of polyethylene oxide crystallized from the melt.   Several types of spiral growths are present.   (Fischer (57).)

j. Polyisoprene

(1) Rubber (cis-Polyisoprene)

Solution (24) and melt (26) grown spherulites of purified rubber were described during the 1930's. Those precipitated from solution, at temperatures below −35°C, consist of radiating aggregates of needlelike crystals. The molecular axes are apparently normal to the needles, "the individual needles in each spherulite having optical properties that closely approach those of a uniaxial crystal with negative elongation" (26). Spherulites crystallized from molten thin films, at −25°C, display a well-formed Maltese cross.

(2) Gutta-Percha (trans-Polyisoprene)

The optical properties of various types of gutta-percha spherulites are described in detail by Schuur (3). The polymer should be of interest since, as in the case of polypropylene, two different types of spherulites, consisting of different crystal modifications, can grow simultaneously. No electron micrographic studies have been reported. Several early authors (22,23) also worked with gutta-percha, crystallizing the spherulites from solution.

k. Poly-4-methyl-pentene-1

Spherulites of poly-4-methyl-pentene-1 crystallized from the melt are nonbanded. Although tending to be obscured by noncrystalline material, Cr replicas of the surface indicate that the spherulites are composed of lamellae (Fig. IV-62). Numerous cracks between the lamellae develop as the sample cools.

l. Poly-oxa-1,4-bis (dimethylsilylene) benzene (18)

Spherulites of this polymer, crystallized from the melt, are nonbanded. When crystallized at elevated temperatures they appear coarsely fibrillar, internal reflection occurring. Presumably, although electron micrographs were not presented, the spherulites are lamellar, the internal reflection occurring from cracks between the lamellae. At lower crystallization temperatures the fibrous structure is finer. Although Price discusses (18) his results in terms of a spherulite being composed of radial aggregates of regularly twisting, anisotropic crystallites imbedded in an amorphous matrix, it is believed that his observations on this polymer can also be explained in terms of irregularly twisted (as in polyoxymethylene) lamellae with numerous cracks between the lamellae.

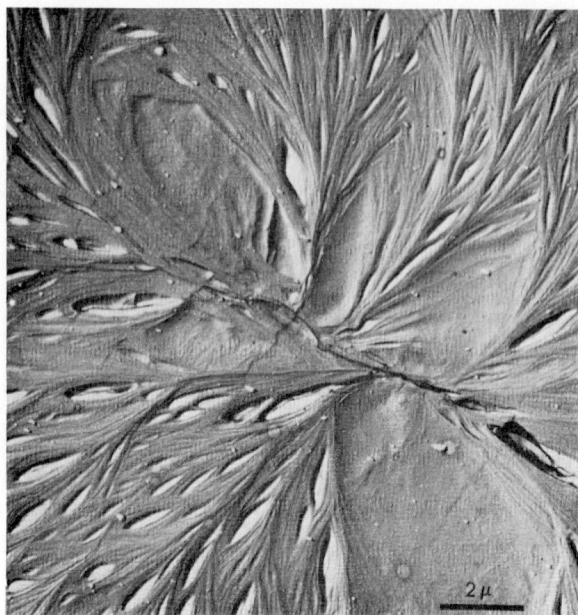

Fig. IV-62. Complex spherulite (Cr replica) of poly-4-methyl-pentene-1 crystallized from a molten thin film. The elongated white regions are cracks between the lamellae that developed either during growth or subsequent cooling. Several sheaves of lamellae appear to have been nucleated at angles to the central sheaf. The lamellae which are parallel to the surface in regions at the sides of the sheaf are obscured by noncrystalline or low molecular weight polymer.

m. Polystyrene

Isotactic polystyrene crystallizes from the melt in the form of nonbanded spherulites with large radial fibrils (103–105). A plot of spherulite growth rate vs. temperature is essentially Gaussian (103), resembling that of nylon 6 (46). The maximum is at 174°C. The material can be easily quenched to a glassy state and since the glass transition temperature (80–100°C) is above room temperature, crystallinity will develop only during subsequent annealing. Spherulites form if the glassy sample is reheated above T_g (105). In this case the spherulites that form are coarsely banded. No electron micrographs have been reported.

n. Polytetrafluoroethylene

Polytetrafluoroethylene spherulites can be grown by slow cooling from above the melting point (106) and by crystallization from perfluorokerosene

Fig. IV-63. Spherulite of polytetrafluoroethylene crystallized from a perfluorokerosene solution. The widths of the individual primary dendrites varies considerably in different samples. (Symons (107).)

solutions (107). As might be expected from its unusual physical and molecular properties, the morphology of the spherulites differs to some extent from that of other polymers, the primary difference appearing to be that of the thickness of the lamellae, i.e., the fold period.

The spherulites crystallized from solution consist of radial arrays of dendritic fibers (Fig. IV-63). The molecules are parallel to the long axis of the primary dendrite and thus normal to the axis of branches. Electron micrographs of these spherulites have not been obtained.

The spherulites that develop during the slow cooling of polytetrafluoroethylene melts are generally incompletely formed. The use of lower molecular weight polymer, long residence times in the melt and slow cooling lead to more perfect spherulites. When well formed, polytetrafluoroethylene spherulites have a zigzag Maltese cross (31). The zigzag can fill an entire quadrant resulting in a bandlike structure (Fig. IV-64). Electron micrographs of spherulites as well formed as in Figure IV-64 have not been published.

Incipient spherulites that are formed from high molecular weight polymer appear to resemble sheafs (Fig. IV-65), being composed of one or more long "bands." These bands, which are about 0.1 to 1 micron wide, branch as they leave the central region. Each band is seen to have a series of parallel striations, more or less well defined, running the length of the band (Fig. IV-66). Although Bunn, Cobold, and Palmer (106) indicate that this structure can be seen on the fracture surface of a sample prepared

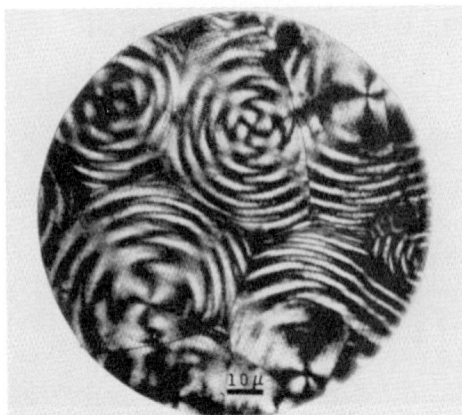

Fig. IV-64. Spherulites of polytetrafluoroethylene.   The method of preparation is not stated in the original reference (crossed polaroids).   (Point (31).)

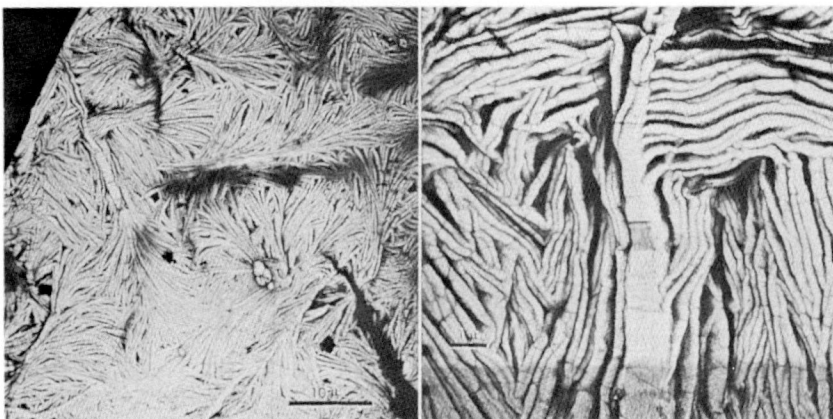

Fig. IV-65. Incompletely formed spherulites of polytetrafluoroethylene crystallized by slowly cooling (0.2°C/min.) a slab of dispersion-based polymer from 380°C (Cr replica). The fibrils on this and subsequent micrographs result from polymer adhering to the Cr replica as it is removed from the sample.   Two different magnifications are shown.

from dispersion based polymer (see Section VIII-1-c) and broken at liquid nitrogen temperatures, we have never observed it on such a sample. The exterior surfaces, however, of all samples from dispersion-based and granular polymer that we have examined have a structure similar to that in Figures IV-64 and IV-65. In some cases, stereomicrographs suggest that the bands are approximately as thick as they are wide. In other regions of the same sample, however, it is obvious that the bands are composed of a few lamellae, each striation corresponding to one lamella (Fig. IV-67). In Figure IV-65b each band may correspond to a single lamella.

When a fractured surface is examined, from a sample prepared from dispersion or granular polymer, the bands are again observed (Fig. IV-68). In this case, however, parallel striations several hundred angstroms apart are observed, oriented normal to the long axis of the band. This structure is typical of all highly crystalline samples of polytetrafluoroethylene that we

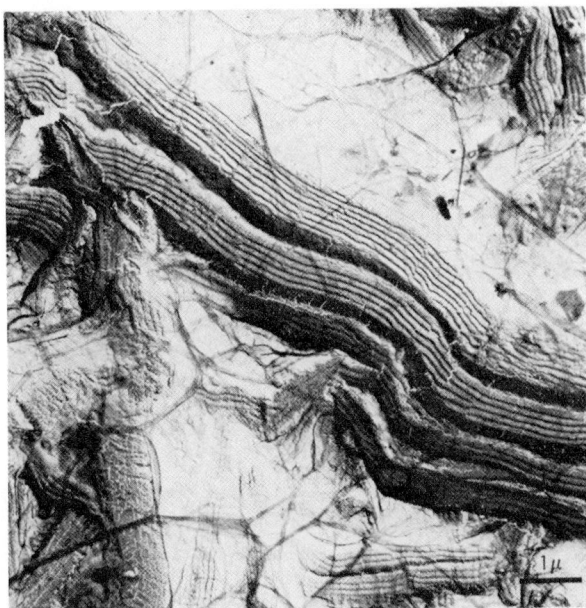

Fig. IV-66. Parallel striations on bands of a polytetrafluoroethylene sample crystallized under the same conditions as that shown in Figure IV-64 (Cr replica).

have observed, regardless of source. Bunn, Cobbold, and Palmer (106) report it, however, only for samples prepared from granular polymer. The bands in the interior of the sample appear to be as thick as they are wide; all bands on the fractured surface are of about the same width. Although some stereomicrographs of the long bands suggest they may be polyhedral, their cross section is not known. It has not been possible to identify a band fractured normal to its axis. In fact, a peculiar feature is that in many portions of a highly crystalline sample most of the fracture surface is occupied by bands which appear to be parallel to the surface.

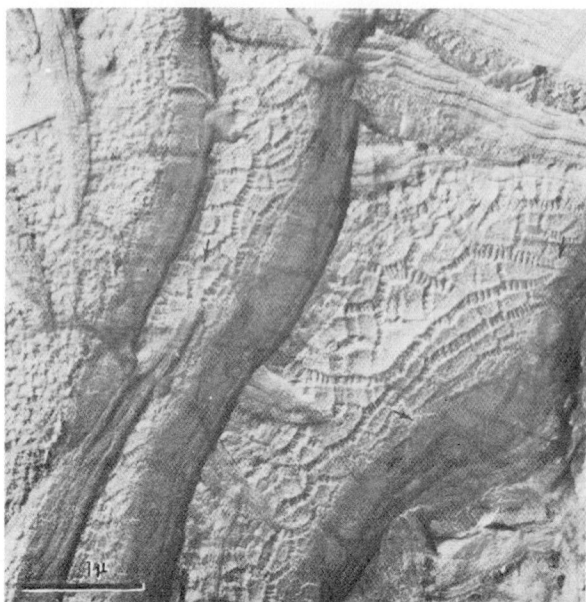

Fig. IV-67. Bands of polytetrafluoroethylene composed of several lamellae. Along the edge of the band, lamella edges are observed to bend, curving down the side of the band (arrows). In these regions, parallel striations are observed normal to and between the lamellae.

Fig. IV-68. Fracture surface (Cr replica) of a slowly cooled (0.2°C/min.) sample of dispersion-based polytetrafluoroethylene. (a) Low magnification. (b) Higher magnification of a region in which a band may have been fractured normal to its axis.

Fig. IV-69. Replica of the fractured surface of a sample of "granular" polytetrafluoro-ethylene heated to 500°C and slowly cooled (Bunn, Cobbold, and Palmer (106)).

Using somewhat higher melt temperatures (500°C instead of 380°C) which, it is indicated, caused some degradation of their sample, Bunn, Cobbold, and Palmer (106) were able to grow nearly symmetrical spherulites. Electron micrographs of fractured surfaces from these samples (Fig. IV-69) have a band structure. As the authors suggest, each band has the appearance of being the edge of a sheet, i.e., the bands consist of individual lamellae. The molecular weight should remain relatively high after the indicated heat treatment. Although no lamellae apparently were observed parallel to the exterior surface of these samples, the authors did observe easily recognizable lamellae, about 200 A thick, on the exterior surfaces of samples crystallized from very low molecular weight polymer.

Optical measurements (106) indicate that the molecules are normal to the long axis in the bands of the type shown in Figures IV-65 to IV-67. The magnitude of the birefringence, however, appears relatively low (74). Although originally it was suggested that the width of the bands corresponds to the molecular length (106), the average length of the molecules requires that they be folded (108). Considerably more work is needed, however, before the different types of striations and bands that are observed can be correlated with each other and with the position of the molecules.

The relationship of the perpendicular striations on the fractured bands, the longitudinal striations on the exterior bands, and a lamellar structure is not known. The most direct evidence that there is a relationship has

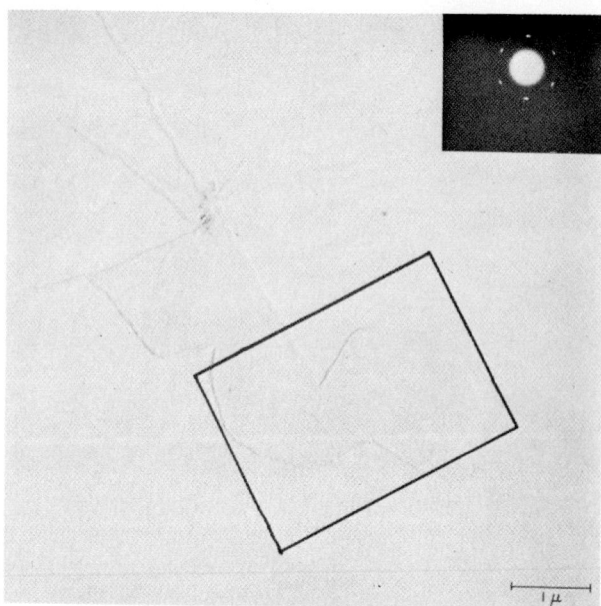

Fig. IV-70. Single crystal of polytetrafluoroethylene.   The sample was heated for 2 hours at 350°C and then cooled to below the melting point of 327°C in about 15 minutes. A crystal such as that shown here would contain the polymer from a cluster of about 100 dispersion particles.   The diffraction pattern from the 150 A thick lamella is shown in the insert.   (Symons (109).)

resulted from studies of the morphology of samples crystallized from the melt by Symons (109). He sprayed dilute dispersion particle suspensions onto glass slides, melted the polymer at 350°C and then slowly crystallized the resulting sample. From highly dispersed preparations he obtained individual single crystal lamellae (Fig. IV-70) and regions in which bands and crystals are intermingled (Fig. IV-71). Electron diffraction patterns indicated the polymer molecules are normal to the 150 A thick, hexagonal lamellae and therefore must be folded. Degradation should not have occurred under the conditions used in their preparation. In Figure IV-70 the lamellae appear to merge into the bands. The same type of structure appears in Figure IV-71 where the long ribbons appear to be extensions of the edges of the crystals. Many of the crystals pictured by Symons (109) have this thickened-edge effect. Symans suggests that the longitudinal striations on his bands are the result of edges of clusters of lamellae, in agreement with the suggestions based on observations of thicker samples.

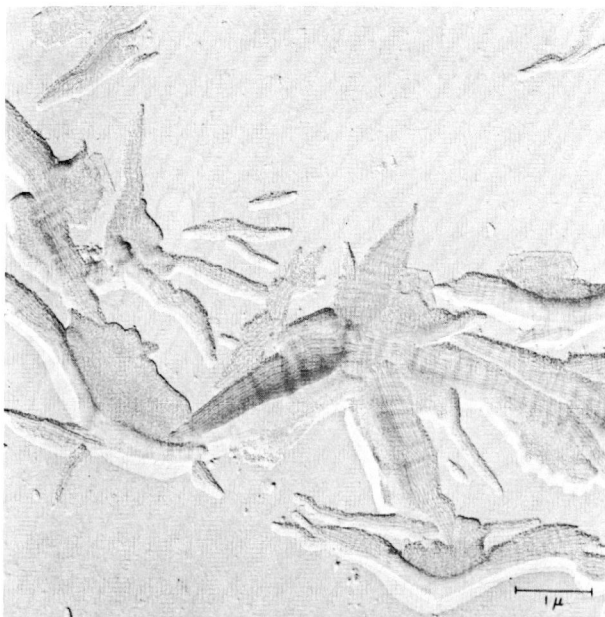

Fig. IV-71. Bands and lamellae of polytetrafluoroethylene prepared in the same manner as the sample in Figure IV-69. The individual bumps on the polymer are probably due to aggregation of the chromium shadowing material. (Symons (109).)

However, in the case of Figure IV-70 and possibly Figure IV-71 it appears as if the striations may result instead from some structure within the lamellae themselves and may even represent a different form of crystallization similar to the ribbons and lamellae observed in polyhexamethylene terephthalate (Figs. IV-58 to IV-60).

If more concentrated samples are held for a number of hours at temperatures just above the melting point, single crystal lamellae as thick as 900 A, as well as banded needles, are formed which are large enough to be observed in the optical microscope (109). Figure IV-72 shows an area in such a sample. The single crystals are isotropic in polarized light, as would be expected from the molecular orientation and hexagonal unit cell. The banded structures, as observed by Bunn, Cobbold, and Palmer (106), are anisotropic; the birefringence indicates that the molecules are normal to the long axis of the needles. However, even when extinguished, some bright striations are observed in the needles (Fig. IV-72c). Although at present there still remain a number of questions çoncerning the details of the

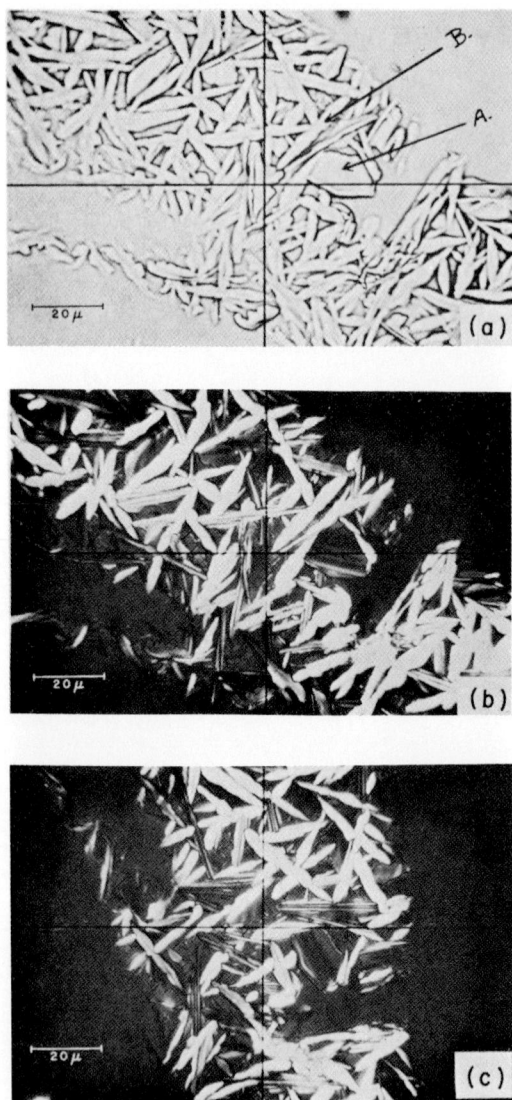

Fig. IV-72. Optical micrographs of a somewhat less dispersed sample of polytetra-fluoroethylene than in Figures IV-69 and IV-70. This sample was held at 350° for 2 hours and then at 334°C for 20 hours before being cooled. The crystal marked A in (a) is isotropic in polarized light, whereas the needles, for instance B, are birefringent. (a) Transmitted light. (b) Crossed polaroids. (c) Crossed polaroids but with sample rotated by 45°. (Symons (109).)

morphology of polytetrafluoroethylene, it is believed that a continuation of the studies of Symons should lead to considerable clarification of the situation. Similar observations would probably also be useful with other polymers.

o. Polyurethane

The only published micrograph of a polyurethane is of a sample crystallized from the glassy state by annealing at 150°C for 2 hours (melting point = 215°C) (76). The authors indicate that the material has a fibrillar structure. Small angle scattering indicates the presence of an 84 A spacing. As in the case of polyethylene terephthalate, polyurethane can be quenched to a glassy state with spherulites developing during subsequent annealing.

p. Synthetic Polypeptides

Robinson (110,111) reported that poly-γ-benzyl-L-glutamate can be crystallized in methylene chloride in the form of spherulites. A spiral structure, with spacings large enough to cause diffraction of visible light, can be observed in the three dimensional spherulites by natural and polarized light. Their morphology may be considerably different than the polymer spherulites discussed here. They probably can best be described as spherical aggregates of liquid crystals.

q. Proteins

Coleman, Allan, and Vallee (112) have shown that at low concentrations (<5 mg/ml) the enzyme, carboxypeptidase, precipitates from 1 M NaCl solutions during dialysis with buffered solutions of decreasing salt concentration, in the form of spherulites. The enzyme is insoluble in distilled water. At concentrations greater than 5 mg/ml it comes out of solution in the form of polyhedral crystals which are considerably larger than the needle like crystals making up the spherulites. It is not known if the crystal structure in the spherulites and the polyhedral crystals is the same, nor is the orientation of the molecules in the spherulite known. In the polyhedral crystal each molecule is folded and twisted into a globule, held together by intrachain hydrogen bands. Optical micrographs are presented by the authors in their paper (112).

r. Starch

Weigel (25) found that spherulites with a radiating, fibrous appearance were formed during the precipitation of starch from aqueous solutions of various alcohols. Optical micrographs, presented in this paper, resemble those of spherulites of synthetic polymers.

REFERENCES

1. C. W. Bunn and T. C. Alcock, "The Texture of Polythene," *Trans. Faraday Soc.,* **41**, 317 (1945).
2. H. Staudinger and R. Signer, "Über den Kristallbau hochmolekularer Verbindungen. 17. Mitteilung über hochmolekulare Verbindungen," *Z. Krist.,* **70**, 193 (1929).
3. G. Schurr, "Some Aspects of the Crystallization of High Polymers," Rubber-Stichting Communication #276, Delft (1955).
4. H. A. Stuart, *Die Physik der Hochpolymeren,* Vol. III, Springer, Berlin, Goettingen and Heidelberg (1955).
5. A. Keller, "Morphology of Crystalline Polymers, a Review," in *Growth and Perfection of Crystals,* edited by R. H. Doremus, B. J. Roberts, and D. Turnbull, Wiley, New York (1958).
6. M. Levy and Munier-Chalmas, "Memoire sur diverses formes affectees par le reseau elementaire du quartz," *Bull. soc. franc. mineral,* **15**, 159 (1892).
7. O. Lehman, *Molecularphysik,* Vol. 1, Wilhelm Engelmann, Leipzig (1888).
8. F. Wallerant, "Sur les enroulements helicoidaux dans les corps cristallises," *Bull. soc. franc. mineral,* **30**, 45 (1907).
9. P. Gaubert, "Sur le pseudopolychroisme des spheroletes," *Compt. rend.,* **149**, 456 (1909), and other by the same author.
10. Boris Popoff, "Spherolithenbau und Strahlungs kristallization," *Latv. Farm. Zurn.,* **1934**, 1 (1934). A description of the postulated growth mechanism is given in ref. 11 below.
11. H. W. Morse and J. D. H. Donnay, "Optics and Structure of Three-Dimensional Spherulites," *Am. Mineralogist,* **21**, 391 (1936). (See for other papers of series.)
12. A. V. Shubnikov, "On the Initial Form of Spherulites," *Sov. Phys., Cryst. (Eng. trans.),* **2**, 578 (1959).
13. H. W. Morse, C. H. Warren, and J. D. H. Donnay, "Artificial Spherulites and Related Aggregates," *Am. J. Sci.,* **23**, 421 (1932).
14. W. Jansen, "Röntgenographische Untersuchung über die Kristallorientierung in Spharolithen," *Z. Krist.,* **85**, 239 (1933).
15. See for instance, F. R. N. Nabarro and P. J. Jackson, "Growth of Crystal Whiskers" in *Growth and Perfection of Crystals,* edited by R. H. Doremus, R. W. Roberts, and D. Turnbull, Wiley, New York (1958).
16. F. Bernauer, "Gedrillte Kristalle" in *Forschungen zur Kristallkunde,* Heft. 2, Borntrager, Berlin (1929).
17. A. Keller, "The Spherulitic Structure of Crystalline Polymers Part I. Investigations with the Polarizing Microscope," *J. Polymer Sci.,* **17**, 291 (1955).
18. F. P. Price, "The Structure of High Polymer Spherulites," *J. Polymer Sci.,* **37**, 71 (1959).

19. A. Keller and J. R. S. Waring, "The Spherulitic Structure of Crystalline Polymers Part III. Geometrical Factors in Spherulitic Growth and the Fine Structure," *J. Polymer Sci.*, **17**, 447 (1955).
20. A. Keller, "Investigations on Banded Spherulites," *J. Polymer Sci.*, **39**, 151 (1959).
21. V. S. Yoffe, "About the Structure and Properties of Real Crystalline Substances," *Uspekhi Khimii*, **13**, 144 (1944) (in Russian).
22. A. Tschirch and O. Muller, "Über die Guttapercha von Deutsch Neu-Guinea," *Arch. d. Pharm.*, **243**, 114 (1905).
23. F. Kirchhof, "Über die Kristall-Struktur der Tjipetir-Gutta Percha," *Kautschuk*, **5**, 175 (1929).
24. W. H. Smith, C. P. Saylor, and H. J. Wino, "The Preparation and Crystallization of Pure Ether-Soluble Rubber Hydrocarbons: Compositions, Melting Point, and Optical Properties," *J. Research Natl. Bur. Standards*, **10**, 479 (1933).
25. E. Weigel, "Uber die kristallisierte Ausscheidung von Starke aus Losungsmittelgemischen," *Kolloid-Z.*, **102**, 145 (1943).
26. W. H. Smith and C. P. Saylor, "Optical and Dimensional Changes Which Accompany the Freezing and Melting of Hevea Rubber," *J. Research Natl. Bur. Standards*, **21**, 257 (1938), and *Rubber Chem. Technology*, **12**, 18 (1939).
27. H. D. Keith and F. J. Padden, Jr., "The Optical Behavior of Spherulites in Crystalline Polymers. Part II. The Growth and Structure of the Spherulites," *J. Polymer Sci.*, **39**, 123 (1959).
28. W. M. D. Bryant, "Polythene Fine Structure," *J. Polymer Sci.*, **2**, 547 (1947).
29. M. Herbst, "Röntgenographische Untersuchung an Spharolithen in Polyamid-Spritzgussmassen," *Z. Elektrochem.*, **54**, 318 (1950).
30. A. Keller, "The Spherulitic Structure of Crystalline Polymers Part II. The Problem of Molecular Orientation in Polymer Spherulites," *J. Polymer Sci.*, **17**, 351 (1955).
31. J. J. Point, "Structure fibreuse et phenomenes de cristallisation rayonnante dans les hauts polymeres," *Bull. Acad. roy. Belg.* (*Classe Sci.*), **41**, 974 (9155).
32. J. J. Point, "Enroulement helicoidal dans les spherolithes de polyethylene," *Bull. Acad. roy. Belg.* (*Classe Sci.*), **41**, 982 (1955).
33. J. J. Point, "Spherolithes de polyadipate de glycol, type α," *Bull. Acad. roy. Belg.* (*Classe Sci.*), **39**, 435 (1953).
34. J. J. Point, "Recherches sur l'etat solide de hauts polymeres spherolithiques," *Soc. des sci., des arts et des lettres du Hainaut, Mem. et publ.*, **71**, 65 (1958).
35. H. D. Keith and F. J. Padden, Jr., "The Optical Properties of Spherulites in Crystalline Polymers. Part I. Calculation of Theoretical Extinction Patterns in Spherulites with Twisting Crystalline Orientation," *J. Polymer Sci.*, **39**, 101 (1959).
36. F. P. Price, "On Extinction Patterns of Polymer Spherulites," *J. Polymer Sci.*, **39**, 139 (1959).
37. P. H. Geil, "Lamellar Crystallization of Low Density Polyethylene," *J. Polymer Sci.*, **51**, S10 (1961).
38. R. J. Clark, R. L. Miller, R. S. Stein, and P. R. Wilson, "The Scattering of Light from Polyethylene Samples Having Ringed Spherulites," *J. Polymer Sci.*, **42**, 275 (1960).
39. R. S. Stein and M. B. Rhodes, "Photographic Light Scattering by Polyethylene Films," *J. Appl. Phys.*, **31**, 1873 (1960).
40. R. Dauscher, E. W. Fischer, and H. A. Stuart, "Lichtzerstreung an kristallisiertin Hochpolymeren," *Z. Naturforsch.*, **15a**, 116 (1960).

41. H. Hendus, "Zur Deformation von Polyathylan," *Kolloid-Z.*, **165**, 32 (1959).
42. S. W. Hawkins and R. B. Richards, "Light Transmission and the Formation and Decay of Spherulites in Polythene," *J. Polymer Sci.*, **4**, 515 (1949).
43. F. P. Price, "The Development of Crystallinity in Polychlorotrifluoroethylene," *J. Am. Chem. Soc.*, **74**, 311 (1952).
44. W. M. D. Bryant, R. H. H. Pierce, Jr., C. R. Lindegren, and R. Roberts, "Nucleation and Growth of Crystallites in High Polymers. Formation of Spherulites," *J. Polymer Sci.*, **41**, 131 (1955).
45. M. Herbst, quoted by W. Brenschede, "Spharolithische Struktur synthetischer Hochpolymeren," *Kolloid-Z.*, **114**, 35 (1949).
46. B. B. Burnett and W. F. McDevitt, "Growth of Spherulites from Supercooled Polymer Melts," *J. Polymer Sci.*, **20**, 211 (1956).
47. G. Schuur, "Mechanism of the Crystallization of High Polymers," *J. Polymer Sci.*, **11**, 385 (1953).
48. F. J. Padden, Jr. and H. D. Keith, "Spherulitic Crystallization in Polypropylene," *J. Appl. Phys.*, **30**, 1479 (1959).
49. H. D. Keith, F. J. Padden, Jr., N. M. Walter, and H. W. Wyckoff, "Evidence for a Second Crystal Form of Polypropylene," *J. Appl. Phys.*, **30**, 1485 (1959).
50. P. H. Geil, "Morphology of an Acetal Resin," *J. Polymer Sci.*, **47**, 65 (1960).
51. H. D. Keith and F. J. Padden, Jr., to be published.
52. H. D. Keith, "The Crystallization of Long Chain Polymers" in *The Physics and Chemistry of the Organic Solid State*, edited by A. Weissberger, D. Fox, and M. M. Labes, Interscience-Wiley, New York (1963).
53. C. F. Hammer, T. A. Koch, and J. W. Whitney, "Fine Structure of Acetal Resins and Its Effect on Mechanical Properties," *J. Appl. Polymer Sci.*, **1**, 169 (1959).
54. V. Peck and W. Kaye, "Behavior of Crystallites in Polyethylene," paper presented at Elec. Micro. Soc. Am. meeting, abstract in *J. Appl. Phys.*, **25**, 1465 (1954).
55. G. C. Claver, Jr., R. Buchdahl, and R. L. Miller, "Spherulitic Fine Structure in Polyethylene," *J. Polymer Sci.*, **20**, 202 (1956).
56. E. W. Fischer, "Stufen-und spiralformiges Kristallwachstum bei Hochpolymeren," *Z. Naturforsch*, **12a**, 753 (1957).
57. Micrographs courtesy of E. W. Fischer (personal communication).
58. Y. Fujiwara, "The Superstructure of Melt Crystallized Polyethylene I. Screwlike Orientation of Unit Cell in Polyethylene Spherulites with Periodic Extinction Rings," *J. Appl. Polymer Sci.*, **4**, 10 (1960).
59. E. W. Fischer, "Thermodynamical Explanation of Large Periods in High Polymer Crystals and Drawn Fibers," *Ann. N. Y. Acad. Sci.*, **89**, 620 (1961).
60. F. R. Anderson, "Internal Morphology of Bulk Crystallized Polyethylene," paper presented at Inter. Cong. Elec. Micro., Philadelphia (1962).
61. L. Mandlekern, A. S. Posner, A. F. Diorio, and D. E. Roberts, "Low-Angle X-ray Diffraction of Crystalline Nonoriented Polyethylene and Its Relation to Crystallization Mechanisms," *J. Appl. Phys.*, **32**, 1509 (1961).
62. S. S. Pollack, W. H. Robinson, R. Chiang, and P. J. Flory, "X-ray Diffraction of Linear Polyethylene Crystallized at 131°C," *J. Appl. Phys.*, **33**, 237 (1962).
63. H. Hendus, "Neuere physikalische Untersuchungen an Hochpolymeren. IV. Roentgenographische und eletronenmikroskopische Strukturuntersuchungen," *Ergeb. exakt. Naturwiss.*, **31**, 331 (1959).
64. C. Sella, "Etude des polyethylenes par diffraction des rayons X aus petit angles," *Compt. rend.*, **248**, 1819 (1959).

65. H. Hendus, personal communication.
66. P. H. Geil, "Application of Small Angle X-ray Scattering to the Determination of the Structure of Macromolecules," Thesis, Univ. of Wis., 1956.
67. J. D. Hoffman and J. I. Lauritzen, Jr., "Crystallization of Bulk Polymers with Chain Folding: Theory of Growth of Lamellar Spherulites," *J. Research Natl. Bur Standards*.
68. A similar suggestion has been made by E. Orowan—private communication. He suggests internal strains, due to defects, may also contribute to the irregular twisting observed in some polymer lamellae.
69. A. Schram, "Zur Struktur des Polyathylens," *Kolloid-Z.*, **150–151**, 18 (1957).
70. E. Jenckel, E. Teege, and W. H. Inrichs, "Transkristallsation in hochmolekularen Stoffen," *Kolloid-Z.*, **129**, 19 (1952).
71. H. D. Keith and F. J. Padden, Jr., "Ringed Spherulites in Polyethylene," *J. Polymer Sci.*, **31**, 415 (1958).
72. D. H. Reneker, unpublished data.
73. H. A. Stuart, "Kristallisationsbedingungen und morphologische Strukturen bei Hochpolymeren," *Kolloid-Z.*, **165**, 3 (1959).
74. P. H. Geil, unpublished data.
75. G. Kampf, "Zur Ausbildung morphologischer Strukturen am Polykohlensaureester des 4,4,-Dioxydiphenyl-2,2-Propane (Polycarbonat)," *Kolloid-Z.*, **172**, 50 (1960).
76. R. Eppe, E. W. Fischer, and H. A. Stuart, "Morphologische Strukturen in Polyathylen, Polyamiden und andern kristallsierenden Hochpolymeren," *J. Polymer Sci.*, **34**, 721 (1959).
77. R. L. Miller, "On the Existence of Near Range Order in Isotactic Polypropylenes," *Polymer*, **1**, 135 (1960).
78. F. J. Padden, Jr., and H. D. Keith, "Microstructure and Growth Mechanisms in Spherulitic Crystallization," paper presented at Am. Phys. Soc. meeting, Monterey, Calif., March (1961).
79. H. D. Keith and F. J. Padden, Jr., "Deformation Mechanisms in Crystalline Polymers," *J. Polymer Sci.*, **41**, 525 (1959).
80. W. A. Tiller, "Alloy Crystal Growth," in *Growth and Perfection of Crystals*, edited by R. H. Doremus, B. W. Roberts, and D. Turnbull, Wiley, New York (1958).
81. V. F. Holland, "Crystalline Morphology of Polyacrylonitrile," *J. Polymer Sci.*, **43**, 572 (1960).
82. P. Kassenbeck, "Neue praparatus Methoden der Elektronenmikroskopie und ihre Ergebnisse auf dem Gibiet der Faserforschung," *Melliand Textilberichte*, **39**, 55 (1958).
83. H. A. Stuart and B. Kahl, "Beobachtungen über die morphologische Struktur in Hochpolymeren," *J. Polymer Sci.*, **18**, 143 (1955).
84. H. A. Stuart, U. Veiel, and M. Hartman-Fahnenbrock, "Über die morphologische Struktur bei festen Korpern mit Fadenmolekular," *Naturwissenschaften*, **40**, 339 (1953).
85. R. J. Barriault and L. F. Gronholz, "Formation of Spherulitic Structure in Polyhexamethylene Adipamide (Nylon 66). I. Structure and Optical Properties of Spherulites at Room Temperature," *J. Polymer Sci.*, **18**, 393 (1955).
86. E. H. Boasson and J. M. Woestenenk, "Some Aspects of the Crystallization of Nylon 66 (Polyhexamethylene Adipamide)," *J. Polymer Sci.*, **24**, 57 (1957).
87. F. Khoury, "The Fibrillar Structure of Spherulites in Polyhexamethylene Adipamide," *J. Polymer Sci.*, **26**, 114 (1957).

88. F. Khoury, "The Formation of Negatively Birefringent Spherulites in Polyhexamethylene Adipamide (Nylon 66)," *J. Polymer Sci.*, **33**, 389 (1958).
89. F. D. Hartley, F. W. Lord, and L. B. Morgan, "Crystallization phenomena in polymers, Part IV, the course of the primary crystallizations in polyhexamethylene adipamide," *Intern. Symp. Macromol. Chem., Ricerci sci. Suppl.*, **1**, 577 (1954).
90. C. M. Langkammerer and W. E. Catlin, "Spherulite Formation in Polyhexamethylene Adipamide," *J. Polymer Sci.*, **3**, 305 (1949).
91. P. H. Geil, "Nylon Single Crystals," *J. Polymer Sci.*, **44**, 449 (1960).
92. M. J. Schick, unpublished data.
93. A. Keller, "Electron Microscope-Electron Diffraction Investigations of the Crystalline Texture of Polyamides," *J. Polymer Sci.*, **36**, 361 (1959).
94. R. G. Scott, "Structure of Spherulites as Revealed by Selected Area Electron Diffraction and Electron Microscopy," *J. Appl. Phys.*, **28**, 1089 (1957).
95. J. H. Magill and P. H. Harris, "Single Crystals from Polyamide Melts," *Polymer*, **3**, 252 (1962).
96. J. D. Hoffman and J. J. Weeks, "Melting Process and the Equilibrium Melting Temperature of Polychlorotrifluoroethylene," *J. Research Natl. Bur. Standards*, **66A**, 13 (1962).
97. J. D. Hoffman, personal communication.
98. K. Masuzawa, "Studies on the Spherulite of Polyethylene Adipate, 1. On the Process of Growth of the Spherulites and Occurence of Cracking in Them," *Chem. High Polymers*, **14**, 1 (1957) (in Japanese).
99. M. Takayanagi and T. Yamashita, "Growth Rate and Structure of Spherulite in Fractionated Poly(ethylene adipate)," *J. Polymer Sci.*, **22**, 552 (1956).
100. A. Keller, G. R. Lester, and L. B. Morgan, "Crystallization Phenomena in Polymers. I. Preliminary Investigation of the Crystallization Characteristics of Polyethylene Terephthalate," *Phil. Roy. Soc. Trans. (London)*, **247**, 1 (1954).
101. L. B. Morgan, "Crystallization Phenomena in Polymers. II. The Course of the Crystallization," *Phil. Trans. Roy. Soc. (London)*, **247**, 13 (1954).
102. F. D. Hartley, F. W. Lord, and L. B. Morgan, "Crystallization Phenomena in Polymers. III. Effect of Melt Conditions and the Temperature on the Course of the Crystallization in Polyethylene Terephthalate," *Phil. Trans. Roy. Soc. (London)*, **247**, 23 (1954).
103. A. S. Kenyon, R. C. Gross, and A. L. Wurstner, "Kinetics of Spherulite and Crystallite Growth in Isotactic Polystyrene," *J. Polymer Sci.*, **40**, 159 (1959).
104. G. Natta, "Polymeres isotactiques et autres polymeres stereoisomeres," *Chimie Industrie*, **77**, 1009 (1957).
105. F. Danusso and F. Sabbioni, "Strutture sferulitiche e poliedritiche nel polisterolo isotattico," *Rend. Inst. Lomb. Sci. e Letteres*, **A92**, 435 (1958).
106. C. W. Bunn, A. J. Cobbold, and R. P. Palmer, "The Fine Structure of Polytetrafluoroethylene," *J. Polymer Sci.*, **28**, 365 (1958).
107. N. K. J. Symons, "Solution Grown Crystals of Polytetrafluoroethylene," *J. Polymer Sci.*, **51**, S21 (1961).
108. A. Keller, "The Morphology of Crystalline Polymers," *Macromol. Chem.*, **34**, 1 (1959).
109. N. K. J. Symons, "The Growth of Single Crystals of Polytetrafluoroethylene from the Melt," *J. Polymer Sci.*, in press.

110. C. Robinson, "Liquid-Crystalline Structures in Solutions of a Polypeptide," *Trans. Faraday Soc.*, **52**, 571 (1956).

111. C. Robinson, "Spherulites and Paracrystalline Structures in Solutions of Synthetic Polypeptides," abstract of paper presented at Faraday Society Discussion, July 15, 1954, Maidenhead, England, in *Disc. Faraday Soc.*, **50**, 1011 (1954).

112. J. E. Coleman, B. J. Allan, and B. L. Vallee, "Protein Spherulites," *Science*, **131**, 350 (1960).

113. J. H. Magill and P. H. Harris, personal communication.

114. B. J. Spit, "Nylon 6 Staining with Phosphotungstic Acid," paper presented at Inter. Cong. Elec. Micro. meeting, Philadelphia, Pa., Sept. 1962.

115. F. R. Anderson, "Morphology of Isothermally Bulk-Crystallized Polyethylene," *J. Appl. Phys.*, in press.

V. ANNEALING

We consider, in this chapter, the effect of annealing polymers at temperatures below their equilibrium melting points. In previous chapters it has been shown that the thickness of a lamella, whether crystallized from the melt or solution, depends on the crystallization temperature. The lamellae are thicker when crystallized at higher temperatures. In this chapter it is shown not only that annealing at temperatures between the crystallization temperature and the equilibrium melting point can result in an increase in thickness of the lamellae but that the annealing of samples originally quenched to the glassy state, at temperatures above the glass transition temperature, can lead to crystallization in the form of lamellae. In the following sections we consider the effect of annealing on the morphology of single crystals, hedrites, and samples crystallized and quenched from the melt and also the effect of annealing on melting-point determinations. Unfortunately, as will become obvious, there is still a great deal of uncertainty concerning the details of the effect of annealing.

1. ANNEALING OF POLYMER SINGLE CRYSTALS

Shortly after recognition of the relationship between the fold period and the long period observed with x-rays in polyethylene single crystal mats, Keller and O'Conner (1) showed that the usual spacing of 120 A nearly disappears and is replaced by a new spacing of 200–300 A if the mats are annealed near the melting point. They further indicated that the molecular axes rotate about the **b** axis of the crystals (see later discussion, p. 329). Subsequently, Randy, Morehead, and Walter (2) also showed that the long period spacing of a mat of polyethylene crystals, composed of crystals grown at 60°C, increases following annealing of the mat in glycerine or air at 110°C for 4 hours.

Electron microscopic studies of the effect of annealing followed soon after the small angle x-ray studies. Frank, Keller, and O'Connor (3) showed that when individual crystals of poly-4-methyl-pentene-1 on a glass slide are heated almost to the equilibrium melting point, holes develop within the lamellae (see Fig. V-19). The polymer tends to concentrate on the edges and diagonals of the square crystals. Electron and small angle x-ray diffraction studies were not reported for these crystals. Subsequent results from the

Fig. V-1. Truncated single crystal of polyethylene annealed near 130°C. Much of the material in the {100} fold domain has recrystallized with a larger fold period. (Keller and Bassett (5).)

same laboratory (4) indicated that the {100} fold domains in truncated polyethylene crystals become "distinct" when individual crystals are heated to 130°C, 7°C below the melting point. The {100} fold domains melt or are transformed at a lower temperature than {110} fold domains (Fig.V-1). This is in agreement with the suggestion that {100} fold domains have a different type of fold and, therefore, a different thermal stability than {110} fold domains.

Considerably more extensive results on the effect of annealing polyethylene crystals were reported by Statton and Geil (6). Using small angle x-ray scattering and electron microscopy they showed that during annealing the molecules refold, the lamellae thickening and developing holes. These results and subsequent work are described below. The effect of annealing crystals in suspension was discussed in Chapter II.

The crystals used by Statton and Geil were obtained by ambient cooling of 0.1% tetrachloroethylene solutions of linear polyethylene (type A, Section II-4). Mats were prepared by evaporation of the solvent, the crystals forming a dense, layered aggregate. The three orders of small angle x-ray diffraction obtained from the mats indicate that the lamellae, as sedimented, are 104 A thick. Following annealing in an air oven for various times and at various temperatures the samples were cooled to room temperature and the small angle scattering again determined photographically. If the annealing

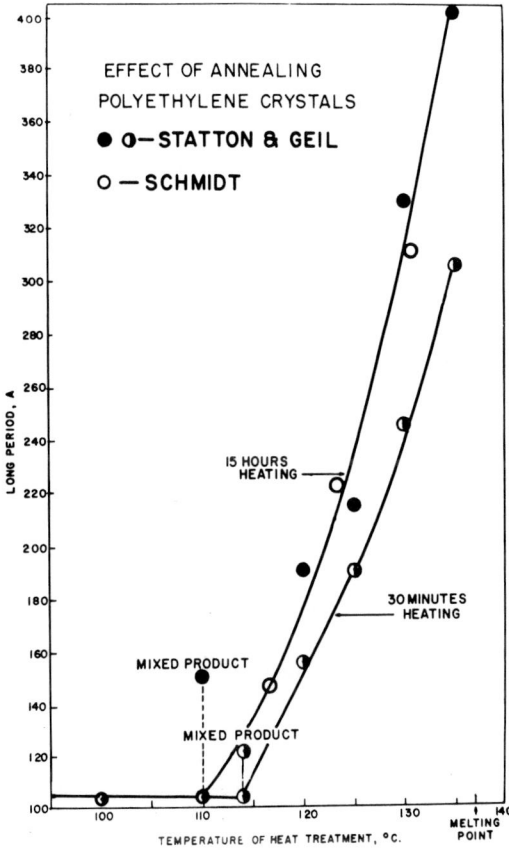

Fig. V-2. Dependence of the long period of mats of polyethylene crystals on the annealing temperature. The data of Schmidt (7) were obtained at the annealing temperature, whereas that of Statton and Geil (6) were obtained after the samples had been cooled to room temperature. Schmidt's time of annealing is believed to be on the order of one day.

is at temperatures above 110°C and is continued for a sufficient length of time, two orders of diffraction are found in the annealed samples corresponding to a larger long period (Fig. V-2). The rate of cooling of the crystals after annealing appeared to have no effect. Subsequent work by Schmidt (7), in which the small angle patterns were obtained at the annealing temperature, is in agreement with the above results (his data are included in Figure V-2).

Although the value of 110°C as the transition temperature above which recrystallization takes place may merely be related to the fold period of the

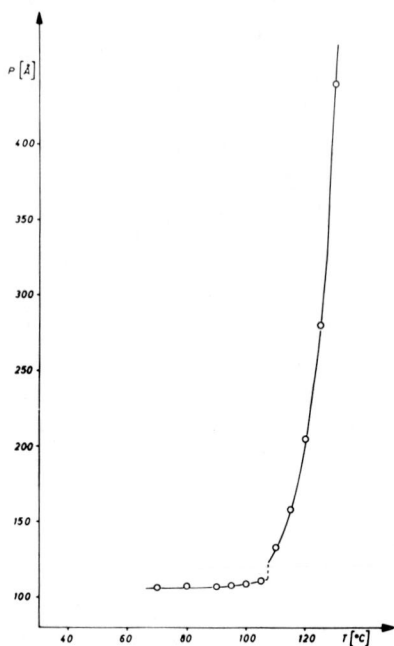

Fig. V-3. Long period from polyethylene single crystal mats annealed for 24 hours.   The crystals were grown in tetrachloroethylene at 70°C.   (Fischer and Schmidt (8).)

original crystals (i.e., it would be lower if thinner crystals were used), it may also be related to the critical temperature predicted by the theory of Peterlin *et al.* (Chapter VI, Section 3) above which the crystals have no equilibrium length. Fischer and Schmidt (8), using single crystal mats, find a very small (Fig. V-3) increase in fold period with a 24-hour anneal at temperatures below 110°C whereas at higher temperatures the period rapidly increases with increasing temperature. In the same paper, however, they found significant increases in the long period from quenched bulk polyethylene films at temperatures as low as 100°C.

The time of annealing in air, as well as the temperature, is seen to affect the results (6). The long period after annealing for 15 hours is greater than that after annealing for 30 minutes. Similar increases in thickness with time have been found by Hirai *et al.* (9). Although it was suggested that a stable value is reached after 15 hours, no further change occurring with times up to 65 hours (6), Hirai *et al.* (9) and Fischer and Schmidt (8) indicate that the long period from single crystal mats, after an initial abrupt increase, continues to increase linearly with the logarithm of the time of annealing

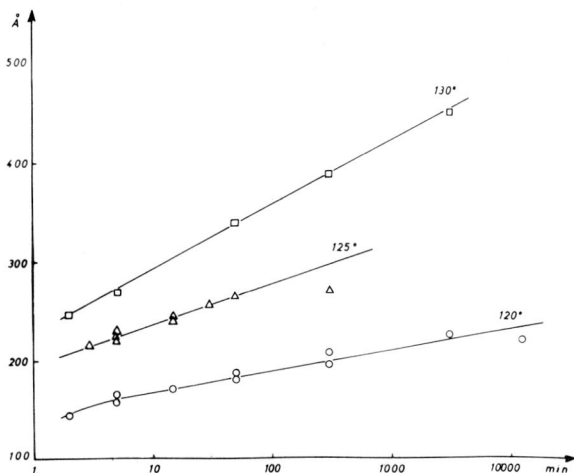

Fig. V-4. Dependence of the long period from polyethylene single crystal mats on annealing time. The mats were quenched in ice water following the annealing treatment. The crystals were grown from a dilute xylene solution at 80°C and had an initial long period of 118A. (Fischer and Schmidt (8).)

(Fig. V-4). However, whereas Fischer and Schmidt (8) find that the rate of increase in thickness with the logarithm of the time increases rapidly with increasing temperature, Hirai *et al.* (9) find that it is almost constant. Reneker has shown (10) that considerable recrystallization can take place in times of the order of seconds if a crystal is rapidly heated (see Figure V-16). Statton has found (11) that annealing in a liquid greatly speeds up the refolding process (Fig. V-5). Results equivalent to those obtained after 15 hours of annealing in air are found after less than 1 minute in baths such as ethylene glycol and Wood's metal. The cause of the difference in rate is at present unknown. Heat conduction and transfer probably play a part. The crystal mats, however, should be up to temperature within at least 30 minutes in the air oven. The liquid baths, at least in the case of Wood's metal, have no solvent or swelling action which might loosen the lattice. One factor that may be important is the amount of retained solvent in the mats. This factor, which was not measured and may have differed from sample to sample in all of the above work, will be discussed later.

Hirai, Mitsuhata, and Yamashita (9) have noted that when a crystal mat is heated in xylene at 90° for 35 minutes a long period of 270 A is seen as soon as the sample is removed from the xylene. However, as the sample stands at room temperature, this long period is replaced by another, at

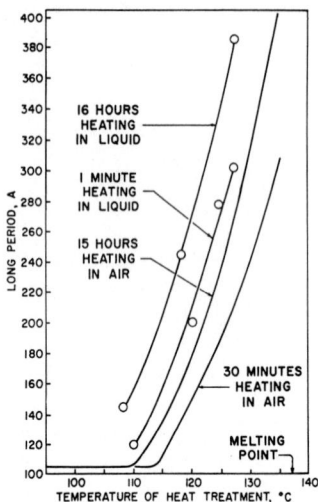

Fig. V-5. Comparison of long periods obtained from polyethylene crystal mats when annealed in air and in liquids for various periods of time. The curves for annealing in air are from Figure V-2. (Statton (11).)

140 A, within about 30 minutes. Apparently no sign of the original long period remains. The fact that this does not occur if the sample is heated in air suggests that the larger period is somehow associated with solvent between the lamellae. These authors also find that an increase in the long period occurs at temperatures above 110° when the samples are heated in air. When heated in xylene, as would be expected since the melting point (solubility temperature) is lower, the increase in long period can occur at a lower temperature. They indicate that the results (change of long period and rate of change) obtained during annealing at 80°C in xylene are similar to those obtained at 120°C in air. Whether or not solution and reprecipitation took place in these latter experiments is not known.

Statton and Geil reported (6) that when a mat is annealed in air or vacuum at the lower annealing temperatures, a so-called "mixed product" is obtained; two distinct sets of long period spacings are obtained (Fig. V-2). The smaller period corresponds to the initial lamella thickness. The results suggest that a certain degree of molecular motion is required before refolding can occur. In the linear polyethylene crystals used some parts of the lattice apparently loosen sufficiently at 110°C to permit this motion. The lamellar thickness corresponding to crystallization at this temperature may be considerably greater than that of the original crystals so that a large

increase in fold period occurs in the crystals or portions thereof which recrystallize. The fact that, at higher temperatures, at least, the fold period
increases with the logarithm of the time suggests that the refolding occurs
through a number of small increases in thickness resulting in a more or less
uniform increase in lamellae thickness throughout the sample, a fact of
significance for theoretical treatments of the problem.

TABLE V-1
Characteristics of the Polyethylene Crystalline Aggregate (6)

Parameter	Before heating	After heating at 125° C
Thickness of cake (through direction), mm[a]	0.751	0.772 (3% expansion)
Width of cake (edge direction), mm[a]	1.197	1.209 (1% expansion)
Density, g/cc	0.979	0.978
Melting point, °C (Kofler hot stage)	137	No change
Crystallite width, A[b]	230	290
Crystallite length, A[b]	120	300
Direction of x-ray beam to produce long-period diffraction	Parallel to platelets	No change
Crystallite orientation in aggregate	Chain axis well aligned perpendicular to surface of platelets	No change

[a] Microscope measurements with calibrated eyepiece.

[b] Standard line-broadening measurements from appropriate diffractions in the wide-
angle x-ray pattern; calculated by use of the Warren approximation of the Scherrer
formula, ±10%. Crystallite width is probably more correctly interpreted as the extent of
crystallite perfection in a mosaic, since the individual single crystal lamellae are several
orders of magnitude larger.

Several physical characteristics of the mats used by Statton and Geil,
i.e., dimensions, density, melting point, and molecular orientation, did not
change during the annealing (Table V-1). The lack of a difference in the
initial and final density would suggest that the mats in this experiment were
solvent free, or else that there was a compensatory change in density and
solvent concentration. A major change occurred in the crystallite length,
as measured by x-ray wide angle peak broadening, corresponding to the
change in x-ray small angle long period. The lateral perfection of the crystals
appeared to increase moderately.

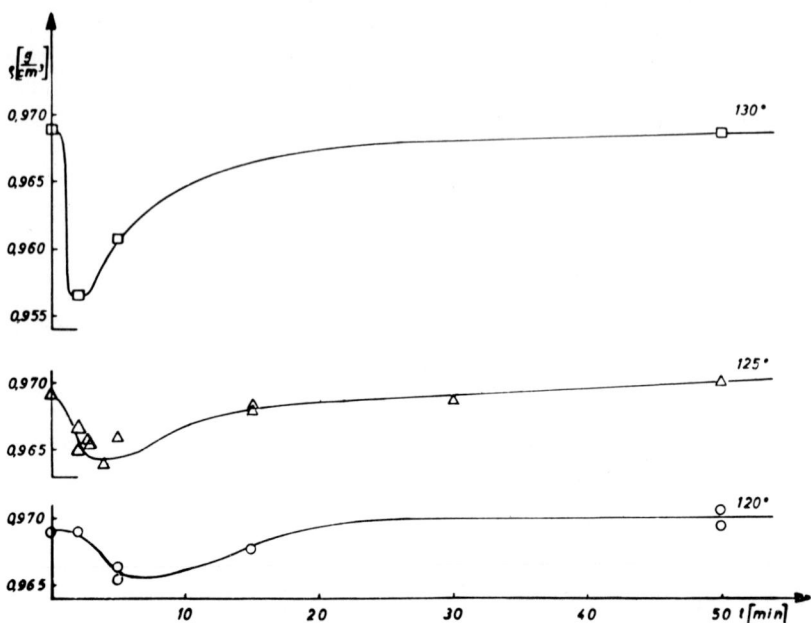

Fig. V-6. Dependence of the density of polyethylene single crystal mats on the time and temperature of annealing. The samples were quenched in ice water at the times shown, and the density measured at 30°C. (Fischer and Schmidt (8).)

Fischer and Schmidt (8) have shown, by optical birefringence, density and x-ray diffraction techniques, that the refolding process during annealing is accompanied by an initial rapid loss of order and, depending on the temperature, relatively slow recovery of order. The density results shown in Figure V-6 were obtained after quenching the specimen in ice water following its anneal at the temperature and for the time given. The decrease in density is seen to be considerably larger at the highest annealing temperature. Similar results have also been published for bulk polyethylene terephthalate and polyethylene crystallized from the melt (Section V-3). In addition to the density changes, if a crystal mat is viewed from on edge (position of maximum birefringence) the birefringence is seen to rapidly decrease and then increase again as recrystallization takes place (Fig. V-7). The rate of recovery of the birefringence appears to decrease rapidly with increasing annealing temperature. Using wide angle x-ray diffraction at the annealing temperature they found (8) that the crystalline reflections disappear almost completely within a few minutes when a mat is heated to 132°C (Fig. V-8). With continued annealing at this temperature the crystalline peaks re-

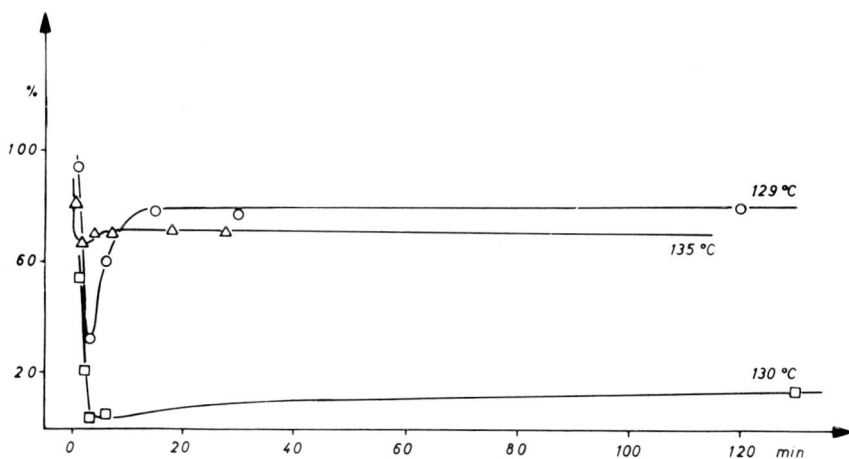

Fig. V-7. Percentage change in birefringence during annealing of polyethylene single crystal mats (at 129°C and 130°C) and drawn polyethylene (at 135°C) as a function of time (Fischer and Schmidt (8).)

appear and gradually increase in magnitude. Similar results were also found when a mat previously annealed was then reannealed at higher temperatures (Fig. V-9). The weak, broad {110} reflection remaining in both cases corresponds to crystals with lateral dimensions of 40 A or less, suggesting that a near "liquid" structure develops. The long period spacings, as shown by samples quenched at various times (Fig. V-4), increase during the annealing. The low intensity of the small angle diffraction did not permit measurement of the pattern as a function of the time of annealing at the annealing temperature. These results of Fischer and Schmidt (8) all suggest a major disordering or melting of the crystals when these temperatures are raised. However, the fold period must be increasing with time (as seen in the quenched specimens) even though the sample is still considerably disordered. These results are discussed further at the end of this section.

Bassett reports (12) that if a rapid rate of heating is used (i.e., ambient to 130°C in 1 to 2 minutes) the wide angle pattern is present at the shortest time for which he can obtain a pattern (5 minutes). However, if the temperature is increased slowly or the sample is heated at 125°C, for instance, for 30 minutes or more and the temperature then increased to 130°C he obtains results similar to those above of Fischer and Schmidt.

Ranby and Brumberger (13) reported that the long period can be decreased by annealing at temperatures below the initial crystallization temperature, as well as increased by annealing at higher temperatures. We believe these results were affected by solution and reprecipitation of the crystals during annealing. The crystals were annealed in solvents at temperatures equal to or above those at which the polymer will crystallize from the solvent (2). Reannealing of crystal mats in air for 18 hours at tempera-

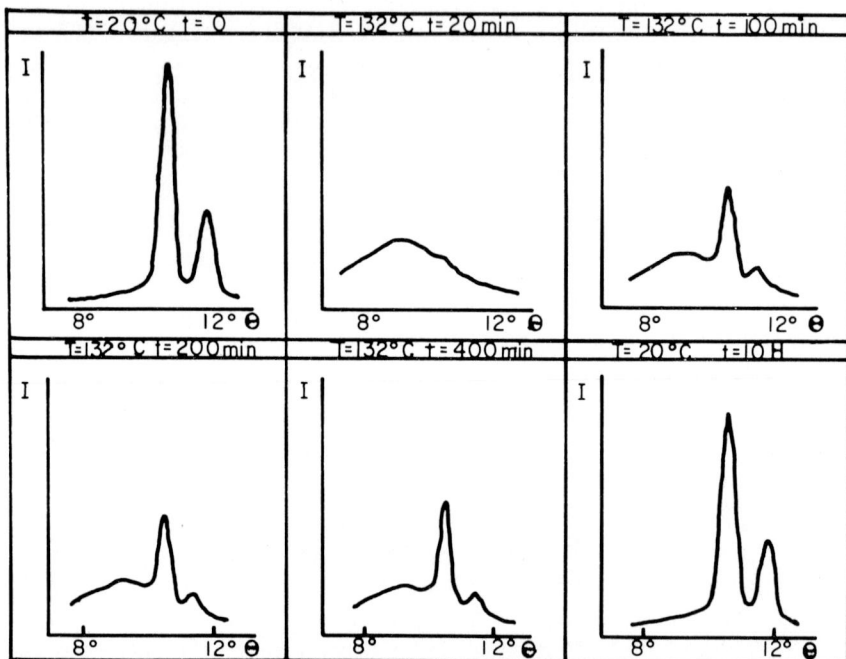

Fig. V-8. Wide angle diffraction patterns from a polyethylene crystal mat while being annealed, as a function of time (Fischer (16); see also ref. 8).

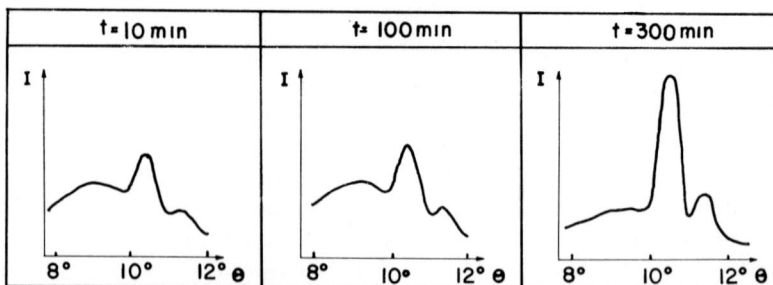

Fig. V-9. Wide angle x-ray diffraction patterns from a polyethylene crystal mat which had previously been annealed at 125°C, as a function of time at 132°C (Fischer (16)).

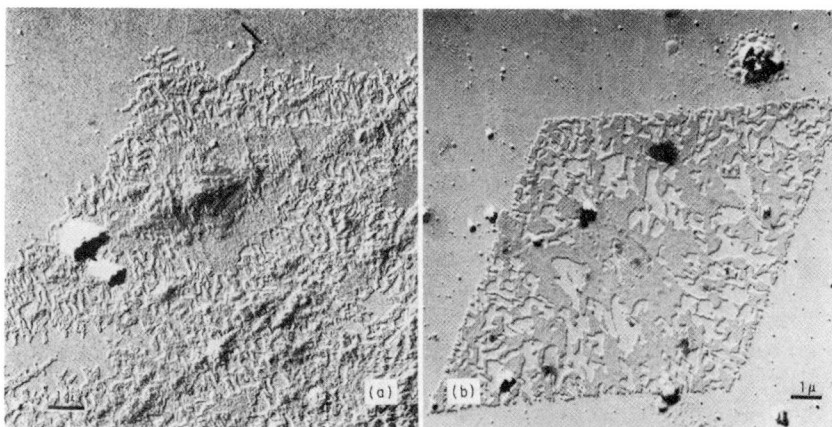

Fig. V-10. (a) Polyethylene single crystal precipitated from tetrachloroethylene and annealed at 125°C for 30 minutes (Statton and Geil (6)). (b) Polyethylene single crystal precipitated from p-xylene at 85°C and annealed at 125.7°C for 35 minutes.

tures below that of the initial anneal, but above 110°C, did not result in any change in the long period (6).

Balta, Bassett, and Keller (14) have recently indicated that if polyethylene is crystallized from xylene at 90°C (type D crystals of Section II-4) and then annealed at 110° to 120°C for 24 hours the long period decreases from 150 to 140 A and also sharpens. This, they suggest, may be due to a rearrangement of the folds, similar to that which Bassett, Frank, and Keller (15) suggest (see Chapter VI, Section 2) takes place during the original crystallization. They suggest that the fold period in the original crystals as the molecule is laid down on the growth face is variable. Isothermal or higher temperature annealing could then result in the fold period becoming more uniform with the lamellae then packing together more closely. While the average thickness of the lamellae might stay the same or increase during this process, the separation and therefore the long period could decrease. It may also result from a related type of motion resulting in restaggering of the folds and fold planes to form a new pyramidal modification such that the fold period remains constant (or may even increase somewhat) but the tilt of the molecular axes with respect to the fold surface increases and therefore the thickness normal to the fold surface decreases. Wide angle x-ray patterns indicate an increase in molecular tilt.

Fischer and Schmidt showed (8) that if one plots the density of the mats versus the reciprocal of the long period, the values for the annealing tem-

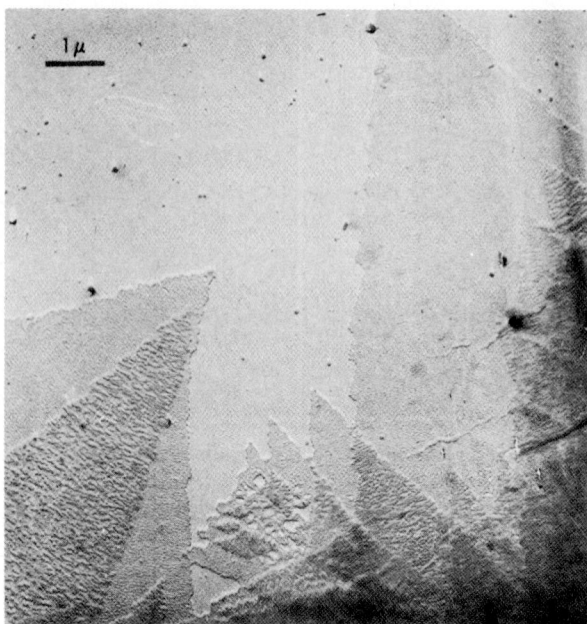

Fig. V-11. Polyethylene crystals from tetrachloroethylene solution annealed at 125.7°C for 35 minutes.   A small ridge of recrystallized material surrounds each hole in regions which are only one lamella thick.   These crystals were precipitated more slowly than that in Figure V-10a.   They are from the same preparation as those in Figure II-2.

peratures of 120°, 125°, and 130°C extrapolate at infinite fold period to the calculated (from unit cell constants) perfect crystal density. However, the different fold periods at a given annealing temperature were obtained by varying the annealing time and then quenching the sample (Fig. V-4). As suggested by the disorder present in the sample at the time it is quenched (Figs. V-6, V-7, and V-8) the fold period may not directly represent the annealing temperature. Fischer and Schmidt (8) interpret their data in terms of "amorphous" surface layers on the lamellae. This author has reservations concerning this interpretation and prefers an alternative possibility (16) involving a 3 to 4% void content within the lamellae.

Observations by Statton and Geil (6) of the annealed crystals in the electron microscope show that the refolding process, resulting in a larger fold period and long period, occurs within the single crystal lamella. The crystal shown in Figure V-10a was from the same type of preparation as those

Fig. V-12. Polyethylene crystal in which the edges have thickened during annealing. {100} Sectors and the uppermost layer at the top of the micrograph have also been transformed. The time and temperature of annealing was not given. (Keller and Bassett (5).)

used for the crystal mats in their study. It was annealed on a carbon-coated glass slide in an air oven and subsequently shadowed. The crystal in Figure V-10b, precipitated from p-xylene, was annealed under nitrogen. A number of holes of approximately equal size developed in regions in which the crystal was originally one lamella thick (no spiral overgrowths). The increase in thickness of the lamellae, as measured in the electron microscope, agrees with the increase in long period.

The appearance of the crystal in regions originally several lamellae thick differs from one sample to another. In some preparations the resulting structure is obscure (Fig. V-10a). The edges of the spiral growths tend to disappear and a coherent, nonlamellar appearing mound remains. In other samples, apparently those in which the recrystallization is not as extensive (Fig. V-11), it is found that the size of the holes depends on the number of lamellae present, larger holes being found in regions two or more thick. Since the bottoms of the holes could not be seen, it is not known if the larger holes extend through two or more lamellae. On the basis of this and other micrographs, and the lack of dimensional change of the mats, it is believed that in those regions more than one lamella thick and in the crystal mats,

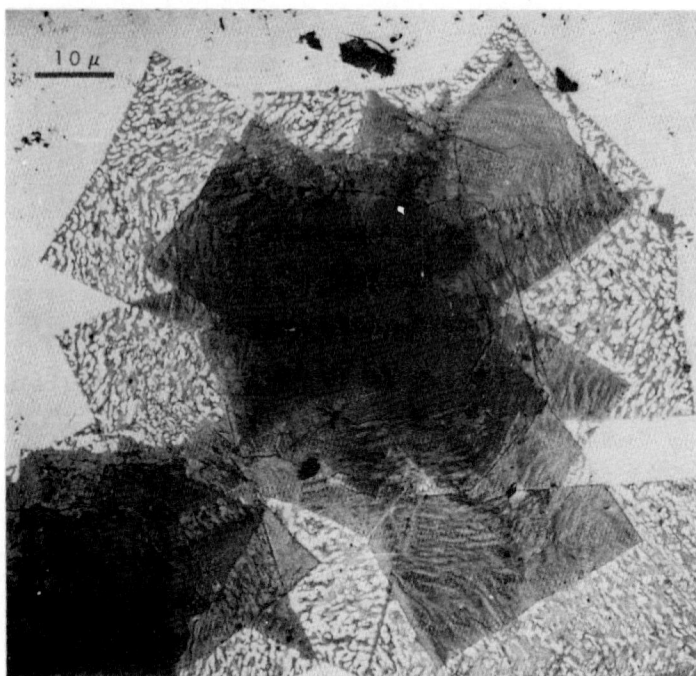

Fig. V-13. Polyethylene crystal heated on a glass slide for a few minutes. Only the lamellae in contact with the glass have recrystallized, even in those portions which have overlying, untransformed lamellae. The holes are more or less parallel to the *b* axis in this crystal. The lathlike growth is due to twinning. (Bassett (12).)

the holes in one lamella tend to be filled in by thickening of adjacent lamellae. Keller and Bassett (5) suggest that if more than one lamella is present, as in spiral growths or when one crystal is on top of another, striations more or less parallel to the **b** axis develop in the uppermost lamellae at a lower temperature than the ones beneath. In Figure V-12, large transformed regions can be seen in the second lamella of the large crystal as well as numerous striations in the lamellae in the spiral growth. However, the (100) fold domain in the crystal at the bottom of the figure has been completely transformed even though it is in contact with the substrate. Subsequently, Bassett has indicated (12) that recrystallization may also occur first in the lamella in contact with a glass substrate (Fig. V-13).

In some cases a craterlike structure develops at each site of recrystallization in the crystal (Fig. V-11). Surrounding each hole is a circular ridge. Apparently only material in the immediate vicinity of each hole has re-

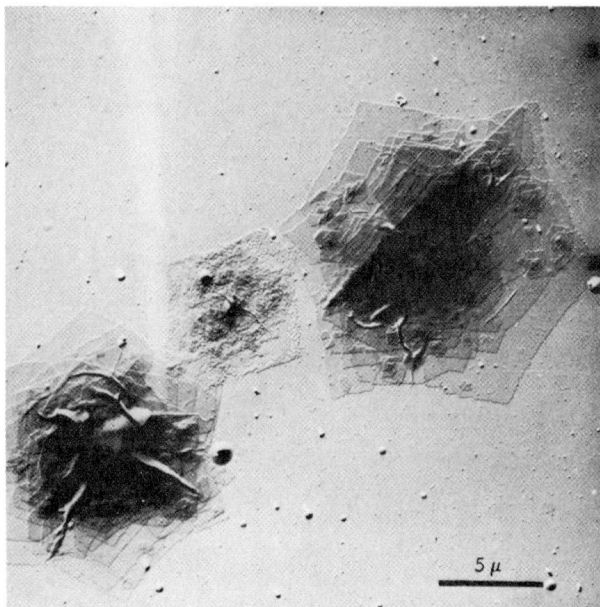

Fig. V-14. Polyethylene crystals from an octane solution, annealed at 120°C for 12 hours. Only the smaller crystals have recrystallized (Cr shadowed at $\tan^{-1} 4/7$.) (Statton and Geil (6).)

crystallized with a longer fold period. It may be that each hole is associated with a defect in the original crystal, the lamella being less stable in the vicinity of the defect. Defects within the lattice disrupt the lattice forces in their vicinity and thus reduce the energy required for molecular motion. This type of recrystallization is probably related to the "mixed product" type of small angle patterns obtained after low temperature anneals.

Another possible cause of the two sets of spacings is related to the fact that the crystals are often grown by ambient cooling of a solution (6). The crystals may be nucleated and start growing at different times and, therefore, temperatures. Furthermore, the edges of crystals grown by ambient cooling of solutions are thinner and, presumably, less perfect than the central portions. Two features have been observed during the annealing of such crystals, either of which could lead to the "mixed product" type of small angle x-ray diffraction pattern.

1. The edges would be expected to undergo recrystallization at a lower temperature than the central portion. Because the new fold period is considerably larger, the edges, after annealing, may be thicker than the central

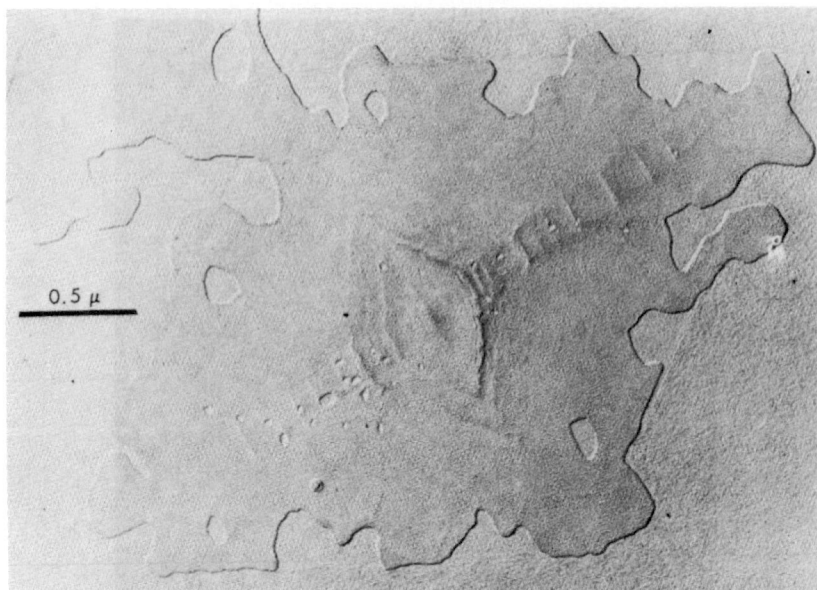

Fig. V-15. Linear polyethylene crystal precipitated from xylene and annealed at 120°C for 5 minutes. The surface area is within 5% of that indicated for the original crystal by the imprint on the collodion substrate (Cr shadowed). (Hirai (62).)

portion of the crystal. Such effects apparently have been observed (9,17,18) (Fig. V-12).

2. In a suspension of polyethylene crystallized from octane, hexagonal crystals with various lateral dimensions were found (6). The variation in dimensions is greater than that which can be attributed to competition for dissolved polymer during growth by neighboring lamellae, as in a spiral growth. (This effect, for instance, can lead to concave edges on polyethylene crystals in the vicinity of a spiral growth but it does not seem to greatly affect the overall size of the crystal.) The central portion of the largest and, presumably, first nucleated crystals should be thicker than that of the smaller crystals. The edges and all regions crystallizing at the same temperature should have the same thickness and therefore recrystallize at the same temperature. However, as shown in Figure V-14, the smallest crystals, presumably nucleated at the lowest temperature, recrystallize entirely at a lower temperature than the larger crystals. Although this process would lead to two sets of long periods if it also occurred in the crystal mats, the cause of the preferential recrystallization of the smaller crystals in their entirety is not known. In both cases, however, care should be used in ex-

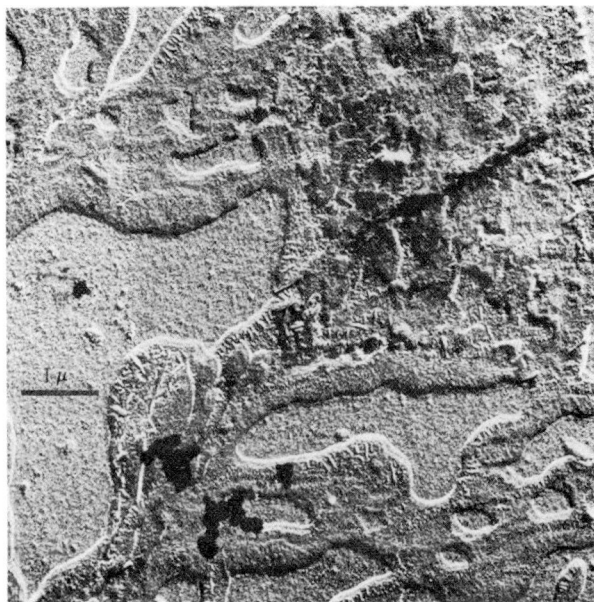

Fig. V-16. Portion of a polyethylene crystal similar to those in Figure V-14 except annealed at 132°C for 30 minutes. The individual lamellae appear doubled, the total thickness being about 500 A. Small rods are also seen on some of the surfaces (Pt-Pd shadowed at $\tan^{-1} 4/8$).

tending observations on isolated crystals to their behavior in mats. The effect of local defects is believed most likely to lead to the "mixed period" type of diffraction in the mats.

Hirai et al. (9) have indicated that an "amebalike" motion occurs when crystals are annealed at 120°C for 10 minutes on a collodion film substrate (Fig. V-15). They indicate that no change in total surface area or thickness occurred, the original outline of the crystal being visible from the imprint on the substrate. One also notes that ridges or corrugations are present in the {100} fold domains.

At relatively high annealing temperatures the holes that develop in the crystals may be of micron dimensions (Fig. V-16). A peculiar feature of the resulting lamellae is that they often appear to have a thin layer of material on their surface as well as some small rods and other debris. This material may result from low molecular weight or other poorly crystallizible material being exuded during the recrystallization at the annealing temperature and crystallizing later when the temperature is reduced.

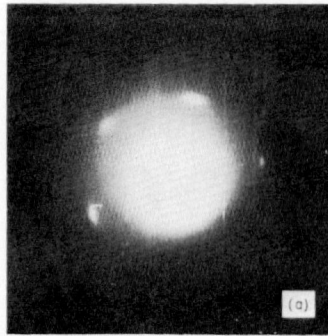

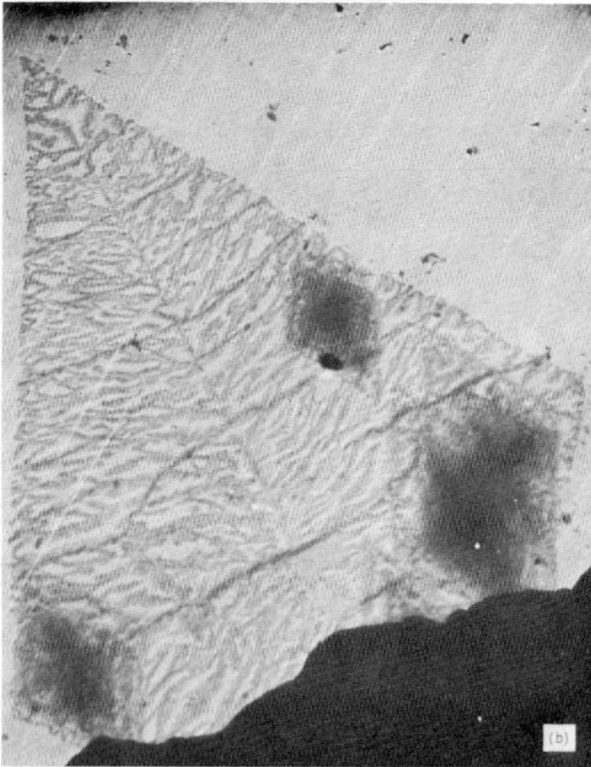

Fig. V-17. Electron diffraction pattern (a) from a polyethylene crystal (b) precipitated from tetrachloroethylene and annealed for 8 seconds in the focused rays of the sun. Further heating resulting in melting of the crystal followed by formation of structures similar to those in Figure V-18.   (Reneker (10).)

Wide angle diffraction patterns from the annealed crystal mats used for the data shown in Figure V-2 indicate (6) that the orientation of the c axis was retained during annealing. The molecules before and after annealing are essentially normal to the plane of the mats. The Bragg x-ray reflections, due to the increase in fold period and, presumably, lateral perfection, are sharper after annealing. The observed lateral dimensions of the lamellae (spacings of the holes), after annealing, are larger than the calculated lateral dimensions (from the peak broadening) of "perfect crystals."

Electron diffraction patterns from individual, annealed crystals show that the alignment of the a and b axes is also maintained during annealing (Fig. V-17). The crystal shown was annealed for 8 seconds in focused sunlight (10). Statton and Geil showed (6) that single crystal patterns were observed after annealing for 12 hours at 132°C. The Bragg electron reflections in both cases are only slightly wider than those from the original crystals. It was suggested (6) that this is due to a slight, random rotation of the a and b axes about the c axis in various portions of the lamellae. Dark field micrographs also show that the orientation of the molecules is maintained during the annealing. The entire crystal, excepting the holes, is visible in the dark field micrographs. This suggests that, if the crystal was originally pyramidal (other crystals in the same preparation used for the dark field micrographs formed pleats while collapsing during solvent removal), the molecules in the various fold domains may have rotated toward the normal to the substrate during annealing. These results indicate that either the lattice is essentially maintained while recrystallization is taking place or else only some portions of the crystal melt at one time, the molecules taking on some degree of randomness, followed by recrystallization on the remaining lattice with the original orientation but a new fold ·period.

Bassett and Keller (17) have reported that electron diffraction patterns indicate that during annealing of solvent cast continuous films (Fig. II-67) and isolated single crystals, the molecular axes rotate about the b axis by amounts up to at least 56°. In the case of a crystal mat, x-ray diffraction indicated that the rotation about the b axis could be sufficient to cause the a axis to become normal to the film (the molecular axes would thus lie in the film). The amount of rotation in all cases, they indicated, increases with increasing annealing temperature. The rotation is similar to that which occurs when drawn polyethylene is annealed (see Chapter VII). Balta et al. have recently extended these observations (14).

Electron microscopic observations, they report (17) show that corrugations or ridges parallel to the b axes gradually develop near the center of

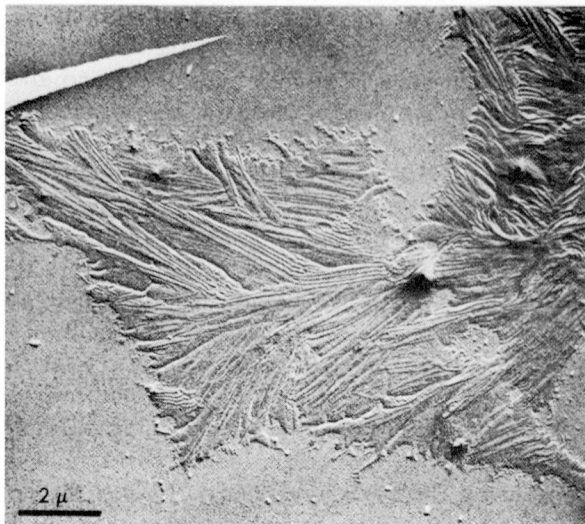

Fig. V-18. Polyethylene crystal annealed at 132°C for 30 minutes, apparently in the presence of a small amount of the tetrachloroethylene solvent. The solvent often is trapped under the hollow pyramids.

the crystals. In some cases the edges of the crystals developed a ragged appearance (Fig. V-12). They indicated that similiar corrugations, caused by buckling of the lamellae, occurred during annealing of the continuous film. Subsequently (19) they indicated that these ridges, which they attribute to the hollow pyramidal nature of polyethylene crystals, are often roof-shaped, i.e., they have planar sides. The ridges are best developed in crystals which are heated in the solvent but are also enhanced by heating in air (19). It may be that the formation of the corrugations and some of the molecular rotation, which they suggest are related, result from a change in the fold and fold-plane staggering during the annealing treatment.

Statton and Geil (6) indicated that wrinkles parallel to the **b** axis develop during the annealing of crystals precipitated from octane (Fig. V-14). They attributed the wrinkles to thermal expansion effects. These wrinkles, limited to the {100} fold domains, have since been shown to result from the pyramidal structure of the crystals (Chapter II). They are frequently found on crystals before as well as after annealing.

The results of Bassett and Keller (5,14,17) as well as those of other authors may have been affected by the presence of retained solvent. Several instances where retained solvent has affected annealing observations will be discussed shortly. The samples used by Bassett and Keller, for instance, were crystallized from xylene and annealed for a few minutes (14) in air on a temperature gradient bar at temperatures between 125°C and 130°C. Xylene boils at a temperature above those used for annealing and thus may easily stay trapped in or under the crystals (particularly since the crystals are pyramidal) for considerable periods of time. Bassett (12) indicates, however, that he has obtained similar evidence for rotation of the axes from crystals which have been stored for several months exposed to the air.

When crystal mats are prepared by filtration or evaporation of the solvent, a considerable quantity of solvent is retained which may be an important factor in the annealing experiments. Infrared measurements indicate (20) that a polyethylene crystal mat prepared by evaporation of the tetrachloroethylene solvent (b.p. 122°C) at room temperature was 9% solvent, by weight, when air dried, and was still 6% solvent after drying in a vacuum oven at 50°C for 200 hours. The location of the solvent in the mats is not known. Likewise mass spectrometry measurements on mats prepared from crystals precipitated from xylene show that at least several percent xylene is retained (21).

The presence of solvent, in one instance at least, has been shown (22) to affect the annealing process. A slide was placed in a vacuum oven at 132°C shortly after it was coated with a tetrachloroethylene suspension of polyethylene crystals. On part of the slide, typical annealed single crystals are found, i.e., during annealing holes developed and the lamellae thickened. On another part of the slide, however, a considerably different structure developed. In places where the crystals had been isolated, they retained their outline (Fig. V-18). The internal structure of these crystals resembles that described by Keller (23) as developing when polyethylene is crystallized as a film by evaporation of the solvent at an elevated temperature (see Figure II-67). It also resembles, to a considerable extent, the micrographs of Bassett and Keller (17) of annealed films in which molecular rotation has taken place. Apparently the orientation of the axes of the lattice in the various polygonal areas in Figure V-18 is not related to that in the original crystal. In other regions in which numerous crystals had originally been deposited on top of each other, all semblance of the original structure disappeared, and banded spherulites formed. It was thus concluded that the polymer molecules had essentially dissolved and then reprecipitated as the retained solvent evaporated.

Recrystallization, similar to that described previously, apparently also occurs when crystals of other polymers are annealed. As indicated pre-

Fig. V-19. Single crystal of poly-4-methyl-pentene-1 crystallized from xylene and annealed just below the melting point (Frank, Keller, and O'Connor (3).)

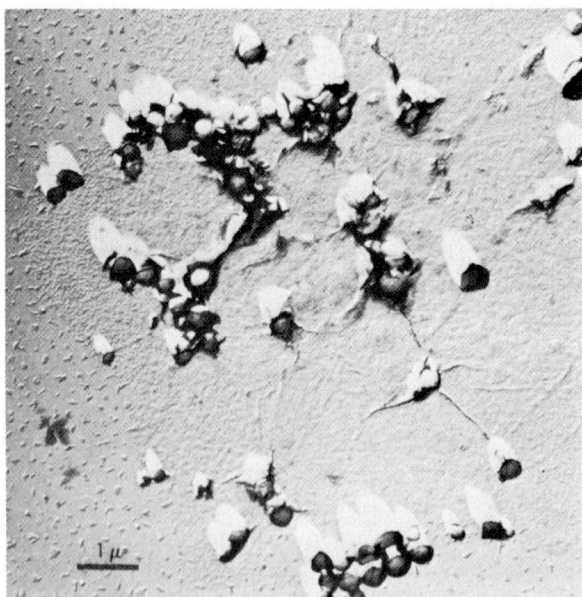

Fig. V-20. Portion of a single crystal of polyoxymethylene crystallized from cyclohexonal and annealed at 165°C for 150 minutes.

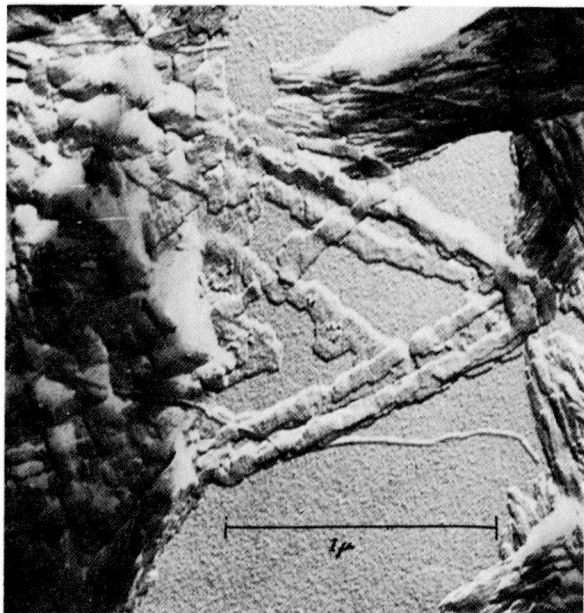

Fig. V-21. Single crystal of nylon 6 crystallized from glycerin and heated at 150°C. Compare with Figure II-1. (Holland (24).)

viously, poly-4-methylpentene-1 crystals develop holes if heated almost to the melting point (3) (Fig. V-18). Holes have also been observed in crystals of polyoxymethylene annealed at 165° (22) (Fig. V-20). Nylon 6 crystals annealed at 150°C apparently develop large holes in the center, the molecules forming a thickened periphery (24) (Fig. V-21). The striations seen on nylon crystals (Chapter II) may also result from the heating required to remove the solvent. No small angle x-ray diffraction results are available for the crystals of these polymers.

Nuclear magnetic resonance (NMR) measurements have shown (25–27) that in polyethylene crystal mats the number of mobile segments and their relative freedom is greatly increased after annealing. Linear polyethylene crystallized from the melt has, at room temperature, a compound NMR absorption spectrum (Fig. V-22). In hydrocarbon polymers the narrow component is attributed to protons in segments undergoing large scale oscillation and rotation whereas protons in segments in which the oscillations are small contribute to the broad band. At low temperatures, below the glass transition temperature, only the broad component is seen.

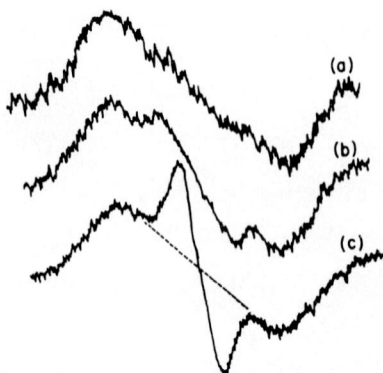

Fig. V-22. Effect of heat treatment on the shape of the derivative curves of the nuclear magnetic resonance of linear polyethylene (Marlex 50) crystallized at 90°C. from an 0.5% xylene solution. (a) Original crystals, (b) after heating at 130°C (between $1/2$ and 25 hours), and (c) after heating at 135°C. The curve from sample c resembles that of melt crystallized material. (Slichter (26).)

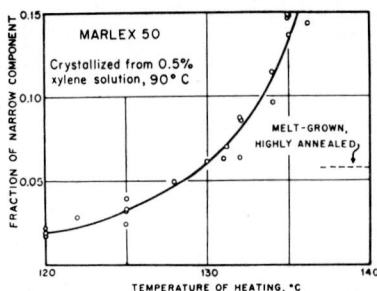

Fig. V-23. Variation of the ratio of the area of the narrow component of the total area in the nuclear resonance envelope (derivative curve) with temperature of heat treatment. The "melt grown" sample was crystallized at 128°C. (Slichter (26).)

In contrast to the melt-crystallized material the spectrum of *as*-crystallized, crystal mats consists essentially of only a broad component (Fig. V-22), suggesting that few chain segments have any great degree of mobility. However, when the mats are heated to a high enough temperature, a narrow component of measurable intensity develops, as shown in Figure V-22 and Figure V-23. Slichter (26) found, in agreement with the small angle x-ray results, that a minimum temperature of about 110°C was required before the segmental mobility in the crystals was sufficient to permit a nonreversible change in the NMR spectrum. Thurn (28), who measured the spectrum at the annealing temperature, reports a value of 108°C for the temperature

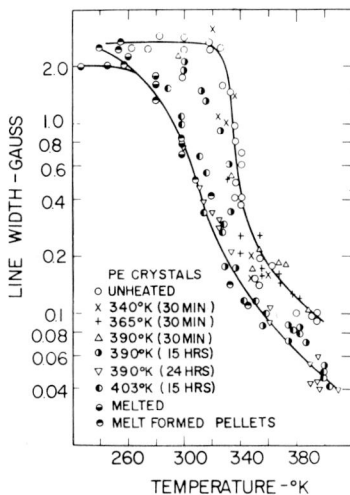

Fig. V-24. Line width (narrow component) versus temperature for polyethylene single crystals annealed at various temperatures and melt crystallized polyethylene (Odajima, Sauer, and Woodward (27).)

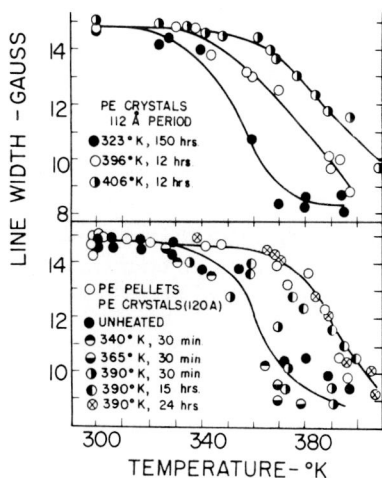

Fig. V-25. Line width (broad component) versus temperature for polyethylene single crystals with original long period spacings of 112 A and 120 A (Odajima, Sauer, and Woodward (27)).

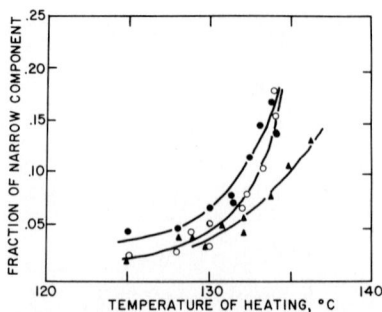

Fig. V-26. Variation of the fraction of the narrow component of the nuclear resonance curve from polyethylene crystals (as in Figure V-23) with concentration and temperature of crystallization. ● crystallized from 2% xylene solution at 70°C; ○, crystallized from 2% xylene solution at 90°C; and ▲, crystallized from 0.1% xylene solution at 90°C. (Slichter (26).)

transition. The data of Peterlin *et al.* (25), on the other hand, suggest that the narrow component is present after the mats have been heated above 70°C. The maximum ratio of the narrow component to the whole signal after annealing reported by Slichter was about twice that reported by Peterlin *et al.* Kedzie has recently shown that the difference in the two results is probably due to retained solvent (29).

Odajima, Sauer, and Woodward (27), in agreement with the results of Slichter, found that annealing single crystal mats results in an increase in area of the narrow component, indicating an increase in the number of mobile segments. They found (Fig. V-24) that the line width of the narrow component depends on temperature and that increasing the time or temperature of prior annealing treatments resulted in the transition in the narrow component shifting to lower temperatures. At the same time, the decrease in line width of the broad component, usually associated with melting, shifts to higher temperatures (Fig. V-25). Since this narrowing of the broad component takes place at temperatures as low as 70°C, it is probable, as in the case of the data of Peterlin *et al.* (25), that solvent was present in the mats.

Slichter indicates (26) that concentration, type of solvent, temperature of crystallization, and molecular weight have little or no effect on the results when the crystals are precipitated from dilute solution. At higher concentrations, in the range in which hedrites develop, concentration and temperature do result in discernible effects (Fig. V-26). Apparently defects and interlamellar links incorporated into the lattice during crystallization affect the results.

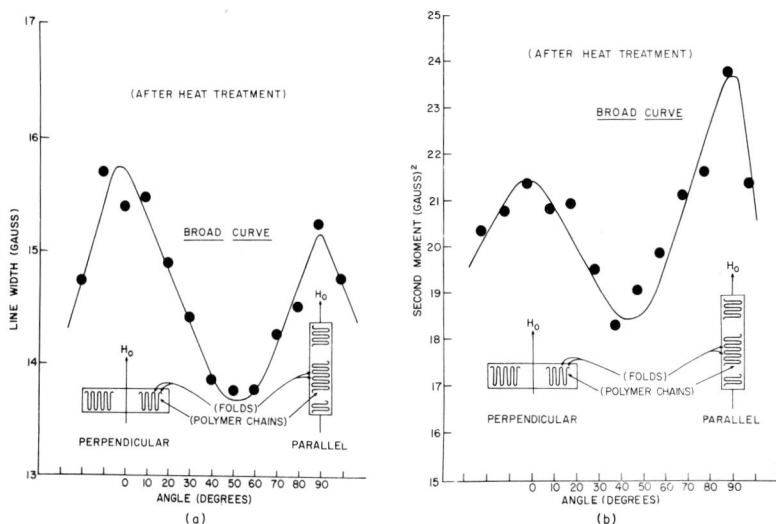

Fig. V-27. Angular dependence of the line width and second moment of polyethylene single crystal mats after annealing. A portion of the maximum at 90° is assumed to be due to the motion of proton pairs in the folds. The curves before heat treatment are quite similar to those shown here. (Kedzie (29).)

The results presented above, of which those of Slichter are believed to be the least complicated by the effects of retained solvent, suggest (25–27) that during annealing defects are introduced into the crystal mats, possibly similar to those inherent in melt crystallized samples, which permit segmental mobility in the vicinity of the defects at relatively low temperatures. The data of Odajima et al. (27) suggest that the nonmobile segments remain so at high temperatures as the fold period is increased. The increase in mobility of the already mobile segments in annealed samples, which Odajma et al. find to occur at lower temperature the longer or higher the annealing temperature, may be related to some type of aggregation of the defects such as occurs in grain boundaries. These defects are expected to be similar to those in low molecular weight solids, such as dislocations, as well as those peculiar to polymers, as for instance those resulting from chain ends, branches, improper folds, etc.

Thurn (28) has suggested that the shape of the NMR curve for polyethylene crystallized from xylene varies with the length of time the crystals are stored at room temperature, approaching the form of the curve from melt crystallized samples. Although again the influence of retained solvent needs investigation, particularly in view of the shape of the NMR curves

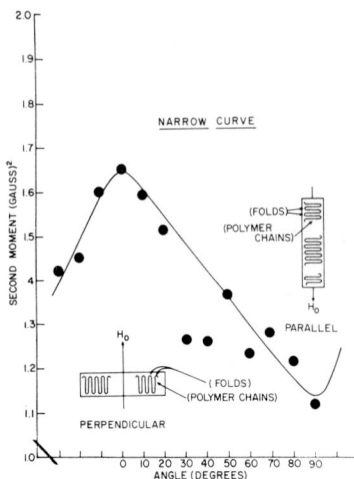

Fig. V-28. Angular dependence of the second moment of the narrow portion of the NMR curve from annealed polyethylene single crystal mats. The angular dependence is interpreted as being due to motion in defects within the lamellar crystal. A plot of the angular dependence of the area of the narrow curve (first moment) resembles that shown here. (Kedzie (29).)

Thurn observes, it may be that some form of annealing can take place at room temperature. The shape of Thurn's curves are close approximations to the theoretical line shape whereas the curves of other investigators deviate considerably.

Recent work by Kedzie (29) indicates that these defects have a specific orientation with respect to the lamellae and therefore must be associated with molecules in the lamellae. He has studied the NMR spectra from oriented polyethylene crystal mats before and after heat treatment. He finds that before and after annealing both the line width and second moment of the broad curve have maxima of about equal size when the lamellae in the mats are parallel and perpendicular to the applied field with a minimum when they are oriented at 45° to the field (Fig. V-27). This indicates, as would be expected, that the nonmobile segments are oriented. One would expect the line width when the molecules are parallel to the field to be twice that when they are perpendicular to the field, whereas the protons associated with the folds will produce the opposite effect. Kedzie suggests that one can explain the relative sizes of the maxima and the effect of annealing on the broad curve by considering the protons in both the main chain and folds.

In contrast to the type of anistropy in the broad line, Kedzie finds (29) that the second moment and the area of the narrow line found in annealed samples have a maximum when the lamellae are oriented normal to the applied field and a minimum when they are parallel to the applied field (Fig. V-28). No anisotropy could be observed in the line width. The anisotropy of the second moment indicates that the mobile protons giving rise to the narrow line are oriented with respect to the crystals and, it is suggested, are located at defects within the lamellae.

Prior to the work on single crystal mats it was shown (30) that annealing of bulk samples also results in a change in the room temperature NMR spectrum. The narrow component decreases in intensity. This change was attributed to a change in crystallinity, the molecules within the crystal being immobile by NMR standards. Slichter's results (26) (Fig. V-23) suggest that an equilibrium volume of defects are introduced into the lattice during annealing that are retained during cooling. When crystallized from the melt by quenching, a large volume of defects are frozen into the lattice resulting in the intense narrow line. During annealing at temperatures where there is sufficient segmental mobility, some of these defects are eliminated and the structure would appear to approach that obtained by the introduction of defects, during annealing, into the crystal mats. However, Slichter has recently shown (31) that the type of defects in the two cases probably is different. Samples crystallized from the melt and then annealed have a broader distribution of correlation times, i.e., rates of segmental reorientation, than the annealed crystal mats. Retained solvent may again be important, however.

The results presented above suggest that the increase in fold period during annealing occurs through two distinct processes. If, for a given fold period, ℓ, the crystal is heated rapidly to above its melting point

$$T_m\,(\ell) \;=\; T_m \left(1 \,-\, \frac{2\,\sigma_e}{\ell \Delta h_f} \right)$$

(T_m is the melting point of a crystal of infinite fold period, σ_e is the end surface free energy, and Δh_f the heat of fusion) (see eq. VI-36), it will essentially melt. Although the molecules probably do not become entirely misaligned and randomized, the low degree of birefringence (Fig. V-7) suggests they approach that state fairly closely at the highest annealing temperatures. The retention of alignment of the axes of the lattice in isolated single crystals (and also the uniformity of the rotation of these axes in some cases) suggests that nuclei are retained, at least on an average basis, which serve as seed crystals for the recrystallization. Further investigation of the lattice orientation in single crystals heated rapidly and annealed at relatively high temperature is needed however to test the above suggestion.

If heated slowly and also during the isothermal logarithmic increase in long period, it is believed that the fold period is increasing by a mechanism involving molecular motion along the backbone of the molecule while the overall orientation of the molecules in the lattice is maintained. This may involve either the motion of point dislocations of the types suggested by Reneker (32) and Kedzie (33) (described below) or the motion of entire segments of the molecule. At the same time defects incorporated into the crystal during either the initial crystallization or recrystallization can be ejected or corrected. One notes (Fig. V-7) that the birefringence at 129°C (and presumably lower temperatures also) rapidly recovers and becomes essentially constant although the density (Fig. V-6) and long period (Fig. V-4) continue to increase.

The situation at the higher annealing temperature (greater than 130°C) is unclear. Fischer and Schmidt (8) find (Fig. V-4) that the long period is increasing (as determined from quenched specimens) even though there is considerable disorder in the sample. As shown in Section V-3, Geil has found (34) that if a sample crystallized from the melt is quenched while in a similar disordered state, the resulting long period is characteristic of the quench temperature and may be shorter than the original long period (see Fig. V-51). In Fischer and Schmidt's experiments different heating rates were used for the small angle x-ray and the birefringence and density measurements. A sufficiently slow heating rate would permit a continuous increase of fold period in the small angle experiment with the material never becoming disordered.

Reneker (32) has suggested a plausible model, involving the motion along the chain of a point dislocation or crowdion, that will result in mass transport along the backbone of a polymer molecule in a crystal lattice. A related point defect, for helical molecules, has been suggested by Kedzie (33). This type of motion may be related to the logarithmic increase of the fold period during annealing, as well as occurring during mechanical deformations. The type of defect considered by Reneker is that in which a short portion of the chain in the crystal is compressed in the direction of the chain axis. The compression is localized (considerably less than 1 fold period) and is large enough that the portions of the chain on either side of the point dislocation are in their proper lattice positions. If this compressed region starts at one end of a segment within a crystal and moves to the other, the entire segment is transported within and with respect to the

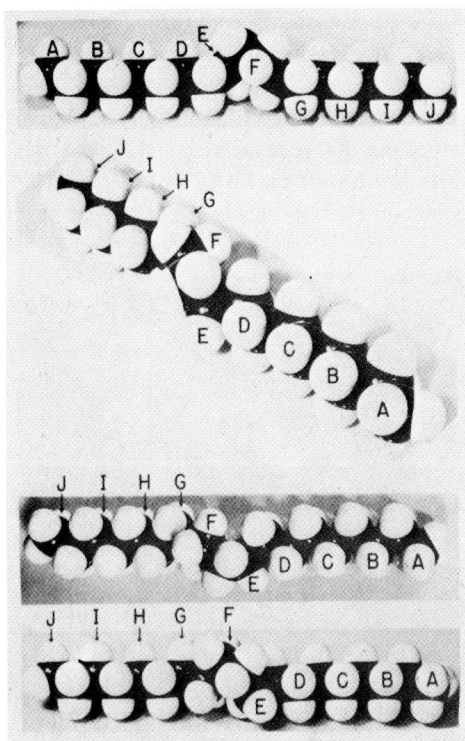

Fig. V-29. Various views of a point dislocation in a polyethylene molecule (Reneker (32)).

crystal as a whole by a distance equal to the excess length included within the dislocation. The energy required for the motion of such a defect may, as in the case of dislocations in crystals of low molecular weight materials, be considerably less than that required to move the entire segment an equivalent distance at one time.

In polyethylene a suitable point dislocation, involving one excess CH_2 group, can form relatively easily. Figure V-29 is a photograph of a Stuart-Briegleb model of such a dislocation. To form it, one section of the chain was rotated by 180° and shifted toward the other section by half a repeat distance. The extra CH_2 group resulting from the shift is included within the distorted region. The periodicity in the remaining sections of the segment is in phase with, and the chain would fit in, the lattice. The cross-sectional area of the defect is slightly larger than that of the remainder of the chain. Similar twisting of the molecule had previously been suggested for short-chain molecules to explain dielectric properties and crystallographic transitions (for review see Daniel (35)). Some of these suggestions, however, did not consider the compression that must accompany the twisting.

Reneker (32) estimates that at 100°C in polyethylene there is approximately one such defect per 400 carbon atoms, i.e., about 1 defect per every 4 folds. These results are based on an estimate that the energy involved in the strain in the carbon-carbon bands in the defect and as a result of the

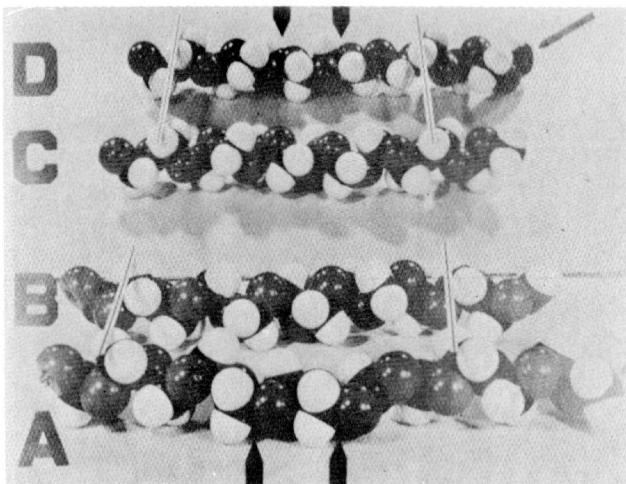

Fig. V-30. Model of a point defect in a polyoxymethylene molecule. One repeat distance of the molecule is indicated by the lines. In the molecule in front (and its mirror image at the top of the photograph) there is one less chemical repeat unit (CH_2O group) between the arrows than in the other molecule. By untwisting the helix for a short distance the molecules can remain in lattice register over the remainder of their length. (Kedzie (33).)

distortion of the lattice in its vicinity is about 0.2 electron volts. The number would increase rapidly with increasing temperature. Other similar defects permitting molecular motion may also occur in polyethylene. Reneker suggests, however, that the above type of defect is the simplest and has the lowest energy of creation.

An increase in fold period during annealing requires that the point dislocations be created and move along the chain. Thermal oscillations of the end of the chain, which must be in a distorted region of the lattice, could serve as the source of the twist and compression. These dislocations would then have to move along the folded chain and disappear at one of the folds. Some of the defects must pass through one or more folds before disappearing. The folds at which it disappears is raised by one CH_2 unit with respect to its neighbors for every two point dislocations that it absorbs.

The point defect considered by Kedzie (33) for helical molecules, in particular polyoxymethylene, involves a local straightening out of the helix (Fig. V-30). One chemical repeat unit is removed, in contrast to the interstitial nature of the defect suggested by Reneker. In this case an increase in fold period would involve formation of the defects at the fold and disappearance at the ends of the molecules. Kedzie (33) interprets his NMR results to indicate that defects similar to these are present in polyoxymethylene crystallized from the melt but are not present, in detectable amounts, in single crystals grown from solution. He did not anneal his single crystals but it is believed that the NMR spectrum characterizing the defects would be found in annealed specimens.

It is obvious from many of the above comments that further work on the effect of annealing single crystals is required. In these and other experiments, care must be taken to eliminate or at least recognize, control, and measure the amount of retained solvent and also to define the rate of heating.

2. ANNEALING OF HEDRITES

The lamellae of solution-grown hedrites apparently recrystallize in a manner analogous to those of the single crystals. It is likely that the materials used by Ranby, Morehead, and Walter (2), described as a "light cake," and by Ranby and Brumberger (13) were composed of porous, solution-grown hedrites. Mats prepared by filtering or evaporating solutions of polyethylene single crystals are characteristically dense (Table V-1) and coherent, although flaking like mica when stressed.

The annealing of melt-grown polyoxymethylene hedrites has been described in Chapter III. In contrast to solution-crystallized material as well as spherulitic, melt-crystallized material (next section), recrystallization apparently did not take place during the annealing treatments. Lamellae crystallized at the lowest temperature melt and remain molten, although in contact with still crystalline lamellae formed at higher temperatures, for at least several hours at temperatures below the polymer's melting point. No wide or small angle diffraction results are available, nor have similar studies with hedrites of other polymers been reported.

3. ANNEALING OF SPHERULITIC POLYMER

A number of authors have indicated (8,36–40) that the long period of bulk, unoriented polyethylene increases if it is annealed near the melting point. Some of their results are given in Tables V-2 and V-3. None of these authors combined their small angle x-ray measurements with electron microscope observations.

The results of Mandlekern *et al.* (36,37) (Table V-2), as in the case of single crystals, show both a time and temperature effect (see also Fig. V-49). One notes that in several cases the two periods measured are not related as first and second orders, as is also found for this and other polymers as crystallized from the melt (Chapter IV). In addition, the difficulty in locating the maxima of the diffraction pattern is suggested by the variable changes noted as a function of time for a given annealing temperature.

The data shown in this table were taken from samples 1.0 mm thick and, therefore, the long periods and specific volumes of the samples crystallized at temperatures below about 125°C actually correspond to somewhat higher crystallization temperatures (Chapter IV). In all cases, the long period obtained by isothermal crystallization at a given temperature is larger than that obtained even by prolonged annealing at this temperature. Since the melting point is a function of the fold period, this suggests that the highest melting point can be obtained by isothermal crystallization at a given temperature rather than by crystallization at a lower temperature followed by annealing (see Section V-5).

Sella's results (38,39), Table V-3, suggest that molecular weight has only a small effect on the long period but that the degree of branching has a significant effect. The branched polyethylene was apparently annealed at 107°C for 48 hours while the other samples were annealed at temperatures in the neighborhood of the melting point for a sufficient period of time to obtain a "stable" long period (41). The annealing temperature depended on the sample. In the case of the low molecular weight branched polymers the long period did not increase on annealing. The number average length of molecules in these samples is of the same order as the long period.

Sella's work (38,39) suggests that the transition temperature for recrystallization in branched polyethylene quenched from the melt may be near room temperature or below (Fig. V-31). He indicates that the long period, originally 150 A, of a sample of branched polyethylene quenched to -196°C increases after 400 hours at room temperature to 180 A and, after 3 hours at 100°C, to 220 A. It should be noted, in this connection,

TABLE V-2

Changes in Long Period and Specific Volume of Linear Polyethylene During Annealing (Mandlekern et al. (36,37))

	(a)ᵃ		(a)		(b)ᵇ		(a)		(b)	
Crystallization Conditions	0°		0°		75°		0°		0°	
Annealing temperature	120°		125°		125°		130°		130°	
Time of annealing	$\bar{V}$	"d"	$\bar{V}$	"d"	$\bar{V}$	"d"	$\bar{V}$	"d"	$\bar{V}$	"d"
0 days	1.0562	230	1.0562	230	1.0422	280	1.0562	230	1.0562	230
3 days					1.0323	315			1.0281	210ᶜ
6 days	1.0308	322	1.0282	370,182	1.0312	330	1.0249	460,198	1.0278	225ᶜ
12 days					1.0304	330			1.0222	510,220
36 days	1.0287	330,145	1.0253	380,180			1.0222	480,220		
66 days	1.0274	335,160	1.0248	405,180			1.0239	430,215		
96 days	1.0271	310	1.0239	364,173			1.0210	440,206		

Isothermal crystallization yielded the following values:

Crystallization temperature	$\bar{V}$	"d"
120°	1.0276	400,190
125°	1.0227	223ᶜ
130°	1.0182	355ᶜ

ᵃ Data from reference 36.
ᵇ Data from reference 37.
ᶜ Very likely these are second orders of a longer spacing. It is indicated that long periods greater than 400–450 A could not be resolved with their equipment.

TABLE V-3

Long Period of Polyethylene as a Function of Branching and Temperature (Sella (38))

Type	Grade	$[\eta]^a$ (dl/g)	$\overline{M}_n{}^b$ ($\times 10^{-3}$)	CH₃ per 1,000 C	C=C per 1,000 C	"Completely annealed" P^c	C%ᵈ	Quenched at −196° P	C%	Drawn at 20° P
Branched	100,000	0.2	2.0	60	3	220	21	220	20	—
	2,000	0.42	5.6	45	2	200	48	220	44	—
	500	0.56	9.0	35	1.4	210	50	200	45	—
	20	0.81	16.0	28	0.8	220	53	180	47	110
	7	1.01	23.0	20	0.5	230	56	170	49	—
	2	1.1	30.0	15	0.3	250	59	150	52	—
	0.9	1.3	35.0	10	0.2	260	62	160	54	—
Linear (a) Ziegler	2	1.10	40	7	0.9	320	77	220	70	—
	0.4	1.7	70	5	0.8	360	79	250	71	170
	0.25	2.0	90	4	0.8	370	80	250	71	—
	0.11	2.7	140	4	0.7	360	79	240	71	—
	—	5.8	400	3	0.7	350	77	240	70	—
	—	11	1,000	3	0.6	340	75	240	68	200
(b) Phillips	0.55	1.6	70	2	1.5	420	88	280	76	—

ᵃ $[\eta]$: Intrinsic viscosity in tetraline at 80°C for the branched polyethylene and at 130°C for the linear polyethylene.

ᵇ $\overline{M}_n$: Number average molecular weight calculated from the distribution curve for the branched polyethylene and from the relation $[\eta] = 5.1 \cdot 10^{-4}M^{0.725}$ for the linear polyethylene.

ᶜ P: Period in angstroms deduced from the small angle x-ray diffraction pattern.

ᵈ C %: Crystallinity deduced from the wide angle diffraction pattern (by transmission with monochromatic x-rays).

that the spacings of quenched linear polyethylene have not yet been measured at the quenching temperature.

One of the most interesting results reported by Sella was the effect on the long period of mixing branched and linear polyethylene. If simply mixed, two separate long periods are observed, corresponding to the periods of the two components. However, if the sample is melted, homogenized, and recrystallized slowly, one obtains a single long period intermediate to the original values (Fig. V-32). As will be seen in the next chapter, the number of defects within a crystal and, to a lesser extent, the molecular weight (terminal groups introduce defects) should affect the fold period and thermal stability. These factors affect the crystallization temperature. Sella's results agree, at least qualitatively, with the expected dependence of fold period on defect concentration.

It may be, as shown for polymers quenched to a glassy state (see Section V-4), that crystallization and recrystallization can take place at temperatures not far above the glass transition temperature. In such a case, the apparent 110° transition temperature seen for polyethylene single crystals

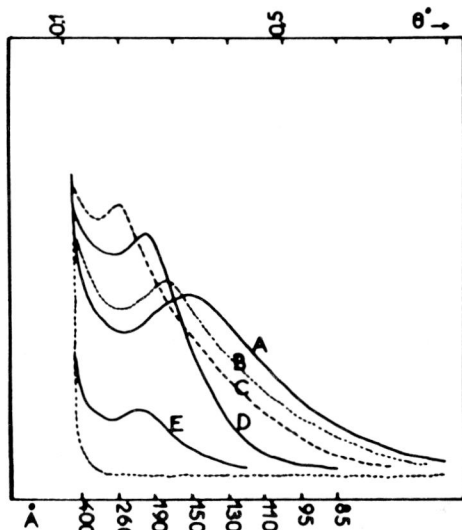

Fig. V-31. Small angle diffraction curves of a branched polyethylene (15 CH_3/1000 C atoms, melting point 115°C) quenched from the melt and annealed at various temperatures. It was not stated whether the patterns were obtained at room temperature or the annealing temperature. A, melted at 160°C and quenched to −196°C; B, after 400 hours at 20°C; C, after melting, holding at 107°C for 48 hours and cooling at 10°C per hour; D, sample A after 3 hours at 100°C; and E, sample D at 110°C. (Sella (39).)

may be due entirely to the fact that they already have a fold period of 100 A or more.

Belbeoch and Guinier (40) have also shown that the long period of a branched polyethylene increases when heated at temperatures above room temperature. Using a sample originally cooled rapidly from the melt to 20°C (42) the long period increased from 200 A at 20°C to 210 A at 62°, 230 A at 85°, and 260 A at 95°. The patterns were taken at the annealing temperature (42).

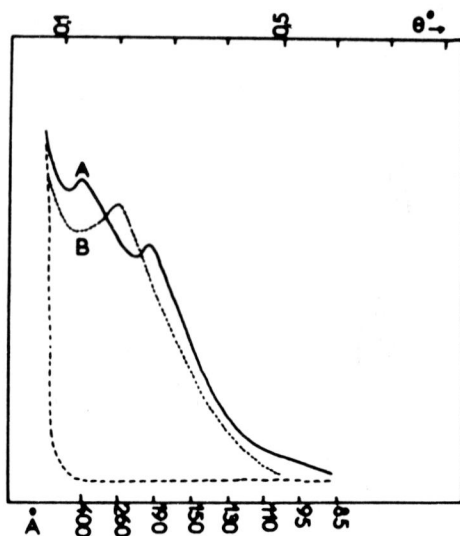

Fig. V-32. Small angle diffraction scans from a 50-50 mixture of branched and linear polyethylene before (a) and after (b) melting, homogenization and recrystallization. (Sella (39).)

As indicated in Chapter IV (Table IV-1), Hendus (43) found two sets of spacings in polyethylene samples. He indicates that during annealing of a quenched sample the larger of these two periods, as well as the apparently related width of the {110} reflections, approaches the values for the slow-cooled specimens. Because of its low intensity he was unable to determine the effect of annealing on the smaller of the two periods (44).

Fischer and Schmidt repeated many of their single crystal annealing experiments using quenched bulk polymer (8). The variation of long period with the logarithm of the time is shown in Figure V-33 for a linear polyethylene. Note that the initial values for the samples annealed at 130°C are less than those for samples annealed at 120 and 125°C. The samples were presumably quenched following the annealing treatment. Fischer and Schmidt's density and wide angle x-ray diffraction results for polyethylene crystallized from the melt are discussed later (p. 358) in this section.

Electron microscope studies have shown that the lamellae in spherulitic polyethylene thicken during annealing in a manner similar to that for single crystals (34). Small angle x-ray patterns and electron microscopy surface replicas were obtained from thin films crystallized with an unrestrained surface on a glass slide. Micrographs of replicas from the original films are

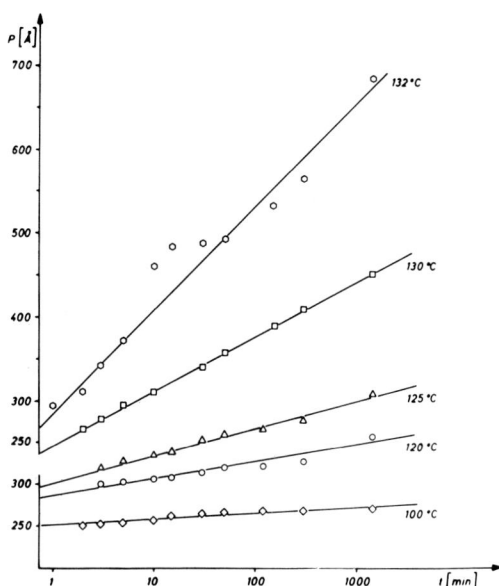

Fig. V-33. Dependence of the long period of a linear polyethylene film on the annealing time. The sample was originally quenched from the melt and was also quenched after the given annealing time. (Fischer and Schmidt (8).)

shown in Figures IV-21 to IV-25. The method of film preparation is also given in Chapter IV. Some compression molded samples of greater thickness were also studied by small angle x-ray diffraction. Optical microscopic observations indicate that the spherulitic structure of the compression molded films corresponds to that of the thin films, i.e., the dry ice-acetone quenched films are essentially nonspherulitic, the ring spacing and spherulite sizes are equivalent in the "rapidly cooled" compression molded films and the thin films quenched on a table top and the spherulites in the slowly cooled films are of equivalent size and are not banded.

The samples of varying thickness and thermal history were simultaneously annealed in nitrogen in a tube inserted in an oil bath at 125°C or 131°C (130°C for the thickest sample) for 100 hours. The long periods for the various samples are shown in Table V-4.

In general, the smaller of the two spacings listed is not a second order of the larger. Although slit-smearing corrections would reduce the larger period more than the smaller, the change in the relative values would not be sufficient to explain this effect. This would suggest that the two sets of periods may be due to different lamella thicknesses on the surface and the

TABLE V-4
Long Period of Annealed Samples of Linear Polyethylene

| Sample treatment | 0.25 mm Films | | Thin films | | | | | |
| | Quenched | Slow cooled | Quenched | | Rapidly cooled | | Slow cooled | |
			Small angle diffraction	Electron microscopy	Small angle diffraction	Electron microscopy	Small angle diffraction	Electron microscopy
As crystallized (A)	216, 95	325, 147	208, 65	60–70	252	100–120	294, 130	130–150
Annealed 100 hours at 125°C (A)	321, 158	342, 156	338, 165	300	340, 152	[a]	340, 155	350[b]
Annealed 100 hours at 131°C (A)	570, 265	510, 226, 143	430, 230	450	575, 240	500	615, 265	500

1.5 mm Films

Sample treatment	Quenched	Slow cooled
As crystallized (A)	210, 85	330
Annealed 100 hours at 130°C (A)	340, 150	380, 170

[a] No change observed in electron microscope.
[b] Observed in regions which had been deformed—see text.

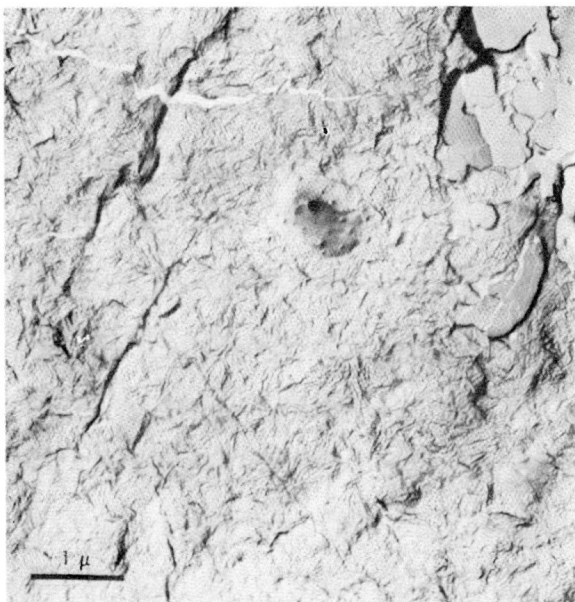

Fig. V-34. Linear polyethylene quenched from the melt into dry ice-acetone and annealed at 125°C for 100 hours in nitrogen. The material in the upper right is believed to consist of platelets of paraffin (Cr replica).

interior. Hendus, however, has measured (44) the relative intensity of the two periods from samples of varying thickness and found no difference. One notes that the deviation from the "expected" ratio of 2:1 for the long periods is largest for the quenched specimens. It is in these samples that the lamellae have the smallest lateral dimensions (as well as thickness) and are the most poorly organized (see Chapter IV). One might expect that diffraction from such a sample should be interpreted in terms of a dense collection of randomly oriented small clusters of lamellae. The number of lamellae per cluster, at least in the case of the quenched samples, may be too few to give rise to Bragg scattering. However, as discussed below, the lamella thickness as measured in the electron microscope corresponds closely to the smaller of the two periods.

The electron micrographs show that the morphological structure of the annealed samples differs considerably among themselves and with the original samples (Fig. V-34 and subsequent). These figures should be compared with micrographs of similar samples before annealing in Figures IV-16, IV-18, IV-23, and IV-25, and the changes in long periods noted in

Fig. V-35. Different portion of the same sample as in Figure V-34.   The arrows indicate regions in which the lamellae are essentially parallel to the surface.

Table V-4 of this chapter. As discussed below, they show that the melt crystallized material recrystallizes in a manner similar to that of the single crystals, i.e., the lamellae thicken by a refolding of the molecules. There is no gross change in structure and orientation, however, and thus the physical properties will be affected by the original crystallization conditions as well as the recrystallization.

The greatest morphological change occurs, as might be expected, during the annealing of the quenched samples. The original sample is non-spherulitic, having lamellae approximately 60 A thick randomly oriented on the surface (Fig. IV-23). Following annealing at 125°C for 100 hours, the lamellae are still randomly oriented but have thickened to about 300 A. In Figure V-34 they appear to be oriented essentially normal to the surface whereas in Figure V-35 there are regions (marked with arrows) in which they are essentially parallel to the surface. The cause of the rough surface in Figure V-35 is unknown. The smooth platelets in the upper right of Figure V-34 are believed to be composed of paraffin that exuded to the surface during annealing and crystallized during cooling. These platelets are seen on many of the subsequent micrographs also. Following annealing at 131° for 100 hours the lamellae, as suggested by the small angle x-ray results, are even thicker (~450 A) than those obtained by annealing at 125°. Much of the sample resembles that shown in Figure V-36 and, at higher magnification, Figure V-37. The lamellae retain a random orientation. In some areas, however (Fig. V-38 and V-39), the orientation of a number of neighboring lamellae appears to be essentially the same, resulting in ribbonlike structures composed of lamellae oriented

Fig. V-36. Linear polyethylene quenched from the melt and annealed in nitrogen at 131°C for 100 hours (Cr replica).

normal to the surface. It is not known if these structures developed spontaneously during the annealing or if they are due to some orientation imposed on the surface of the sample before annealing (see discussion of the effect of scratches below). They may possibly be related to the ribbons shown in Figure IV-26.

The major observed result of the 125° anneal of the rapidly cooled sample is the exudation of paraffin. The small globules seen on Figures V-40 and V-41, as well as the platelets, are believed to be paraffin. A possible change in lamella thickness, as suggested by the small angle scattering results, cannot be determined from the micrographs.

Fig. V-37. Higher magnification of a portion of Figure V-36.

Fig. V-38. Rows of lamellae oriented normal to the surface as seen in portions of a quenched sample of linear polyethylene annealed in nitrogen at 131°C for 100 hours. The white line is a crack in the Cr replica.

Fig. V-39. Higher magnification of a portion of Figure V-38.

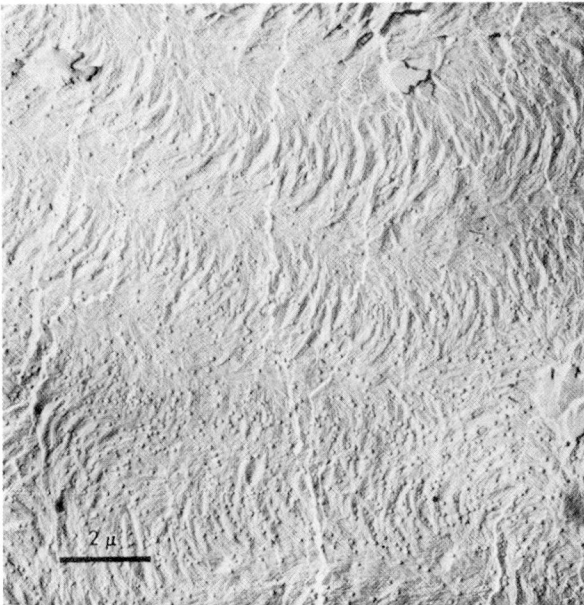

Fig. V-40. Portions of two spherulites in a rapidly cooled film of linear polyethylene after annealing in nitrogen at 125°C for 100 hours. The globules as well as the platelets are believed to be paraffin. Cracks in the replica are present (Cr replica).

355

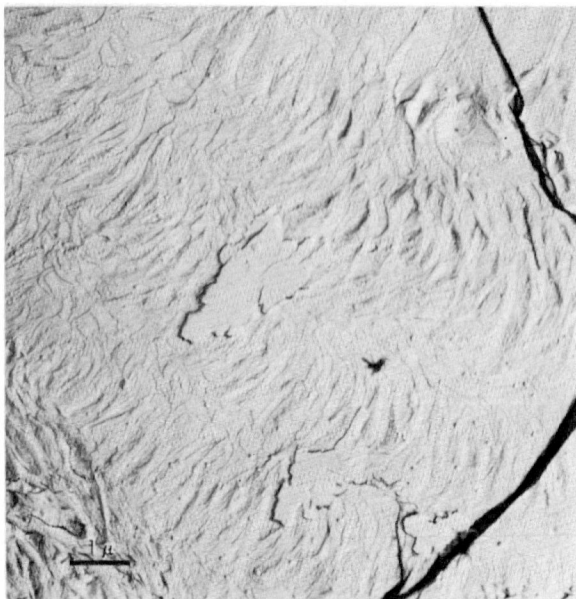

Fig. V-41. Higher magnification of a portion of the sample in Figure V-40.

Following annealing of the rapidly cooled samples at 131°, however, there is no doubt that recrystallization and a thickening of the lamellae occurred (Figs. V-42 and V-43). The lamellae, in agreement with the small angle x-ray results, are about 500 A thick. It is to be noted that the size and period of banding of the spherulites did not change during annealing. Apparently, as in the case of the single crystals, the molecules recrystallized without a gross change in orientation.

Contrary to a report by Reding and Walter (45) the spherulites did not change in size during the annealing. Using an etching technique, they reported erratic changes in size and morphology in spherulites of branched polyethylene during annealing. It is believed that these results are due to the "etching" process itself and do not represent the actual structure of the polymer. A subsequent paper by Mackie and Rudin (46) would tend to confirm this opinion, these authors finding that Reding and Walter's technique does not result in etching. Direct observation (47) of spherulitic films of branched polyethylene during annealing on a hot stage microscope at temperatures equal to or above those used by Reding and Walter show no change in size or optically visible structure. Likewise, as suggested by

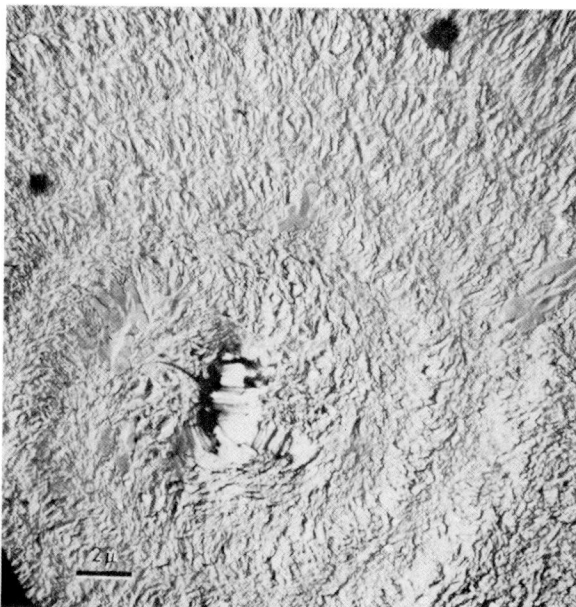

Fig. V-42. Spherulites of linear polyethylene in a rapidly cooled film after annealing at 131°C for 100 hours. At higher magnification the 50–100 A diameter fibers drawn across the void at the center of the spherulite are seen to have a beaded structure. (Cr replica).

Figure V-42, there is no gross change in size or structure of linear polyethylene spherulites. The lamellae increase in thickness, accompanied in many cases by the development of large voids at the spherulite centers, but the size of the spherulite does not change. Growth of some spherulites at the expense of their neighbors would require gross changes in orientation of the molecules and lamellae at the boundaries of the growing spherulites.

In a personal communication Reding indicated (48) that an increase in spherulite size as a result of annealing could be observed in microtomed sections of thick samples. Likewise Hoffman (49) indicates that he has observed the disappearance of birefringence followed by recrystallization with a new spherulite structure during the annealing of linear polyethylene. It is possible that the rate of heating and maximum temperature are significant, as discussed in connection with the results on single crystals.

Recrystallization of the slowly cooled samples, following annealing at 125°C, is evident only in regions in which the sample's surface was scratched or rubbed prior to annealing (Figs. V-44 and V-45). The lamellae formed

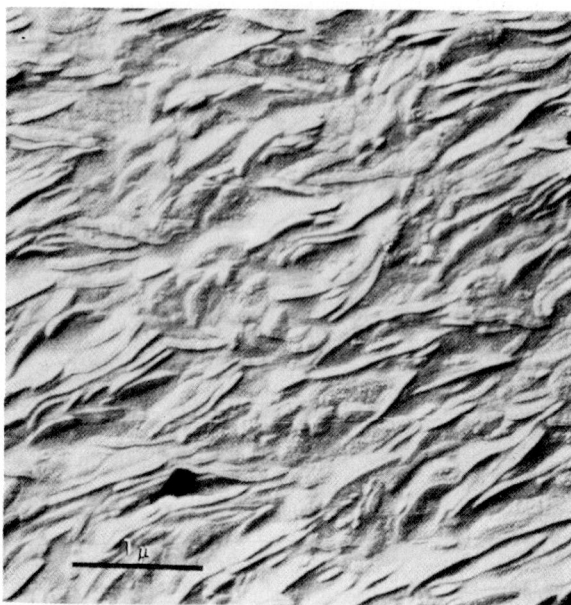

Fig. V-43. Portion of the spherulite shown in Figure V-42. Some of the lamellae have
a line parallel to their edge, resulting in a double appearance.

in these areas, oriented essentially normal to the direction of the scratch,
are about 350 A thick. No visible change in lamella thickness in unstressed
areas can be seen. As with the rapidly cooled samples, annealing at 131°C
results in an increase in lamella thickness, to about 500 A, but no gross
change in orientation of the molecules or size of the spherulites (Fig. V-46).
No holes are observed in the lamellae of melt-crystallized samples following annealing. In many cases, however, the surfaces are obscured by what
would appear to be degraded polymer or wax and thus it is likely that any
fine structure within the lamellae is similarly obscured.

    In agreement with the results obtained with single crystals, Fischer
and Schmidt (8) have shown that during the annealing of melt crystallized
material, the wide angle Bragg reflections first nearly disappear and then
reappear as the sample is held at temperature (Fig. V-47). Likewise, Gubler,
Rabesiaka, and Kovacs (50), Matsuoka (51), Peterlin (52), and Fischer
and Schmidt (8) have shown that during annealing the density first decreases and then increases again. In Figure V-48 is shown the difference
between the specific volume (density^{-1}) of the undercooled melt (at the
respective temperature) and the sample, expressed as mm^3/gram of poly-

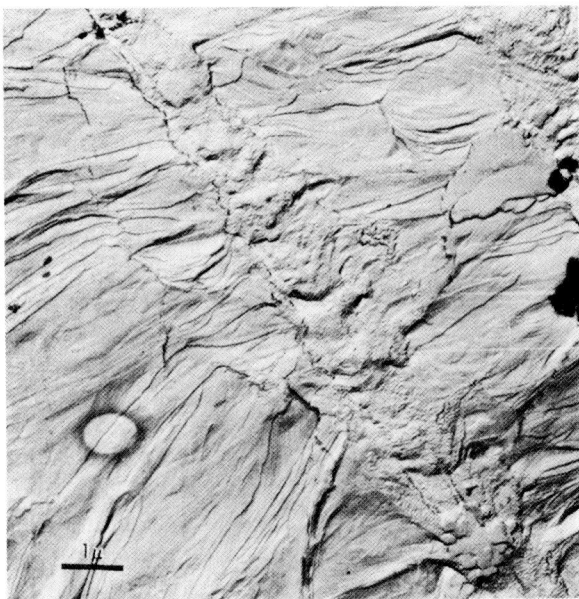

Fig. V-44. Portion of a spherulite of linear polyethylene in a slowly cooled film whose surface was scratched or rubbed slightly before being annealed at 125°C for 100 hours. Recrystallization has apparently occurred only in those areas which had been distorted (Cr replica).

mer. At all temperatures below 133° the density does not reach that of the liquid whereas at 133° the sample has essentially the density of the liquid state for a considerable period of time before starting to recrystallize. Likewise if a sample is first equilibrated at an elevated temperature (i.e., held at temperature until the density is essentially constant) and then heated to a still higher temperature, a similar process occurs (50). The specific volume will again increase slightly and then decrease with time.

Fischer and Schmidt's (8) density data are shown in Figure V-49. The density is seen to increase linearly with the logarithm of the time following an initial disordering and recrystallization. The same increase in density can be observed during the "secondary crystallization" period when a sample is crystallized isothermally from the melt (53). As in the case of the single crystals this isothermal increase in density (and long period—Fig. V-33) is believed to result from motion in the lattice of the molecules along their backbone. The data of Mandlekern et al. (36) are also shown on Figure V-49 and are seen to agree well with those of Fischer and Schmidt.

Zachman and Stuart (54) have obtained similar evidence for polyethylene terephthalate originally crystallized at 140°C and then annealed at temperatures between 200°C and 247°C (Fig. V-50).

When a polymer sample is annealed at a temperature greater than that corresponding to its small angle periodicity, all memory of its original fold period is lost. It may recrystallize with either a larger or smaller fold period depending upon how it is recrystallized (34). As in the case of the single crystal mats, reannealing of a melt crystallized sample of linear polyethylene at a higher temperature results in a corresponding further increase in

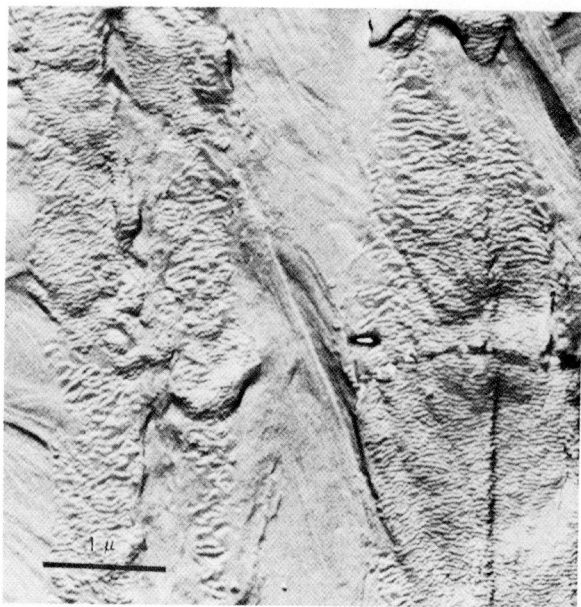

Fig. V-45. Higher magnification of a portion of the sample shown in Figure V-44.

spacing. However, if a sample is first heated to a temperature above that corresponding to its fold period and then quenched within the time that the density measurements (Figs. V-48 and V-49) suggest that it is in a disordered state, the resulting long period corresponds to the quench temperature (Fig. V-51). Thus, if a sample annealed at 125°C for 100 hours and having a long period of 320 A is heated to 132.6° and quenched from this temperature, the resulting long period after as much as 40 hours of annealing at 132° is characteristic of the quench temperature. The slightly greater upswing of the 40-hour anneal curve in Figure V-51b suggests that a slight amount of recrystallization has occurred, in agreement with the density observations. From the data in Figure V-51 it appears as if recrystallization with a larger long period occurs more readily if the original long period is close to the long period corresponding to the annealing temperature. The disordering may be less or even absent in this latter situation.

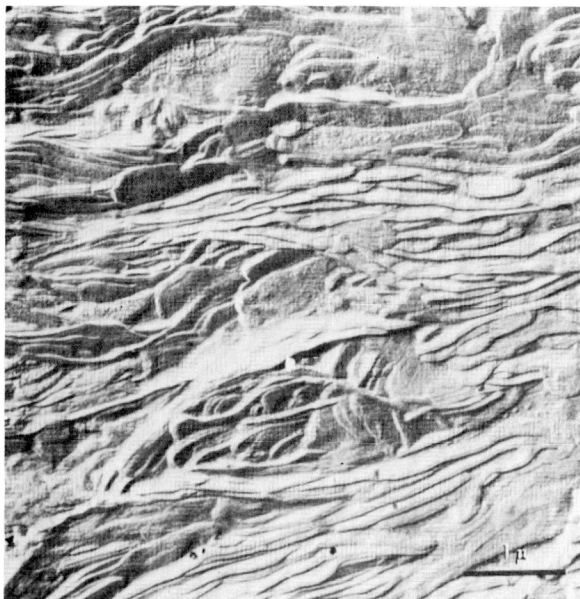

Fig. V-46. Portion of a spherulite of linear polyethylene in a slowly cooled film after annealing at 131°C for 100 hours. The doubled appearance of the lamellae is even more obvious here than in Figure V-43. As in the case of single crystals, it may be due to exuded material on the surface of the lamellae.

Gubler, Rabesiaka, and Kovacs (50) have followed a similar process dilatometrically. A Marlex 50 sample was first quenched from 170° to 25° and then annealed at temperatures near the melting point (Fig. V-52) for a time sufficient to reach the maximum specific volume (see Fig. V-48). The sample was then quenched to 129°C and the crystallization process followed as a function of time. The curve marked 170° on Figure V-52 corresponds to a sample cooled directly from the melt (170°C) to 129°C. The sample apparently becomes more disordered, the higher the initial annealing temperature. However, it is only at the highest annealing temperature that the disordered state closely approaches that of the true melt. At lower temperatures, nuclei must be retained which serve as seed crystals, permitting faster crystallization.

It is well known (see, for instance, reference 8) that as the temperature of bulk polyethylene is raised the intensity of the crystalline peaks, measured at the annealing temperature, decreases. The effect is more marked

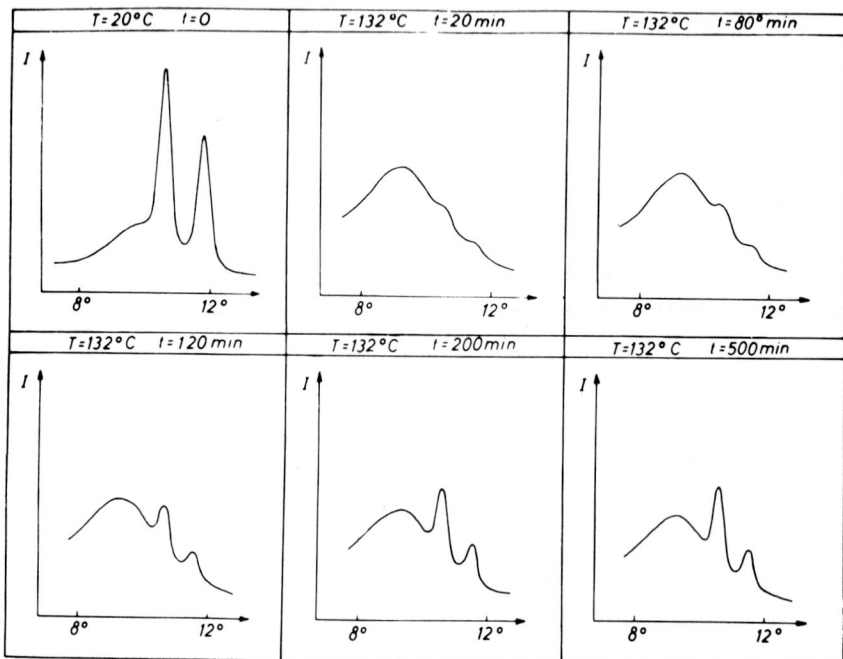

Fig. V-47. Wide angle x-ray diffraction patterns obtained during the annealing of a sample of linear polyethylene quenched from the melt (Fischer (16); see also ref. 8).

for a branched polyethylene than for the linear polymer. Other evidence for this so-called premelting has resulted from infrared studies, density measurements, NMR studies, etc. As Stuart points out (55), the rate of decrease in "crystallinity" with rising temperature as determined by the different techniques for measuring crystallinity differs greatly. At 120°C, for instance, linear polyethylene is much more "crystalline" by density than by x-ray measurements. It is, perhaps, only fortuitous that in many polymers at room temperature there is reasonable aggreement between crystallinity values calculated from the results of different techniques.

It is obvious from the x-ray studies of Fischer and Schmidt (8) that the x-ray patterns at elevated temperatures need to be obtained as a function of time as well as temperature. For temperatures above the minimum recrystallization temperature and up to a limiting temperature, the intensity of the crystalline peaks should increase, resulting in a better correlation with density measurements, if the samples are held at temperature sufficiently long.

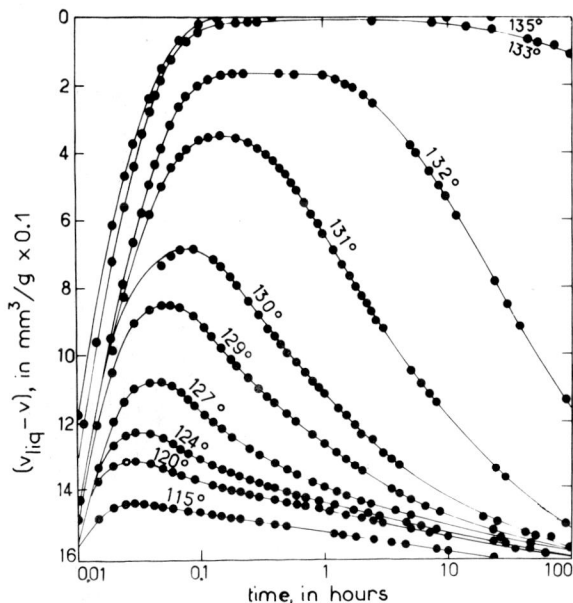

Fig. V-48. Difference in specific volume of undercooled liquid and sample for a linear polyethylene (Marlex 50) crystallized from the melt and annealed at various temperatures (Gubler, Rabesiaka, and Kovacs (50)).

Results similar to those obtained with polyethylene are obtained for the time and temperature dependence of the annealing of polyoxymethylene (Table V-5) (22). Ice water-quenched, 0.25 mm thick samples were annealed in copper foil jackets in an oil bath for various periods of time and then quenched into ice water. At temperatures of 171°C and below, recrystallization takes place in less than 2 minutes whereas at 175°C it takes on the order of 1 hour. Recrystallization takes place, in these samples, at temperatures as low as 150°C and possibly 130°C. The small angle peaks in melt-crystallized polyoxymethylene samples are much more clearly

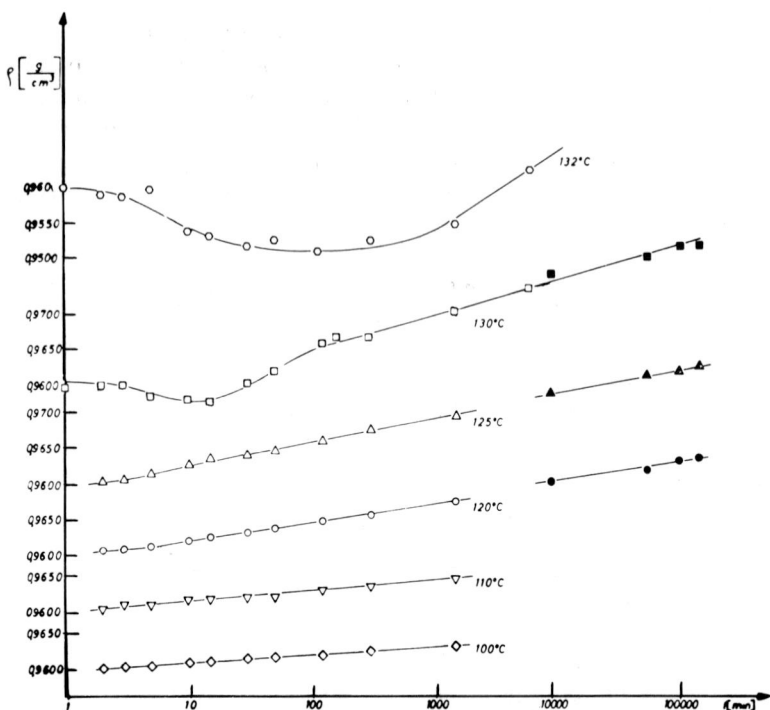

Fig. V-49. Density of linear polyethylene quenched from the melt as a function of annealing time and temperature. The solid points are data of Mandlekern *et al.* (36). (Fischer and Schmidt (8).)

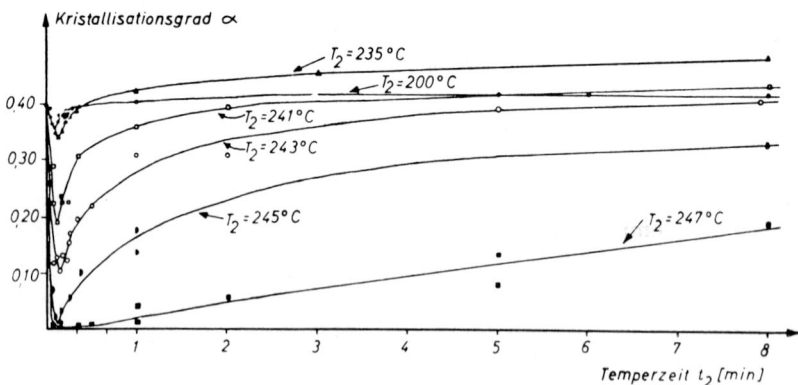

Fig. V-50. Crystallinity of polyethylene terephthalate obtained from density measurements as it depends on the annealing temperature and time. The original sample was crystallized at 140°C. (Zachmann and Stuart (54).)

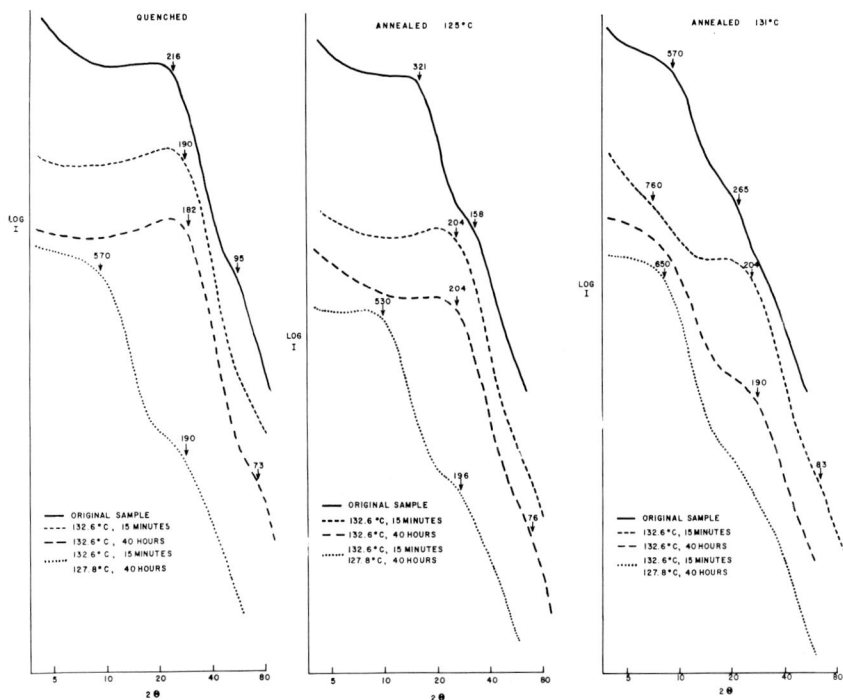

Fig. V-51. Small angle diffraction curves (slit collimation) from linear polyethylene films with different crystallization histories. The annealed samples had been annealed for 100 hours at either 125 or 131°C. All samples were quenched into dry ice-acetone following the further heat treatment listed on the figure.

defined than in the case of linear polyethylene, permitting relatively precise measurement. No electron microscope observations of the effect of annealing melt crystallized polyoxymethylene have been reported.

TABLE V-5
Long Period (A) of Annealed Polyoxymethylene[a]

Time of annealing	Temperature of annealing						
	130°	150°	160°	165°	168°	171°	175°
5 Minutes		168	180	200	228	283	165
2 Hours	163						275

[a] Small angle period of original sample, 155 A. Values are not corrected for slit effects. The samples were quenched in ice water following the annealing.

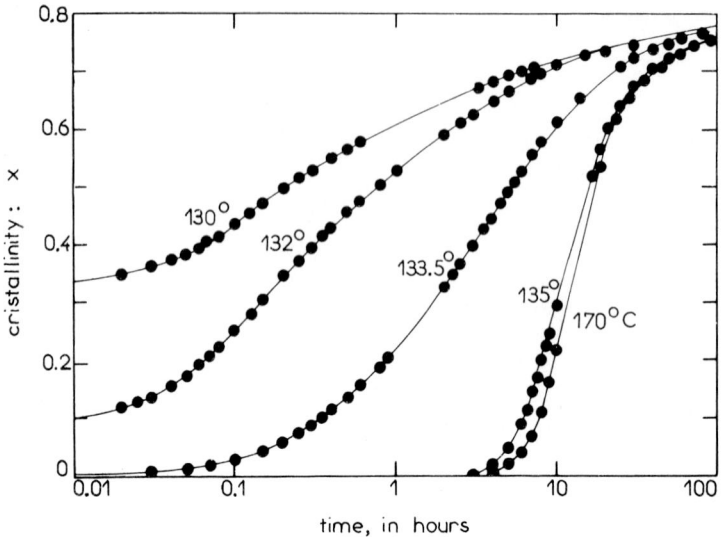

Fig. V-52. Crystallinity of polyethylene samples originally quenched from the melt and then heated to the temperatures indicated. When the maximum specific volume was attained at this temperature (Fig. V-48) the sample was quenched to 129°C and the specific volume measured as a function of time (Gubler, Rabesiaka, and Kovacs (50).)

4. ANNEALING OF GLASSY POLYMERS

As shown in Chapter IV, several polymers can be quenched from the melt to a glassy state, crystallization and spherulite growth occurring during subsequent annealing at temperatures above the glass transition temperature. Eppe, Fischer, and Stuart have shown (56) that discrete small angle x-ray diffraction maxima develop during the annealing of glassy polyurethane coincidentally with the development of crystallinity (Fig. V-53). The spherulites of polyurethane that are formed during annealing may be composed of fibrous or ribbonlike structures. In the case of polyethylene terephthalate, Fischer (7) has obtained similar small angle diffraction results for short annealing times. In both polymers the long period increases with the annealing temperature. As shown in Chapter IV, polycarbonate (57) can crystallize from the glassy state in the form of well-developed spherulites composed of lamellae.

A significant result of Fischer's annealing experiments with polyethylene terephthalate is his report (16) that the long period, after an initial increase, decreases with time at a given annealing temperature. He explains

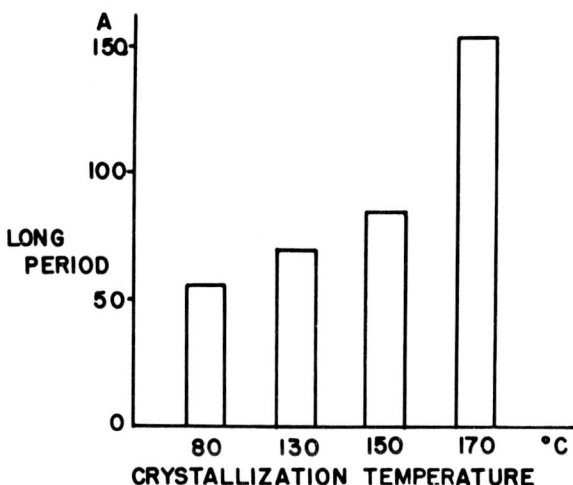

Fig. V-53. Dependence of the long period of a sample of polyurethane on the annealing temperature (Eppe, Fischer, and Stuart (56)).

these results as being due to an increase in perfection of the crystals during the annealing. This increase in perfection would, in terms of the theory of Peterlin and Fischer (Chapter VI), result in a decrease in the equilibrium length of the crystal but the results would appear to be compatible with the theory of Lauritzen and Hoffman (Chapter VI) as well.

The experimental results described in this chapter show conclusively that the thickness of lamellae can be increased by annealing at temperatures between the crystallization temperature and the melting point. This change in fold period can be followed microscopically or by means of small angle x-ray scattering. Further experimental work is needed, as explained in the text, to extend and clarify many of the factors involved in the changes produced by thermal treatment. It is particularly obvious that the relationship between the small angle results and the electron microscope observations for bulk polyethylene needs clarification. Somewhat similar changes in long period occur during the annealing of drawn polymers. These results will be described in Chapter VII.

5. EFFECT OF ANNEALING ON MELTING POINT DETERMINATIONS

Knowledge of the limiting or equilibrium melting point, T_m, of a polymer is essential in determining many of the thermodynamic properties of polymers. T_m corresponds to the melting point of an "infinite" crystal (so large that surface effects are negligible) with a defect concentration that is in equilibrium at the melting point. Until recently it has been believed that the best way to determine T_m was to slowly raise the temperature of a dilatometric specimen, waiting after each temperature rise for the specific volume to become constant. On a plot of this limiting specific volume versus temperature there is a reasonably sharp break at the melting point. X-ray and birefringence measurements have also been used to determine T_m. By slowly raising the temperature the most perfectly ordered crystalline structures were believed to develop, and the melting point determined in this manner was believed to closely approximate T_m (58). As pointed out previously in this chapter and as assumed by the authors as also occurring with fringed micelle-type crystallites (59), a slow heating process permits recrystallization to take place. The thinnest lamellae or smallest crystallites "melt" and recrystallize with a larger fold period or micelle size. The problem with this approach is that a considerable portion of the time alloted for the experiment on a given sample is used in reaching the next to highest temperature. Similar "nearly perfectly ordered" or, in the lamellar model, lamellae with as long a fold period as possible, should be obtainable with less total expended time and possible sample degradation by crystallizing from the melt at as high a temperature as possible, probably seeding the melt to induce crystallization more rapidly. Even in this case, however, it is expected that the observed melting points will be somewhat less than the equilibrium melting point.

Hoffman and Weeks (60) have developed an extrapolation technique, based on knowledge of the recrystallization process, to determine T_m. They have shown this to be a more desirable method for at least polychlorotrifluoroethylene and polyethylene. In addition, while testing the technique, they prepared samples by slow cooling that melted at or above the temperatures previously obtained for T_m by the stepwise heating technique.

The basic objective of the extrapolation technique is to obtain the observed melting point, T_{obs}, as a function of the fold period and then extrapolate to infinite fold period. As developed by Hoffman and Weeks it is not necessary to measure the fold period of the various samples. They determine the observed melting point, usually dilatometrically, as a function of the isothermal crystallization temperature T_{xl}, and determine the intersection of a line through these points with the line defined by $T_{obs} = T_{xl}$. T_m will lie on this line since for an "infinite" crystal T_m, T_{obs} and T_{xl} are the same.

The results obtained by Hoffman and Weeks for polychlorotrifluoroeth-

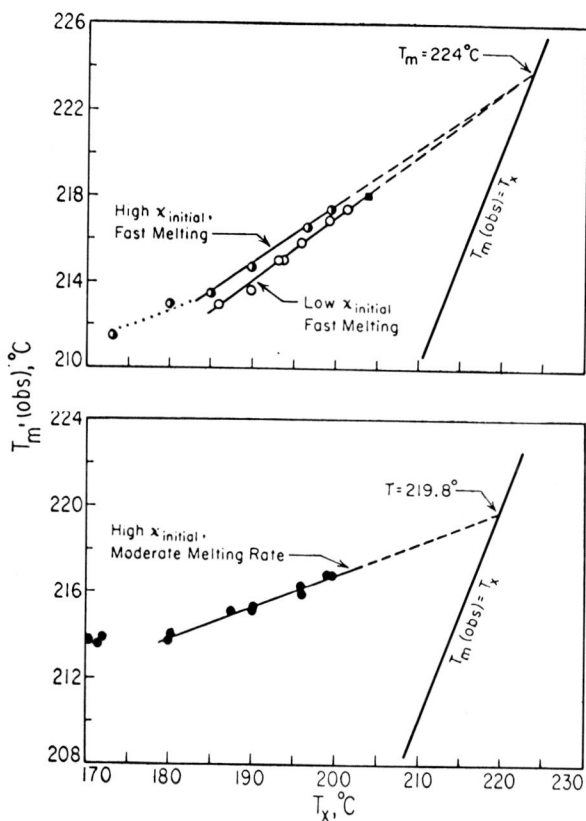

Fig. V-54. Observed melting point (T_m (*obs*)) versus crystallization temperature (T_x) for polychlorotrifluoroethylene of various degrees of crystallinity (χ) (Hoffman and Weeks (60)).

ylene samples with high ($\sim$50%) and low($\sim$10%) initial crystallinity and two heating rates are shown in Figure V-54 (60). For the moderate heating rate the sample was heated above 205°C until it melted, the time for this process being on the order of 5 hours. For the rapidly heated specimens the samples were placed directly in oil baths operating near the previously estimated melting point and it was determined whether or not the sample melted at the particular temperature of each bath. Depending on the number of baths available one can closely bracket the melting point in this way. Hoffman and Weeks estimate that this procedure is equivalent to a normal melting experiment of only about 20 minutes. The samples with higher crystallinity were prepared by holding the sample at the crystallization temperature for longer periods of time (at 199°C the crystallinity varied from 9% after 15.6 hours to 65% after 34 days).

The moderate heating rate curve is believed to show the effect of recrystallization. The points for the lower values of T_{rl} will be raised because the lamellae thickened during the heating and therefore melted at a higher temperature than that corresponding to their original thickness. If the heating had been carried out slowly enough, one would expect to get a horizontal line corresponding to T_m as determined by the previously accepted technique. By the latter, stepwise technique, Hoffman and Weeks obtained a melting point of 216.4°C, the experiment lasting 35 days (60). Not only is this value significantly lower than the extrapolated values for both heating rates but is also below that which they were able to obtain by seeding a specimen and crystallizing it at 203.9°C for 7 days (T_{obs} = 218.0°C).

It is seen that both the high and low initial crystallinity data, with fast melting, extrapolate to nearly the same value for T_m. Hoffman and Weeks suggest (60) that the difference is due to an increase in fold period during the isothermal crystallization. This would agree with the findings of Fischer and Schmidt (8) that during annealing treatments the long period increases linearly with the logarithm of the time of annealing. As indicated previously such results require that the molecules have a considerable degree of mobility in the crystal lattice; probably the motion is along the backbone by some mechanism involving point defects similar to that suggested by Reneker (32). Melting and recrystallization with a longer fold period is not expected during isothermal treatments. An increase in perfection of the crystals might also contribute to the increased values of T_{obs} at the higher crystallinities. Even with rapid heating the curvature of the plot for crystallization temperatures below 180°C suggests that recrystallization occurred while these samples were heated.

In Chapter VI theoretical aspects of crystallization and melting will be discussed. It will be shown that the following relation can be applied to the melting of lamellae crystals, $T_{obs} = T_m - (T_m - T_{rl})/2\beta$ where β is a parameter related to the difference in thickness of the lamellae at the time of melting and the thickness of the primary nucleus at the time of crystallization. The value of β can be determined from the slope of the T_{obs} versus T_{rl} plot and is expected to be slightly larger than unity if no increase in fold period occurs either as a result of recrystallization during heating or with storage at T_{rl}. For the three sets of data shown in Figure V-54, Hoffman and Weeks list β as being 1.69 (low initial crystallinity, fast melting rate), 1.85 (high initial crystallinity fast melting rate), and 3.36 (high initial crystallinity, moderate melting rate).

Initial results for polyethylene, using the extrapolation technique, yield a value for T_m of $143 \pm 2°C$ (60). This is considerably above the previously accepted value of $137.5 \pm 0.5°C$ obtained by a stepwise melting process (58), and agrees favorably with the value for the limiting melting point ($141.1 \pm 2.4°C$) of n-paraffins (limit as n approaches infinity) with a unit cell similar to polyethylene (61). β was found to be approximately unity for polyethylene.

REFERENCES

1. A. Keller and A. O'Connor, "Study of Single Crystals and their Associations in Polymers," *Disc. Faraday Soc.*, #25, 114 (1958).
2. B. G. Ranby, F. F. Morehead, and N. M. Walter, "Morphology of n-Alkanes, Linear Polyethylene and Isotactic Polypropylene Crystallized from Solution," *J. Polymer Sci.*, **44**, 349 (1960).
3. F. C. Frank, A. Keller, and A. O'Connor, "Observations on Single Crystals of an Isotactic Polyolefin: Morphology and Chain Packing in Poly-4-Methyl-Pentene-1," *Phil. Mag.*, **4**, 200 (1959).
4. D. C. Bassett, F. C. Frank, and A. Keller, "Evidence for Distinct Sectors in Polymer Single Crystals," *Nature*, **184**, 810 (1959).
5. A. Keller and D. C. Bassett, "Complementary Light and Electron Microscope Investigations of the Habit and Structure of Crystals with Particular Reference to Long Chain Compounds," *J. Royal Micro. Soc.*, **79**, 243 (1960).
6. W. O. Statton and P. H. Geil, "Recrystallization of Polyethylene during Annealing," *J. Applied Polymer Sci.*, **3**, 357 (1960).
7. E. W. Fischer, "Thermodynamical Explanation of Large Periods in High Polymer Crystals and Drawn Fibers," *Ann. N. Y. Acad. Sci.*, **89**, 620 (1961).
8. E. W. Fischer and G. F. Schmidt, "Über die Langperioden von verstrecktem Polyathylen," *Angew. Chem.*, **74**, 551 (1962).
9. N. Hirai, T. Mitsuhata, and Y. Yamashita, "Thickening of Polyethylene Single Crystals under Heat Treatment," *Chem. High Poly. (Japan)*, **18**, 33 (1961) (in Japanese).
10. D. H. Reneker, personal communication.
11. W. O. Statton, "Rate of Recrystallization of Polyethylene Single Crystals," *J. Appl. Phys.*, **32**, 2332 (1961).
12. D. C. Bassett, personal communication.
13. B. G. Ranby and H. Brumberger, "Observations on the Chain Folding in Crystalline Polymers," *Polymer*, **1**, 399 (1960).
14. F. Balta-Calleja, D. C. Bassett, and A. Keller, "A Study of X-ray Long Periods Produced by Annealing Polyethylene Crystals," to be published.
15. D. C. Bassett, F. C. Frank, and A. Keller, "Some New Habit Features in Crystals of Long Chain Compounds IV," to be published.
16. E. W. Fischer, personal communication.
17. D. C. Bassett and A. Keller, "Similarities Between the Behaviour of Single Crystals and That of Bulk Material on Annealing in Polyethylene," *J. Polymer Sci.*, **40**, 565 (1959).

18. D. C. Bassett and A. Keller, "On the Habits of Polyethylene Crystals," *Phil. Mag.*, **7**, 1553 (1962).

19. D. C. Bassett and A. Keller, "Some New Habit Features in Crystals of Long Chain Compounds Part II. Polymers," *Phil. Mag.*, **6**, 345 (1961).

20. R. Brown, personal communication.

21. R. Salovey, personal communication.

22. P. H. Geil, unpublished data.

23. A. Keller, "A Note on Single Crystals in Polymers: Evidence for a Folded Chain Configuration," *Phil. Mag.*, **2**, 1171 (1957).

24. V. F. Holland, "Electron Microscope Studies of Crystalline Structures Precipitated from Dilute Polyamide Solutions," paper presented at Elec. Micro. Soc. Am. Meeting, Milwaukee, Wis. (1960).

25. A. Peterlin, F. Krasovec, E. Pirkmajer, and I. Levstek, "Dilatometric and Nuclear Magnetic Resonance Studies of Polyethylene with Different Branching and Crystallinity," *Makromol. Chem.*, **37**, 231 (1960).

26. W. P. Slichter, "Defects in Polyethylene Crystals," *J. Appl. Phys.*, **31**, 1865 (1961).

27. A. Odajima, J. A. Sauer, and A. E. Woodward, "Proton Magnetic Resonance of some Normal Paraffins and Polyethylene," *J. Phys. Chem.*, **66**, 718 (1962.)

28. H. Thurn, "Kernresonanzmessungen an Polyathylen," *Kolloid-Z.*, **179**, 11 (1961).

29. R. W. Kedzie, "NMR in Gravitationally Oriented Polyethylene Crystal Mats," to be published.

30. R. L. Collins, "Crystalline Recovery of Quenched Polyethylene," *J. Polymer Sci.*, **27**, 75 (1958).

31. W. P. Slichter, "Nuclear Magnetic Resonance Studies of Disordered Regions in Solid Polymers," paper presented at Conference on Physics of Polymers, Bristol, England (Jan., 1961).

32. D. H. Reneker, "Point Dislocations in Crystals of High Polymer Molecules," *J. Polymer Sci.*, **59**, 539 (1962).

33. R. W. Kedzie, "NMR Observations of Molecular Motion in Polyoxymethylene Crystal Lattice Defects," paper presented at Am. Phys. Soc. Meeting, Baltimore, March 1962.

34. P. H. Geil, "Recrystallization of Melt Crystallized Polyethylene during Annealing," paper presented at Am. Phys. Soc. meeting, Baltimore, March 1962.

35. V. Daniel, "The Physics of Long Chain Crystals," *Adv. in Physics*, **2**, 450 (1953).

36. L. Mandlekern, A. S. Posner, A. F. Diorio, and D. E. Roberts, "Low Angle X-Ray Diffraction of Crystalline Nonoriented Polyethylene and Its Relationship to Crystallization Mechanisms," *J. Applied Phys.*, **32**, 1509 (1961).

37. L. Mandlekern, A. S. Posner, A. F. Diorio, and D. E. Roberts, Paper presented at Am. Chem. Soc. Meeting, Boston, April (1959).

38. C. Sella, "Etude des polyethylenes par diffraction des rayons X aux petits angles," *Compt. rend.*, **248**, 1819 (1959).

39. C. Sella, "Etude des polyethylenes par diffraction des rayons X aux petits angles," IUPAC Symposium on Macromolecules, paper IB 7, Weisbaden (1959).

40. B. Belbeoch and A. Guinier, "Structure a Grande Echelle du Polyethylene," *Makromol. Chem.*, **31**, 1 (1959).

41. C. Sella, personal communication.

42. A. Guinier, personal communication.

43. H. Hendus, "Neuere physikalische Untersuchungen an Hochpolymeren. IV. Roentgenographische und electronmikeroskopische Strukturunsuchungen," *Ergeb. exakt. Naturwiss.*, **31**, 331 (1959).
44. H. Hendus, personal communication.
45. F. P. Reding and E. R. Walter, "An Electron Microscope Study of the Growth and Structure of Spherulites in Polyethylene," *J. Polymer Sci.*, **38**, 141 (1959).
46. J. S. Mackie and A. Rudin, "Electron Microscopy of Solvent-Etched Polyethylene Surfaces," *J. Polymer Sci.*, **49**, 407 (1961).
47. P. H. Geil, "Lamellar Crystallization of Low Density Polyethylene," *J. Polymer Sci.*, **51**, S10 (1961).
48. F. P. Reding, personal communication.
49. J. P. Hoffman, personal communication.
50. M. Gubler, J. Rabesiaka, and A. J. Kovacs, to be published.
51. S. Matsuoka, "Dilatometric Observation of Annealing Processes in Bulk Polyethylene," paper presented at Am. Phys. Soc. Meeting, Baltimore, March 1962.
52. A. Peterlin and E. Roeckl, "NMR and Dilatometric Studies of Polyethylene Recrystallization," *J. Applied Phys.*, **34**, 102 (1963).
53. A. Kovacs, "Kinetics of Crystallization of Polymers," *Recerci Sci.*, **25**, 668 (1955).
54. H. G. Zachman and H. A. Stuart, "Schmelz-und Kristallisationserscheinungen bei makromolekularen Substanzen. V. Partielles Schmelzen and Neukristallisieren von Terylen," *Makromol. Chem.*, **41**, 148 (1960).
55. H. A. Stuart, "Problems of High Polymer Crystallinity," *Ann. N. Y. Acad. Sci.*, **83**, 3 (1959).
56. R. Eppe, E. W. Fischer, and H. A. Stuart, "Morphological Structures in Polyethylenes, Polyamides and Other Crystallizable High Polymeric Materials," *J. Polymer Sci.*, **34**, 721 (1959).
57. G. Kampf, "Zur Ausbildung morphologischer Strukturen am Polykohlensaureester des 4,4'-Dioxydiphenyl-2,2 Propans (Polycarbonat)," *Kolloid-Z.*, **172**, 50 (1960).
58. F. A. Quinn, Jr., and L. Mandlekern, "Thermodynamics of Crystallization in High Polymers: Poly-(ethylene)," *J. Am. Chem. Soc.*, **80**, 3178 (1958).
59. L. Mandlekern, "The Dependence of the Melting Temperature of Bulk Homopolymers on the Crystallization Temperature," *J. Polymer Sci.*, **47**, 494 (1960).
60. J. D. Hoffman and J. J. Weeks, "Melting Process and the Equilibrium Melting Temperature of Polychlorotrifluoroethylene," *J. Research Natl. Bur. Standards*, **66A**, 13 (1962).
61. M. G. Broadhurst, "Extrapolation of the Orthorhombic *n*-Paraffin Melting Properties to Very Long Chain Lengths," *J. Chem. Phys.*, **36**, 2578 (1962).
62. N. Hirai, personal communication.

VI. THEORETICAL CALCULATIONS OF FOLDING

1. INTRODUCTION

In the first few years following the discovery of folded chain polymer crystals a number of suggestions were advanced as to the cause of the regularity of folding. Most of these suggestions were not developed beyond the hypothesis stage, being discussed briefly in papers devoted to experimental results, published as discussion to other papers, or presented informally at various meetings (1–6). At present only two theories have been developed to a stage such that they can be compared with experimental results. One, developed by Peterlin, Fischer, and Reinhold (7–11), suggests that the fold period is determined thermodynamically, corresponding to a minimum in the free energy density of the crystal at the crystallization temperature. The other theory, suggested independently by Lin (12), Price (13,14) and Lauritzen and Hoffman (15) and based on a kinetic approach, was subsequently developed and applied to crystallization from both solution and the melt by Lauritzen and Hoffman (16) and Hoffman and Lauritzen (17). In their extensive papers, Lauritzen and Hoffman also show the relationship and theoretical implausibility of the fringed micelle type of crystal growth with respect to the folded chain type for crystallization from both solution and the melt. More recently, some of the finer details of the theory have been discussed by Price (18), Frank and Tosi (19), and Lauritzen (20). It is primarily these two theories which we shall consider in this chapter. At present, insufficient experimental data are available to differentiate between the kinetic and equilibrium approaches.

Both of the present theories are based on free energy calculations. If a process, in this case crystallization, takes place in a system at equilibrium and at constant temperature and pressure, the free energy change accompanying the process is zero. If a system is not in equilibrium, for instance a melt or solution below the crystallization temperature, any process may occur spontaneously at constant temperature and pressure if it results in a lowering of the free energy of the system. A recent review of the thermodynamics of phase changes has been given by Turnbull (21).

In considering the development of polymer crystals it is necessary to consider both nucleation and crystal growth analogous to the way the theory has been developed for growth of crystals of low molecular weight materials. The general theory of nucleation is based on the assumption that thermal fluctuations in the liquid (melt or solution) result in the continual formation and disappearance of "crystalline" clusters of molecules. In these regions, which are assumed to be built up one molecule at a time, the molecular positions are similar to those in the crystalline solid. At any temperature, above or below the crystallization temperature, there is a statistically determined distribution in size of these clusters, many more small "crystalline" clusters than large ones being present. Those of interest for nucleation are assumed to contain a large number of molecules. As the temperature is lowered, thermal vibrations are less effective in over-coming the attractive intermolecular forces and the clusters of any size exist for longer periods of time. The average size of a cluster will also become larger.

Below the melting point the molar free energy or chemical potential of the molecules in a crystal is less than that of the molecules in a liquid; one might thus expect the "crystalline" clusters to be stable and grow at any temperature below the melting point. However, when the cluster is very small the free energy of the system at first increases as more molecules are added. The interfacial surface tension between the liquid and solid results in an interfacial free energy which increases with the surface area of the cluster, whereas the difference in chemical potential of the liquid and solid results in a bulk free energy which decreases with the volume of the cluster. The free energy of formation of a crystalline cluster, i.e., the change in free energy of the system when such a cluster is formed, is given by

$$\Delta F = \sum_i n_i a_i \sigma_i + \frac{n}{N_A} (\mu_s - \mu_l) \qquad \text{(VI-1)}$$

where the sum is over all the faces of the cluster and

n_i = number of molecules on faces of type i
a_i = cross-sectional area of molecules on faces of type i, $n_i a_i$ = area of faces of type i
σ_i = interfacial tension of faces of type i
n = total number of molecules in the cluster
N_A = number of molecules per mole, i.e., Avagadro's number
μ_s = chemical potential of the bulk solid
μ_l = chemical potential of material in the liquid state, solution or melt

The temperature dependence of ΔF in Equation VI-1 is primarily through μ_s and μ_l. For later discussions it must be recognized that μ_s is the change

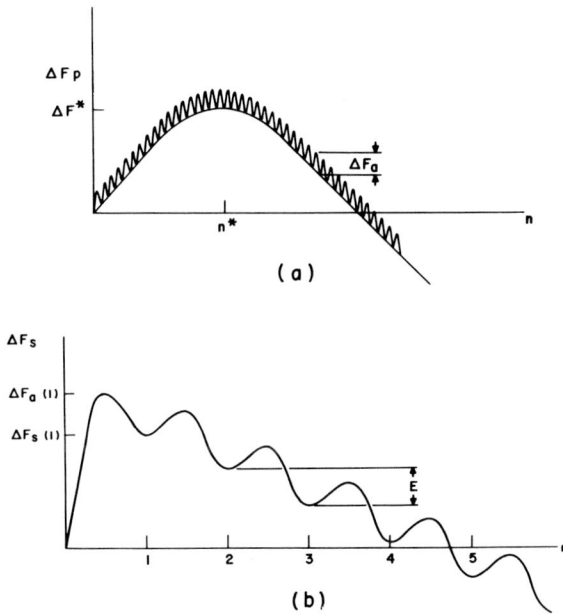

Fig. VI-1. (a) Free energy of formation of a primary nucleus built up through the addition of a large number of elements (16). ΔF_a is the height of the barrier for the addition of each element, i.e., the activation energy. After n^* elements have been added, ΔF_p decreases with further growth. The free energy of the system, however, does not decrease until $\Delta F_p < 0$. (b) Free energy of formation of a secondary nucleus consisting of one fold plane (16). The activation energy or free energy barrier for the laying down of the first segment on the previously completed growth face is $\Delta F_a(1)$. Subsequent addition of segments decreases the free energy of the monosegment nucleus, $\Delta F_s^*(1)$, by equal increments, E.

in free energy of the solid when one mole of molecules (repeating units in the case of polymers) is added to a crystal already so large that no further increase in surface energy occurs. When $\mu_s - \mu_l > 0$, ΔF is always positive and any size cluster will disappear as a result of thermal fluctuations, i.e., crystals will melt or dissolve. When $\mu_s - \mu_l < 0$, ΔF is positive for small n, negative for large n and 0 for $n = 0$ (Fig. VI-1). It thus has a maximum, ΔF^* (often termed a free energy barrier) for a particular n which we call n^* When n is greater than n^*, the free energy of the cluster will decrease as more molecules are added.

A cluster having n^* molecules is known as a nucleus and is said to have a critical nucleus size. The value of n^* will vary somewhat with the shape of the cluster due to the difference in interfacial free energy for different crystal faces. The free energy of a cluster smaller than the critical nucleus size increases as more molecules are added. Such growth is thus thermodynamically unstable and, on the average, thermal fluctuations will result in its disappearance. A few clusters, however, do grow to the critical nucleus size. For $n = n^*$ and even somewhat larger, ΔF is still greater than zero and thus the nucleus is not yet stable; the probability of growth, however, is now greater than the probability of dissolving and with increasing size ΔF soon becomes negative and the free energy of the system decreases. Growth will then continue until the supply of molecules is exhausted or a minimum in the free energy is reached.

The critical nucleus size decreases with decreasing temperature and thus the nucleation rate will rapidly increase with increasing supercooling. Turnbull and Fisher (22,23) have calculated this rate, assuming that the nucleus contains a large number of repeat units and is built up unit by unit.

It can be written as

$$I = K \exp \left\{ -(\Delta F^* + \Delta F_a)/kT \right\} \tag{VI-2}$$

where

$$K = cv_0 \left(\frac{\sum_i n_i a_i \sigma_i}{9\pi kT} \right)^{1/2}$$

c = number of molecules/unit volume of liquid
v_0 = fundamental jump frequency usually set equal to kT/h
ΔF_a = free energy of activation, i.e., the height of the free energy barrier the molecule must cross in jumping from the liquid to the solid

Thus $K \propto T^{+1/2}$. The "jumps" referred to involve the process by which an individual repeat unit crosses the interface and is added to or subtracted from the nucleus.

Once past the critical nucleus size, the growth rate limiting step is assumed to be the nucleation of a new lattice plane of molecules, a secondary nucleus, on a completed growth face. In crystals of low molecular weight materials the occurrence of screw dislocations permits faster growth rates. The perpetual angle between the advancing lattice plane and the underlying plane when a spiral growth pattern is present results in attractive forces in at least two directions for a crystallizing molecule and accounts for the rapid crystallization by spiral growth in crystals of such materials. No secondary nucleation is required. A similar type of defect has not been observed or postulated for the growth of lateral faces of polymer crystals. The spiral growths that are observed on polymer crystals are associated with growth parallel to the molecular axes and permit the crystal to thicken much more rapidly than would occur if thickening resulted only from the spontaneous nucleation of new lamellae on the fold surface of the older lamellae. The only evidence for the nucleation of a lamella on an existing lamella without a screw dislocation is the presence of overgrowths consisting of small crystals on polyethylene crystals which have been partially dissolved and then recrystallized (24) (Figs. II-59 and II-60). It was suggested (24) that the nucleation of the overgrowths was associated with rather long segments of molecules which protrude from the crystal. In polymer crystals the attraction between two fold surfaces is likely to be very small, and, therefore, the presence of the underlying lamella does not affect the growth rate other than through its limiting the solid angle in the liquid from which molecules are added to the growth face. If growth is diffusion-controlled to some extent, the growth rate of the growth faces of the lamellae in a spiral growth would be less than that of the basal lamella. The growth rate normal to the lamellae, resulting in thickening of the crystal by spiral growth, is related to this lateral growth rate.

The formation of a new lattice plane on a surface of any crystal, polymer or not, and therefore the growth of a crystal, involves a free energy change similar to that in Equation VI-1 except for the interfacial free energy term. This differs in that the only new surfaces formed during creation of the secondary nucleus are those which are not parallel to the original crystal face. The parallel face of the secondary nucleus is a projection of the original surface of the crystal. The growth rate can be calculated from an equation similar to Equation VI-2.

The free energy of formation of a small polymer crystal can be given more explicitly than in Equation VI-1. The interatomic forces in a polymer crystal are highly anisotropic. Strong covalent bond forces hold the repeating units together along the molecule whereas much weaker van der Waals forces act between the chains. Thus, the nature of the surfaces and the surface free energies would be expected to differ in directions normal to and parallel to the chains, regardless of whether a lamellar, folded chain type crystal or a fringed micelle, bundlelike crystal is considered. For a polymer crystal, bundlelike or lamellar, the free energy of formation is

$$\Delta F = 2\nu a \sigma_e + \ell C (\nu a)^{1/2} \sigma_s + \frac{\nu a \ell}{V_s} (\mu_s - \mu_l) \qquad \text{(VI-3)}$$

where

ν = number of segments in a cross section of the crystal

a = cross-sectional area per segment

σ_e = free energy of the surface normal to the chain direction per unit area

ℓ = length of crystal in the direction of the molecules

C = geometrical factor relating perimeter to area, for instance $C = 4$ for a square and $2(12)^{1/4}$ for a hexagon

σ_s = interfacial free energy per unit area (surface tension) for the sides of the crystal (assumed the same for all sides)

μ_s = free energy per mole of polymer repeating unit in the bulk crystalline phase

μ_l = free energy per mole of polymer repeating unit in the liquid (melt or solution)

V_s = volume per mole of polymer repeating unit in the crystalline phase

In a lamellar crystal, ℓ is the fold period and σ_e is the sum of the interfacial tension and work per unit area (actually change in free energy) required to form the folds ($a\sigma_e = \frac{1}{2}$ the work required to form a fold since on a given surface there are 2 segments per fold). For a bundlelike crystal, ℓ will be the length of the bundle and the amount of work necessary to form a unit area of the end of the crystal, σ_e will consist primarily of the amount of work necessary to separate the chains as they leave the end of the crystalline region.

The molar free energy or chemical potential of a crystal, μ_s, as indicated previously, can be defined as the change in free energy of the crystal per mole of molecules added at constant temperature, pressure, surface area, etc.; i.e., all other factors which contribute to the free energy are assumed to be unaffected. In a lamellar crystal, μ_s can be considered as the free energy change per mole of repeating units in the form of straight segments added to a crystal whose lateral extent is so large that the total interfacial energies are not altered. The total change in free energy that occurs would include that due to the folds.

The kinetic theories of polymer crystallization, as developed by Lin (12), Price (14), and Lauritzen and Hoffman (16,17), begin with an equation similar to Equation VI-3, assume that a polymer crystallizes by folding, and determine the rate of nucleation of primary and secondary (new fold planes on a growing crystal) nuclei as a function of their fold period assuming that the chemical potential, μ_s, and the interfacial free energy are independent of the segment length in the crystal. The nuclei which are most probable, reaching critical size most rapidly as molecules are added, are those which correspond to the lowest free energy barrier. A plot of the free energy of formation as a function of ℓ and ν is a saddle shaped surface, the lowest barrier corresponding to the saddle point. Although, as will be shown, nuclei with longer fold periods are thermodynamically more stable under the assumption that chemical potentials and interfacial free energies are segment length independent, those nuclei having long fold periods require a greater free energy of activation and thus fewer are formed than those whose fold period and lateral extent correspond to the saddle in ΔF. With this theory an increase in fold period may occur at any temperature below the melting point at which the molecules have sufficient mobility and thus may occur while the crystals are still at the crystallization temperature or, perhaps, if stored at somewhat lower temperatures. Recrystallization during annealing at temperatures above that corresponding to the crystallization temperature may occur as above or through a melting and renucleation step. Recrystallization at a smaller fold period would be thermodynamically forbidden unless the sample is first heated sufficiently to melt it and then supercooled to a temperature such that a short fold period would be produced by the above arguments.

The theory of Peterlin, Fischer, and Reinhold (9,10), on the other hand, does not consider the nucleation step. They attempt to show that the molar free energy of the crystal, μ_s, depends on the segment length in the crystal. Very likely σ_s would then also depend on the segment length. They suggest that the amplitude of thermal oscillations of the atoms in a polymer chain in a crystal, at a given temperature, increases with increasing segment length. Thus, if the oscillations in neighboring chains are incoherent, the average force of attraction between the chains and, as a result, the molar free energy of the crystal is lowered as the segment length increases. They find that below a certain critical temperature, which is in the range of the crystallization temperatures of interest, there is a minimum in the free energy density of the crystal with respect to the segment length when the free energy due to the ends of the crystal is also considered. The segment length corresponding to the minimum, which increases with increasing temperature, is the equilibrium thickness of the crystal. Above the critical temperature only an infinite crystal is stable and no finite fold period would be expected. This theory does not discriminate between bundlelike crystals and crystals with folded chains, requiring only a decoupling of the thermal oscillations at distances corresponding to the equilibrium segment length. Decoupling can occur by folding or other disordering of the crystalline lattice. Recrystallization during annealing, according to this theory, is an equilibrium process and may result in an increase or decrease of the fold period. However, the development of a shorter fold period, by annealing below the crystallization temperature, may take place so slowly as to be experimentally unmeasurable. In addition, annealing between the crystallization temperature and the critical temperature would be expected to lead to a new, finite fold period whereas annealing above the critical temperature would result in the fold period increasing without limit.

In the following two sections we develop the two theories in more detail. The results of the calculations should be compared with the experimental results, summarized briefly below, which are now available.

1. Many, perhaps all, crystallizable polymers crystallize from solution and from the melt in the form of lamellae or the degenerate form, ribbons, in which the molecules are folded.

2. The fold period depends on the temperature of crystallization, increasing with increasing temperature. For many polymers it is on the order of 100 A.

3. At a given crystallization temperature the number of orders of small angle x-ray diffraction reflections that are observed indicate that the fold period of solution grown crystals differs by 8% or less from crystal to crystal.

4. The development of a hollow pyramidal crystal structure during crystallization suggests that there is a sharply defined fold period in a given single crystal. There are suggestions that the pyramid may develop by a readjustment of an original, somewhat variable, fold period.

5. Discrete changes in crystallization temperature, either up or down, result in corresponding substeps in the growing lamellae.

6. The fold period increases if a lamella crystallized from solution or the melt is heated at temperatures above the original crystallization temperature. There is probably a limiting temperature, near but above the glass transition temperature, below which refolding does not take place. In addition there may be a critical temperature above which the fold period will increase indefinitely with log time but below which it will approach a finite value.

7. Drawn polymers often have a well-defined periodicity, on the order of 100 A, which may or may not be related to the same cause as the long period from the lamellae. The long periods in drawn polymers also increase during annealing. This is discussed in Chapter VII.

8. Lamellae develop when previously quenched, glasslike polymers crystallize due to heating above the glass transition temperature. The thickness of the lamellae depends on the maximum temperature of annealing, provided the melting point is not exceeded. If it is exceeded, the thickness depends on the manner in which the sample is cooled.

2. KINETIC OR NUCLEATION DETERMINED GROWTH

The derivation presented here, in which μ_s and σ_s are assumed independent of ℓ, is primarily based on the papers of Lauritzen and Hoffman (16,17). Various aspects of the derivation were also considered by Lin (12) and Price (14). Although in their first paper (16) Lauritzen and Hoffman concluded that during crystallization from the melt or concentrated solution, a bundlelike crystal is energetically the most favorable, in the second paper (17) they indicate that this is not so. Calculations, based on a density gradient model for the end of a bundlelike crystal, show that the magnitude of the end free energy of a bundlelike crystal is considerably larger than the end free energy including the fold energy of a lamellar crystal. It requires more energy to separate all the molecules at the end of a bundlelike crystal and form some sort of strained, disordered region than it does to fold the molecule back on itself. In the following pages we shall present only the essential parts of the derivation; for further details the reader is referred to the original papers. In their derivation Lauritzen and Hoffman (16,17) assume, once a given fold plane is nucleated with a certain fold period, that that fold period is maintained during the fold plane's growth. In subsequent papers, whose results we shall discuss briefly, Price (18), Frank and Tosi (19), and Lauritzen (20) have considered the effect of letting the fold period vary one or more times during the growth of a specific fold plane. The results of these refinements are in essential agreement with the original results of Lauritzen and Hoffman.

a. Primary Nucleation

The free energy of formation of a primary or homogeneous nucleus from melt or solution may be written, as in Equation VI-3, as

$$\Delta F_p = 2\nu a \sigma_e + \ell_p C \nu^{1/2} a^{1/2} \sigma_s - \nu a \ell_p \Delta\mu \tag{VI-4}$$

where $\Delta\mu = (\mu_l - \mu_s)/V_s$ and ℓ_p refers to the length of the primary nucleus.* Equation VI-4 defines, for a given temperature, and as a function of ℓ_p and νa, a saddle-shaped free energy surface (Fig. VI-2). The primary effect of a change in temperature is to change the value of $\Delta\mu$ so that ΔF decreases with increasing temperature and equals zero at the melting point (or the solubility temperature in a solvent). The variation in σ_e and σ_s with temperature is neglected. For any given ℓ_p the maximum in ΔF_p, which is the free energy barrier for the formation of a nucleus of fold period ℓ_p, is given by

$$(\Delta F_p)_{\max} = \frac{C^2\ell_p{}^2\sigma_s{}^2}{4(\ell_p\Delta\mu - 2\sigma_e)} \qquad (\text{VI-5})$$

and is at

$$(\nu a)^{1/2} = \frac{C\ell_p\sigma_s}{2(\ell_p\Delta\mu - 2\sigma_e)} \qquad (\text{VI-6})$$

With increasing temperature, $\Delta\mu \to 0$ and thus the magnitude of the free energy barrier for any ℓ_p increases. Furthermore it is seen that ℓ_p must be greater than $2\sigma_e/\Delta\mu$, since ΔF will increase indefinitely with νa for $\ell_p \lesssim 2\sigma_e/\Delta\mu$.

The free energy surface shown in Figure VI-2 is basically the same for both bundlelike and lamellar crystals, only the definitions of ℓ_p and σ_e differing. In both cases the most probable nucleation path is that which passes over the lowest point of the barrier. In the case of a bundlelike nucleus, the nucleus would presumably be built up by a series of additions of repeating units, both along the molecules already incorporated as they straighten out and, to an extent depending on the polymer concentration, as additional molecules are added laterally. In a folded chain, lamellar crystal, Lauritzen and Hoffman assume that the elementary process in nucleus development is the addition of "step elements" of length equal to the fold period, stating (16) that "the paths by which nuclei with folds are formed are characterized by a length that is invariant" as the embryo

---

* Lauritzen and Hoffman (16) also introduce an edge free energy term proportional to the perimeter of the fold surfaces to account for the energy required to form an edge of the crystal. In a growing crystal it accounts for the energy involved during secondary nucleation if the new fold plane has a different fold period than the crystal on which it is being formed. It does not, we believe, apply to a primary nucleus, the energy involved in regular or irregular displacements of folds being proportional to the area of the fold surface and, therefore, included in σ_e. In the following, as in their derivation, the edge free energy for a primary nucleus will not be considered.

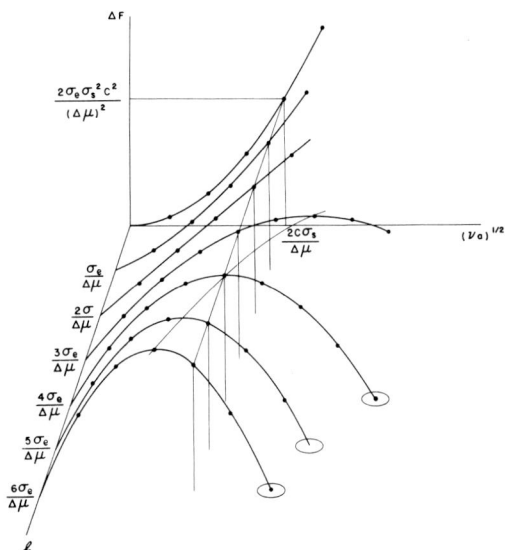

Fig. VI-2. Free energy of formation of a primary nucleus of a polymer crystal as a function of its thickness and lateral size, $(\nu_a)^{1/2}$. The minimum free energy barrier $\Delta F_p{}^*$ corresponds to the saddle point, occurring at

$$\ell_p{}^* = \frac{4\sigma_e}{\Delta\mu} \quad \text{and} \quad [(\nu a)^*]^{1/2} = \frac{C\sigma_s}{\Delta\mu}$$

develops. Although this is not necessarily true, the fold period perhaps being variable during nucleus formation, it is not believed to affect the subsequent derivation for primary nuclei.

The most probable value of ℓ_p, corresponding to the lowest free energy barrier at the given temperature, is

$$\ell_p{}^* = \frac{4\sigma_e}{\Delta\mu} \tag{VI-7}$$

The free energy of a nucleus with such a fold period will decrease with the addition of further step elements when

$$\nu a \geqslant \frac{C^2\sigma_s{}^2}{(\Delta\mu)^2} \tag{VI-8}$$

The minimum free energy barrier for a primary nucleus is thus

$$\Delta F_p{}^* = \frac{2C^2\sigma_s{}^2\sigma_e}{(\Delta\mu)^2} \tag{VI-9}$$

and

$$V_p^* = (va\ell)_p^* = \frac{4C^2\sigma_s{}^2\sigma_e}{(\Delta\mu)^3} \tag{VI-10}$$

At this point we can consider the equilibrium value of ℓ, ℓ_e, for a crystal of a given volume, corresponding to a minimum in ΔF (12). The variation of μ_s with ℓ, as treated by Peterlin, Fischer, and Reinhold (9,10), is not involved in this calculation. For any crystal, regardless of its size, Equation VI-4 applies, and can be written as

$$\Delta F = V\left(-\Delta\mu + \frac{2\sigma_e}{\ell} - \frac{C\ell^{1/2}2\sigma_s}{V^{1/2}}\right)$$

where V is the volume of the crystal. $\partial\Delta F/\partial\ell = 0$ for

$$\ell_e = \left(\frac{4\sigma_e}{C\sigma_s}\right)^{2/3} V^{1/3} \tag{VI-11}$$

but

$$V > (va\ell)_p^* = \frac{4C^2\sigma_s{}^2\sigma_e}{(\Delta\mu)^3} \tag{VI-12}$$

therefore, substituting Equation VI-12 in Equation VI-11

$$\ell_e > \frac{4\sigma_e}{\Delta\mu} = \ell_p^*$$

Thus, the equilibrium value of ℓ for a grown crystal is greater than the fold period of the primary nucleus. The influence of a mosaic structure on V, perhaps limiting the size of the crystal that can be grown at a given temperature and therefore the maximum fold period that can be attained during subsequent annealing is not known.

According to Hoffman (25), $\Delta\mu$, for crystallization from the melt, is given approximately by

$$\Delta\mu = \Delta h_f \frac{T(T_m - T)}{T_m{}^2} = \frac{\Delta h_f T \Delta T}{T_m{}^2} \tag{VI-13}$$

where $\Delta T = T_m - T$ and Δh_f is the heat of fusion per unit volume of the crystal at the equilibrium ($\Delta\mu = 0$) melting point of the crystal, T_m. In portions of their derivation Lauritzen and Hoffman (16,17) also use the approximation, suitable for crystallization from solution,

$$\Delta\mu = \frac{\Delta h_f \Delta T}{T_m} \tag{VI-14}$$

Unless otherwise indicated we shall use the latter approximation throughout this derivation. In the presence of a solvent, T_m is the equilibrium solubility temperature in the presence of a large volume of the solvent and thus refers to the solubility temperature for a crystal with an "infinite" fold period. T_m, in the case of solutions, is greater than the temperature at which a crystal of finite fold period will dissolve but is lower than the equilibrium melting point.

On the basis of Equation VI-14

$$\ell_p{}^* = \frac{4\sigma_e T_m}{\Delta h_f \Delta T} \tag{VI-15}$$

$$(va)_p{}^* = \frac{C^2 \sigma_s{}^2 T_m{}^2}{(\Delta h_f)^2 (\Delta T)^2} \tag{VI-16}$$

and

$$\Delta F_p{}^* = \frac{2C^2 \sigma_s{}^2 \sigma_e T_m{}^2}{(\Delta h_f)^2 (\Delta T)^2} \tag{VI-17}$$

From Equations VI-5 and VI-9 it is seen that the height of the free energy barrier for some ℓ_p other than $\ell_p{}^*$ is

$$(\Delta F_p)_{\max} = \Delta F_p{}^* \left(1 + \frac{\ell_p/\ell_p{}^* - 1}{1 + 2(\ell_p/\ell_p{}^* - 1)} \right) \tag{VI-18}$$

Using a derivation similar to that used by Turnbull and Fisher (22) in calculating Equation VI-2, Lauritzen and Hoffman (16) show that the number of nuclei formed per unit time per unit volume with fold periods between ℓ_p and $\ell_p + d\ell_p$ is given by

$$i(\ell_p)d\ell_p = i(\ell_p{}^*)d\ell_p \exp \left[\frac{-\Delta F_p{}^*}{kT} \left(\frac{(\ell_p/\ell_p{}^* - 1)^2}{1 + 2(\ell_p/\ell_p{}^* - 1)} \right) \right] \tag{VI-19}$$

where $i(\ell_p{}^*)$ is the nucleation rate corresponding to the minimum in the free energy barrier. Thus, according to Equation VI-19 incipient crystals with a fold period $\ell_p{}^*$, having the lowest barrier to surmount, reach the critical nucleus size in fewer steps and are thus most probable. The distribution in fold periods about $\ell_p{}^*$ becomes sharper as $\Delta F_p{}^*$ increases, i.e., for a given polymer, as ΔT decreases.

In their papers, Lauritzen and Hoffman (16,17) show that the over-all nucleation rate for primary nuclei with folds, assuming a rectangular parallelepiped shape, is

$$I_{\text{folds}} = I_{0\,\text{folds}} \exp \frac{-\Delta F_a}{kT} \exp \frac{-32\sigma_s{}^2\sigma_e}{(\Delta\mu)^2 kT} \qquad \text{(VI-20)}$$

The nucleation rate for bundlelike crystals of similar shape is given by

$$I_{\text{bundle}} = I_{0\,\text{bundle}} \exp \frac{-\Delta F_a}{kT} \exp \frac{-32\sigma_s{}^2\sigma_e}{(\Delta\mu)^2 kT} \exp \frac{16\sigma_s{}^2 \ln v_2}{a(\Delta\mu)} \qquad \text{(VI-21)}$$

where v_2 = volume fraction of polymer in the liquid. Equations VI-20 and VI-21 differ primarily in the value of σ_2 and the exponential term containing v_2; variations of ΔF_a and I_0 are less significant and $\Delta\mu$ and σ_s are probably nearly the same. In their first paper (16) they show that regardless of the values of σ_e, $I_{\text{fold}} > I_{\text{bundle}}$ if v_2 is small enough. The change in entropy involved in collecting together a large number of molecules to form a bundle is a much greater barrier to nucleus formation than the energy involved in the folds. In crystallization from the bulk ln v_2 is zero. However, near the melting point they point out (17) that if $(\sigma_e)_{\text{fold}} < (\sigma_e)_{\text{bundle}}$, then lamellar type nuclei will be formed preferentially. Their calculations of $(\sigma_e)_{\text{bundle}}$, based on a qualitative model (17), suggest, as might be expected due to the energy required to separate and disorder the molecules at the boundary of the crystal, that $(\sigma_e)_{\text{bundle}}$ is greater than $(\sigma_e)_{\text{fold}}$.

b. Secondary Nucleation

Growth of the crystal requires the addition of new molecules to the growth faces. At various stages in this process a new fold plane must be nucleated on a completed growth face, for example when a molecule reaches a corner and has to turn back or continue around the corner or when a new molecule is added to the crystal at some point along a completed growth face. Although the secondary nucleus may consist of a number of step elements in several fold planes this is considered less likely than the formation of a single fold plane.

The free energy of formation of a coherent, monomolecular layer type nucleus is given by

$$\Delta F_s = 2va_0 b_0 \sigma_e + 2b_0 \ell \sigma_s + 2va_0 \epsilon - va_0 b_0 \ell \Delta\mu \qquad \text{(VI-22)}$$

where va_0 is the width of the nucleus on the growth face containing step elements of thickness b_0.

In this case we have added the term for the edge free energy, ϵ, to take into account the possibility that the new fold plane may have a smaller fold period than the substrate. Still further terms involving σ_s must be added if the new fold period is larger than that of the substrate. Equation VI-22 differs from Equation VI-4 in that no term $2\nu a_0 \ell \sigma_s$ appears.

As in the case of a primary nucleus ℓ must be greater than

$$\ell_0 = 2\sigma_e/\Delta\mu + 2\epsilon/b_0\Delta\mu \qquad (\text{VI-23})$$

for ΔF_s to become negative for some value of ν. For any possible ℓ the highest value of ΔF_s corresponds to $\nu = 1$; as more elements are added ΔF_s decreases (Fig. VI-1b). Thus, Lauritzen and Hoffman conclude (16) that the formation of the nucleus for a new fold plane is a one-step process involving the attaching of a segment of length ℓ onto the completed fold plane. It is assumed that the laying down of a segment of length ℓ involves only one step, i.e., the fact that the segment probably is laid down a few chemical repeat units at a time is neglected. Frank and Tosi (19) show that this assumption does not introduce a significant error. After a number of further step elements are added ΔF_s will become negative.

The free energy surface defined by Equation VI-22, for $\nu = 1$ and possible ℓ, does not have a minimum. The change in free energy when one step element is added to a completed growth face is

$$\Delta F_s(1) = 2a_0(b_0\sigma_e + \epsilon) + b_0\ell(2\sigma_s - a_0\Delta\mu) \qquad (\text{VI-24})$$

Lauritzen and Hoffman point out (16) that $2\sigma_s > a_0\Delta\mu$ (numerical values of these and other factors are given near the end of this section) in the temperature range of interest and, therefore, $\Delta F_s(1)$ increases with increasing ℓ. The addition of each further step element changes the free energy by a constant amount

$$E = -a_0 b_0 \left(\ell\Delta\mu - 2\sigma_e - \frac{2\epsilon}{b_0} \right) \qquad (\text{VI-25})$$

Thus the decrease in free energy with further growth is faster, the larger ℓ. For the relationship between these factors see Figure VI-1b.

Lauritzen and Hoffman calculate (16) the net rate of formation of new fold planes of fold period ℓ, assuming that once nucleated the fold plane will rapidly become completed with the same fold period, as a function of ℓ. The growth rate limiting step is thus assumed to be that of secondary nucleation.

The rate at which a new fold plane becomes stable (ΔF becomes negative, see Figure VI-1) is derived as a function of the fold period ℓ. The value of ℓ.

corresponding to those fold planes which most rapidly nucleate and become stable will be the most probable. They assume that once a fold plane begins growing all additional segments laid down in that plane have the same period. It is in this assumption that the fluctuation theories of Price (18), Frank and Tosi (19), and Lauritzen (20) differ, the latter theories permitting one or more fluctuations in fold period during the growth of a given fold plane. The rate of steady state nucleation of new fold planes of length ℓ is given by

$$r \approx K' \exp \frac{-\Delta F_a(1)}{kT} \exp \frac{-\Delta F_s(1)}{kT} \sinh \frac{E}{2kT} \qquad \text{(VI-26)}$$

where $\Delta F_s(1)$ and E depend on ℓ, and $\Delta F_a(1)$ is the free energy barrier to the addition of one step element. K' is a constant depending on factors similar to K in Equation VI-3. This rate of growth of a fold plane is derived by comparing the rate at which fold planes with ν step elements of length ℓ add 1 element to become a layer with $\nu + 1$ elements and the reverse process in which a layer with $\nu + 1$ elements loses 1 element. The average or most probable value for ℓ is then determined

$$\bar{\ell} = \frac{\int_{2(\sigma_e + \epsilon/b_0)/\Delta\mu}^{\infty} \ell r d\ell}{\int_{2(\sigma_e + \epsilon/b_0)/\Delta\mu}^{\infty} r d\ell}$$

$$= \frac{2(\sigma_e + \epsilon/b_0)}{\Delta\mu} + \frac{2kT}{4b_0\sigma_s - 3a_0b_0\Delta\mu} + \frac{2kT}{4b_0\sigma_s - a_0b_0\Delta\mu} \qquad \text{(VI-27)}$$

where the lower limit of the integration is set at

$$2(\sigma_e + \epsilon/b_0)/\Delta\mu$$

since for ℓ less then this, E is positive (Eq. VI-25).
When

$$\sigma_s \gg a_0\Delta\mu$$

$$\bar{\ell} = \frac{2(\sigma_e + \epsilon/b_0)}{\Delta\mu} + \frac{kT}{b_0\sigma_s}$$

$$= \frac{2\sigma_e T_m}{\Delta h_f \Delta T} + \frac{2\epsilon T_m}{b_0\Delta h_f \Delta T} + \frac{kT}{b_0\sigma_s} \qquad \text{(VI-28)}$$

The mean square deviation in ℓ is given, to the same approximation, by

$$< (\ell - \bar{\ell})^2 > = \frac{1}{2} \frac{(kT)^2}{b_0\sigma_s} \qquad \text{(VI-29)}$$

Comparison with Equation VI-7 indicates that

$$\bar{\ell} = \frac{\ell_p^*}{2} + \frac{2\epsilon}{b_0 \Delta\mu} + \frac{kT}{b_0 \sigma_s} \tag{VI-30}$$

Therefore $\bar{\ell}$ lies between $\ell_p^*/2$ and ℓ_p^* since $kT/b_0\sigma_s$ is found to be less than $\ell_p^*/2$ (see Section VI-2d) and ϵ is nonzero only for $\bar{\ell} < \ell_p^*$.

The difference between the equation for $\bar{\ell}$ and ℓ_p^* arises from the fact that the primary nucleation process is considered to involve a large number of activated steps, i.e., the addition of a large number of step elements of the same length, the free energy of the system increasing with each addition until the critical size is attained, whereas in the secondary nucleation process the critical size is attained in one step and all further additions of step elements of the same length result in a lowering of the free energy (compare Figs. VI-1a and VI-1b).

The lateral growth rate of a crystal, assuming that a fold plane is rapidly completed following nucleation, is equal to the product of the rate of nucleation of new fold planes on the advancing face and the lattice spacing, i.e., $G = rb_0$. The macroscopically observed growth rate is assumed equal to the rate of nucleation for $\ell = \bar{\ell}$, thus

$$G = G_0 \exp\frac{-\Delta F_a(1)}{kT} \exp\frac{-\overline{\Delta F_s(1)}}{kT} \tag{VI-31}$$

where G_0 is a function of ℓ and T. However from Equations VI-24 and VI-27

$$\overline{\Delta F_s(1)} = \frac{4\sigma_s}{\Delta\mu}(b_0\sigma_e + \epsilon) + kT\left(2 - \frac{a_0\Delta\mu}{\sigma_s}\right) \tag{VI-32}$$

Therefore,

$$G = G_0 \exp\frac{-\Delta F_a(1)}{kT} \exp\frac{-4\sigma_s(b_0\sigma_e + \epsilon)}{kT\Delta\mu} \exp\frac{a_0\Delta\mu}{\sigma_s} \tag{VI-33}$$

Since, from Equation VI-13

$$\Delta\mu = \frac{T}{T_m}\frac{\Delta h_f \Delta T}{T_m}$$

where the factor T/T_m is not present for crystallization from dilute solution, the temperature dependence of the growth rate is given by

$$G = G_0(T, \Delta T) \exp\frac{-\Delta F_a(1)}{kT} \exp\frac{-4\sigma_s(b_0\sigma_e + \epsilon)T_m^2}{kT^2\Delta h_f\Delta T} \exp\frac{a_0 T \Delta h_f \Delta T}{T_m^2}$$

$$\tag{VI-34}$$

The growth rate is thus essentially proportional to exp $(\Delta T)^{-1}$ since the last term is negligible under normal conditions. This law should apply for lamellar growth from solution or the melt down to a temperature such that nucleative collapse occurs; i.e., the critical size is so small that nuclei appear throughout the sample. This temperature may be above or below that at which $G_0(T)$ becomes so small that growth would cease due to lack of motion of the molecule, the material becoming glassy. In using this equation for growth rates as a function of ΔT the exp T^{-1} term must be considered.

It is to be noted that the same equations apply to crystallization from the melt and solution. The only difference is in the values of the parameters, the major effect being seen in the values of μ_l, σ_s, and K' and G_0.

It is appropriate at this point to consider what happens when the end of a molecule is reached as it is folding within a fold plane. Hoffman and Lauritzen (17) suggest that the last segment of the molecule, being shorter than the fold period, will not attach to the crystal face but would protrude like a cilia from the fold surface. Renucleation with a new molecule would be more probable at this jog in the growth face than elsewhere. They suggest that a few ends may be incorporated as defects within the lattice but that in this case the step height locally would be larger. The local change in fold period is credited to a local deficit in Δh_f, i.e., the defect due to the terminal group of the molecule creates a strain in the surrounding lattice resulting in a change in the local free energy density.

No direct evidence of "cilia" or localized changes in fold period due to defects has been reported. Probably, however, they would not be directly visible in the electron microscope. Experience with branched polyethylene indicates that the lattice can accommodate a large number of defects. We believe that many, perhaps most, of the terminal groups are included within the lattice. It is likely that the molecule lies down on the growth face one or a few repeat units at a time and therefore the end would not greatly affect the folding unless the end is close to the fold. It is only when the presence of the end affects the folding that terminal segments might be required to protrude as cilia. Furthermore, it is not considered likely that the first segment of a new molecule to attach itself will be a terminal segment. Rather, it may be anywhere along the chain, resulting in two long ends, only one of which can be included in the then growing fold plane. The left-over portion of the molecule would have to be accommodated in the next fold plane. In a rapidly grown crystal the result may be a large number of improper folds and probably a higher than predicted growth rate due to secondary nucleation at various places along the growth face. In a slowly grown crystal surface diffusion probably occurs, permitting readjustment of the molecule and inclusion within one fold plane. Considering the rate at which the fold period can increase during annealing treatment (Chapter V), such surface diffusion, probably through point defect motions as postulated by Reneker (26), may play an important role in the growth of lamellar crystals from both the melt and solution. An increased concentration of defects within a crystal of given thickness should result in a lowering of its melting point. This effect is found in the comparison of branched and linear polyethylene.

Lattice strain, resulting from an accumulation of defects and impurities or due to the presence of folds, may cause the growth rate of the lamellae to decrease with time (and size). Hoffman and Lauritzen (17) indicate that in such a case growth may be reactivated by the formation of a noncoherent nucleus on the growth face of the lamella being considered. Nucleation here is favored over spontaneous nucleation elsewhere, the growth face serving as a wettable substrate. The nucleus that is formed will be three dimensional in character, similar to the primary nucleus and in contrast to the monomolecular layer or single fold plane type of secondary nucleus. Following nucleation, growth occurs coherently by the formation of new fold planes as described above until sufficient lattice strain has again accumulated. If the formation of the noncoherent nucleus is the radial growth rate determining step, the growth being spasmodic in time, they indicate that a $(\Delta T)^{-2}$ law will apply. The noncoherent nucleus may form epitaxially, having preferred angles of attachment to the original growth face. In this case, if the rotation is always in the same direction, a twisted lamellar structure will form with, probably, a nearly regular period to the twist. Regular twisting of the coherently grown lamellae may also occur. This type of twisting is discussed later (p. 401).

c. Melting Behavior

The free energy per unit volume of a crystal of volume $\nu a \ell$ with respect to the liquid state at any temperature, is given by (see Eq. VI-4)

$$\frac{\Delta F}{\nu a \ell} = \frac{2\sigma_e}{\ell} + \frac{C\sigma_s}{\nu^{1/2} a^{1/2}} - \frac{\Delta h_f \Delta T}{T_m} \qquad \text{(VI-35)}$$

where the factor T/T_m is again neglected in the following.

For a crystal of given ℓ the equilibrium melting temperature or solubility temperature $T_m(\ell)$ can be found by setting $\Delta F = 0$, remembering that T_m is the equilibrium melting temperature for an infinite crystal ($\Delta\mu = 0$).

$$T_m(\ell) = \frac{T_m}{\Delta h_f}\left(\Delta h_f - \frac{2\sigma_e}{\ell} - \frac{C\sigma_s}{\nu^{1/2}a^{1/2}}\right)$$

$$\approx T_m\left(1 - \frac{2\sigma_e}{\ell\Delta h_f}\right) \qquad\qquad \text{(VI-36)}$$

since νa will be large. From Equation VI-30 $\ell_p{}^*/2 < \ell < \ell_p{}^*$ where $\ell_p{}^*$ is given by Equation VI-15 ($\Delta T = T_m - T_{xl}$) and T_{xl} is the temperature of crystallization corresponding to $\ell_p{}^*$ and $\bar{\ell}$. Therefore, if $\ell = \bar{\ell}$, i.e., there is no change in ℓ after crystallization, then $T_m(\ell)$ is within the limits;

$$T_{xl} < T_m(\ell) < \frac{T_m + T_{xl}}{2} \qquad\qquad \text{(VI-37)}$$

In any experimental determination of the melting point of a collection of crystals, the observed melting point T_{obs} is likely to be higher than the melting point of the crystals with the average fold period. There will be a distribution of fold periods among the crystals and possibly within each crystal. Most experimental methods determine some type of maximum melting point, i.e., the temperature at which the last traces of crystallinity, as defined by the experiment, disappear. Not only will T_{obs} vary with the sensitivity of the experiment but it will also be dependent on the tail of the fold period distribution. To allow for this feature and also for the possibility that the fold period may increase after formation, while the crystal remains at the crystallization temperature, as a result of chain mobility in the lattice (Chapter V), Hoffman and Weeks (27) have introduced a parameter $\beta = \ell_{obs}/\ell_p{}^*$. ℓ_{obs} signifies that fold period corresponding to the observed melting point, T_{obs}.

From Equation VI-36 and VI-15, T_{obs} is given by

$$T_{obs} = T_m - \frac{1}{2\beta}(T_m - T_{xl}) \qquad\qquad \text{(VI-38)}$$

As $\beta \to \infty$, $T_{obs} \to T_m$ while for $\beta = 1$, $\ell_{obs} = \ell_p{}^*$ and $T_{obs} = \dfrac{T_m + T_{xl}}{2}$.

One should note that for $\beta = 1$, ℓ_{obs} is larger than $\bar{\ell}$ (i.e., $\ell_p \approx 2\bar{\ell}$). Because of the contribution of the edge free energy and $kT/b_0\sigma_s$ one expects β to be between $1/2$ and 1 if the fold period remains constant after formation and to be larger than 1 if it increases between the time of formation and the melting of the crystal. As indicated in Chapter V, Section 5, Hoffman and Weeks (27) find that in experiments in which recrystallization is avoided, β is approximately 1 for polyethylene and about 1.7 for polychlorotrifluorcethylene, suggesting that at least for the latter polymer the fold period increases after formation of the lamellae.

d. Numerical Values of Parameters and Comparison with Experimental Data

Unfortunately, even for polyethylene which is probably the best characterized polymer, several of the parameters in the previous equations, in particular σ_e, σ_s, and T_m in solution, have not yet been adequately determined. Those which are reasonably well known are listed below.

$$\Delta h_f = 2.8 \cdot 10^9 \text{ ergs/cm}^3 \text{ (16)}$$
$$a = 18.7 \text{ A}^2 \text{ (90°C) or } 19.0 \text{ A}^2 \text{ (130°C) (28)}$$
$$T_m = 415° \text{ in bulk polyethylene}^*$$
$$\bar{\ell} = 140 \text{ A for crystallization from xylene at 90°C}$$
$$\approx 300 \text{ A for crystallization from the melt at 131°C (29)}$$
$$kT = 1.38 \cdot 10^{16} \, T \text{ ergs}$$
$$C = 4.13 \text{ for a rhombus with an acute angle of 70°}$$
$$b_0 = 4.1 \text{ A}$$

σ_s can be estimated from the data of Turnbull and Cormia (23) for the nucleation rates of n-paraffin droplets. Averaging over lateral and end surfaces they obtain $\sigma_s = 9.6$ ergs/cm² for a paraffin with a crystal structure similar to polyethylene. Averaging in the end surfaces probably makes this value too large. In addition, Hoffman and Weeks (31) have estimated σ_e and σ_s (as discussed below) from growth rate and nucleation rate data of Price (32) and Cormia, Price, and Turnbull (33). They suggest $\sigma_s = 12.2$ ergs/cm². Thus, $kT/b_0\sigma_s \approx 10$ A.

A number of authors have estimated σ_e, with values, even when using the same data, ranging from about 50 to 150 ergs/cm.² Using Price's polyethylene growth rate data and Equation VI-34, Hoffman and Weeks (31) estimate the product $\sigma_s\sigma_e$ to be about 600 ergs²/cm⁴. From polyethylene droplet data (33), assuming the homogeneous nucleation rate to be of the order of 10^9 for $\Delta T = 56°C$, a temperature at which homogeneous nucleation becomes extremely rapid, they estimate $\sigma_s^2\sigma_e$ to be 7320 ergs³/cm⁶. This yields a value of $\sigma_s = 49$ ergs/cm². Cormia, Price and Turnbull (33), using the change in rate of nucleation with temperature from their droplet data find $\sigma_s^2\sigma_e = 1.55 \cdot 10^4$ ergs³/cm⁶, $2^1/_2$ times larger than Hoffman and Weeks from the same data. Combining this with the average value of σ

---

* 414.3° by extrapolation from linear paraffins (30); 416° by extrapolation of a fold period-melting point relationship (27).

(9.6 ergs/cm²) from paraffin droplet data (23) they obtain $\sigma_e = 168$ ergs/ cm². Most other values that have been suggested lie between these extremes with Frank and Tosi estimating $\sigma_e = 70$ to 100 ergs/cm² (19).

One can also estimate σ_e from ℓ_{obs}, using Equation VI-28 and neglecting the edge free energy term. To use the data for crystallization from xylene, one needs to know T_m, the solubility temperature for an "infinite" crystal. This we estimate to be between 100°C and 110°C. For $T_m = 110$°C and $\ell = 140$ A, $\sigma_s = 90$ ergs/cm² while for $T_m = 100$°C and $\ell = 140$ A, $\sigma_s = 50$ ergs/cm². Any increase in step height following crystallization would make this value too large. A more reliable method of calculating σ_e from values of ℓ_{obs} is to use Equation VI-36 which involves only straightforward thermodynamics. For this one needs to know either $T_m(\ell) \approx T_{obs}$ or β. From Equation VI-38 and Hoffman and Week's (27) value of $\beta = 1$ for melt crystallized polyethylene and the values listed above one finds $\sigma_e = 57$ ergs/cm².

All of these values thus fall in the range predicted and also are reasonable in terms of the energy required to produce a fold. Approximately 3 non-*trans* bonds are formed in producing a fold in a $\{110\}$ plane. In addition, the neighboring CH_2 groups will be slightly twisted. Infrared measurements (34) and studies of the temperature dependence of the retractive force in polyethylene (35) indicate that the lowest energy non*trans* bond has an energy of about $3.5 \cdot 10^{-14}$ ergs more than a *trans* bond. De Santes *et al.* (36) and McCullough indicate (37) that this lowest energy, non*trans* bond corresponds to a rotation of 90° about the C—C bond, rather than the 120° that corresponds to a gauche bond and as is present in the diamond lattice. The value of $3.5 \cdot 10^{-14}$ ergs per non*trans* bond is considerably lower than the value of $5.6 \cdot 10^{-14}$ ergs per bond used by Frank and Tosi (19) in their estimation of σ_e. The value used here corresponds to that determined from infrared measurements on the longer hydrocarbons. Three such non*trans* bonds per fold correspond to 28 ergs/cm² at 90°C; adding an interfacial energy of about 10 ergs/cm² and allowing for some additional distortion suggests that σ_e should be near 50 ergs/cm².

Some data has also been obtained for polychlorotrifluoroethylene. Hoffman and Weeks (31) estimate σ_s to be about 5 ergs/cm²; combining this value with their spherulite growth rate data for $\sigma_e\sigma_s$ (184 erg²/cm⁴) yields $\sigma_e = 36$ ergs/cm². This value is somewhat lower than for polyethylene despite the fact that polychlorotrifluoroethylene is a stiffer molecule. However, the cross-sectional area is about twice as large and thus the energy per fold is larger than in their calculation of σ_e for polyethylene. Using step heights determined from electron microscope (Section IV-5) and allowing for the increase in step height during crystallization by incorporating the factor β, they obtain a value of 35 ± 7 erg/cm^2, in good agreement with that obtained from the growth rate data.

There are a number of questionable assumptions in the derivation of Hoffman and Lauritzen as given in their first two papers (16,17). Attempts to overcome the most questionable, the assumed constancy of the fold period in a fold plane once it has started growth, are considered in more detail in the next section. In addition they have assumed that σ_s and σ_e are independent of temperature and that μ_s is independent of ℓ (the dependence of μ_s on ℓ forms the basis of the thermodynamic approach of Peterlin, Fischer, and Reinhold (7–11)). Furthermore, one may question the applicability and validity of the nucleation rate calculations with respect to polymer crystals. In the derivations of Equation VI-19 for the nucleation rate of the primary nuclei it is assumed that all step elements, i.e., segments, in a nucleus have the same fold period and that the nucleus, before reaching the critical size, is built up by the addition of a large number of such elements. Using the parameters listed above for crystallization from xylene at 90°C, the critical number of segments ν^*, Equation VI-16, is about 160, a reasonably large number. Since the average length of a polyethylene molecule corresponds to only about 10–20 folds, the above value, however, is not in agreement with the suggestion elsewhere by Lauritzen and Hoffman (16) that the nucleus reaches critical size before an entire molecule is incorporated. In fact, it suggests that a number of molecules are involved in forming the primary nucleus.

However, the fact that consistent values of σ_e are obtained from spherulite growth rate and homogeneous nucleation rate data, and melting point and lamellae thickness observations for two polymers and also the energy involved in bending a molecule to produce a fold for one of the polymers (31), suggests that the basic assumptions and resulting calculations are at least reasonable.

Simultaneously with Lauritzen and Hoffman, Price (14), by a somewhat different calculation of the average fold period during growth, arrived at a similar, but different, expression than Equation VI-27 for the thickness of the growing crystal. Considering only the free energy change during secondary nucleation, he finds, neglecting edge free energy considerations, that

$$\bar{\ell} = \frac{2\sigma_e}{\Delta\mu} + \left(\frac{kT}{\pi b_0 \Delta\mu} \right)^{1/2}$$

$$= \frac{\ell_p^*}{2} + \left(\frac{kT}{\pi b_0 \Delta\mu} \right)^{1/2} \tag{VI-39}$$

The numerical value of the second term in this equation and $kT/b_0\sigma_s$ in Equation VI-28 are of the same order at the temperature of interest.

In this derivation, Price assumes that ΔF_s for a fold plane nucleus is a continuous, saddle-shaped surface in terms of ℓ and νa_0. Lauritzen and Hoffman (16), on the other hand, assuming that secondary nucleation is a one-step process, do not consider the saddle-shaped nature of the surface and, instead, travel over the pass at a much higher elevation than the saddle point. In the region considered by Lauritzen and Hoffman ΔF_s is a maximum for $\nu = 1$ and decreases with further increase in ν. In the case of Price's calculation, however, ΔF_s is a maximum (at the saddle point) for a value of $\nu > 1$. Price also calculated the fold period of the original nucleus, pointing out that it was equal to $4\sigma_e/\Delta\mu$, i.e., nearly twice as large as the fold period of a fold plane nucleus.

e. Fluctuation Theory

Three attempts have been made to overcome the restriction in the Lauritzen and Hoffman theory that a fold plane, once nucleated, will grow with a constant fold period (18–20). Whereas Frank and Tosi (19) and Lauritzen (20) use the same basic approach as that presented in the previous sections, Price uses a related but somewhat different approach. Since the results of the calculations in all three papers are nearly the same as those discussed previously we will only summarize their treatments in this section.

Price (18) pictures the development of a polymer single crystal, after nucleation, as occurring by the successive addition of fold planes, each fold plane acting as a substrate for the next one. In order to perform the calculations he assumes that each preceding fold plane is of uniform fold period when a new one is forming. Although the fold period in the preceding one may, of course, vary, this being the subject of the calculation, it is not believed that this assumption greatly affects the calculation. He then calculates the free energy involved in (a) laying down on this substrate a new fold plane composed of individual molecular segments, all of length greater than $2\sigma_e/\Delta\mu$, but whose length may vary one from another, (b) the molecular segments coming together to form a lattice, and (c) forming the folds. The segments are originally considered as individual entities, but the energy of scission of the polymer molecule to form the segments is regained when they are rejoined by the folds. The equilibrium distribution of molecular segment lengths is then calculated.

Price finds that the average fluctuation in fold period is less than one repeat distance in the case of polyethylene, thus justifying the assumption of the substrate being smooth. He also suggests that at low crystallization temperatures the bulk free energy ($\Delta\mu$, which increases with increasing segment length) overwhelms the surface energy and the crystals increase in thickness without limit. This increase in fold period with decreasing crystallization temperature, which is also predicted by the other two fluctuation theories (19,20), has not been observed experimentally.

If the temperature of crystallization is changed during growth of the crystal, the calculations predict that the fold period of new layers will approach the "equilibrium" (as determined by this theory) length for the new temperature. They may either increase or decrease depending on whether the temperature is raised or lowered. Price finds that his results agree best with experimental, single crystal fold periods if he assumes $\sigma_s = 13.3$ ergs/cm^2, $\sigma_e = 110$ ergs/cm^2, and T_m (in xylene) = 113°C. The surface energies would appear to be somewhat too large.

Frank and Tosi (19), assuming for the purpose of calculation that the probability of a change in fold period during growth of a given fold plane is small, attempt to determine the effect of one such change during growth of each fold plane. Although they find that the probability of a change or fluctuation in fold period is quite large, nevertheless they indicate that the results should be a useful approximation. More important, they and Price indicate that if fluctuations occur during growth they will be such as to converge to an average thickness which is dependent on the temperature of crystallization. Fluctuations of both larger and smaller values than the average are permitted. This average fold period is similar in magnitude to that calculated by Lauritzen and Hoffman (16).

The theory, as developed by Frank and Tosi, suggests three different types of crystallization behavior corresponding to three different temperature ranges. For small supercoolings (high crystallization temperatures) they suggest that the fold period, regardless of the thickness of the nucleus, will tend toward ℓ_0 (Eq. VI-23), the minimum fold period for which the free energy will decrease during crystal growth. The initial fold planes must be thicker than ℓ_0 if growth is to occur at all. They suggest that crystal growth in this temperature range is not possible unless molecular rearrangement within the crystal permits the fold period to increase following which new fold planes can again be deposited. The growth rate would then be dependent on this rearrangement rate. This restriction of crystallization at low supercooling is not found by either Price (18) or Lauritzen (20). The results for the middle range of supercooling are those discussed in the previous paragraph. Regardless of the initial thickness of the substrate, crystallization proceeds with the fold period converging to a value $\ell > \ell_0$, where ℓ is dependent on the temperature. Frank and Tosi suggest that each of these two ranges extend over 10–20°C.

According to this theory, as well as those of Price (18) and Lauritzen (20), a temperature is reached with increasing supercooling (decreasing fold period) at which the fold period suddenly begins to increase rapidly. Frank and Tosi suggest (19) that before this temperature is reached another process, intramolecular crystallization, sets in. Because of the small lateral size of the nucleus at low temperatures, only a few folds are required to produce a stable nucleus and thus a number of nuclei may form in an individual molecule. This will obviously alter the crystallization process. Although Frank and Tosi suggest this type of intramolecular crystallization may be related to the formation of dendritic crystals, it is fairly obvious that it is not associated with the type of dendrites pictured in Chapter II-5. The individual fold domains in the dendrites pictured in that section may contain 10^4 or more molecules. They also suggest that as one approaches this temperature range the magnitude of the fluctuations increases rapidly and that this may lead to dendritic growth.

Using only the experimental results of Keller and Bassett (38) for comparison with the calculations, Frank and Tosi find that values of $T_m = 110°C$ (in xylene), $\sigma_s = 6.4$ ergs/cm², $\sigma_e = 110$ ergs/cm² or $T_m = 115°C$, $\sigma_s = 7.2$ ergs/cm², $\sigma_e = 130$ ergs/cm² give about equally good fits. However, if one also includes the data for crystal thickness obtained by Price (18) (see Fig. II-6) at temperatures as low as 50°C it would appear that considerably larger values of all three parameters are needed.

Frank and Tosi state that the nature of the crystals they describe differs from those of Lauritzen and Hoffman in that in the latter's model "the segment length is uniform within each crystal, but there is a variation among crystals, all grown at the same temperature." Actually, however, the two models are not that dissimilar; Lauritzen and Hoffman (16) permit a fluctuation in fold period at the time of nucleation of each new fold plane (they suggest the average fluctuation in fold period is about $kT/2/^{1/2}b_0\sigma_s$) (Eq. VI-29) whereas Frank and Tosi permit one additional fluctuation during the growth of each fold plane. In both models all crystals grown at the same temperature have the same average fold period. Frank and Tosi also state that their picture implies "that the crystal is likely to change after growth, by creeping displacements of the molecular chains tending to even out the segment lengths." Bassett and Keller (39) suggest that this is what happens when hollow pyramids (Section II-3) form, the crystals first forming with a variable fold period and having, perhaps, a grossly planar topography with the folds and fold periods later readjusting to form the pyramids.

Several of the inconsistencies between experiment and theory in Frank and Tosi's approach have been corrected by Lauritzen (20). The high temperature (low supercooling) range, for which crystallization supposedly does not occur, is not found if one permits the molecules to fold only at a finite number of points, these points corresponding to the actual atomic distances. In addition, Lauritzen shows that the effect at high supercooling, the rapid increase of fold period with decreasing temperature, while still evident in the theory, occurs at experimentally inaccessibly low temperatures. Intramolecular crystallization as suggested by Frank and Tosi or the formation of a glass would probably occur first.

f. Lamellar Twisting

In a number of polymers, the lamellae appear to twist regularly during spherulite growth (see Chapter IV). The period of the twist usually depends on the temperature of crystallization, increasing with increasing temperature. In the case of spherulites of low molecular weight materials, which also often consist of twisted ribbons, it has been suggested (40) that the twisting arises because of a surface stress. In a thin crystal the interfacial surface tensions would give rise to such a force, placing the surface in compression. In a polymer crystal, as Hoffman and Lauritzen suggest (17), the presence of the folds may contribute an additional surface stress. The unstrained lattice dimensions of the folds is expected to be larger than the dimensions of the crystal lattice with which it is continuous. A somewhat different suggestion, the stress developing on the lateral edges of the lamellae due to the incorporation of impurities, has been suggested by Keith and Padden (41) in connection with their cellulation model for spherulite growth (Chapter IV, Section 5-a).

The surface stress in a polymer lamella resulting from the presence of the folds may be accommodated in two ways. During growth from solution, as discussed in Chapter II, the fold planes may be displaced resulting in the development of hollow pyramids and a lowering of the stress. If the fold planes are not displaced, the surface stress can, if large enough, result in a twisting of the lamella. This twist will be gradual, the lamella developing by coherent nucleation of new fold planes. Hoffman and Lauritzen (17) indicate that a ribbon of width x and thickness ℓ, having a surface stress of f dynes cm^{-2} will twist when

$$f \geqslant \frac{S\ell^2}{2tx^2}$$

where S is the appropriate shear modulus and t is the thickness of the stressed surface layer. For polyethylene crystallized from the melt ℓ/x is of the order of $1/10$ to $1/100$ and t is about 5 A. Thus, twisting will result for values of f/S on the order of 10^{-3}, a value which they indicate (17) is quite reasonable to expect. The period of the twist should increase with decreasing ℓ, as is observed experimentally. The dependence of x on temperature is less significant.

During growth, all the lamellae in a spherulite would be expected to twist in the same direction. The sense of twist of a primary lamellae will be propagated as it grows. The secondary and tertiary lamellae as defined in Chapter IV will, in general, twist in the same direction since, when first formed, they are in contact with the primary lamella. If the spherulite is nucleated heterogeneously different sectors may result from different acts of primary nucleation on the foreign substance. The sense of twist in the different sectors may then be of opposite sign, resulting in radial faults.

3. THERMODYNAMIC EQUILIBRIUM THEORY

Peterlin, Fischer, and Reinhold (7–11) attempt to calculate the free energy density of a polymer chain in a lattice in terms of its length and inter- and intrachain forces. They point out that a polymer lattice is different than the lattice of most low molecular weight materials, such as metals, in that the binding forces are highly anisotropic. In a polymer crystal such as polyethylene, the binding forces that oppose axial translation or rotation of one segment with regard to its neighbors are much less than the forces within the chain, due to primary valance bonds, that oppose bending, stretching, or partial lateral translation of the chain. Furthermore, on the basis of the motion observed in NMR experiments, they suggest that at temperatures near the melting point the oscillations in neighboring chains should be essentially incoherent.

Following the initial treatment by Peterlin and Fischer (7–9) of the effect on the free energy density of a polymer molecule (free energy per backbone atom) of longitudinal vibrations along the axis of a given chain and its neighbors in a crystal, Peterlin, Fischer, and Reinhold (10,11) considered the effect of librations or torsional oscillations. They suggest that librations should have a greater influence on the morphology than linear vibrations. In paraffin crystals, for instance, X-ray investigations have shown that the amplitude of the librations is much larger than the amplitude of linear vibrations along the molecular axis (42). Since the derivations are quite similar we will consider both types of vibrations simultaneously, calculating the energy of a free chain undergoing the vibration in question, considering the effect of placing the chain in a lattice in which the neighbors are at rest, the reduction of the potential caused by the motion of the neighbors and finally calculating the free energy density as a function of the length of the segment of the molecule in the crystal.

It is found that there is a minimum in the free energy density for some finite value of the number of backbone atoms N in the segment. This value of N, N^*, which depends on temperature, is thermodynamically stable and the length of the crystal will be restricted to it. Although not required by the theory, the most likely means of restricting the length, in view of the known morphology of crystalline polymers, is by folding. However, a fringed-micelle-type model for either unoriented or drawn polymer in which the crystalline regions are limited to a length corresponding to N^*, with some form of intervening disordered regions, would also be in agreement with the theory.

In the following discussion equations and quantities referring specifically to linear vibrations or librations will be labeled (v) or (r) respectively. Different symbols are used for some factors than those used in the original papers, in part to illustrate the resemblance between the two treatments and in part to agree with those used elsewhere in this text. The treatment here and in the original papers is restricted to polyethylene; the lattice constants which are needed in particular for the libration calculations are known most precisely for polyethylene and it is a relatively simple molecule.

a. Free Chain

In order to simplify the calculation of the total energy of a vibrating or librating free chain $(U_{fc,v}$ or $U_{fc,r})$ Peterlin et al. (9–11) assume a model for the zigzag polyethylene chain. For the vibration calculation N CH_2 groups are treated as being equally spaced at a distance equal to $c/2$ along a single line whose center of gravity is fixed but whose ends are free. The elements of mass m are assumed to be bound by linear elastic forces of strength f_v (Fig. VI-3a). For the libration calculation (11) N pairs of CH_2 groups are spaced at a distance c along opposite sides of a line whose ends are free (Fig. VI-3b). Their moment of inertia is θ and a restoring force f_r opposes rotation of the N pairs with respect to their neighbors. Note that the model chain for the libration calculation for a given value of N is twice as long as that for the vibration calculation.

(a)

(b)

(c)

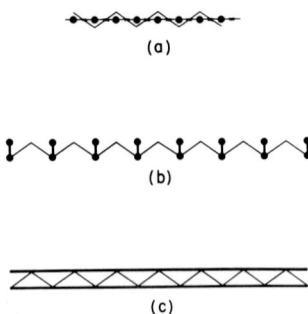

Fig. VI-3. Diagram of the models used by Peterlin, Fischer, and Reinhold (9–11) for their calculations of the free energy density of a polyethylene molecule. (a) Model used for linear vibrations of a free chain, (b) model used for liberations of a free chain, and (c) model used for liberations of a chain in lattice. In a and b the masses of the CH$_2$ groups (m) are localized at the points shown, whereas in c the mass is assumed to be evenly distributed along the lines.

For these models U_{fc} is given by (9-11)

$$U_{fc,v} = \frac{1}{2m} \sum_{j=1}^{N} p_{j,v}^2 + \frac{m}{2} \sum_{j=1}^{N-1} \omega_{j,v}^2 \, a_{j,v}^2$$

and   (VI-40)

$$U_{fc,r} = \frac{1}{2\theta} \sum_{j=1}^{N} p_{j,r}^2 + \frac{\theta}{2} \sum_{j=1}^{N-1} \omega_{j,r}^2 \, a_{j,r}^2$$

where

$$\omega_j = \omega_0 \sin \frac{\pi j}{2N}$$

$$\omega_0 = 4f_v/m \text{ or } 4f_r/\theta$$

$$p_j = m\dot{a}_{j,v} \text{ or } \theta \dot{a}_{j,r}$$

and where a_j is defined by the following transformation from the linear z_i or angular ϕ_i coordinate

$$z_i = \left(\frac{2}{N}\right)^{1/2} \sum_{j=1}^{N-1} a_{j,v} \cos \frac{\pi j(2i - 1)}{2N}$$

and   (VI-41)

$$\phi_i = \left(\frac{2}{N}\right)^{1/2} \sum_{=1}^{N-1} a_{j,r} \cos \frac{\pi j(2i - 1)}{2N}$$

The partition function Q is of significance in the following calculations of both the amplitude of the vibration or libration and the free energy density $(F = -kT \ln Q)$.

Assuming all frequencies of vibration to be equally excited, it is given by

$$Q = \left(\frac{1}{h}\right)^{N-1} \int_{-\infty}^{\infty} \cdots \int_{-\infty}^{\infty} \exp(-U/kT) dp_1 \ldots dp_{N-1} da_1 \ldots da_{N-1} \quad \text{(VI-42)}$$

$$= \prod_{j=1}^{N-1} \frac{kT}{\hbar\omega_j} = \left(\frac{kT}{\hbar\omega_0}\right)^{N-1} \prod_{j=1}^{N-1} \frac{1}{\sin \pi j/2N} \quad \text{(VI-43)}$$

Therefore the free energy per chain element of a free chain in libration or vibration is given by

$$\frac{F_{fc}}{N} = \frac{-kT}{N} \left[(N-1) \ln \frac{KT}{\hbar\omega_0} - \sum_{j=1}^{N-1} \ln \sin \frac{\pi j}{2N} \right] \quad \text{(VI-44)}$$

For large N, $\sin \frac{\pi j}{2N} \approx \frac{\pi j}{2N}$ and Peterlin and Fischer indicate that F_{fc} can be approximated by

$$\frac{F_{fc}}{N} = -kT \left(\frac{N-1}{N} \ln \frac{2KT}{\hbar\omega_0} - \frac{\ln N}{2N} \right) \quad \text{(VI-45)}$$

where ω_0 will be different for the vibration and librational modes.

The mean square amplitude of vibration or libration of the mass elements in the free chain $\langle S_{fc}^2 \rangle$ is determined from the following relations:

$$\langle S_{fc}^2 \rangle = \sum_{i=1}^{N} \frac{\langle z_i^2 \rangle}{N} \text{ or } \sum_{i=1}^{N} \frac{\langle \phi_i^2 \rangle}{N} \quad \text{(VI-46)}$$

$$= \sum_{j=1}^{N-1} \frac{\langle a_j^2 \rangle}{N}$$

but

$$\langle a_j^2 \rangle = \frac{\int_{-\infty}^{\infty} \cdots \int_{-\infty}^{\infty} a_j^2 \exp(-U_{fc}/kT) dp_1 \ldots dp_{N-1}, da_1 \ldots da_{N-1}}{Q_{fc}}$$

and a_j^2, from Equation VI-40, can be shown to be given by

$$a_j^2 = \frac{2kT}{mQ_{fc}} \frac{dQ_{fc}}{d\omega_j^2} = \frac{-2kT}{m} \frac{\partial \ln Q_{fc}}{\partial \omega_j^2}$$

Therefore,

$$\langle S_{fc}^2 \rangle = \frac{-2kT}{mN} \sum_{j=1}^{N-1} \frac{\partial \ln Q_{fc}}{\partial \omega_j^2}$$

$$= \frac{kT}{mN} \sum_{j=1}^{N-1} \frac{1}{\omega_j^2} \approx \frac{NkT}{\pi^2 f} \sum \frac{1}{j^2} \qquad \text{(VI-47)}$$

from Equation VI-43.

For large N, $\sum 1/j^2 \approx \pi^2/6$ and therefore

$$\langle S_{fc}^2 \rangle = \frac{NkT}{6f} \qquad \text{(VI-48)}$$

where f will be different for the vibrational and librational modes.

$\langle S_{fc}^2 \rangle$ increases with the number of mass elements in the chain. Using a reasonable value of f for polyethylene [$f_v = 2.10^5$ dyne/cm (43) and $f_r = 0.51 . 10^{-12}$ ergs (from Szigetis' measurement of the frequency ω_0 of libration (44)] one finds that

$$S_{fc,v} = 1.15 \, NT \cdot 10^{-5} \, \text{A}^2 \, {}^{\circ}\text{K}$$

$$S_{fc,r} = 4.5 \, NT \cdot 10^{-5} \, \text{radians}^2/{}^{\circ}\text{K}$$

At 100°C the mean linear amplitude of vibration of a free polyethylene segment 100 A long ($N_v = 80$) would be more than 20% of the axial separation of the atoms while its mean amplitude of libration ($N_r = 40$) would be more than 45°. Even near the glass transition the values of S_{fc} for a 100 A long segment are about 70% of those given above, whereas in metals the amplitude of vibration does not even reach $1/10$ of the lattice spacing until near the melting point (45). These values should apply to straight portions of a molecule in a liquid where the time average potential due to neighboring molecules is constant. As indicated in previous chapters, it is likely that there are extended, nearly ordered segments in polymer melts and solutions, suggesting that F_{fc}, for instance, may be related to μ_l.

b. Chain in a Lattice

Peterlin and Fischer suggested (9) that if the molecular segment is placed in a polymer lattice the thermal oscillations of the chain along its axis would not be greatly affected. The same assumption has been used for the interpretation of infrared bands (46). Likewise, it is assumed that the librational motions are not greatly affected. In both cases the major effect is the addition of a periodic potential (along or about the axis), due to the periodic force field of the neighbors, to the inherent potential of the single chain. In the case of linear vibrations, if the periodic potential is assumed to be a cosine function whose amplitude is independent of the length of the chain, as originally assumed by Fischer (7), the free energy per chain element is independent of the length of the chain (47). However, as pointed out by Peterlin and Fischer (9), the atoms in the neighboring chains in the lattice, as well as in the chain under consideration, will be vibrating with an amplitude depending on N. As a result the periodic potential is "smeared" out; i.e., its amplitude is reduced. The vibrations in the chain of interest will therefore be larger, approaching those of a free chain.

We shall consider, below, first the effect of the linear vibrations and then the librations of the neighboring molecules on the potential and free energy of a molecule in a lattice. Although the results are similar the treatment differs in a few respects.

If all the chains in a lattice are at rest the potential energy of a given chain due to its neighbors can be written as

$$\Phi_v(z) = \phi_{0,v} - \phi_v \cos 2\pi z/c \qquad \text{(VI-49)}$$

where

$\phi_{0,v}$ = average value of the potential ($\phi_{0,v}$ varies with temperature as the lattice spacing changes)

ϕ_v = amplitude of the periodic portion of the potential (ϕ_v will vary with temperature as a result of the motion of the molecules. It has its maximum value $\phi_v(\text{max})$ when the neighboring chains are at rest).

If the positions of the atoms on the neighboring chains can be considered random, due either to large oscillations or a liquidlike structure, $\phi_v = 0$ and $\Phi_v(z)$ is constant and equal to $\phi_{0,v}$. The definitions given here differ somewhat from those in the papers by Peterlin and Fischer (8,9) but are believed correct in terms of Figure 1 of Reference 9 and agree with those used later in the calculations based on a libratory type of motion.

The total energy of the vibrating chain in a lattice in which the neighbors are at rest is given by adding the potential due to the neighbors (Eq. VI-49) to the intrinsic energy of the single chain (Eq. VI-40).

$$U_{xl,v} = \frac{1}{2m} \sum_{j=1}^{N} p_{j,v}^2 + \frac{m}{2} \sum_{j=1}^{N-1} \omega_{j,v}^2 a_{j,v}^2 + N\phi_{0,v}$$

$$- \sum_{j=1}^{N} \phi_v(\text{max}) \cos 2\pi z_j/c \qquad \text{(VI-50)}$$

Substituting this value of $U_{xl,v}$ into the partition function, as in Eq. VI-42, results in the following expression for Q_{xl} of a chain undergoing linear vibrations in a lattice in which the neighbors are at rest.

$$Q_{xl,v} = \left(\frac{1}{\hbar}\right)^{N-1} \left(\frac{kT}{2\pi}\right)^{\frac{N-1}{2}} \exp\left(\frac{-N\phi_{0,v}}{kT}\right) \int_{-\infty}^{\infty} \ldots \int_{-\infty}^{\infty}$$

$$\exp\left(-\frac{m}{2kT} \sum_{j=1}^{N-1} \omega_{j,v}{}^2 a_{j,v}{}^2 + \phi_v(\max) \sum_{j=1}^{N} \cos 2\pi z_j/c\right) da_1 \ldots da_{N-1}$$

$$(VI-51)$$

Peterlin and Fischer show (9) that this can be simplified and approximated by

$$Q_{xl,v} = \left(\frac{kT}{\hbar}\right)^{\frac{N-1}{2}} \left(1 + \frac{4N^2\phi_v(\max)}{fa^2}\right)^{-1/2} \exp\left\{\frac{-N}{kT}(\phi_0 - \phi_v(\max))\right\}$$

$$(VI-52)$$

The free energy can then be calculated, if desired, from Eq. VI-52. However, the neighboring chains as well as the chain under consideration will be vibrating and therefore ϕ_v will be reduced. At any point along the chain the periodic portion of $\Phi_v(z)$ (assuming that the oscillations of the neighboring chains are independent) will be given by

$$\phi_v \cos 2\pi z/c = \int_{-\infty}^{\infty} \phi_v(\max) |\cos 2\pi(z | + S)/c \frac{\exp - S^2/\langle S_{xl,s}{}^2\rangle}{2\pi\langle S_{xl,s}{}^2\rangle} ds$$

$$(VI-53)$$

Therefore

$$\phi_v = \phi_s(\max) \exp\left(-2\pi^2 \langle S_{xl,s}{}^2\rangle/c^2\right) \qquad (VI-54)$$

where the mean square amplitude of vibration in the crystal $\langle S_{xl,s}{}^2\rangle$ can be calculated, as before (Eq. VI-47), from the partition function.

$$\langle S_{xl,s}{}^2\rangle = \frac{-2kT}{mN} \sum_{j=1}^{N-1} \frac{\partial \ln Q_{xl,v}}{\partial \omega_{j,v}{}^2} \qquad (VI-55)$$

Peterlin and Fischer indicate (9) that $\langle S_{xl,v}{}^2\rangle$ is given by

$$\langle S_{xl,v}{}^2\rangle = \frac{NkT}{6f_v}\left(1 - \frac{2/5}{1 + \dfrac{fc^2}{4N^2\phi_v}}\right) \qquad (VI-56)$$

$$= \langle S_{fc}{}^2\rangle \left(\frac{2.4\,N^2\phi_v + fc^2}{4\,N^2\phi_v + fc^2}\right)$$

The approximations involved in obtaining Eq. VI-52 result in $\langle S_{x1,v}{}^2 \rangle$ being larger than it should be (the factor $2/5$ should be larger) (9). The results indicate, however, that the interaction potential depends exponentially on the chain length as well as the forces between the chains.

For the calculation of the periodic portion of the potential of a librating chain in a crystal due to its neighbors, Peterlin, Fischer, and Reinhold (10,11) start with the Lennard-Jones potential

$$u = -2\epsilon \left(\frac{R_{\min}}{R} \right)^6 + \epsilon \left(\frac{R_{\min}}{R} \right)^{12} \tag{VI-57}$$

where R is the distance between CH_2 groups on neighboring molecules and $R_{\min}$ is the separation when the potential is a minimum; i.e., $u_{\min} = -'\epsilon$. They attempt to calculate the numerical value of the amplitude, as a function of temperature, of the periodic portion of the potential Φ_r due to neighboring chains, first for the case in which the neighboring chains are at rest and then for the case in which they also undergo libration.

For this calculation a different model is used than for the free chain. The zigzag molecule is replaced by two parallel lines passing through the carbon atoms and having a continuous uniform density and a separation of $2r = 0.88$ A (the separation of the carbon atoms normal to the axis as determined by Bunn (48)). Because of the rapid decrease in u with R only the 6 nearest neighbors in the polyethylene lattice will produce a significant contribution to the potential energy of a given molecule. For a molecule in the center of the unit cell these are the 4 molecules at the corners of the cell and the 2 molecules at the centers of the adjoining unit cells along the **b** axis direction (see Fig. I-8).

For the model used in their calculation, Peterlin et al. (10,11) state that

$$\Phi_{ij,r} = \int_{-\infty}^{\infty} \frac{u}{2.53} \, dz$$
$$= \frac{0.931\epsilon R_{\min}{}^6}{s_{ij}{}^5} + \frac{0.306\epsilon R_{\min}{}^{12}}{s_{ij}{}^{11}} \tag{VI-58}$$

where $\Phi_{ij,r}$ is the contribution to the potential energy per 2.53 A repeat distance along one of the central molecular lines, i, from an infinite neighboring line j, at rest and situated at a distance s_{ij}; z is the coordinate along the chain and $R^2 = s_{ij}{}^2 + z^2$. The distance s_{ij} will vary somewhat with the angles of inclination, ψ_i and ψ_j, of the molecules (pairs of molecular lines one each of which are i and j) with the line connecting both molecules in the **a**, **b** plane of the lattice.

$$s_{ij} = B_{ij} + 0.44 \,(\cos \psi_j - \cos \psi_i) \tag{VI-59}$$

where B_{ij} is the lattice distance between the corresponding molecular axes and 0.44 A is the distance of one of the lines from its molecular axis.

The total contribution to the potential per repeat distance, of a given infinite molecular line in a molecule oriented at an angle α with the **a** axis, is then found by adding up the contributions of the 6 neighboring molecules (6 pairs of lines). In order to calculate R_{min} and s_{ij} Peterlin *et al.* (10,11) used the lattice constants measured as a function of temperature by Cole and Holmes (28) (both R_{min} and B_{ij} increase with temperature) and the angles ψ measured at room temperature by Bunn (48). The value of ϵ used ($1.135 \cdot 10^{-14}$ erg) was determined from the heat of sublimation of paraffins (49). Inserting these values in Equation VI-58 they find that at $-150°C$

$$\Phi(\alpha)_{r,-150°C} = [-30.875 + 1.652 \sin (2\alpha + 7.7°)$$

$$+ \ 0.129 \sin (4\alpha - 56.6°) + \ldots] \cdot 10^{-14} \text{ ergs.} \quad \text{(VI-60)}$$

At $100°C$, due to the change in R_{min} and B_{ij} it is given by

$$\Phi(\alpha)_{r,100°C} = [-30.091 + 0.818 \sin (2\alpha + 7.4°) +$$

$$0.0544 \sin (4\alpha - 52.6°) + \ldots] \cdot 10^{-14} \text{ ergs.} \quad \text{(VI-61)}$$

For this calculation all of the molecules are assumed fixed at the angles determined by Bunn.

As in the case of the vibrations Φ_r can be written in terms of an average value $\phi_{0,r}$ plus a periodic factor, whose amplitude is ϕ_r, in this case periodic with rotation of the given chain while the neighboring chains are motionless. It is the coefficient of the $\sin (2\alpha + \delta)$ term that is of most significance in the periodic factor and correspondingly is labeled ϕ_r. By calculating the value of $\Phi_r(\alpha)$ at a number of temperatures Peterlin *et al.* (10,11) found that ϕ_r can be approximated by the following expressions over the temperature ranges indicated:

$$\phi_r = 2.1 \ kT \exp (-0.007T) \qquad 223°K < T < 373°K$$

$$\phi_r = 0.64 \ kT - 0.13 \cdot 10^{-2} kT^2 \qquad 353°K < T < 403°K$$

As in the vibration calculation, the constant term $\phi_{0,r}$ also varies with temperature.

The values of ϕ_r given above actually correspond to $\phi_r(max)$, ϕ_r being reduced if the libration of the neighboring chains are considered. The average potential energy for the case where the central chain is oriented at an angle α (with the **a** axis) and the neighboring chains are librating is given as (10,11)

$$\langle \Phi_r(\alpha) \rangle = \int_{-\infty}^{\infty} \Phi(\alpha, S) W(S) dS \qquad \text{(VI-62)}$$

$W(S)$ is the probability of one of the neighboring molecules making an angle S with its rest position. For small incoherent librations the distribution is assumed to be Gaussian with the mean square being the same as for a free chain. Peterlin *et al.* indicate it can be represented by:

$$W(S)dS = (2\pi \langle S_{fc,r}{}^2 \rangle)^{-1/2} \exp \left(\frac{-S^2}{2\langle S_{fc,r}{}^2 \rangle} \right) dS \qquad \text{(VI-63)}$$

Writing $\Phi_r(\alpha)$ as a Fourier series in $\sin 2nS$ and $\cos 2nS$ and averaging over these terms Reinhold indicates (11) that at $100°C$ $\langle \Phi_r(\alpha) \rangle$ can be written as

$$\langle \Phi_r(\alpha) \rangle = [(-30.1206 - 0.0171e(-1)$$
$$+ 0.0478e(-2) - 0.0010e(-3) + \cdots)$$
$$+ \cos 2\alpha(0.1296 - 0.0531e(-1) + 0.0268e(-2) - 0.0001e(-3) + \cdots)$$
$$+ \sin 2\alpha(\qquad 0.8198e(-1) - 0.0090e(-2) + 0.0005e(-3) + \cdots)$$
$$+ \cos 4\alpha(-0.0495 + 0.0037e(-1) + 0.0025e(-2) + 0.0001e(-3) + \cdots)$$
$$+ \sin 4\alpha(\qquad 0.0334e(-1) - 0.0004e(-2) + 0.0001e(-3) + \cdots)$$
$$+ \cdots \cdots] \cdot 10^{-14} \text{ ergs.} \qquad \text{(VI-64)}$$

where $e(-n)$ is written for $\exp(-2n^2 \langle S_{fc,r}{}^2 \rangle)$. If $S^2 = 0$, the neighboring molecules being at rest, $e(-n) = 1$ and Equation VI-64 reduces to Equation VI-61.

Peterlin *et al.* point out (10,11) that the most significant term in Equation VI-64, with respect to the smearing of the angularly periodic potential, is the first coefficient of the $\sin 2\alpha$ term, i.e., $0.8198 \exp(-2\langle S_{fc,r}{}^2 \rangle)$. Although the 0.1296 coefficient of $\cos 2\alpha$ is not negligible, it is independent of the amplitude of oscillation. They then state that the potential can be written as

$$\Phi_r(\alpha) = \phi_{0,r} - \phi_r \cos(2\alpha - \delta) \qquad \text{(VI-65)}$$

where ϕ_r is proportional in the first approximation to

$$\phi_r(\text{max}) \exp(-2\langle S_{fc,r}{}^2 \rangle).$$

As in the case of linear vibrations the total energy for a chain containing N mass points is found by adding the potential energy due to the neighbors $N\langle \Phi_r(\alpha)\rangle$ to the inherent total energy (Eq. VI-40) of the chain.

$$U_{xl,r} = \frac{1}{2\theta}\sum_{j=1}^{N} p_{j,r}^2 + \frac{\theta}{2}\sum_{j=1}^{N-1} \omega_{j,r}^2 a_{j,r}^2 + N\phi_0 - \sum_{j=1}^{N}\phi_r\cos(2\alpha_j - \delta)$$

(VI-66)

which is mathematically nearly identical to that obtained for linear vibrations of a chain in a lattice in which the neighbors, in that case, are at rest (Eq. VI-50). Using the same simplifications and approximations as in the case of Equation VI-52, one obtains

$$Q_{xl,r} = \left(\frac{kT}{\hbar^2}\right)^{\frac{N-1}{2}}\left(1 + \frac{8N^2\phi_r}{f_r\pi^2}\right)^{-1/2}\exp\left\{\frac{-N}{kT}(\phi_0 - \phi_r)\right\}$$ (VI-67)

Equation VI-67 differs from Equation VI-52, however, in that ϕ_r (max) in Equation VI-52 refers to the potential with the neighbors at rest (Peterlin and Fischer (9) calculate ϕ_v and $\langle S_{xl,v}^2\rangle$ from this Q_{xl} in Equations VI-54 and VI-56 whereas in Equation VI-67 ϕ_r has already been reduced by the smearing effect of the librations of the neighboring molecules).

As before (Eq. VI-56) $\langle S_{xl,r}^2\rangle$ is calculated from

$$\langle S_{xl,r}^2\rangle = \frac{-2kT}{N\theta}\sum_{j=1}^{N-1}\frac{\partial\ln Q_{xl,r}}{\partial\omega_j^2}$$

$$= \frac{NkT}{6f_r}\left[1 - 0.4\left(1 + \frac{f_r\pi^2}{8N^2\phi_r}\right)^{-1}\right]$$ (VI-68)

Again the factor 0.4 should be larger. Thus it is found that $U_{xl,r}$ (Eq. VI-66), $Q_{xl,r}$ (Eq. VI-67) and $\langle S_{xl,r}^2\rangle$ (Eq. VI-68) are nearly identical in form to $U_{xl,v}$ (Eq. VI-50), $Q_{xl,v}$ (Eq. VI-52) and $\langle S_{xl,v}^2\rangle$ (Eq. VI-56) where ϕ is interpreted as the amplitude of the smeared periodic portion of the potential. For N in the range of interest the terms $fc^2/4N^2\phi_v$ and $f\pi^2/8N^2\phi_r$ are small compared with unity and therefore

$$\langle S_{xl}^2\rangle \approx 0.6\langle S_{fc}^2\rangle = \frac{NkT}{10f}$$ (VI-69)

where f is evaluated for the appropriate mode of oscillation.

The free energy density for a chain in a crystal, $F_{xl/N} = \frac{-kT}{N}\ln Q$ is given by

$$\frac{F_{zl,v}}{N} = \phi_{0,v} - \phi_v - kT\left[\frac{N-1}{N}\ln\frac{2kT}{\hbar\omega_{0,v}} - \frac{1}{2N}\ln N\left(1 + \frac{4N^2\phi_v}{f_vc^2}\right)\right]$$

(VI-70)

$$\frac{F_{zl,r}}{N} = \phi_{0,r} - \phi_r - kT\left[\frac{N-1}{N}\ln\frac{2kT}{\hbar\omega_{0,r}} - \frac{1}{2N}\left(1 + \frac{8N^2\phi_r}{f_r\pi^2}\right)\right]$$

The total free energy per chain element is obtained by adding the surface free energy of the ends of the crystal (the sides are apparently neglected under the assumption that the growing crystal is already large in lateral dimension). For N of interest ($N > 20$) and under the assumptions made in the derivation, the total free energy is given as

$$\frac{F_{zl,v}}{N} = kT\ln\frac{\hbar\omega_0}{2kT} + \phi_0 - \phi_v\,(\text{max})\exp\left(\frac{-\pi^2NkT}{5f_vc^2}\right) + \frac{2a\sigma_e}{N}$$

$$+ \frac{kT}{2N}\ln\left\{1 + \frac{4N^2\phi_v(\text{max})}{f_vc^2}\exp\left(\frac{-\pi^2NkT}{5f_rc^2}\right)\right\}$$

(VI-71)

$$\frac{F_{zl,r}}{N} = kT\ln\frac{\hbar\omega_0}{2kT} + \phi_{0,r} - \phi_r\,(\text{max})\exp\left(\frac{-NkT}{5f_r}\right) + \frac{2a\sigma_e}{N}$$

$$+ \frac{kT}{2N}\ln\left\{1 + \frac{8N^2\phi_r\,(\text{max})}{f\pi^2}\exp\left(\frac{-NkT}{5f_r}\right)\right\}$$

c. Comparison with Experimental Results

In the case of $F_{zl,r}$ all of the parameters in Equation VI-71 have been measured or calculated except σ_e. In their earlier paper on linear vibration Peterlin and Fischer (9) had to assume a value for the periodic potential. Over much of the temperature range of interest $F_{zl,r}/N$ has 2 minima, one at a finite value of $N_r = N_r^*$ and the other at $N = \infty$. A crystal of length N^* or of infinite length would be stable. The minimum at finite N arises from the opposite effect of the third term and the last two terms. With increasing N and at a given temperature, the third term in Equation VI-71 becomes rapidly smaller (as e^{-N}, F_{zl} increases) while the fourth and fifth terms become smaller more gradually (as $1/N$, F_{zl} decreases). $F_{zl,r}$, for $T = 80°C$ and several values of σ_e, is plotted in Figure VI-4 (a plot of $F_{zl,v}$ would be similar in appearance).

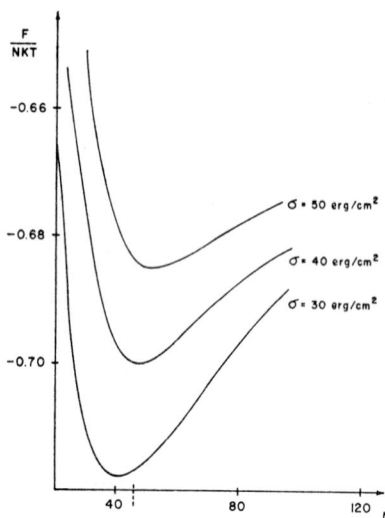

Fig. VI-4.   Free energy density of a liberating polyethylene chain in a lattice as a function of the length of the ordered segment and for various values of the end free energy (Peterlin, Fischer, and Reinhold (10)).

With increasing T and/or σ_e the minimum at $N_r{}^*$ moves to higher values and at the same time becomes flatter. Peterlin et al. (10,11) indicate that the minimum for $F_{xl,r}/N$ completely disappears at a critical temperature T^*, which depends on σ_e, above which only an infinite crystal length would be stable (Fig. VI-5). Likewise, for linear vibrations it was found (9,10) that above a certain temperature only an infinite crystal is stable.

The value of $N_r{}^*$ as a function of temperature for various values of σ_e is shown in Figure VI-6. Also plotted are experimentally determined values for the thickness of polyethylene crystals grown from dilute xylene-butyl acetate solutions (50) which agree reasonably well with those shown in Figure II-6. The values are seen to agree well with the curve for $\sigma_e = 40$ erg/cm^2 which is somewhat less than the value determined by Hoffman and Weeks (31) (Section VI-2-d). This agreement for librations, as might be expected from the greater influence of librations, is somewhat better than was found for the case of vibrations (9). Of interest, also, is the fact that for $\sigma_e = 40$ erg/cm^2, $T^* = 110°C$. Above this temperature the theory predicts that the crystals have no stable thickness but instead continually increase in thickness when annealed, stored or crystallized at these temperatures. This would agree with the findings of Fischer and Schmidt (50) that the thickness of polyethylene crystals increases with log time during annealing. Since

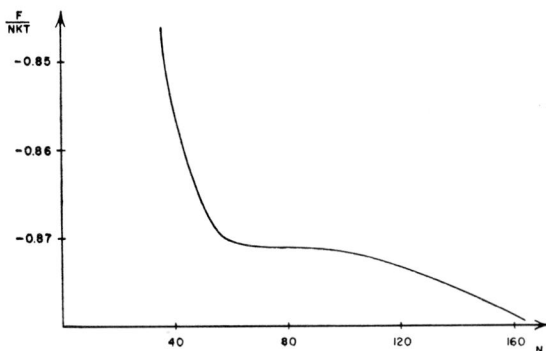

Fig. VI-5. Free energy density of a librating polyethylene chain in a lattice at 120°C. σ_e is assumed to be 40 ergs/cm². Note that there is no minimum in $F_{x,lr}/NkT$ for a finite value of N. (Peterlin, Fischer, and Reinhold (10).)

polyethylene can be crystallized above 110°C (and nearly always does so in the case of crystallization from the melt) with a well-defined fold period, this would suggest that the kinetic theory must apply above 110°C and therefore probably below as well, but that annealing treatments below 110°C might permit the crystals to approach the thermodynamically stable length. In general, however, recrystallization is not observed below 110°C

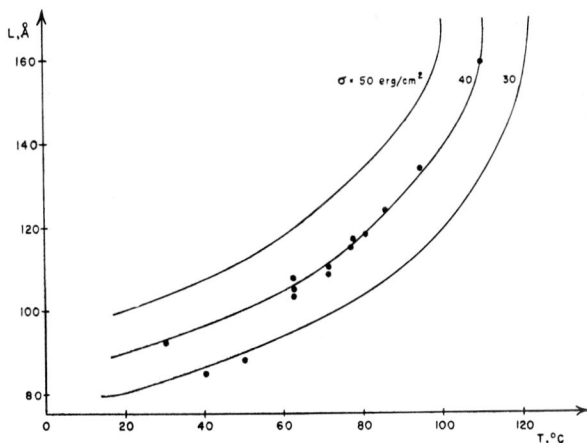

Fig. VI-6. Fold period or crystalline segment length corresponding to N_r^* as a function of temperature for various values of σ_q. Also plotted are Fischer and Schmidt's (50) values of the thickness of polyethylene crystals grown in dilute xylene-butyl acetate solutions. (Peterlin, Fischer, and Reinhold (10).)

(Chapter V). It must also be recognized that the thermodynamic theory applies equally as well to crystals in drawn material as to crystallization from solution and unoriented melt. Fischer and Schmidt have found that the long period of drawn materials increases with the temperature of drawing (50) as well as subsequent annealing treatments (see Chapter VII).

It should be noted that Peterlin *et al.* say nothing about the size or fold period of the nucleus of a lamellar crystal and, furthermore, nothing in the theory requires that the chains fold. After nucleation the crystals grow in the direction of the molecular axes (as well as laterally) until the minimum in the free energy is reached. At this point, sufficient distortion of the molecule must occur so that the oscillations in the segment already incorporated in the crystal and the remainder of the molecule are decoupled. During crystallization from the melt or solution, decoupling presumably occurs by folding whereas in oriented polymers the chain must also be periodically distorted in some fashion. The length of completely aligned chains would be restricted, the oscillations becoming large enough to rupture the chain.

As indicated previously there is still insufficient data to completely rule out either the thermodynamic or the kinetic theories. Both appear to agree reasonably well with the data that is available although at the time of this writing the kinetic theory appears, to this author, to be in better agreement with all known data than the thermodynamic theory.

4. OTHER SUGGESTIONS CONCERNING CRYSTALLIZATION THEORIES

Huggins (6) has suggested that it is possible that intrachain forces tend to twist the backbone of a linear molecule and that a combination of interchain and intrachain forces cause this twisting to have a regular period. This period, on the order of 1000 A or more, should depend on the temperature. When the tendency to twist reaches a certain limiting value, which will depend on the lattice forces opposing the twisting, he suggests that a sudden alteration in structure must occur, i.e., the chain folds in a lamellar crystal. No quantitative or qualititative calculations have been presented as to the cause of the twisting nor does there appear to be any valid experimental evidence suggesting that such a twist occurs in polymer molecules.

Suggestions by Müller and Hess (5) provide a different type of cause for the observed periods. Müller suggested that coupling between the lattice and long wave length phonons might lead to the preferred formation of standing waves. The long periods would correspond to the wave length of these standing heat waves. Hess has apparently suggested a similar idea to explain the long periods found in fibers. These suggestions are related to a suggestion by Born (51) that the anharmonic nature of the thermal vibrations of the atoms in a crystal will set an upper limit on the distance at which long-range order can be maintained. Bresler (3) indicates that Born's results, which are not complete, lead to a thickness of 150 to 200 A for a paraffinlike chain.

Flory (52) has recently suggested a "switchboard" model for polymer lamellae crystallized from solution and the melt. He suggests that, although many of the chains in a given crystal must fold back and reenter the same crystal (the density of chains in the "amorphous" regions is less than that in the crystal), these chains need not be regularly folded as is assumed throughout this volume and as is pictured in Figures II-37 and II-38 for instance. Instead, he assumes a fringed micelle type structure with the added feature of nonadjacent reentry, i.e. the molecule traverses the observed lamella, then enters a disordered, amorphous region and may or may not reenter the same lamella at some nonadjacent position. The number of such folds and the number of molecules which do not reenter the same crystal but pass through the disordered region and enter another crystal (i.e., a sort of "tie" molecule) is suggested to increase with concentration, being a considerable fraction of the total in melt crystallized material. Primary and secondary nuclei are presumed to be bundlelike as are the crystals in copolymers and drawn polymers.

The result of such a model, the surfaces of the lamellae consisting of disordered layers of considerable thickness (as indicated by any observation of models of the surface of such a crystal using atomic models), retains almost all of the defects of the original fringed micelle model. (It does accept the presence of lamellae, i.e., crystals with two dimensions that are considerably larger than the third.) The surface must consist of a diffuse phase boundary, the end surface free energies will be much larger than for a regularly folded array, and it is impossible, we believe, to explain the formation of hollow pyramidal crystals in solution or the growth of single crystals or lamellae with regular growth faces from the melt. In addition the model is not in agreement with, among others, the following observations.

1. Branched polyethylene, which can be considered a copolymer, crystallizes from the melt (Chapter IV) and solution (Chapter II) in the form of lamellae. Flory states that copolymers should not crystallize as lamellae.

2. Polymers have been observed (for instance polycarbonate, Chapter IV, Section 4) to crystallize in the form of lamellae whose thickness corresponds to only a few repeat distances. There is insufficient thickness for

a significant number of molecules to pass through a disordered region and enter the crystal at nonadjacent locations. In addition, polycarbonate crystallizes as lamellae from the glassy state, the one type of crystallization for which a bundlelike crystal should be most suitable if it is at all realistic.

3. Work with oligomers indicates that polymer molecules will form crystals in which they are evidently regular folded (Chapter II) for degrees of polymerization corresponding to lengths just over a fold period and larger. Adjacent reentry and juxtaposition of the ends must be occurring in these crystals. If a few repeat units on the ends of the molecules were in a disordered surface layer, they would crystallize, the resulting crystal thickness corresponding to the total length.

REFERENCES

1. A. Keller and A. O'Connor, "Study of Single Crystals and Their Associations in Polymers," *Discussions Faraday Soc.*, #**25**, 114 (1958).
2. F. C. Frank, "General Discussion," *Discussions Faraday Soc.*, #**25**, 208 (1958).
3. S. E. Bresler, "General Discussion," *Discussions Faraday Soc.*, #**25**, 205 (1958).
4. H. A. Stuart, "Kristallizationsbedingungen und morphologische Strukturen bei Hochpolymeren," *Kolloid-Z.*, **165**, 3 (1959).
5. F. H. Muller, "Weitere Diskussionsbemerkungen" from Symposium Kunstaffe und Kautschuk, Bad Nauheim, 1958, *Kolloid-Z.*, **165**, 38 (1959).
6. M. L. Huggins, "Effect of Intrachain and Interchain Interactions on the Structure of Crystalline Regions in Linear Polymers," *J. Polymer Sci.*, **50**, 65 (1961).
7. E. W. Fischer, "Thermodynamische Deutung der grossen Perioden in kristallinen Hochpolymeren," *Z. Naturforsh.*, **14a**, 584 (1959).
8. A. Peterlin, "Chain Folding and Free Energy Density in Polymer Crystals," *J. Appl. Phys.*, **31**, 1934 (1960).
9. A. Peterlin and E. W. Fischer, "Thermodynamische Stabilität makromolekularer Kristalle I. Der Einfluss der Longitudinalschwingungen der Kettenmoleküle," *Z. Physik*, **159**, 272 (1960).
10. A. Peterlin, E. W. Fischer, and Chr. Reinhold, "Thermodynamic Stability of Polymer Crystals, II. Torsional Vibrations of Chain Molecules," *J. Chem. Phys.*, **37**, 1403 (1962).
11. Chr. Reinhold, "Thermodynamische Stabilität makromolekularer Kristalle," Thesis, University of Mainz, 1962.
12. T. P. Lin, unpublished calculations.
13. F. P. Price, "Growth Habit of Single Polymer Crystals," *J. Chem. Phys.*, **31**, 1679 (1959).
14. F. P. Price, "The Growth Habit of Single Polymer Crystals," *J. Polymer Sci.*, **42**, 49 (1960).
15. J. I. Lauritzen, Jr., and J. D. Hoffman, "Formation of Polymer Crystals with Folded Chains from Dilute Solution," *J. Chem. Phys.*, **31**, 1680 (1959).
16. J. I. Lauritzen, Jr., and J. D. Hoffman, "Theory of Formation of Polymer Crystals with Folded Chains in Dilute Solution," *J. Research Natl. Bur. Standards*, **64A**, 73 1960.

17. J. D. Hoffman and J. I. Lauritzen, Jr., "Crystallization of Bulk Polymers with Chain Folding: Theory of Growth of Lamellar Spherulites," *J. Research Natl. Bur. Standards*, **65A**, 297 (1961).

18. F. P. Price, "Markoff Chain Model for Growth of Polymer Single Crystals," *J. Chem. Phys.*, **35**, 1884 (1961).

19. F. C. Frank and M. Tosi, "On the theory of polymer crystallization," *Proc. Roy. Soc. (London)*, **A263**, 323 (1961).

20. J. I. Lauritzen, Jr., to be published.

21. D. Turnbull, "Phase Changes" in *Solid State Physics*, Vol. 3, edited by F. Seitz and D. Turnbull, Academic Press, New York (1956).

22. D. Turnbull and J. C. Fisher, "Rate of Nucleation in Condensed Systems," *J. Chem. Phys.*, **17**, 71 (1949).

23. D. Turnbull and R. L. Cormia, "Kinetics of Crystal Nucleation in Some Normal Alkane Liquids," *J. Chem. Phys.*, **34**, 820 (1961).

24. D. C. Bassett and A. Keller, "On the Habits of Polyethylene Crystals," *Phil. Mag.*, **7**, 1553 (1962).

25. J. D. Hoffman, "Thermodynamic Driving Force in Nucleation and Growth Processes," *J. Chem. Phys.*, **29**, 1192 (1958).

26. D. H. Reneker, "Point Dislocations in Crystals of High Polymer Molecules," *J. Polymer Sci.*, **59**, S39 (1962).

27. J. D. Hoffman and J. J. Weeks, "Melting Process and the Equilibrium Melting Temperature of Polychlorotrifluoroethylene," *J. Research Natl. Bur. Standards*, **66A**, 13 (1962).

28. E. A. Cole and D. R. Holmes, "Crystal Lattice Parameters and the Thermal Expansion of Linear Paraffin Hydrocarbons, Including Polyethylene," *J. Polymer Sci.*, **46**, 245 (1960).

29. F. R. Anderson, "Internal Morphology of Bulk Crystallized Polyethylene," paper BB4, presented at 5th Inter. Cong. Elec. Micro., Philadelphia (1962).

30. M. Broadhurst, "An Analysis of the Solid Phase Behavior of the Normal Paraffins," *J. Research Natl. Bur. Standards*, **66A**, 241 (1962).

31. J. D. Hoffman and J. J. Weeks, "The Rate of Spherulitic Crystallization with Chain Folds in Polychlorotrifluoroethylene," *J. Chem. Phys.*, **37**, 1723 (1962).

32. F. P. Price, "Spherulite Growth Rates in Polyethylene Crosslinked with High Energy Electrons," *J. Phys. Chem.*, **64**, 169 (1960).

33. R. L. Cormia, F. P. Price, and D. Turnbull, "Kinetics of Crystal Nucleation in Polyethylene," *J. Chem. Phys.*, **37**, 1333 (1962).

34. S. Mizushima, "Structure of Molecules and Internal Rotation," Academic Press, New York 1954 (Chapter V).

35. P. J. Flory, C. A. J. Hoeve and A. Ciferri, "Influence of Bond Angle Restrictions on Polymer Elasticity," J. Polymer Sci., **34**, 337 (1959).

36. P. De Santis, E. Giglio, A. M. Liquori and A. Ripamonti, "Stability of Helical Conformations of Simple Linear Polymers," *J. Polymer Sci.*, **1A**, 1383 (1963).

37. R. McCullough, to be published.

38. D. C. Bassett and A. Keller, "On the Habits of Polyethylene Crystals," *Phil. Mag.* **7**, 1553 (1962). Additional data given in reference 19.

39. D. C. Bassett, F. C. Frank, and A. Keller, "Some New Habit Features of Long Chain Compounds, Part IV," *Phil. Mag.* (in press).

40. V. S. Yoffe, "About the Structure and Properties of Real Crystalline Substances," *Uspekhi Khim.*, **13**, 144 (1944) (in Russian).
41. H. D. Keith and F. J. Padden, Jr., "Interpretation of Melt Crystallization in High Polymers and Related Systems," *J. Appl. Phys.* (in press).
42. A. E. Smith, "The Crystal Structure of the Normal Paraffin Hydrocarbons," *J. Chem. Phys.*, **21**, 2229 (1953).
43. W. Kuhn and H. Kuhn, "Statistische und energiestastische," Rückstellkraft bei stark auf Dehnung beansprцchten Fadenmolekeln., *Helv. Chim. Acta*, **29**, 1095 (1946).
44. B. Szigeti, "Torsional vibrations of long-chain hydrocarbons.," *Proc. Roy. Soc. (London)*, **264A**, 198 (1961).
45. G. Liebfried, "Gittertheorie der mechanischen und thermischen Eigenschaften der Kristalle," *Handbuch der Physik*, *Vol. VII/1*, Springer-Verlag, Verlin, Gottingen, Heidelberg, 1955.
46. S. Krimm, C. Y. Liang, and G. B. B. H. Sutherland, "Infrared Spectra of Polymers II. Polyethylene," *J. Chem. Phys.*, **25**, 549 (1956).
47. F. C. Frank and H. L. Pryce, quoted in reference 9.
48. C. W. Bunn, "The Crystal Structure of the Long Chain Normal Paraffin Hydrocarbons. The 'Shape' of the $\diagup\!\!\!\diagdown CH_2$ Group," *Trans. Faraday Soc.*, **35**, 482 (1939).
49. A. Muller, "The van der Waals Potential and the Lattice Energy of a n—CH_2 Chain Molecule in a Paraffin Crystal," *Proc. Roy. Soc. (London)*, **154A**, 624 (1936).
50. E. W. Fischer and G. Schmidt, "Über die Langperioden von verstrecktem Polyäthylen," *Angew. Chem.* **74**, 551 (1962).
51. M. Born, "Die Gultigkeitsgrenze der Theorie der idealen Kristalle und ihre Uberwindung," *Festschrift Gottingen*, **1**, 1 (1951).
52. P. J. Flory, "On the Morphology of the Crystalline State in Polymers," *J. Am. Chem. Soc.*, **84**, 2857 (1962).

VII. ORIENTATION

Many of the commercial applications of polymers involve the utilization of the basic anisotropy of their molecular forces. As indicated in Chapter I, the strength of the bonds along a polymer molecule are considerably greater than the forces between molecules. By orienting the polymer, i.e., aligning the polymer molecules parallel to the direction of greatest anticipated stress, this anisotropy is utilized in such objects as fibers and films. In this chapter we shall consider the morphological aspects of orienting crystalline polymers, including such features as the changes in wide and small angle x-ray diffraction as a polymer is drawn, microscope observations of the drawing of spherulites and single crystals, the effect of annealing a drawn polymer, and the orientation that is produced by rolling and by the crystallization process itself. In general we shall discuss the orientation of polyethylene, since this polymer has been most widely studied. The results, however, are believed to also apply in general to other crystalline polymers.

1. WIDE ANGLE X-RAY DIFFRACTION

When a polymer is drawn, i.e., stressed uniaxially, the molecules tend to become oriented in the draw direction. Linear polyethylene, for instance, can be drawn about 1000% at room temperature before it breaks. Although some rather complex changes in orientation occur during the initial stages of drawing, the molecular axes in a fully drawn fiber are highly aligned in the draw direction. The other two axes are randomly oriented normal to the fiber axis. Wide angle x-ray diffraction patterns from drawn fibers of several crystalline polymers are shown in Figure I-5.

Diffraction patterns from drawn amorphous polymers also indicate that the molecules became aligned during drawing. In some polymers which do not appear to crystallize in the unoriented state, drawing can produce a one or two-dimensional type of order (1). A diffraction pattern from a drawn fiber of polyacrylonitrile, which is representative of a polymer which normally has "lateral" order only (1,2), is shown in Figure VII-1. As indicated in Chapters II and IV, however, this polymer apparently can be crystallized with three dimensional order, or at least can crystallize in the form of lamellae, under appropriate conditions.

421

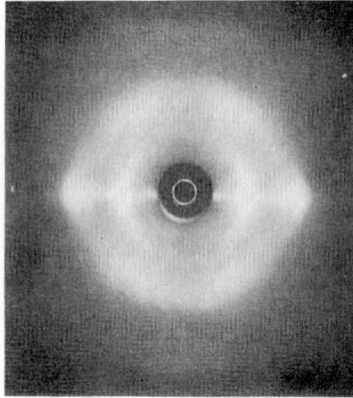

Fig. VII-1. X-ray pattern from a drawn fiber of polyacrylonitrile. Sharp reflections are are observed only on the equator (Bohn, Schaefgen, and Statton (2)).

Wide angle diffraction evidence for the change in orientation that takes place when polyethylene is drawn has been summarized by Aggarwal and Sweeting (3,4). They indicate that the following steps take place (3): "(a) As polyethylene is stretched, the **a** axis first becomes oriented perpendicular to the direction of stretch. The planes containing the **b** and **c** axes are perpendicular to the **a** axis but otherwise randomly distributed. (b) As stretching proceeds the directions $\langle 01\ell \rangle$ with higher and higher index of ℓ become aligned parallel to the direction of stretch. In highly stretched samples the limiting direction $\langle 001 \rangle$ is parallel to the stretching direction." Although the experiments on which these conclusions are based were performed with branched polyethylene, similar changes are believed to occur with linear polyethylene.

The drawing of polymers can take place in two different ways on a macroscopic scale; i.e., they can either draw uniformly or line draw (neck). These types will be discussed in more detail in Section VII-3. The changes in orientation that occur in the necking region (a narrow band normal to the draw direction across which most of the deformation takes place) are believed to be essentially the same as those that occur more gradually and throughout a uniformly drawing sample. Various aspects of the drawing process in polyethylene, many of which are still applicable in terms of present views of morphology, have been discussed by a number of authors (5–9). The results of Aggarwal et al. (4) were obtained from the necking region using an x-ray microcamera. Considerably more detailed x-ray studies are needed, however, of the drawing process and the effect on it of temperature and original morphology, not only for polyethylene but other polymers as well.

2. SMALL ANGLE X-RAY DIFFRACTION

Drawn fibers of most crystalline polymers have one or more orders of small angle x-ray diffraction. A review of this subject and its interpretation in terms of a fringed micelle type structure has recently been published (10). At the present time this type of structure has not yet been disproved for highly drawn polymers, although, as will be seen, questions are beginning to be raised.

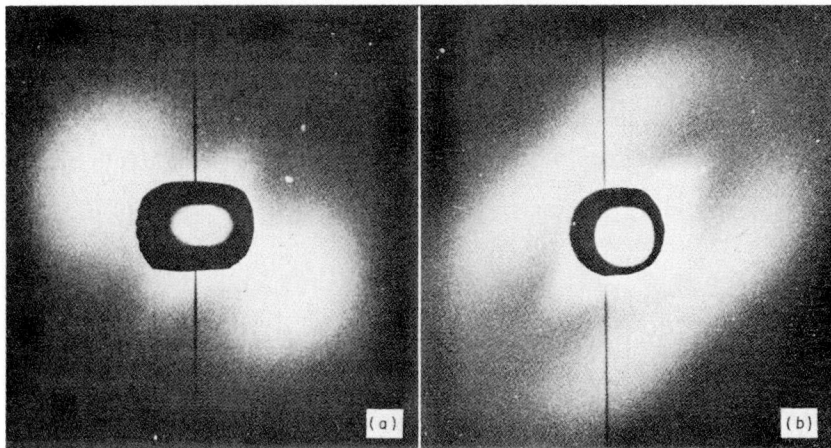

Fig. VII-2. Small angle diffraction patterns from nylon fibers. (a) Meridional small angle diffraction. (b) Quadrant small angle diffraction. The fiber axis is at $-45°$. In both samples there is diffuse scattering elongated in the equatorial direction. (Statton (11).)

The discrete small angle x-ray diffraction patterns from fibers, in general, have the form of lines or streaks normal to the meridian of the pattern (Fig. VII-2a). In some cases quadrant diffraction is observed (Fig. VII-2b). The line pattern is due to a periodic spacing along the fiber axis, i.e., a linear lattice. The lateral extent of the lines is generally interpreted as being inversely proportional to the lateral extent of the diffracting structure. As in the case of wide angle diffraction, however, perfection may be important. The quadrant or four-point diagram, on the other hand, suggests a three-dimensional arrangement of the diffracting structures.

424 POLYMER SINGLE CRYSTALS

Models of the type of fringed micelle structures that could lead to these two types of patterns are shown in Figure VII-3. The diffuse scatter near the center of the pattern probably arises, in large part, from voids in the fiber. As would be expected, the pattern is elongated along the equator indicating that the longest dimension of the voids is along the fiber axis.

The small angle diffraction patterns from drawn polymers are not well

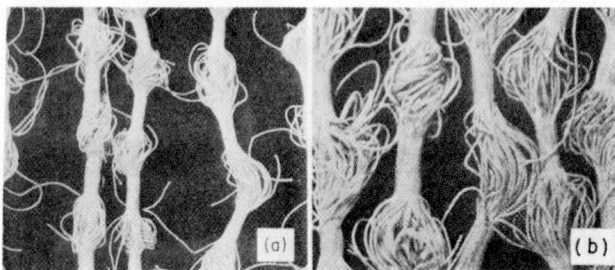

Fig. VII-3. String models of the structure of a fiber. The arrangement of the crystals in (a) would produce meridional diffraction. The mutual compensation of the crystals and amorphous regions in (b) could lead to quadrant diffraction. (Statton (12).)

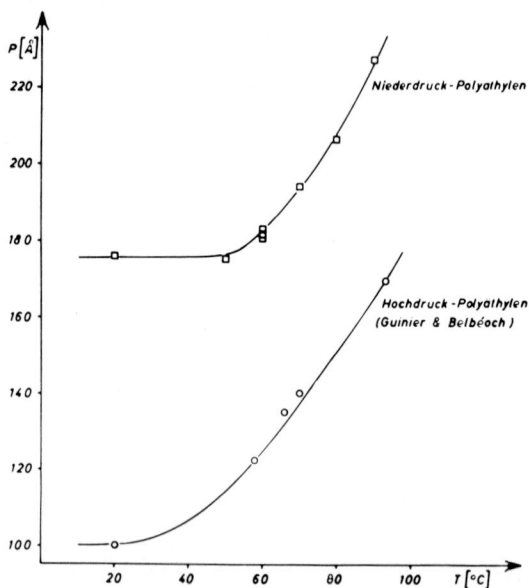

Fig. VII-4. Dependence of the long period of linear (□) and branched (○) polyethylene in the drawing temperature (Fischer and Schmidt (14)).

understood. As Porod indicates (10): "Considerable difficulties are encountered in attempts to interpret the dependence of the long periods on pretreatment, annealing, elongation or swelling. Not only is the location of the reflection changed, but also its distinctness, intensity and the length of the layer line; in the case of four-point diagrams, the angle between the intensity maximum and the meridian is shifted. As a rule, these changes take place irreversibly. Also a layer-line pattern can be converted into a four-pointed diagram." Reversible changes are caused by swelling, whereas elongation may result in reversible or irreversible changes.

In general, temperature increases during or after drawing result in an irreversible increase in the long period. For instance Belbeoch and Guinier (13) report that polyethylene drawn at 58°C has a 125 A period, at 66°C a 135 A period, and at 95°C at 170 A period when measured in the drawn state. If the latter sample is then allowed to relax at 50°C a four-point diagram with a 140 A period along the axis is obtained. These data and some obtained by Fischer and Schmidt (14) are shown on Figure VII-4.

Hosemann (15) indicates that for polyethylene, at least, equatorial diffuse scattering is observed near the main beam (as in Figure VII-2) only for cold drawn samples. If the samples are drawn while hot and then crystallize, the discrete scattering is still present but there is little or no diffuse scattering. Fischer and Schmidt (14) also find that the diffuse equatorial scattering is much more intense from samples drawn at room temperature than from those drawn at 70°C. On the basis of these results Hosemann (15) indicates that a model similar to that in Figure VII-3 cannot apply to hot drawn fibers since the lateral alteration in density would result in some form of scattering along the equator. He suggests that the model shown in Figure VII-5a, involving some folded molecules, is more appropriate. An optical diffraction pattern from such a model is shown in Figure VII-5b and c (see Figure I-9 for comparison purposes). On the other hand, Hosemann suggests that a model similar to that in Figure VII-3a may apply to cold drawn samples. He finds that pressing a cold drawn fiber by hand causes the diffraction pattern to shift from lines along the meridian to a four-point diagram when observed normal to the compression direction, but is unaltered parallel to this direction. This suggests that during the compression the morphology is shifting from that similar to Figure VII-3a to that in Figure VII-3b.

The changes observed in the wide angle pattern during drawing do not have any clear relationship with the changes in the small angle pattern. Hendus has recently obtained (16) some excellent patterns showing the change in small angle diffraction that takes place during drawing (Fig.

VII-6). At low elongations the originally circular diffraction pattern takes on a V-shaped structure, i.e., a quadrant diffraction pattern in which each reflection consists of a line at an angle to the draw direction. The original periodicity is maintained. With further elongation, i.e., 250% or more, an

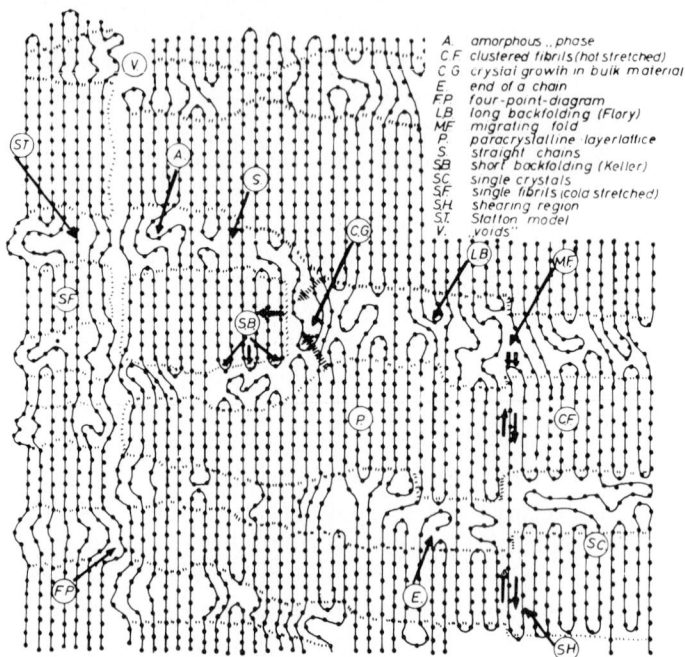

A.	amorphous ..phase
C.F.	clustered fibrils (hot stretched)
C.G.	crystal growth in bulk material
E.	end of a chain
F.P.	four-point-diagram
L.B.	long backfolding (Flory)
M.F.	migrating fold
P.	paracrystalline layerlattice
S.	straight chains
S.B.	short backfolding (Keller)
S.C.	single crystals
S.F.	single fibrils (cold stretched)
S.H.	shearing region
S.T.	Statton model
V.	voids

MODEL OF LINEAR POLYETHYLENE

(a)

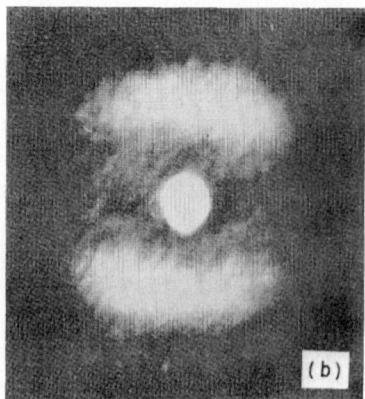

(b)

Faserachse

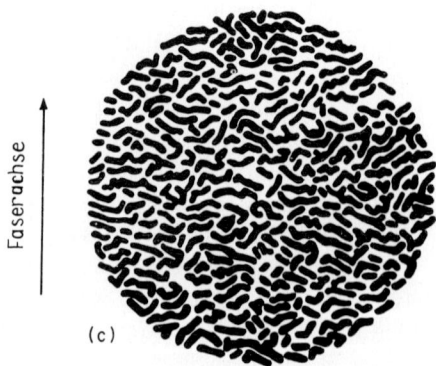

(c)

entirely new periodicity is seen to appear. Although the wide angle pattern shifts gradually, the original long period does not. Somewhat similar results have been described by Kasai *et al.* (17). These results will be discussed further in Section VII-3.

Hendus (16) and Hosemann (15) also investigated the changes in the diffraction pattern that occur when a drawn sample (elongated 200%) is redrawn at an angle of not quite 90° to the original draw direction (Fig. VII-7). Despite an almost complete alignment of the molecular axes in the new draw direction, the small angle pattern remains skewed. The correlation of similar small angle experiments with electron microscope studies, which has not yet been done, should enable one to determine the mechanism of drawing of bulk polymers.

A peculiar feature of the small angle diffraction pattern of drawn polyethylene fibers, noted by Mandlekern, Worthington, and Posner (18), is that 1st-, 2nd-, and 4th-order reflections can be observed. Not only is the 3rd order missing but the 1st-order reflection is less intense than the 2nd. Similar features were not observed by Statton (19).

3. MICROSCOPY OF DRAWN SPHERULITIC SAMPLES

Relatively few papers have been published describing optical or electron microscope observations of the drawing of polymer spherulites (3,5,20–22) and in general the reproduction of the published micrographs is inadequate to permit observation of details of the drawing process. Keith and Padden (20) made the following observations on the basis of their micrographs:

(1) Spherulites of polyethylene in thin films (5–50 microns), grown by very slow cooling and thus having a coarse structure, undergo brittle fracture at very low stresses and negligible elongation. The fracture follows an irregular coarse without regard to the boundaries or internal structure of the spherulites.

←——————————————————————————————

Fig. VII-5. (a) Composite model of linear polyethylene according to Hosemann. On the top left (S.F.) is depicted the structure of cold drawn material, similar to that pictured in Figure VII-3a. When compressed, lower left, the crystallites adjust, resulting in a four-point x-ray diagram (F.P.) and a structure similar to Figure VII-3b. On the right, during hot drawing, the fibrils have clustered (C.F.) to produce lamellaelike structures oriented normal to the draw direction. Near the center of the model is depicted a type of crystal growth (C.G.) that can occur in bulk material, the small crystal gradually disappearing and its chains being incorporated in the larger crystals above and below it. The long backfolding (L.B.) differs from that postulated by Flory (Chapter VI) in that the model shows adjacent re-entry. (b) Optical diffraction model and (c) diffraction pattern corresponding to proposed model of hot drawn fibers. (Hosemann (15).)

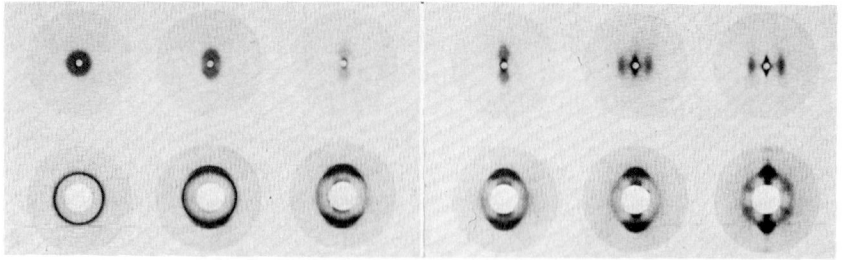

Fig. VII-6. Small and wide angle diffraction patterns from a linear polyethylene as a function of the amount of draw (Hendus (16)).

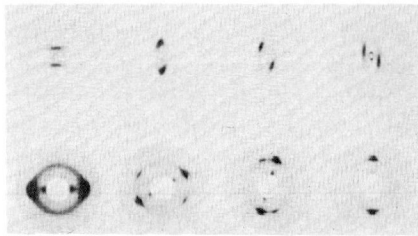

Fig. VII-7. Changes in wide and small angle diffraction patterns as a drawn film of branched polyethylene is redrawn at nearly a right angle to the original draw direction (Hendus (16)).

(2) In more rapidly cooled thin films brittle fracture takes place at the boundaries of the spherulites.

(3) With increasingly more rapidly crystallized spherulites, drawing takes place at the boundaries before rupture and then a stage is reached at which the spherulites themselves begin to yield in place of, or in conjunction with, yield at the boundaries. When the spherulites yield, a neck is often formed across the spherulite diameter. The extinction rings of the spherulite can still be seen in the drawn region at low elongations but tend to disappear as drawing continues (Fig. VII-8).

(4) The resulting drawn material is microfibrillar and splits readily parallel to the fibril axes.

(5) Keith and Padden suggest that the differences in the drawing behavior due to different crystallization rates is a result of differences in the number of tie molecules holding the lamellae together.

In the remainder of this section we shall consider optical and electron microscope evidence for the mechanism of drawing of rapidly cooled spherulites of linear polyethylene similar to those shown in Figures IV-3,

IV-16, and IV-18, and solvent cast spherulites of branched polyethylene. Some micrographs have also been obtained of the drawing process in slowly cooled polyoxymethylene (as in Figures IV-4 to IV-10).

Fig. VII-8. Spherulitic film of polyethylene partially drawn. Voids have developed near the nucleus with drawing taking place along the transverse diameter (polarized light). (Keith and Padden (20).)

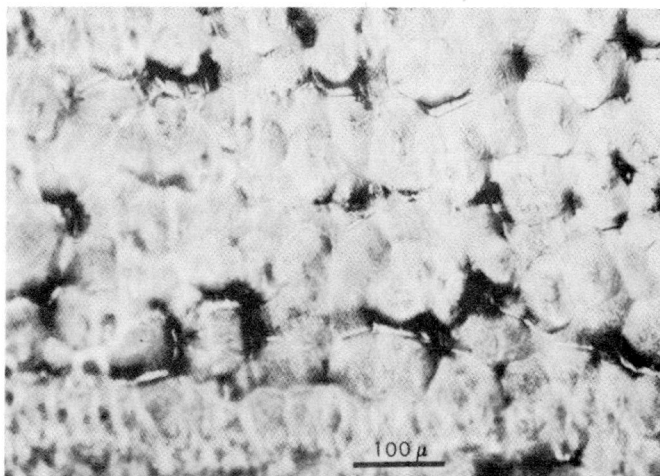

Fig. VII-9. Thin film of linear polyethylene slightly drawn (incident steep oblique illumination).

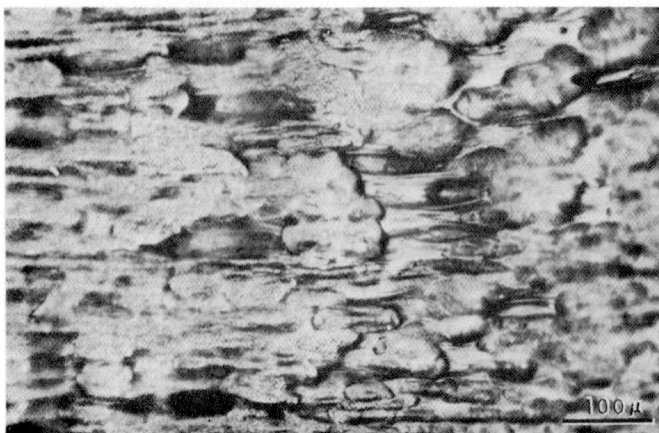

Fig. VII-10. Portion of the deformed region of a film of linear polyethylene that had been drawn almost to the necking stage (incident steep oblique illumination).

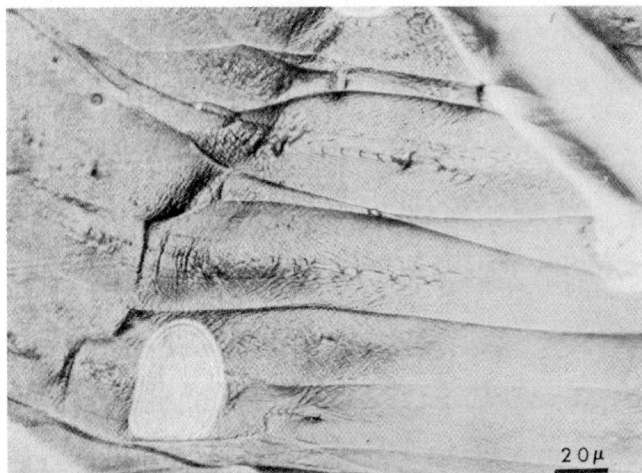

Fig. VII-11. Replica of the necking region of linear polyethylene. The white areas are interference patterns between the polyvinyl alcohol replica and the microscope slide (incident steep oblique illumination).

As indicated by Keith and Padden (20), the first visible elongation of a spherulitic polyethylene film occurs at spherulite boundaries oriented normal to the draw direction (Fig. VII-9). In thin films these boundaries are susceptible to yielding because they are thinner than the rest of the film. During crystallization the polymer contracts toward each spherulite

Fig. VII-12. Drawn spherulitic film of linear polyethylene (incident steep oblique illumination).

nucleus, and actual voids may be produced at the boundaries. It should be noted that since the **b** axis is radially oriented in polyethylene spherulites, the **a** and **c** axes of the portion that draws first will be perpendicular to the draw direction. While this would appear to agree, at least in part, with the wide angle x-ray diffraction results (Section VII-1), the cause of the x-ray results are probably somewhat more subtle. As will be seen in the next section (p. 445) the deformation characteristics of polyethylene lamellae are significantly different in the **a** and **b** axis directions. Furthermore in the published micrograph of Keith and Padden (20) the drawn region is along a spherulite diameter and not along the nearly parallel boundary (Fig. VII-8). In Figure VII-8 the **b** axis in the drawing region is always normal to the draw direction. Surface restraint by a cover glass or similar cover restricted the polymer contraction during the growth of the spherulites in this figure.

With further elongation, linear polyethylene samples usually line draw, i.e., the elongation takes place across a rather sharply defined line. An intermediate stage between that shown in Figure VII-9 and true line drawing is shown in Figure VII-10. Some of the spherulites appear to be virtually unaffected whereas others have elongated. Shear must be taking place along some of the spherulite boundaries parallel to the draw direction.

A replica of a necking region is shown in Figure VII-11. Essentially undrawn spherulites are shown to the left while nearly completely drawn polymer is to the right. A replica is used because the thickness of the film,

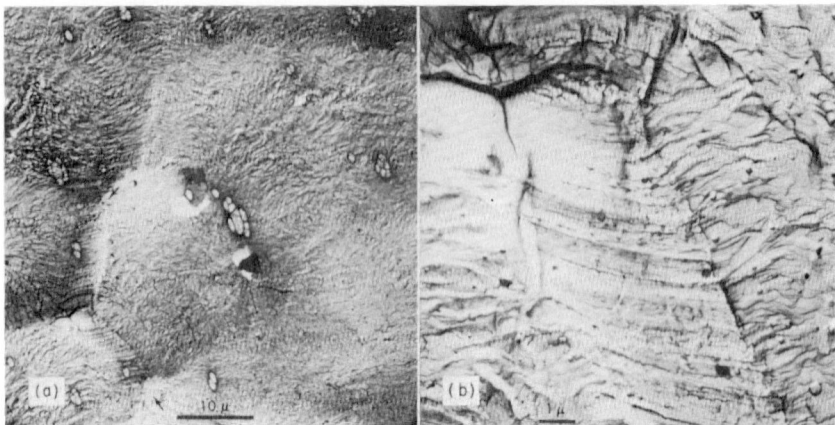

Fig. VII-13. Electron micrograph of slightly drawn linear polyethylene similar to that in Figure VII-9. (a) Drawing has taken place at the spherulite boundaries indicated by the arrow. The spherulites were too small (nuclei too close) to permit the rings to develop. (b) Higher magnification of the boundary region in which drawing is taking place (Cr replica).

Fig. VII-14. Low magnification electron micrograph of drawn linear polyethylene similar to that in Figure VII-12. The remnants of the ring structure can be seen (arrows). The biaxial character of the stress resulted in the void across which fibers have been drawn (Cr replica).

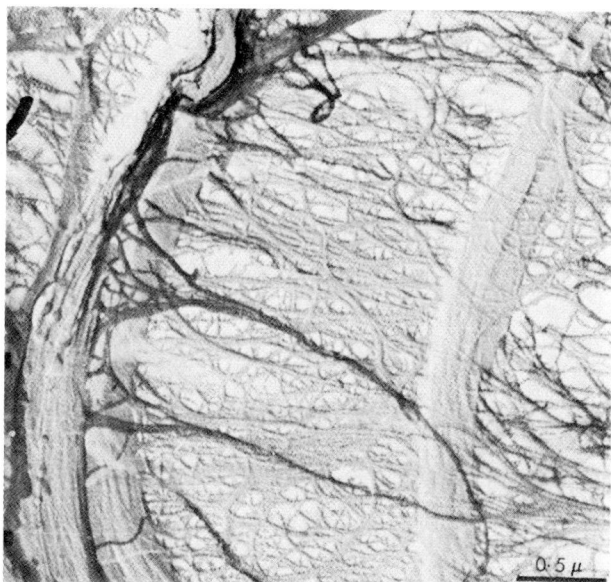

Fig. VII-15. Portion of the void region from Figure VII-14 at higher magnification.

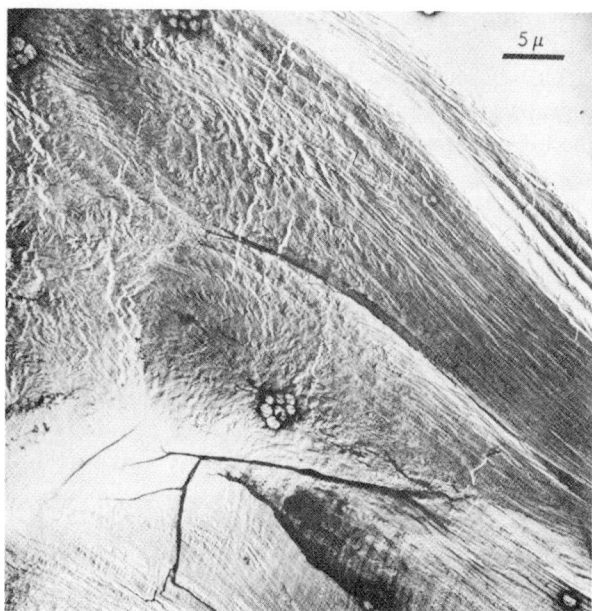

Fig. VII-16. Low magnification electron micrograph of the necking region of a drawn polyethylene film (as in Fig. VII-11) (Cr replica).

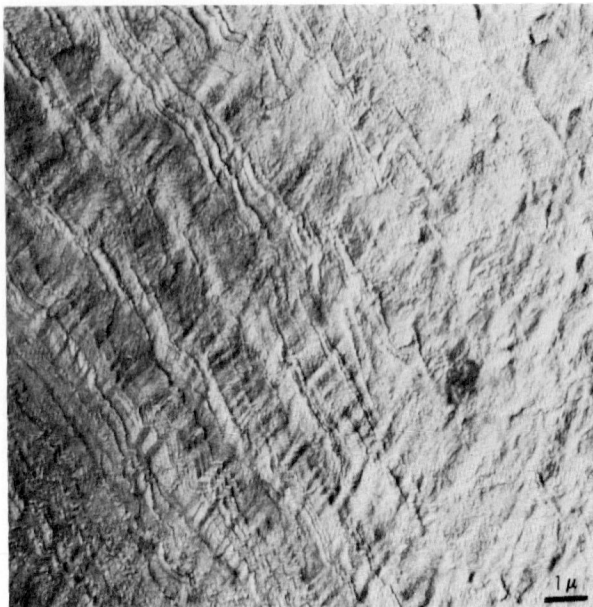

Fig. VII-17. Portion of Figure VII-16 at higher magnification.   Relaxation at room
temperature took place before the film was replicated.

as well as its lateral dimension, decreases abruptly at the neck. Almost the
entire elongation is seen to be restricted to distances of less than a spherulite
diameter. The fully drawn material is shown in Figure VII-12. Spherulites
and their internal ring structure are still visible on the surface and, to an
extent, in polarized light. The surface structure observed here can also be
seen on drawn samples which originally had smooth surfaces, as, for instance, compression molded films. Microcamera x-ray diffraction patterns
from well-defined portions of these drawn spherulites, including the prominent nucleus region, would be useful in defining not only the mechanism
of drawing but also the structure of a fully drawn polymer.

Electron micrographs from a slightly drawn polyethylene (as in Figure
VII-9) indicate that there can be an abrupt change between drawn and
undrawn material (Fig. VII-13). The lamellar structure appears to be
preserved undistorted up to the line of draw. In some cases, fragments of
the lamellae appear to break off and remain as discrete regions in the drawn
material (see later discussion of drawn polyoxymethylene, p. 436).

A biaxial component of the stress results in splitting of the drawn regions
and the formation of voids (Figs. VII-13 and VII-14). Figure VII-14 is

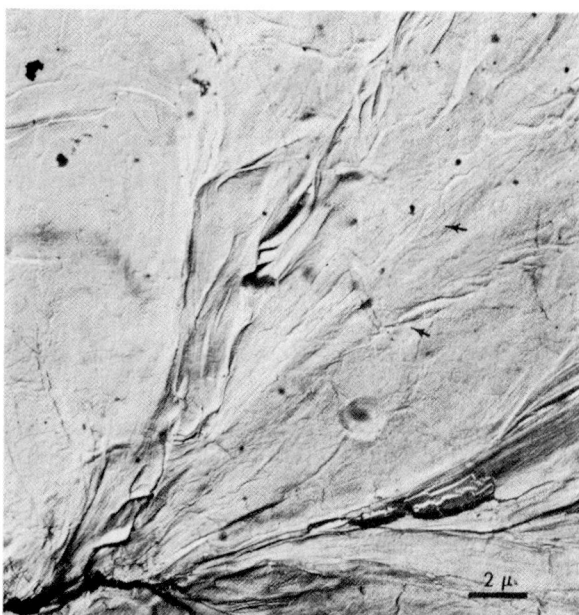

Fig. VII-18. Chromium replica of a slightly drawn film of polyoxymethylene. Drawing has taken place at the regions indicated by the arrows.

from a fully drawn region, similar to that in Figure VII-12. The fibrils that are split off of the edges of the voids are on the order of 100 A in diameter. The beaded appearance in Figure VII-15 is rather typical but may be due to granulation of the replicating metal. Similar beading has been seen on fibrils of polytetrafluoroethylene as well as on some fibrils drawn out of polymer crystals (see the next section, p. 443). Fischer and Schmidt also report a periodicity on fibers drawn from bulk polyethylene and single crystals (14).

Electron micrographs of a neck region similar to that in Figure VII-10 show a gradual transition from the lamellar to the fibrillar structure (Fig. VII-16). Unfortunately those micrographs available to date have not had sufficient resolution, probably because of surface contamination by exuded material, to show the changes in detail. The area in the lower left, shown at higher magnification in Figure VII-17, shows the structure of drawn material which had relaxed before it was replicated. The zigzag structure has been interpreted by Sella and Trillat (22) as consisting of microcrystals which have retained the folded molecule structure throughout the drawing and relaxation treatment.

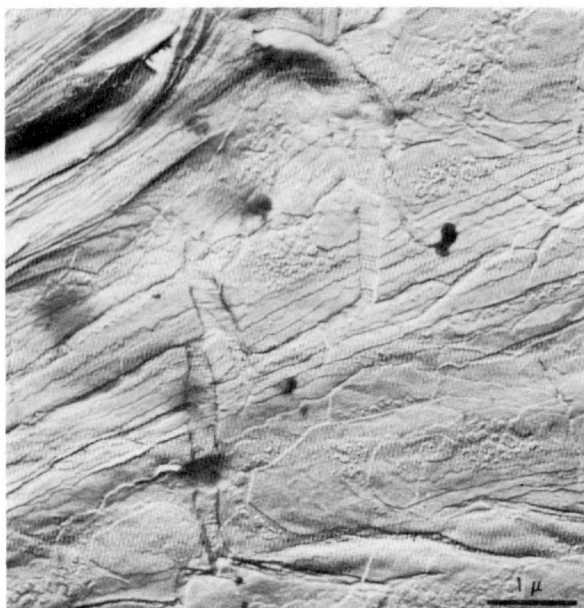

Fig. VII-19. Higher magnification of a portion of Figure VII-18.

Slow cooled films of polyoxymethylene cannot be drawn at room temperature but fracture instead. A line-draw type of elongation can be obtained, however, by heating the samples to temperatures of the order of 100°C or higher. The initial stages of drawing take place within the spherulites, fibers being drawn across small cracks (Figs. VII-18 and VII-19). As in the case of linear polyethylene there is an abrupt change between drawn, fibrillar, and undrawn, lamellar polymers. These observations suggest, but do not prove, that this type of drawing takes place through an unfolding of the molecules.

Apparently slip can take place between the lamellae. As shown in Figure VII-20, broad drawn areas may abruptly terminate with adjacent lamellae appearing undistorted. Although often, as here, the drawn area terminates in a region in which there is a change in orientation of the lamellae, this is not always the case. Lamellar slip is also suggested by the type of structure seen in Figure VII-21 which is formed as the elongation is increased. Sections of individual lamellae or groups of lamellae remain apparently undistorted in the midst of the drawn, fibrillar material. It is not believed that these undistorted sections extend through the sample. The biaxial stress in this case, also, is sufficient to split the drawn fibers and form voids. These voids, in polyethylene at least, result in decreases in density (23) and intense diffuse small angle x-ray scattering.

The morphological structures in polytetrafluoroethylene that correspond to the lamellae are of such a size that their deformation can more easily be viewed in the microscope than is the case with the 100 A thick lamellae in polyethylene and most other crystalline polymers. These bands (see Chapter IV, Fig. IV-68) are on the order of 1000 A or more wide with the molecules apparently aligned parallel to striations that are normal to the band

Fig. VII-20. Portion of partially drawn film of polyoxymethylene (Cr replica).

on fracture surfaces. By following the deformation of the bands and striations when a sample is drawn, Speerschneider and Li (24) have noted two types of deformation (see also Section VII-6 for the effect of rolling on the band structure). They suggest that the initial stages of deformation (Fig. VII-22) involves a relative displacement of the striations within the bands but no deformation of the bands themselves. When the bands are oriented normal to the draw directions, a nonhomogeneous sliding or shear of the striations (and presumably the folded molecules) takes place (A in Figure

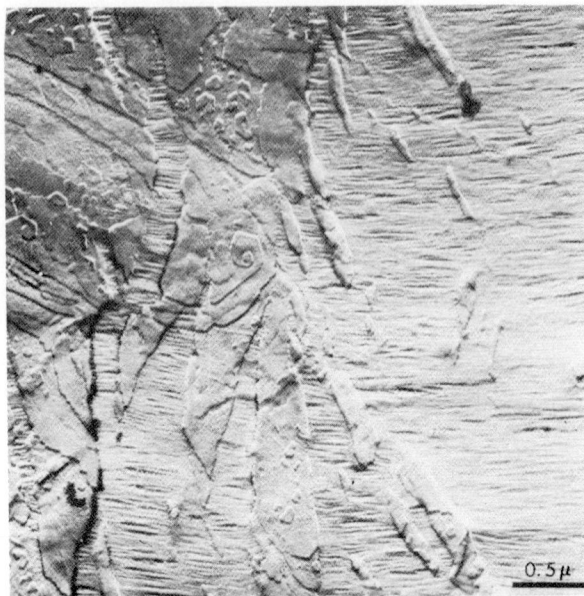

Fig. VII-21. Edge of drawn material in a film of polyoxymethylene. Sections of unde-
formed lamellae are drawn off the edge of the sample as drawing takes place.

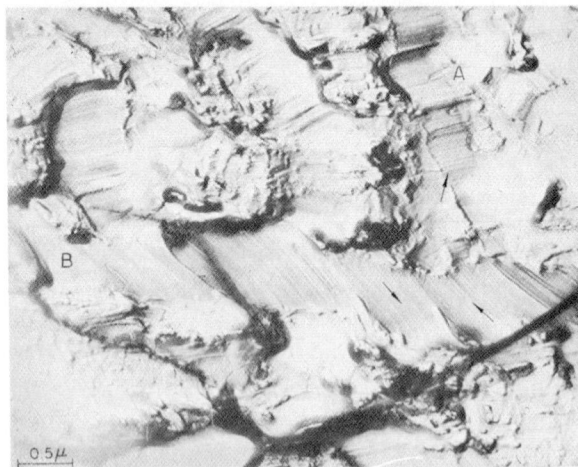

Fig. VII-22. Fracture surface of polytetrafluoroethylene which was drawn 100%
after the surface was produced. Nonhomogeneous shear (A) and homogeneous rotation
(B) of the striations have taken place. (Speerschneider and Li (24).)

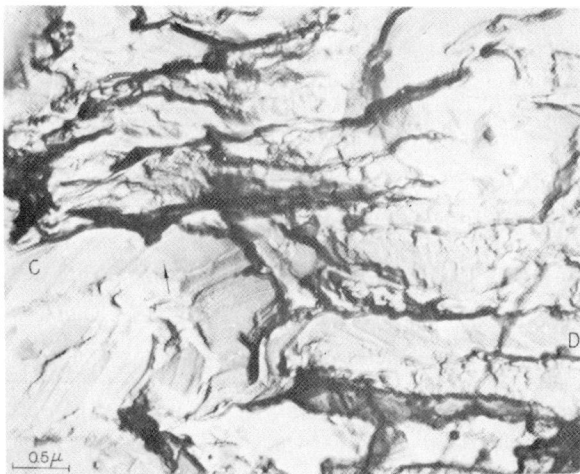

Fig. VII-23. A region from the same sample as in Figure VII-22 in which bowing (C) and kinking (D) of the striations has taken place (Speerschneider and Li (24)).

VII-22) whereas if there is a component of the draw parallel to the band, homogeneous rotation of the striations occurs (B in Figure VII-22).

Speerschneider and Li suggest (24) that if the deformation resulting from the shear of the striations (or molecules) as pictured in Figure VII-22 is locally insufficient to accommodate the macroscopic deformation, then the striations themselves may deform (Fig. VII-23). In a few cases (as in C in Figure VII-23) the striations become bowed, whereas in other regions a kink develops (as at D in Fig. VII-23). The shear and kinking observed by Speerschneider and Li would appear to be related to deformation by dislocation motion and twinning as observed in crystals of low molecular weight materials. Polytetrafluoroethylene, because of the size of its morphological units, should be highly suitable for studies involving the relationship of physical properties and morphology. Speerschneider and Li note (24) that when a drawn sample is annealed at 350°C (23°C above the melting point) the "original" band structure returns, with no trace of the deformation either microscopically or in terms of the shape of the sample. However, particularly since they also note that the new bands are smaller than the original, it may be that recrystallization and the formation of new bands has occurred.

In contrast to linear polyethylene and polyoxymethylene, which are highly crystalline, branched polyethylene, of intermediate crystallinity, can undergo uniform draw rather than line draw. (It should be noted that

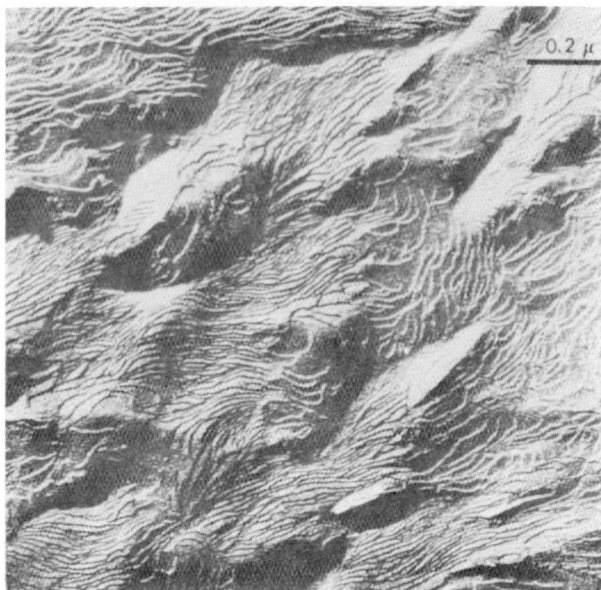

Fig. VII-24. Replica of a film of branched polyethylene drawn 50%
(Peck and Kaye (27)).

whether a polymer draws uniformly or necks is not directly dependent
on whether it is crystalline or not, but rather is related primarily to the
details of the stress-strain relationship for the polymer under the drawing
conditions (25,26).) In their early paper showing the lamellar structure of
polyethylene, which unfortunately was never published, Peck and Kaye
(27) also presented micrographs showing the effect of drawing the films
and then redrawing them at 90°. The lamellar nature of the undrawn polymer is shown in Figure IV-15; Figure VII-24 is a micrograph of a film
drawn 50%. Groups of lamellae, they suggest, have swung as units toward
the draw direction, distorting the surface in the process.

Peck and Kaye indicate (27) that when such a film is then redrawn at
an angle of 90° to the original draw direction the lamellae do not again
twist toward the new draw direction. Instead it appears as if slippage takes
place between the lamellae (Fig. VII-25). The micrograph also suggests
that some sort of fibrillation occurs.

The process of redrawing is even less well understood on a morphological
basis than the process of initial drawing, and it should be obvious from the
preceding that the initial process is not yet well defined. In the next section

Fig. VII-25. Replica of a film similar to that in Figure VII-24, except redrawn at 90° to the original draw direction (Peck and Kaye (27)).

we describe some of the initial experiments on the drawing of polyethylene single crystals. A more complete explanation of the drawing process in bulk samples will probably follow when the deformation of single crystals is understood. Tentatively, however, it appears that the initial stages of uniform draw, in agreement with the small angle diffraction results of Hendus (Section VII-2), consist primarily of an orientation, through twisting and shear, of the lamellae, followed and to some extent accompanied by an unfolding of the molecules. Similar effects also appear to take place in the necking zone during the line draw.

4. DRAWING OF SINGLE CRYSTALS

Fischer (14,28) and Geil (29) have obtained micrographs of polyethylene crystals drawn in various ways. Fischer dispersed the crystals in gutta-percha, drew the gutta-percha and then dissolved it and observed the remnants of the crystals. Dendritic or branched structures were obtained (Fig. VII-26) whose relationship to the original crystals is not yet known. Geil, on the other hand, drew the crystals on elastic substrates of various

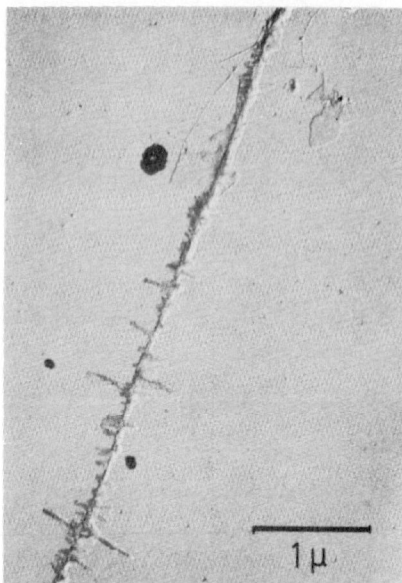

Fig. VII-26. Fiber formed when polyethylene single crystals were drawn while imbedded in gutta-percha. The branches are believed (28) to be residues of the original crystals. (Fischer (28).)

types, afterwards replicating the crystals for microscopy and, in some cases, removing the crystals for electron diffraction. These results are described later in this section. Descriptions of crystals in which drawing took place during sample preparation or which were mechanically scratched have been reported in the literature (14,30–34) and will be considered first.

Mechanical scratching of crystals has shown that fracture takes place most readily between fold planes and between lamellae (30,33). In a dendritic crystal, for instance, fracture planes are found parallel to the growth faces, changing direction across fold domain boundaries (Fig. VII-27). (See also Figure II-16.) In spiral growths each lamella fractures independently; such behavior is expected since it is doubtful that any molecules are incorporated in more than one lamella in solution-grown crystals. When fracture does not occur, fibers on the order of 50 A in diameter and larger are drawn out of the scratched crystal.

As mentioned in Chapter II (see Figure II-23), when hollow pyramidal crystals collapse on a substrate during solvent removal, cracks may develop at the apices of the crystal in addition to the pleats near the center. The

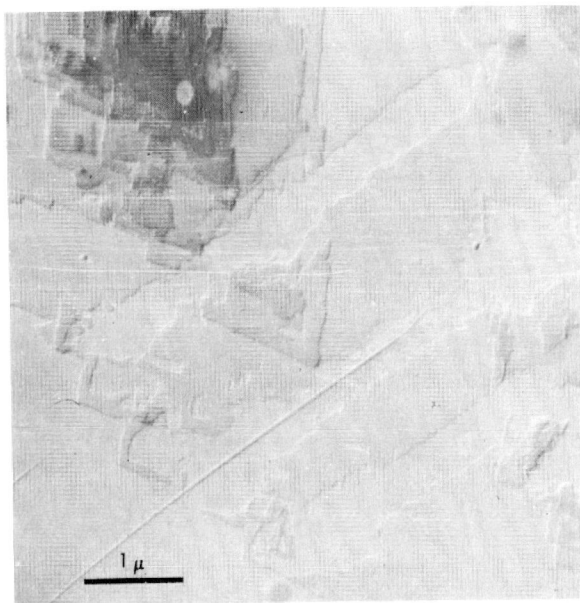

Fig. VII-27. Electron micrograph of mechanically scratched dendritic crystal of polyethylene. The fracture planes are primarily parallel to the fold planes. (Reneker and Geil (33).)

fibers drawn across these tears, which in the figures to be described may have been formed at relatively high temperatures, the solvent evaporating while still warm, are of particular interest. Figure VII-28 shows that the fibers are drawn out of individual lamellae. As in the case of mechanical scratching, each lamella tears independently. The larger fibers have a beaded appearance whereas smaller ones are also present which have a uniform diameter. The beading seen here is not believed to result from granulation of the replicating material since it is visible in the shadows as well as on the fibrils. One would like to interpret the beads in terms of portions of the lamellae which have remained folded and as being related to the small angle x-ray periodicities observed in macroscopic drawn fibers. This interpretation would tend to be confirmed by the structure of the drawn material in Figure VII-29. It appears as if broad thin sections of the lamella have broken off of the edge as the crystal split apart. The large, apparently empty regions in the drawn portion are difficult to interpret; one questions what has happened to the polymer originally present in these areas. Similar, broken-up ribbonlike structures have been observed on long fibrils drawn out of crystals while still in suspension. These ribbons are seen to consist of several more or less individual fibrils periodically held together by the "undrawn" sections of polymer.

On the other hand, structures formed in attempts to reproduce the above type of fibril by vigorously stirring the solution while it was still warm resulted in the formation of apparent exitaxial growth of lamellar crystals on the resulting fibrils (Fig. VII-30) (Also see Section VII-7.) Although these solutions contained dissolved polymer which crystallized out on the fibers during subsequent crystallization, the results suggest that the beading

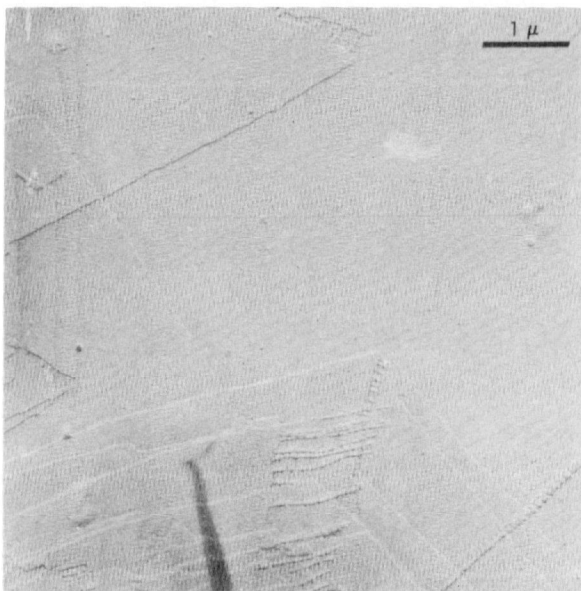

Fig. VII-28. Collapsed pyramidal crystals of polyethylene. The solvent was evaporated from these crystals while they were suspended just above the still hot suspension.

seen on the fibers in Figure VII-28 may be due to similar crystallization of low molecular weight paraffins as the solvent evaporated. Using stereomicroscopy the structures present in Figure VII-30 are seen to be composed of hollow pyramidal crystals standing on edge and oriented normal to the fiber running through and connecting them. Similar structures have been reported by Konstantinopolskaya, Berestneva, and Kargin (35) to form when 0.1 and 0.01% xylene solutions are crystallized on collodion substrates at temperatures in the range of 100 to 110°C. Considerably more work is

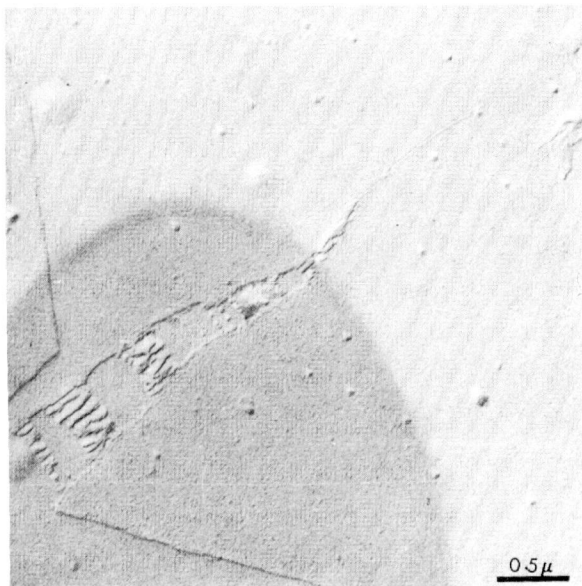

Fig. VII-29. Portion of a collapsed pyramid of polyethylene   A portion of the pleat as well as the tear at the apex can be seen.

needed to define the structure of both these "hedgerow" fibers and the beaded fibers.

Observations of polyethylene crystals drawn on plastic substrates indicated that the type of deformation that occurs is dependent on the direction of draw and on the adhesion to the substrate (29). Significantly different results have been obtained using substrates of "Mylar"* and "Viton."* We shall first consider the results with Mylar substrates, to which the crystals appear to adhere uniformly, and then micrographs of crystals drawn on Viton to which adhesion appears to take place only at various points. Using 0.001-inch Mylar films, elongations up to about 100% can be obtained in some directions (with respect to the axis of the film) whereas elongations of several hundred per cent can be obtained with Viton films.

As long as a component of the **b** axis of the crystal is in the direction of elongation (**a** axis normal to the draw direction), when the crystals are drawn on Mylar, sets of similar, parallel lines of deformation are formed parallel to the **a** axis of the crystal (Figs. VII-31 and VII-32). The appear-

* Registered trademarks of E. I. du Pont de Nemours & Co., Inc.

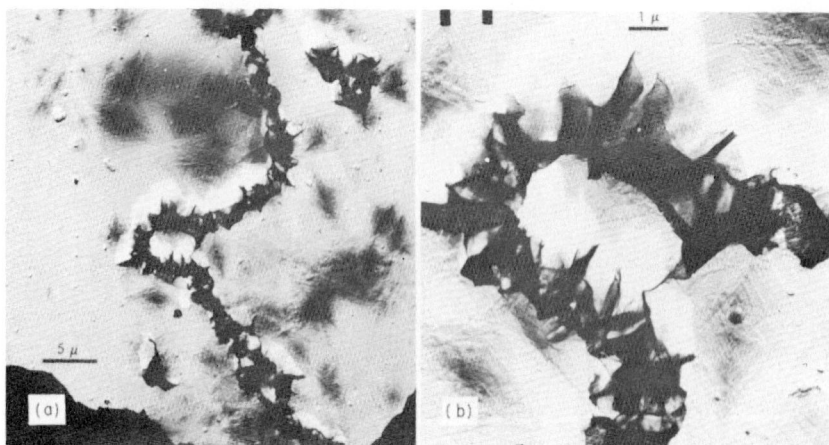

Fig. VII-30. "Hedgerow" type of growth of crystals normal to a drawn fiber.   An 0.1% tetrachloroethylene solution of polyethylene was sprayed into a thin wall tube in an oil bath at 58°C.   A number of the lamellae can be observed normal to the substrate in (b).   Similar structures have been grown (36) by stirring a suspension of crystals at an elevated temperature.

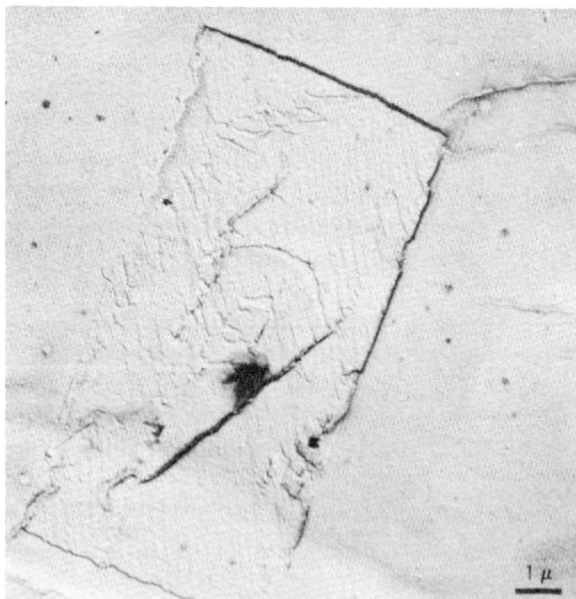

Fig. VII-31. Single crystal of polyethylene drawn 100% on a Mylar substrate.   The **a** axis is nearly vertical in the micrograph.

ance of the lines on one fold surface is different in adjacent fold domains; in fact observations of a number of micrographs suggests that the structure of a particular fold surface in one fold domain, for instance the (001), is similar to that of the opposite fold surfaces, $(00\bar{1})$, in the adjacent fold domains. With samples drawn 50% or more, a few crystals are found in which actual tears form along the lines during the elongation. In general, however, the crystal merely becomes thinner over-all and takes on a rippled structure. It should be noted that both the ripples and the tears cut across fold planes.

Fig. VII-32. Polyethylene crystals drawn 25% on a Mylar substrate. The **a** axes are parallel to the striations on the crystals.

The separation of the lines does not appear to depend on the angle between the draw direction and the **b** axis. The lines can be observed on crystals drawn as little as 15% and are visible on dark field micrographs. It has not been possible, as yet, however, to correlate the various portions of the lines with the Bragg reflection being used to form the dark field image. Occasionally crystals are formed in which similar appearing lines, instead of being parallel to the **a** axis, are parallel to the original {310} planes and may cross each other, forming a network, in the same fold domain.

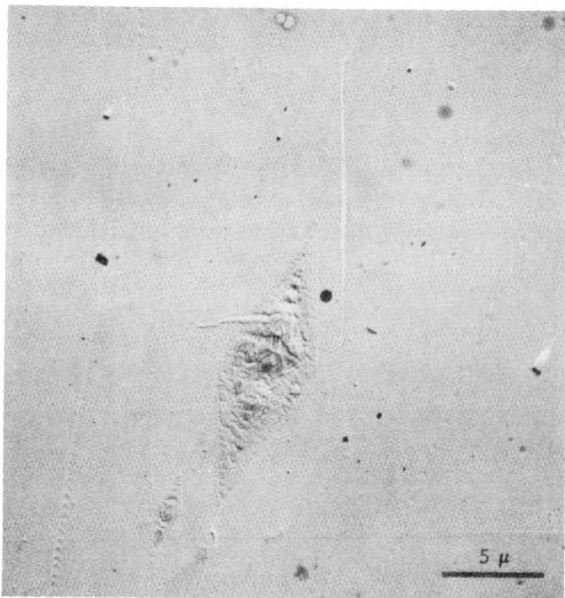

Fig. VII-33. Polyethylene single crystal drawn 100%, nearly parallel to the a axis, on a Mylar substrate.

No deformation lines are observed in crystals in which the **a** axis is parallel, or nearly so, to the direction of elongation (Figs. VII-33 and VII-34). A few irregular fracture lines can be seen on this crystal, which was drawn 100%, but in general only an irregular surface roughness is seen. It would appear that the adhesion to the Mylar substrate is more uniform than that between lamellae; a broad fracture spanned by fibers as well as numerous smaller fractures are seen on spiral growths in this and other crystals. The fibers, which were formed at room temperature, are not beaded. The amount of deviation between the **a** axis and the draw direction that determines whether this more or less uniform draw or the parallel line deformation type of draw seen previously takes place is less than 10°.

Diffraction patterns from crystals drawn various amounts on Mylar have several features in common. (Compare Fig. VII-35 with Fig. II-18.)

1. Some sort of twinning appears to take place as indicated by the presence of 6 {110} reflection and 4 {200} reflections.

2. The **a** and **b** axes rotate to an extent; the diffraction spots have broadened to arcs.

3. Additional reflections, the innermost and outermost on Figure VII-35, appear which are not due to the orthorhombic unit cell of polyethylene. These reflections are believed similar to those seen in extensively drawn

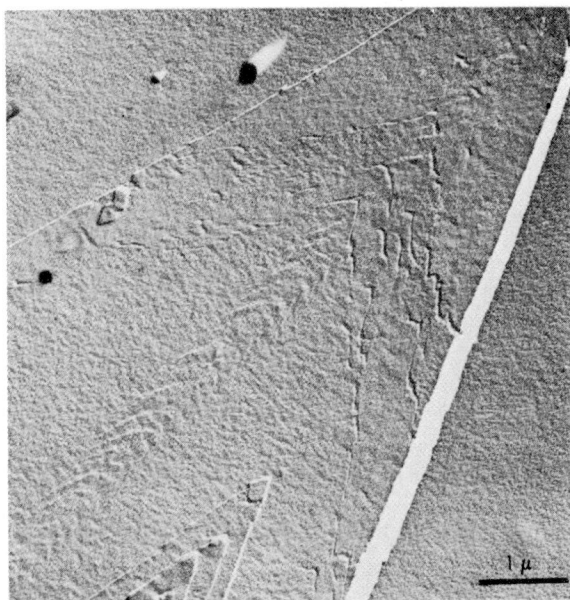

Fig. VII-34. Higher magnification of a portion of the crystal in Figure VII-29. The band down the axis of the crystal is an artifact due to a scratch on the substrate.

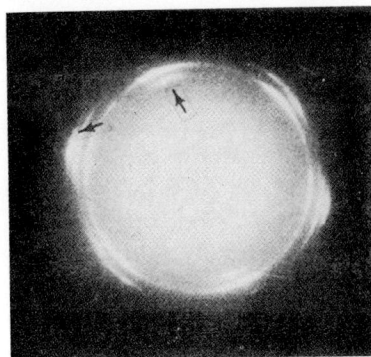

Fig. VII-35. Diffraction pattern from a polyethylene crystal drawn 66% on a Mylar substrate. The reflections indicated by the arrows are not from the normal orthorhombic unit cell of polyethylene.

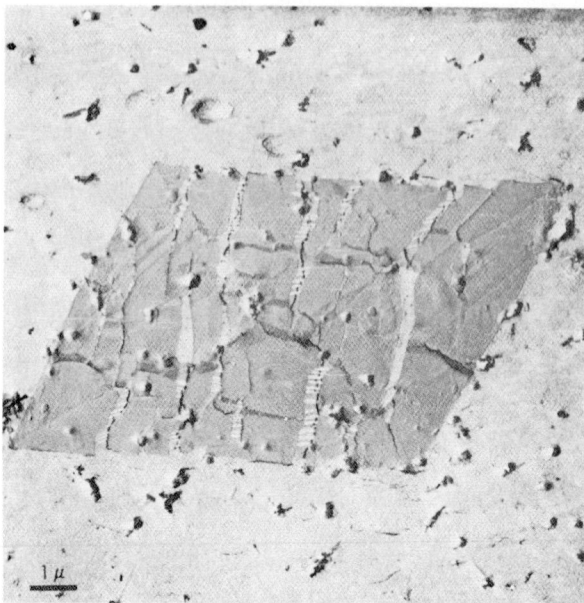

Fig. VII-36. Polyethylene single crystal drawn 50% on Viton. The cause of the black spots is unknown. The a axis is along the longest diagonal.

and redrawn samples of polyethylene film by Pierce, Tordella, and Bryant (37) and subsequently by numerous other authors. The unit cell giving rise to these reflections is not yet known with certainty. Tanaka, Seto, and Hara have recently (52) reported the unit cell of this modification to be monoclinic; all of the planar zigzags of the polyethylene chains are normal to the a axis and therefore parallel to each other (see Section I-4b-10).

Typical fiber diffraction patterns are obtained from those portions of the drawn crystal in which fibers are present, as in Figure VII-33.

The drawn regions in crystals drawn on Viton differ considerably from those in crystals drawn on Mylar. It is believed that the crystal adheres to the Viton in only a few areas. When the substrate is drawn, cracks develop in the crystal, the polymer being drawn across the cracks (Figs. VII-36 and VII-37). When the cracks in a given fold domain make an angle with the growth face, broad ribbons of drawn material are formed. However, if the crack is parallel (or nearly so) to the growth face, only small fibrils are formed, as would be expected, the fold planes being parallel to the growth faces. At a fold domain boundary there is thus an abrupt change in the appearance of a crack. As shown in Figure VII-37, shear can take

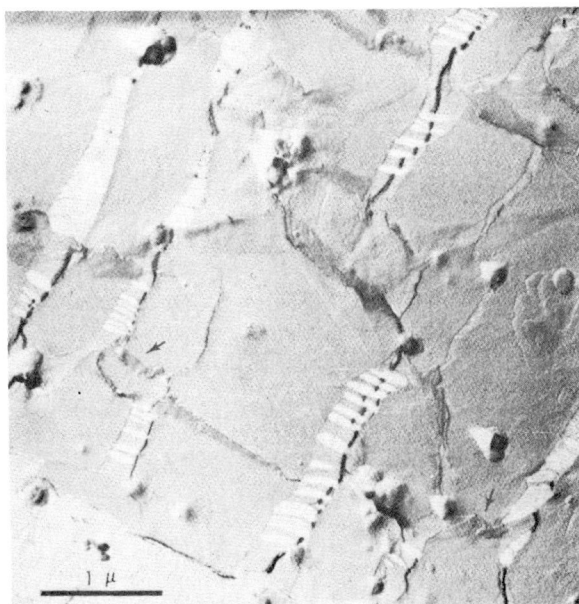

Fig. VII-37. Portion of the crystal in Figure VII-36 at higher magnification. The arrows indicate a region in which shear has taken place between fold planes.

place between fold planes. Other than for the material drawn across the cracks most of the crystal appears relatively undistorted. A peculiar, as yet unexplained, feature is the intense black line on the drawn material in most of the cracks. Diffraction patterns have not yet been obtained from these crystals.

Reneker has drawn polyethylene crystals on a brittle substrate (36). When Mylar is shadowed with carbon and then drawn, the carbon cracks at about 20-micron intervals. Fibers on the order of 100 A in diameter and less are drawn out of crystals which have been deposited on the carbon. In some cases, an 80–100 A periodicity can be seen on the fibers (Fig. VII-38). The fibers appear to have the same structure regardless of the relationship between the directions of crack and fold planes, but the number of fibers formed does depend on this relationship; only a few fibers occur in those fold domains in which the crack is nearly parallel to the fold planes.

The micrographs and diffraction patterns of the crystals drawn on uncoated Mylar suggest that deformation during drawing takes place through a complex combination of shearing and twinning operations. These observations, the fact that the crystals before elongation are collapsed

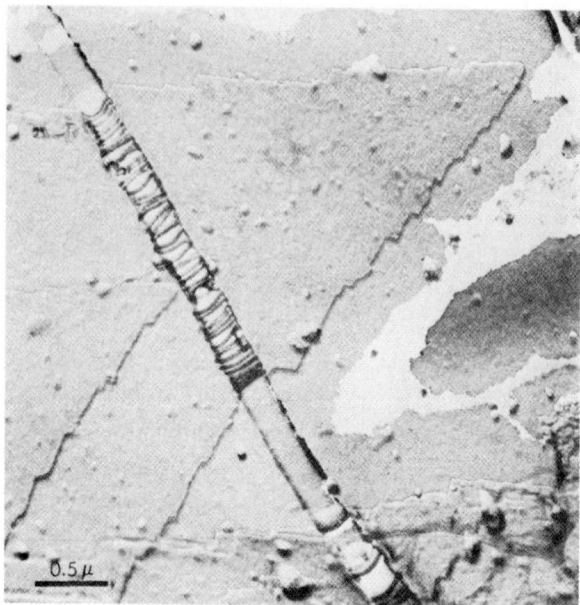

Fig. VII-38. Polyethylene single crystal drawn on a carbon coated Mylar substrate. A 80–100 A periodicity is present on the fibers drawn across the cracks in the carbon. (Reneker (36).)

hollow pyramids in which the molecules are tilted, and the variation of type of draw with substrate, indicate that the mechanism of deformation of polyethylene crystals is highly complex and in need of considerably more research.

5. ANNEALING OF DRAWN MATERIAL

We consider in this section the changes in wide and small angle x-ray diffraction that occur when a drawn polymer is annealed, the annealing being done with and without the restraint of constant length. The changes that occur are complex, depending in a given polymer on time, temperature of annealing and measurement, annealing medium, degree of elongation, and amount of relaxation; they have not yet been adequately defined. Essentially no electron microscope observations have been made (see Section VII-7).

When drawn polyethylene (work reported is on branched but results probably apply to linear as well, although at higher temperature) is an-

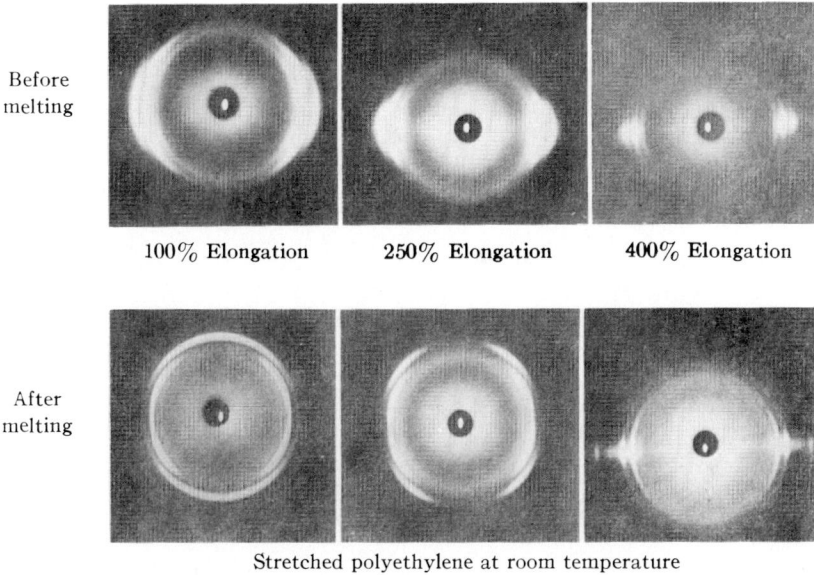

<table>
<tr><td>Before melting</td><td colspan="3"></td></tr>
</table>

100% Elongation   250% Elongation   400% Elongation

After melting

Stretched polyethylene at room temperature

Fig. VII-39. X-ray diffraction patterns from drawn films of lightly crosslinked polyethylene before and after melting and recrystallizing (Judge and Stein (40)).

nealed at temperatures below the melting point and without constraint there is a change in the orientation of the molecules as observed with wide angle x-ray diffraction (4,5,7,8,13,38,39). The results of Belbeoch and Guinier, described below, are typical of the effects found during free relaxation. Related but smaller effects are expected if the annealing takes place with the sample constrained to a constant length (38).

Belbeoch and Guinier found (13) that when a drawn fiber of branched polyethylene (drawn 500% at room temperature) is annealed, contraction begins at about 50°C. The amount of contraction increases with temperature but remains less than 50% up to 95°C. At 105° and 110°C relaxation is complete; i.e., the sample returns to its original length. The polymer melts at 115°C. Related changes take place in the wide angle diffraction patterns. Below 80°C the reflections become sharper, suggesting an increase in crystal perfection. Above 90°C the patterns themselves change, the change being interpreted as indicating that the c or molecular axis is rotating away from the fiber axis. Upon complete relaxation Belbeoch and Guinier suggest that the b axis is normal to the fiber axis with the a and c axes randomly rotated about the b axis. Obtaining diffraction patterns

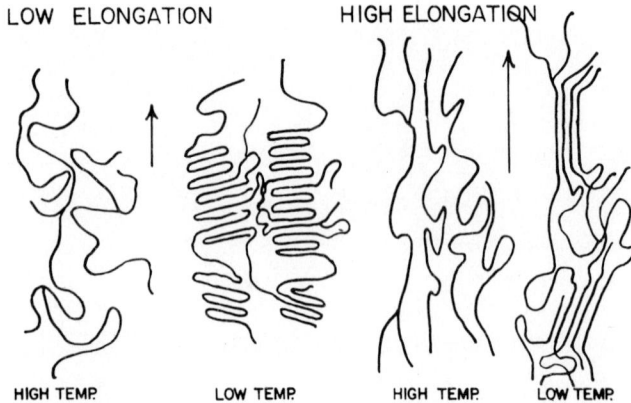

Fig. VII-40. Possible models for the crystallization that takes place when a drawn, slightly crosslinked film of polyethylene is melted and recrystallized (Judge and Stein (40)).

at 105°C during annealing they found a considerable amount of diffuse scatter, with the c axis only somewhat tilted with respect to the fiber axis (similar to that observed after annealing at 50–80°C). The crystals in which the b axis is normal to the fiber axis are apparently formed during cooling to room temperature.

Of interest, with respect to the changes that take place in the wide angle x-ray diffraction patterns from drawn polyethylene during annealing, is the orientation of polymer crystallized from an oriented melt. These results are also of interest because of their implications for the structure of a polymer crystallized as it is being extruded, blown, or spun in commercial operations. Judge and Stein (40), using crosslinked films of linear and slightly branched polyethylene which had been drawn various amounts and then heated above their melting points, have shown that the orientation of the crystals which form on cooling is strongly dependent on the amount of initial elongation. As shown in Figure VII-39, considerably different orientations are obtained above and below 250% elongations. When drawn 500% the initial orientation resulting from the drawing is maintained when the sample is melted and recrystallized. However, below 250% initial elongation, the a axes of the crystals formed during recrystallization are parallel to the original draw direction. A somewhat intermediate type of orientation is produced for a narrow range ($\sim$25%) of elongations in the vicinity of 250%. Similar "a axis-oriented" crystals are found, for instance, in many extruded, noncrosslinked, polyethylene films (3) (see Section VII-7).

The suggestion of Judge and Stein (40) as to the cause of orientation observed in their crosslinked samples is pictured in Figure VII-40. This type of mechanism, or a similar one in which small growing lamellae are oriented due to simultaneous crystallization and orientation, could also explain the **a** axis orientation observed in polyethylene films and similar materials.

A number of papers (10–14, 41–44) have presented some information concerning the effect of annealing on the small angle x-ray diffraction patterns from oriented polymers. As indicated in Section VII-2 the initial long period, in general, increases with increasing draw temperature. Likewise annealing after drawing usually results in an increase in the long period. Some of the available data for various polymers are summarized below. In many cases four-point or quadrant diffraction patterns develop as a result of annealing treatments.

The data of Zahn and Winter (41) on the effect of annealing polyurethane filaments are given in Table VII-1. The effect of heating in water is to reduce the temperature at which the recrystallization takes place.

TABLE VII-1 (41)
Long Period of Polyurethane Filaments as a Function of Annealing Temperature

| | | Long period (A) | |
Treatment		Constrained	Relaxed
Control			75
Heated dry to (°C)	100	75	78
	130	75	78
	140	80	83
	150	88	92
	160	98	102
	170	114	120
Heated in water to (°C)	130	—	104
	145	—	132

Statton's data (11,12) for fibers of polyethylene terephthalate annealed while constrained and while free to shrink are given in Table VII-2. The samples were heated in water (100°C) or mineral oil. It is interesting that in drawn fibers of this polymer, as well as unoriented samples (Chapter V), the long period can decrease upon annealing. The time dependence of the long period was not measured. The intensity of the four-point diffraction patterns increases with increasing temperature although at the highest temperatures the pattern tends to become ringlike.

TABLE VII-2 (11,12)

Long Period of Polyethylene Terephthalate as a Function of Annealing Treatment

Treatment		Long period (A)
Original drawn fiber		112
Constrained (°C)	100	108
	140	117
	220	130
	240	158
Relaxed (°C)	100	100
	140	99
	220	155
	240	167

TABLE VII-3 (13)

Spacings of Four-Point Diagrams from Relaxed, Annealed Polyethylene

Anneal temperature[b]	Long period along fiber axis[c] (A)	Long period normal to fiber axis[c] (A)
Control[a]	100	—
50	104	168
70	110	168
80	135	240
90	145	300
95	165	340
105	170	370
110	170	—

[a] Control was "highly drawn" (probably about 500% at room temperature) sample of branched polyethylene.

[b] Annealing time not given, long periods measured at room temperature.

[c] Separation of maxima in four-point diagram projected on axes.

Belbeoch and Guinier (13) have reported the effect of annealing on the long period spacing of drawn, branched polyethylene (Table VII-3). These samples were not constrained during the annealing. Somewhat similar results would be expected for linear polyethylene. Posner et al. report (42) that the long period first increases and then abruptly decreases as a function of the amount of relaxation. The annealing times and temperatures were not given, however, and it is also possible that the long periods they give for the highest shrinkage ratios are actually second orders.

Fischer and Schmidt, in conjunction with their investigation (14) of the effect of annealing on the long period and other properties of single

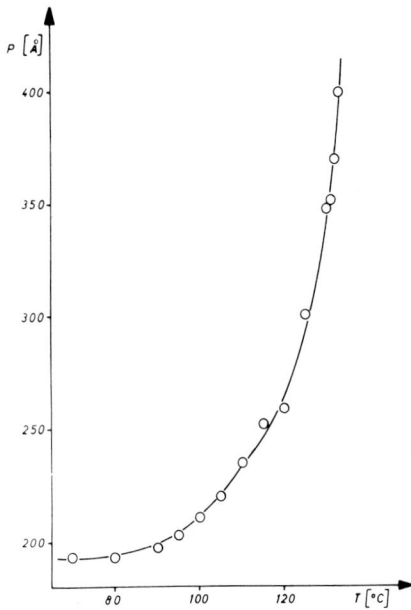

Fig. VII-41. Temperature dependence of the long period of a linear polyethylene sample drawn 1600% at 70°C. The samples were annealed for 24 hours. (Fischer and Schmidt (14).)

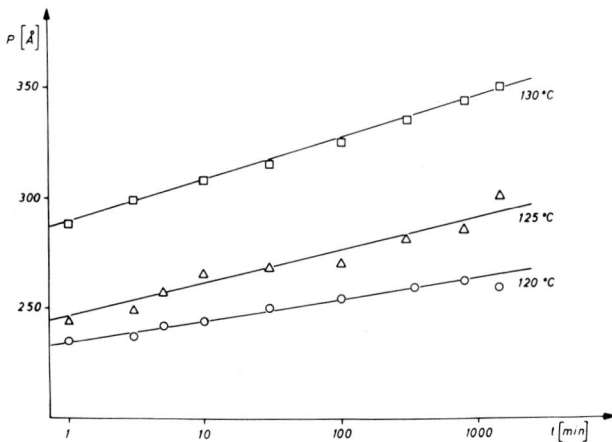

Fig. VII-42. Dependence of the long period of drawn polyethylene on the annealing time (Fischer and Schmidt (14)).

crystal mats and bulk polyethylene, also studied the effect of annealing drawn polyethylene. The temperature and time dependence of the long period are shown in Figures VII-41 and VII-42. The curves, including the logarithmic dependence of the long period on annealing time, are quite similar to those obtained with the single crystals and bulk material. Apparently, however, the disordering attending the increase in long period in single crystals and bulk material is much less or absent in drawn material. There is only a slight dip in the birefringence even at 135°C (Fig. V-7), while the density and wide angle x-ray diffraction patterns show no disordering effects.

On the basis of the discussion presented in this chapter it is obvious that not only the deformation processes for polymers but also the morphology of a drawn polymer are not yet understood, either in general or in detail. It is just as evident, however, that the lamellar structure of the unoriented polymer plays a significant part in the deformation process. Research presently in progress should contribute greatly to our understanding of these processes within the next few years.

6. ROLLING

Rolling, sometimes accompanied by drawing, has long been used to obtain three-dimensionally oriented polymer samples to use in x-ray crystal structure studies. In polyethylene, for instance, Bunn found (45) that rolling of a sample produced an orientation such that the c axes of the crystals are parallel to the roll direction while the $\{110\}$ planes are predominantly parallel to the surface.

Frank, Keller, and O'Connor have discussed (46), in terms of slip and twinning, the expected deformation of polyethylene during rolling. Wide angle x-ray observations (9,46) on samples which were previously oriented by drawing are in agreement with their expectations. They suggest that previously unoriented polyethylene will deform in a similar way when rolled, drawing taking place during the rolling.

On the basis of considerations derived from the deformation of atomic solids and recognition of the long chain nature of polymers they suggest that the most likely slip-mode will involve dislocations with a Burger's vector in the $\langle 010 \rangle$ direction, the dislocation involving extra (010) half planes of molecular segments. (See Fig. VII-43 for diagram of unit cell, the $\langle 010 \rangle$ direction is the shortest lattice translation.) The slip plane would be a $\{100\}$ plane. The second most likely type of slip $\{100\}$ $\langle 010 \rangle$ is not expected to play a significant role in the deformation process since it (a) involves a considerably larger lattice displacement for formation of the

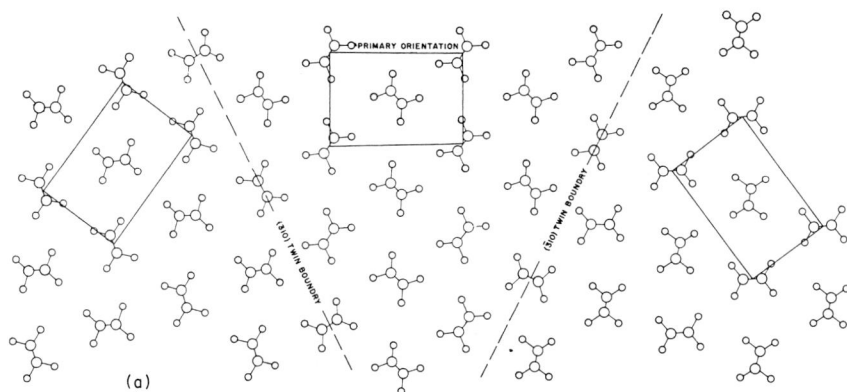

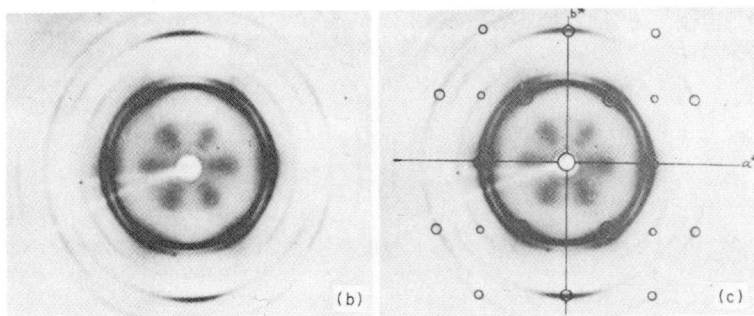

Fig. VII-43. Twinning of the polyethylene unit cell. The composition planes, across which twinning takes place, are {310} planes. A diffraction pattern from rolled linear polyethylene corresponding to this type of twinning is shown in b. The reciprocal lattice due to the material with "primary orientation" is superimposed on the diffraction pattern in c. The additional reflections result from the rotation of this reciprocal lattice by ±53°. Similar diagrams and diffraction patterns have been published by Frank, Keller, and O'Connor (46). (Clark (47).)

dislocation and (b), being orthogonal to the {010} ⟨100⟩ slip system, it is subject to the same driving stresses which can be relieved by the latter system. They indicate (h k 0) dislocations may also play a role in slip, dissociating into pairs of ($\frac{1}{2}$h, $\frac{1}{2}$k, 0) dislocations with {110} slip planes but that the twinning modes discussed below will be in competition with it.

Frank et al. suggest (46) that deformation twinning is most likely to occur with {110} and {310} composition planes; i.e., during the deformation the rearrangement of the molecular segments is such that the twin and the original lattice are mirror images of each other with these planes being the

Fig. VII-44. Slow-cooled (0.2°C/min.), dispersion-based polytetrafluoroethylene deformed 1.5× by multiple rolling at room temperature and fractured at liquid nitrogen temperatures. Fracture surface is parallel to roll direction and normal to roll surface. The micrographs are of similar samples at higher degrees of deformation.

interfaces. The proposed $\{310\}$ twins are shown in Figure VII-43. They indicate that both of these twins, with a slight empirical preference for $\{310\}$, should operate when there are compressive stresses near the $\langle 010 \rangle$ direction or tensile stresses near the $\langle 100 \rangle$ direction.

On the above basis Frank et al. predict (46) the following behavior when drawn polyethylene is rolled. Those crystals with $\{100\}$ and $\{010\}$ planes close to 45° to the axis of compression will deform by $\{010\}$ $\langle 100 \rangle$ slip, undergoing a rotation which brings $\{100\}$ more nearly parallel to the compression plane. Crystals with the $\{010\}$ planes close to normal to the compression axis should twin, resulting in a moderate deformation (25% shear if $\{310\}$ twin) accompanied by a large change in orientation of the unit cell. The new orientation of the unit cell is favorable for $\{010\}$ $\langle 100 \rangle$ slip. Slip and twinning is expected to cease when the $\{100\}$ planes are nearly parallel to the plane of compression.

Frank et al. suggest (46) that with a large amount of deformation retained stresses within the sample induce twinning following removal of the roller

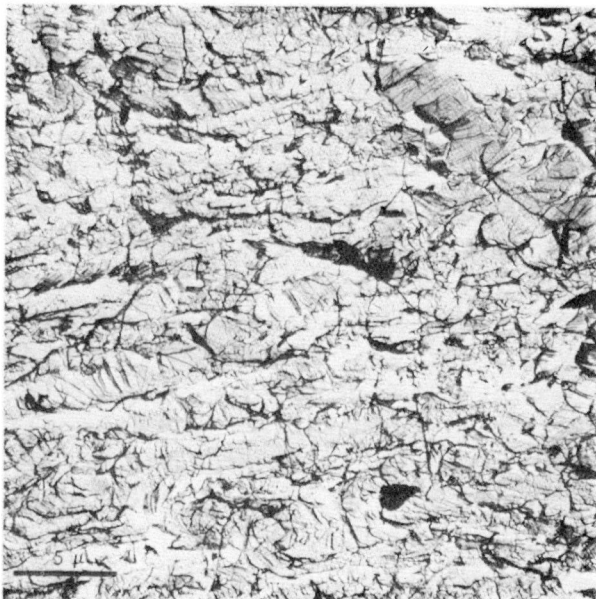

Fig. VII-45. Polytetrafluoroethylene deformed 2×.
Fracture surface as in Figure VII-44.

pressure, such that the $\{100\}$ planes rotate slightly and transform into $\{110\}$ planes. The retained stress, attributed to molecular segments in an amorphous phase and in the folded conformation, is assumed equivalent to a sideways compression and/or to tension parallel to the compression axis. However, it may also be that the compression itself causes this latter twinning. The $\{110\}$ planes, being the planes of closest packing, would be expected to preferentially lie parallel to the compression surface in highly compressed samples. (In Figure VII-43 the axis of compression is horizontal.)

Frank *et al.* (46) are also able to describe observations on the effect of annealing and subsequent redrawing of rolled polyethylene as well as various aspects of the initial drawing process in terms of the proposed deformation mechanism. Although additional, detailed observations are needed to confirm all aspects of the proposed mechanism, this paper is an excellent start toward understanding the deformation of polymers.

Zaukelies has subsequently discussed (53) the deformation of previously drawn ("singly oriented") and drawn and rolled ("doubly oriented") nylon in a similar manner, pointing out that the results can most satis-

Fig. VII-46. Portion of area in Figure VII-45 at higher magnification

factorily be explained in terms of a defect-crystal model rather than a two-phase, crystallite in amorphous matrix, model. He finds that distinct kink bands can be observed both optically and in the electron microscope when the samples are compressed. The formation of the bands involves slip in the molecular axis direction along {010} planes. (These planes contain the H-bonds.) Although microscope observations could be interpreted in terms of slip between and bending of adjacent 100 A diameter fibrils, the fact that definite kink band angles related to the crystallographic lattice are observed indicates that the deformation is occurring within the fibrils. The angle between the molecular axis in the original bristle and in the kink band, which varies with temperature, is found to be related to the number of adjacent molecular planes which slip as a unit when the kink band forms. Zaukelies indicates that at 25°C two adjacent {010} planes act as a unit, whereas at 100°C three adjacent {010} planes are involved.

Zaukelies suggests (53) that the slip occurs through the motion of either an edge dislocation (composed of a linear array of point dislocations related to those described by Reneker in polyethylene (48) except that the long nylon unit cell involves a considerably larger lattice distortion) and/or several types of screw dislocations. He indicates that the formation and

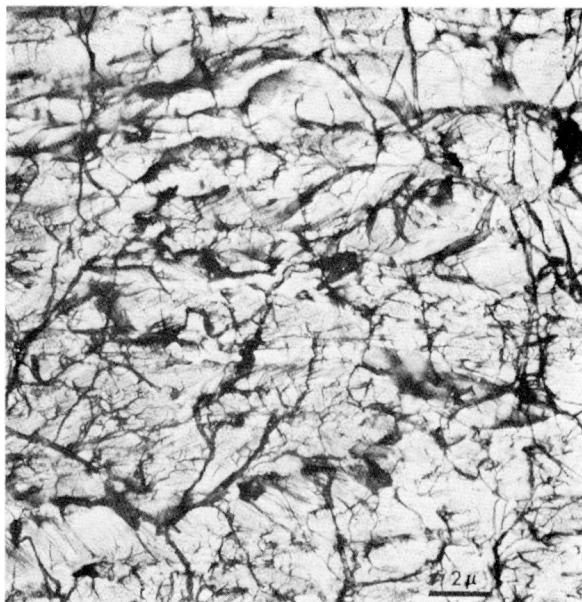

Fig. VII-47. At increasing degrees of deformation (3.5×) the bands become thinner and the tilt of the striations larger. The alignment of the bands in the roll direction also increases. Fracture surface is oriented as in Figure VII-44.

motion of these screw dislocations may also occur during the rolling process itself.

The tensile fracture of a notched doubly oriented nylon is found to depend on the angle between the notch and the {010} planes. Zaukelies indicates (53) that if a doubly oriented bar is notched so that leading edge of the crack is parallel to the {010} planes considerable slip in the ⟨010⟩ direction along these planes takes place before the sample fails. (In his paper Zaukelies uses the convention that the [h k l] direction is normal to the (h k l) plane whereas we use here the convention (Chapter I) that the [h k l] direction is the vector to the h k l real lattice point, thus ⟨001⟩ is parallel to the molecular axis.) However, if the notch is perpendicular to the {010} planes the resulting failure is brittle and no slip occurs.

Electron microscope studies of the deformation of polytetrafluoroethylene during rolling suggest that the lamellar structure of crystalline polymers is retained during rolling up to high degrees of deformation and that slip along and between the molecules and twinning takes place within the lamellae. As shown in Chapter IV (see Fig. IV-68), polytetrafluoroethylene

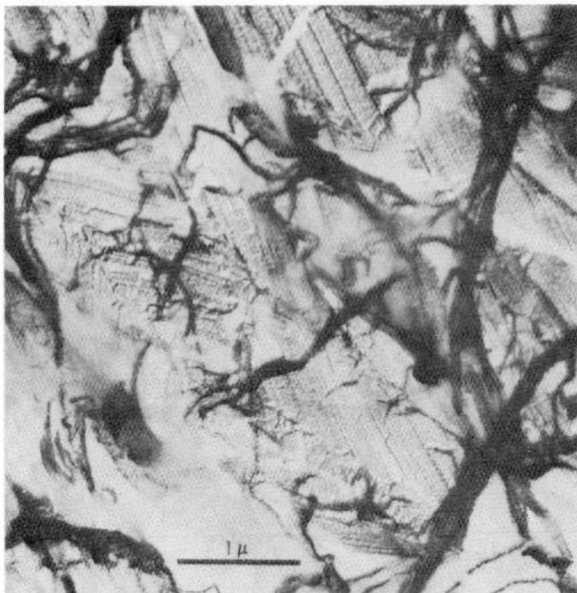

Fig. VII-48. Portion of area in Figure VII-47 at higher magnification. Striations in
neighboring bands may be tilted in opposite directions.

crystallizes in the form of banded structures on the order of 0.1 micron
thick and tens of microns long. Although some question remains concerning
their structure, the molecules are folded within the bands and are parallel
to the observed striations. When such a sample is deformed by rolling, the
bands retain their identity and can be observed up to deformations of at
least 6× (Figs. VII-44 to VII-48). The deformation of the bands is quite
similar to that observed by Speerschneider and Li (24) when similar samples
are drawn (Section VII-3).

   With increasing deformation the bands become more and more aligned
in the roll directions. The striations within each band also tilt with respect to the sides of the bands and approach the roll direction. On surfaces
fractured parallel to the roll surface only a few bands are observed at the
higher deformation ratios. These bands appear to be preferentially aligned
either parallel to the roll direction or normal to it. In all the samples a considerable amount of fibrillation occurred during fracture of the sample.
The number of fibrils appears to increase with increasing deformation of
the sample. X-ray diffraction patterns have not been obtained from these
samples as yet. If one assumes that the molecules remain parallel to the

striations, which appears reasonable, then one must assume that the molecular segments are sliding by each other either individually or in groups corresponding to the width of the striations. The folded conformation is retained but the molecules tilt within the lamellae to a larger and larger degree as the deformation is increased. Twinning, if it takes place, would be on the scale of the width of the striations or less and would be expected primarily in those bands which were originally aligned such that the molecular axes were parallel to the roll surface.

7. CRYSTALLIZATION IMPOSED ORIENTATION

By setting up suitable thermal gradients in a polymer melt it is possible to produce orientation by means of the crystallization process. These effects are of particular importance in many commercial processes. For instance, in injection molding procedures a polymer melt is usually injected into a mold whose surface is well below the crystallization temperature and thus the polymer in contact with the surface is quenched whereas that in the interior of the molded object cools more slowly. In general a transcrystalline layer is produced; i.e., spherulite nuclei formed at the mold surface are so closely spaced that the spherulites can grow in only one direction, inward. Since polymer molecules are in general tangential in polymer spherulites, the result is the formation of a surface layer in which the molecules are more or less parallel to the surface.

An important commercial process for the preparation of a partially oriented polymer film is that of blowing. Molten polymer is extruded

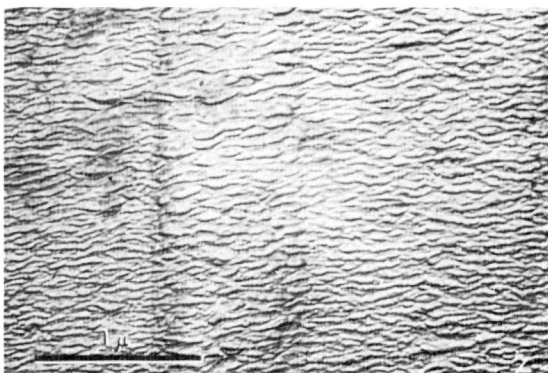

Fig. VII-49. Inner surface of a blown linear polyethylene film. The extrusion direction is vertical. (Kobayashi (50).)

through a circular die, the resulting tube enlarged several times by injecting gas under pressure into its center and then quenched. In the case of polyethylene the resulting material is found to have "**a** axis" orientation; i.e., the **a** axis is preferentially aligned in the extrusion direction whereas the **b** and **c** axes are randomly arranged, on the average, in planes nearly normal to the extrusion direction (3,49). Similar "**a** axis" orientation, but to a lesser degree, is found in flat polyethylene film which is merely extruded and quenched (49).

Fig. VII-50. Slightly stretched blown polyethylene film. Rows of lamellae are particularly apparent in the lower left portion of the micrograph. (Kobayashi (50).)

In both the blown and flat film the thermal conditions during quenching are such that a thermal gradient is set up in the extrusion direction as well as inward from the surface. Such an effect would be expected to produce **b** axis orientation, however, since, in polyethylene spherulites, radial growth is in the **b** axis direction. Aggarwal *et al.* (49) suggested that the **a** axis orientation results from the orientation during the extrusion process of needlelike or, in view of present results, lamellar crystals in which the chains are normal to the long dimension of the crystal. If the film is being drawn in the extrusion direction when it is crystallizing (not transverse as occurs when it is blown), such an effect may occur. Again, however, one would expect **b** axis orientation if the lamellae are elongated in the **b** axis direction.

Recent results of Kobayashi (50) suggest that the effect is more complex. He has investigated the morphology of blown linear polyethylene film,

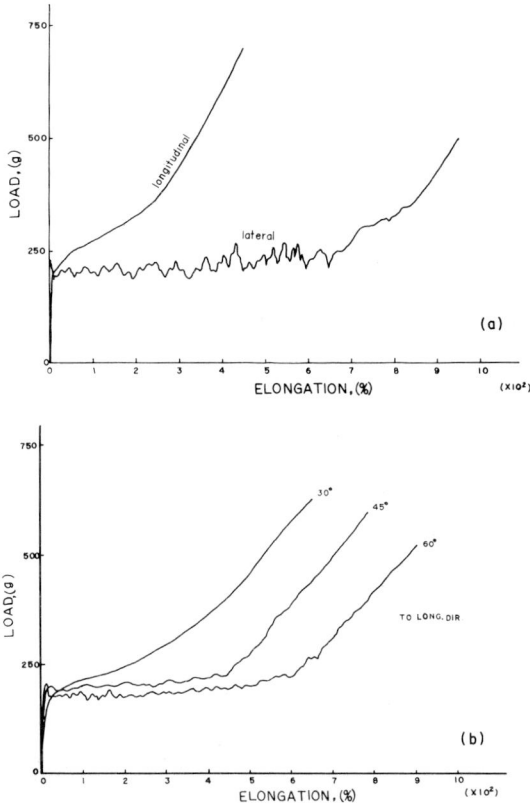

Fig. VII-51. Stress-strain curves for the blown polyethylene film whose morphology is shown in Figure VII-49.   (a) A much smaller stress is required to draw the film normal to the machine direction (parallel to the **b** axis) than parallel to the machine direction. (b) When drawn at an angle to the machine direction, intermediate stress-strain curves are obtained.   (Kobayashi (50).)

the effect of drawing it in various directions and annealing the drawn film, and also the morphology of a polyoxymethylene film which crystallized under a thermal gradient in the machine direction. His results are discussed below.

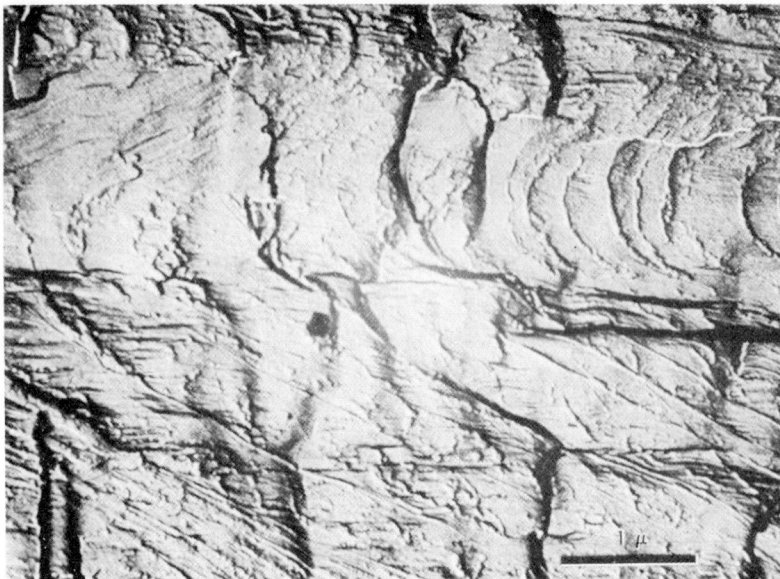

Fig. VII-52. Replica of a blown polyethylene film drawn several hundred per cent normal to the machine direction. The draw direction is horizontal. (Kobayashi (50).)

A replica of an inner surface of a blown film is shown in Figure VII-49. Lamellae are observed oriented normal to the machine direction and tilted with respect to the surface of the film. Kobayashi suggests (50) that the lamella orientation results from numerous spiral growths of lamellae, the axes of the spirals having a component in the machine direction. The **a** axis orientation observed in the films suggests the axes are at a fairly large angle with the surface; this angle probably varies with the quenching conditions. The lateral dimension of the lamellae, parallel to the surface, Kobayashi suggests is parallel to the **b** axis, i.e., it corresponds to the radii of polyethylene spherulites. The lamellar nature, as well as the stacking of lamellae in the machine direction as a result of their development from spiral growths, can be seen more clearly when the film is slightly stretched (Fig. VII-50).

As indicated previously in this chapter, unfolding of the molecules from polyethylene lamellae appears to occur most easily when the **b** axes of the lamellae are aligned in the draw direction. Thus, in these blown films, one would expect draw to take place most easily normal to the machine direction. Such is found to be the case (Fig. VII-51) (50).

Morphological differences in the mechanism of drawing in the two directions are clearly seen in Kobayashi's micrographs of the drawn film. When drawn normal to the machine direction the molecules apparently can easily unfold and form a fibrillar structure (Fig. VII-52). However, when drawn in the machine direction an entirely different effect takes place, the lamellae themselves apparently become distorted (Fig. VII-53). Kobayashi suggests that the linear arrays of white objects in the micrograph are lamellae

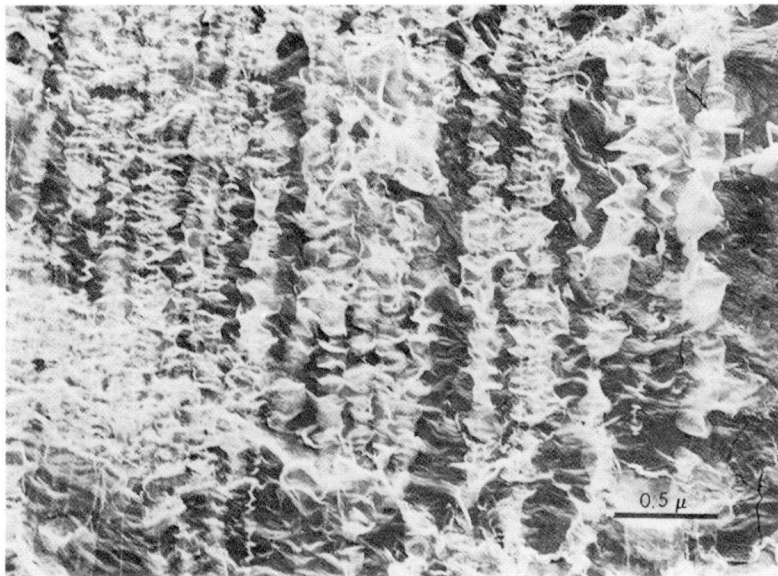

Fig. VII-53. Carbon replica of a blown polyethylene film drawn in the machine direction (vertical). This micrograph is a negative, the white objects consisting of electron dense regions. The lamellae within these regions have probably remained attached to the replica. Their appearance should be compared with the pulled out spiral growth of lamellae in Figure VII-54. (Kobayashi (50).)

oriented normal to the film surface which were originally in spiral growths; i.e., drawing has resulted in a pulling apart of the spiral growths in a fashion similar to that which occurs when a helical spring is stretched. By ultrasonic cavitation Kobayashi produced similar structures from spiral growths of polyoxymethylene single crystals (Fig. VII-54). The stress-strain results in Figure VII-51 suggest that if the film is drawn at an angle to the machine direction molecular unfolding in properly oriented crystals occurs first, followed by distortion of the remaining lamellae.

Fig. VII-54. Ultrasonically dispersed spiral growth of polyoxymethylene. A similar, undistorted polyoxymethylene crystal can also be seen. The hedgerow structures sometimes observed in polyethylene crystal suspensions (Fig. VII-30) may result from a similar effect or from epitaxial growth of crystals on a fiber resulting from disrupting a crystal or a fibrillar impurity. (Kobayashi (50).)

There would appear to be some discrepancy between the length of the stacks of lamellae, suggesting that the axis of the spiral and therefore the c axis is parallel to the machine direction, and the a axis orientation characteristic of blown film. It may be that the a axis orientation is not as complete in the films used by Kobayashi as in many commercial films. It is evident, however, that pole figure diagrams are needed to define the distribution of orientation of the axes in the various films.

Similar morphological effects can be seen even more clearly in polyoxymethylene films extruded in such a fashion that a thermal gradient is set up in the machine direction (51). Kobayashi and Ohta (51) extruded film from a polymer melt through two rolls, one of which was hotter than the crystallization temperature and the other slightly cooler. The extruding film remained attached for awhile to the cooler roll as it was extruded. By adjusting the roll speed it was possible to obtain a situation in which the thermal gradient was in the machine direction, essentially parallel to the surface. The morphology of the resulting film is shown in Figure VII-55. Stacks of lamellae, oriented normal to the surface, resulting (51)

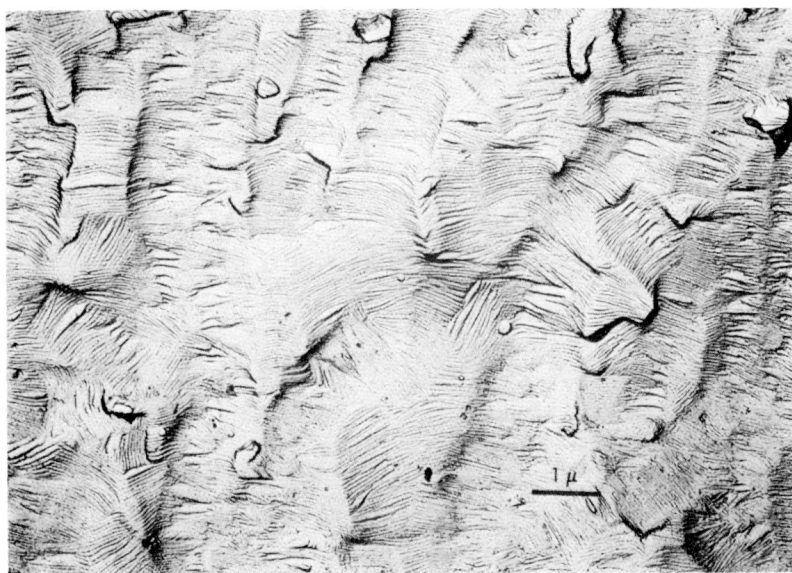

Fig. VII-55. Crystallization oriented film of polyoxymethylene. During crystallization a thermal gradient in the vertical direction was imposed. The rows of lamellae
are believed to result from spiral growth about a screw dislocation axis. (Kobayashi
and Ohta (51).)

from spiral growth with axes primarily parallel to the thermal gradient,
i.e., the machine direction, can be seen.

When such a film is drawn, elevated temperatures being required, the
resulting morphology depends on the angle between the spiral growth axes
and the draw direction (Fig. VII-56). At *A* in Figure VII-56 the lamellae
were originally in the draw direction and unfolding of the molecules is
apparently taking place. In the vicinity of *B* the lamellae are bending,
whereas near *C* they appear to be twisting and slipping toward the draw
direction. In the various regions marked *D*, lamellae oriented normal to
the draw direction appear to be pulling apart. Isolated fibrils appear to
be drawn across the resulting voids.

Kobayashi suggests (50) that when polyethylene is drawn in the **b** axis
direction the fibrils that form are composed of molecules from more than
one lamella (Fig. VII-57). Whereas those molecules drawn from a given
lamella may be in lattice register, it is likely that those from different lamellae will not be in lattice register. Thus, if one passes an x-ray beam through
the fiber parallel to the original lamellae there may be disorder (resulting

Fig. VII-56. Polyoxymethylene film as in Figure VII-55 except partially drawn (draw direction vertical). The symbols are discussed in the text. (Kobayashi and Ohta (51).)

in "amorphous scattering") between the planes of molecules from the various lamellae. When the beam is normal to the original lamella direction the lateral placement of the molecules in each plane of molecules would more likely be in register, Kobayashi suggests, and there should then be less amorphous scattering. Such an effect is found when a blown film is drawn normal to the machine direction (Fig. VII 58). The lamellae in the original film are tilted with respect to the surface, with the **b** axis being in the draw direction.

When the transverse drawn film is viewed by transmission (Fig. VII-58b) only a weak "amorphous" peak is found, whereas the peak is much stronger

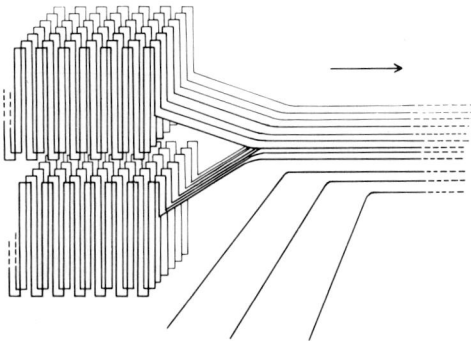

Fig. VII-57. Suggested model for the formation of a fiber by the unfolding of molecules from more than one lamella. The molecules in the ribbon drawn unfolding from lamella, it is suggested, are in lattice register but need not be in phase with the molecules drawn from neighboring lamellae. Stacking faults, dislocations, and other disorders result in intense, diffuse x-ray scattering when the beam is parallel to the plane of the ribbons (Fig. VII-58). (Kobayashi (50).)

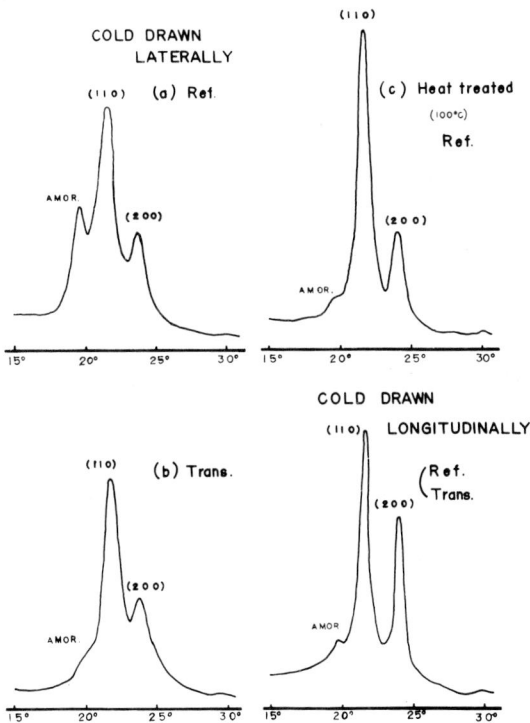

Fig. VII-58. X-ray diffraction patterns from cold drawn, blown polyethylene film as a function of the relative orientation of the film and the x-ray beam (Kobayashi (50)).

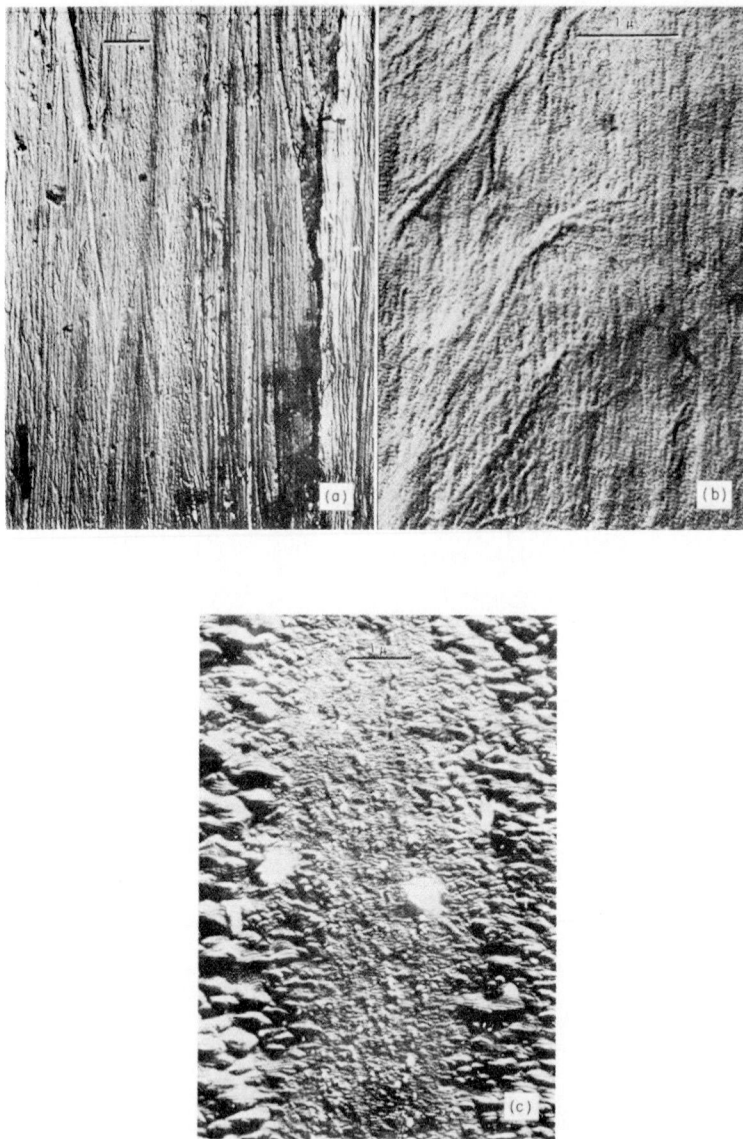

Fig. VII-59. (a) Cold drawn, blown polyethylene film drawn normal to the machine direction. (b) Film as in (a) except annealed at 120°C for 10 minutes in an unrestrained condition. Note difference in magnification. (c) Film as in (a) except annealed at 125°C for 1 hour. (Kobayashi(50).)

when viewed by reflection (essentially equivalent to an edge view) (Fig. VII-58a). When the sample is heated to 100°C, sufficient thermal motion occurs to permit annealing out of the disorder and the "amorphous" peak is significantly reduced (Fig. VII-58c). When cold drawn in the a axis (machine) direction the lamellae themselves are involved as units in the deformation and the "amorphous" peak is the same when viewed from the two directions.

Starting with a fully drawn, blown polyethylene film which had been stretched transverse to the machine direction, Kobayashi investigated the effect of annealing the film for short periods of time in both the free and constrained conditions (50). The initial film has a fibrillar character (Fig. VII-59a). When annealed at 120°C for ten minutes in an unrestrained condition, Kobayashi suggests that the molecules within the fibrils retract and fold to produce a lamellar structure (Fig. VII-59b). The lamellae are observed to have only a small lateral extent and to line up in rows, again suggesting (50) a spiral growth structure. At higher annealing temperatures the lateral extent and perfection of the lamellae increase (Fig. VII-59c). If the samples are annealed while constrained from shrinking, similar effects are found at somewhat longer annealing times and/or higher annealing temperatures. These results tend to agree with, and suggest a mechanism for, the changes in long period during the annealing of drawn polymers discussed previously in this chapter. The question of the origin of the long period in the drawn fibrils themselves, however, still remains.

REFERENCES

1. W. O. Statton, "Directional 'Crystallization' of Polymers," *Ann. N. Y. Acad. Sci.*, **83**, 27 (1959).
2. C. R. Bohn, J. R. Schaefgen, and W. O. Statton, "Laterally Ordered Polymers: Polyacrylonitrile and Poly(vinyl Trifluoroacetate)," *J. Polymer Sci.*, **55**, 531 (1961).
3. S. L. Aggarwal and O. J. Sweeting, "Polyethylene: Preparation Structure, and Properties," *Chem. Revs.*, **57**, 665 (1957).
4. S. L. Aggarwal, G. P. Tilley, and O. J. Sweeting, "Changes in Orientation of Crystallites During Stretching and Relaxation of Polyethylene Films," *J. Polymer Sci.*, **51**, 551 (1961).
5. A. Brown, "X-Ray Diffraction Studies of the Stretching and Relaxing of Polyethylene," *J. Appl. Phys.*, **20**, 552 (1949).
6. W. M. D. Bryant, "Polythene Fine Structure," *J. Polymer Sci.*, **2**, 547 (1947).
7. R. A. Horsley and H. A. Nancarrow, "The stretching and relaxing of polyethylene," *Brit. J. Appl. Phys.*, **2**, 345 (1951).
8. A. Keller, "Unusual Orientation Phenomena in Polyethylene Interpreted in Terms of the Morphology," *J. Polymer Sci.*, **15**, 31 (1955).

9. J. J. Point, "Recherches sur l'etat Solide de Hauts Polymeres Spherolithiques," *Mem. Pub. Soc. Sci. Arts m. c. Lit. du Hainaut*, **71**, 65 (1958).

10. G. Porod, "Anwendung und Ergebnisse der Röntgenklein-winkelstreuung im festen Hochpolymeren," *Fort. Hochpolymeren-Forsch.*, **2**, 363 (1961).

11. W. O. Statton, personal communication.

12. W. O. Statton, "Polymer Texture—The Arrangement of Crystallites," *J. Polymer Sci.*, **41**, 143 (1959).

13. B. Belbeoch and A. Guinier, "Structure a Grande Echelle du Polyethylene," *Makromol. Chem.*, **31**, 1 (1959).

14. E. W. Fischer and G. Schmidt, "Über die Langperioden von verstrecktem Polyäthylen," *Angew. Chem.*, **74**, 551 (1962).

15. R. Hosemann, "Crystalline and Paracrystalline Order in High Polymers," paper *J. Applied Phys.*, **34**, 25 (1963).

16. H. Hendus, personal communication.

17. N. Kasai, S. Fujiwara, S. Morioka, H. Kurose, M. Kakudo, and T. Watase, "The Fine Texture of Cold-Drawn Polyethylene," *Koggo Kagaku Zasshi*, (*J. Chem. Soc. Japan*), **64**, 55 (1961) (In Japanese).

18. L. Mandelkern, C. R. Worthington, and A. S. Posner, "Low Angle X-ray Diffraction of Fibrous Polyethylene," *Science*, **127**, 1052 (1958).

19. W. O. Statton, "Higher Orders of Long Period X-Ray Diffraction in Polyethylene Fibers," *J. Polymer Sci.*, **28**, 423 (1958).

20. H. D. Keith and F. J. Padden, Jr., "Deformation Mechanisms in Crystalline Polymers," *J. Polymer Sci.*, **41**, 525 (1959).

21. C. F. Hammer, T. A. Koch, and J. F. Whitney, "Fine Structure of Acetal Resins and Its Effect on Mechanical Properties," *J. Appl. Polymer Sci.*, **1**, 169 (1959).

22. C. Sella and J. J. Trillat, "Structures periodiques dans les polyethylenes," *Compt. rend.*, **248**, 410 (1959).

23. F. A. Bettelheim and R. S. Stein, "The Change in Density of Low Pressure Polyethylene on Stretching," *J. Polymer Sci.*, **31**, 523 (1958).

24. C. J. Speerschneider and C. H. Li, "Some Observations on the Structure of Polytetrafluoroethylene," *J. Appl. Phys.*, **33**, 1871 (1962).

25. J. S. Lazurkin, "Cold-Drawing of Glass-Like and Crystalline Polymers," *J. Polymer Sci.*, **30**, 595 (1958).

26. P. I. Vincent, "The Necking and Cold Drawing of Rigid Plastics," *Polymer*, **1**, 7 (1960).

27. V. Peck and W. Kaye, "Behaviour of Crystallites in Polyethylene," Paper presented at Elec. Micro. Soc. Am. Meeting, *J. Appl. Phys.*, **25**, 1465 (1954).

28. E. W. Fischer, personal communications.

29. P. H. Geil, unpublished data.

30. H. A. Stuart, "Problems of High Polymer Crystallinity," *Ann. N. Y. Acad. Sci.*, **83**, 3 (1959).

31. P. H. Geil, "Nylon Single Crystals," *J. Polymer Sci.*, **44**, 449 (1960).

32. D. H. Reneker and P. H. Geil, "Morphology of Polymer Single Crystals," *J. Appl. Phys.*, **31**, 1916 (1960).

33. P. H. Geil and D. H. Reneker, "Morphology of Dendritic Polymer Crystals," *J. Polymer Sci.*, **51**, 569 (1961).

34. N. Hirai, H. Kiso, and T. Yasui, "Morphology of Torn Polyethylene Crystals," *J. Polymer Sci.*, **61**, S1 (1962).

35. M. B. Konstantinopolskaya, Z. Ya. Berestneva, and V. A. Kargin, "Helical Structures of Polyethylene. II," *High Mole. Wt. Comps.*, **3**, 1260 (1961) (in Russian).
36. D. H. Reneker, personal communications.
37. R. H. H. Pierce, Jr., J. P. Tordella, and W. M. D. Bryant, "A Second Crystalline Modification of Polythene," *J. Am. Chem. Soc.*, **74**, 282 (1952).
38. M. B. Rhodes and R. S. Stein, "A Light Scattering Study of the Annealing of Drawn Polyethylene," *J. Appl. Phys.*, **32**, 2344 (1961).
39. S. Krimm, "Orientation of Crystallites in Stretched Polyethylene," *J. Appl. Phys.*, **23**, 287 (1952).
40. J. T. Judge and R. S. Stein, "Growth of Crystals from Molten Crosslinked Oriented Polyethylene," *J. Appl. Phys.*, **32**, 2357 (1961).
41. H. Zahn and U. Winter, "Über die Langperiodenreflexe un Röntgenogramm von Polyurethanfaden," *Kolloid-Z.*, **128**, 142 (1952).
42. A. S. Posner, L. Mandelkern, C. R. Worthington, and A. F. Diorio, "Low Angle X-Ray Diffraction of Fibrous Polyethylene," *J. Appl. Phys.*, **31**, 536 (1960).
43. H. Rothe, "Zur Frage der Langperioden—Interferenzen an synthetischen Faserstoffen," *Faserforsch. Textil.*, **8**, 244 (1957).
44. K. Hess, "Über Langperioden—Interferenzen bei synthetischen Fasern und über ein neues Fasermodell," *J. Colloid Sci.*, Suppl. **1**, 135 (1954).
45. C. W. Bunn, "The Crystal Structure of Long Chain Normal Paraffin Hydrocarbons. The Shape of the CH_2 Groups," *Trans. Faraday. Soc.*, **35**, 482 (1939).
46. F. C. Frank, A. Keller, and A. O'Connor, "Deformation Processes in Polyethylene Interpreted in Terms of Crystal Plasticity," *Phil. Mag.*, **3**, 64 (1958).
47. E. S. Clark, personal communication.
48. D. H. Reneker, "Point Dislocations in Crystals of High Polymers," *J. Polymer Sci.*, **59**, 539 (1962).
49. S. L. Aggarwal, G. P. Tilley, and O. J. Sweeting, "Orientation in Extruded Polyethylene Films," *J. Appl. Polymer Sci.*, **1**, 91 (1939).
50. K. Kobayashi, to be published.
51. K. Kobayashi and T. Ohta, to be published.
52. K. Tanaka, T. Seto, and T. Hara, "Crystal Structure of a New Form of High Density Polyethylene, Produced by Press," *J. Phys. Soc. Japan*, **17**, 873 (1962).
53. D. A. Zaukelies, "Observation of Slip in Nylon 66 and 610 and Its Interpretation in Terms of a New Model," *J. Applied Phys.*, **33**, 2797 (1962).

VIII.   *AS*-POLYMERIZED AND AMORPHOUS POLYMERS

1. MORPHOLOGY OF *as*-POLYMERIZED POLYMER

At present very little is known and essentially nothing published concerning the morphology of polymers as they are polymerized. In this section we shall discuss available results for three polymers, polyethylene, polyoxymethylene, and polytetrafluoroethylene.

a. Polyethylene

Niegisch has reported (1,2) the presence of highly dendritic crystals of polyethylene during the early stages of the polymerization of linear polyethylene using a soluble aluminum halide, tetraphenyltin, vanadium halide catalyst in cyclohexane (3) (Figs. VIII-1 and VIII-2). In later stages of polymerization spherulites are found. The **b** axis of the polyethylene unit cell is parallel to the longer axis of the crystals, in contrast to the situation in the typical polyethylene crystals grown from solution (Chapter II).

b. Polyoxymethylene

Formaldehyde was known to polymerize as single crystals consisting of extended chains of polyoxymethylene before 1920 (Chapter II). It is believed that the formaldehyde monomer units add on to growing chains already incorporated in the crystals; the reactive ends of the chain are thus presumably on the ends of the crystals. The crystals that are formed are hexagonal prisms (Fig. VIII-3). The total length of the crystal, in some cases, is considerably greater than the length of the molecules. Fischer (4) has published an electron micrograph of a replica of an *as*-polymerized, polyoxymethylene crystal which shows the presence of parallel striations with an approximately 200 A spacing. The striations are stated to be normal to the molecular axes.

Recently a number of workers have investigated the solid state polymerization of single crystals of trioxane (a ring compound composed of three formaldehyde monomer units) and numerous other cyclic compounds. In the case of trioxane, polymerization has been initiated by irradiation

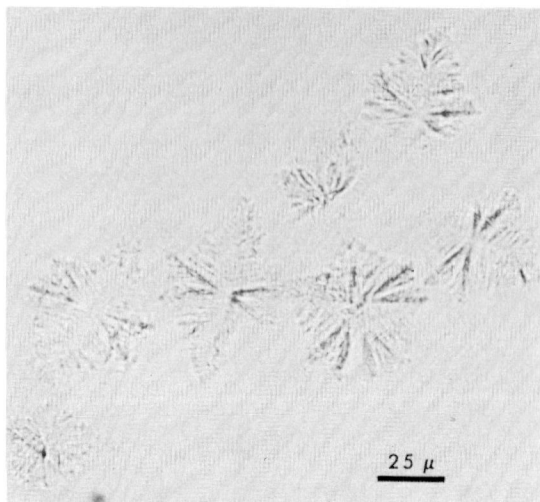

Fig. VIII-1. Optical micrograph of dendritic crystals of linear polyethylene formed during initial $2^1/_2$ minutes of polymerization (Niegisch (1)).

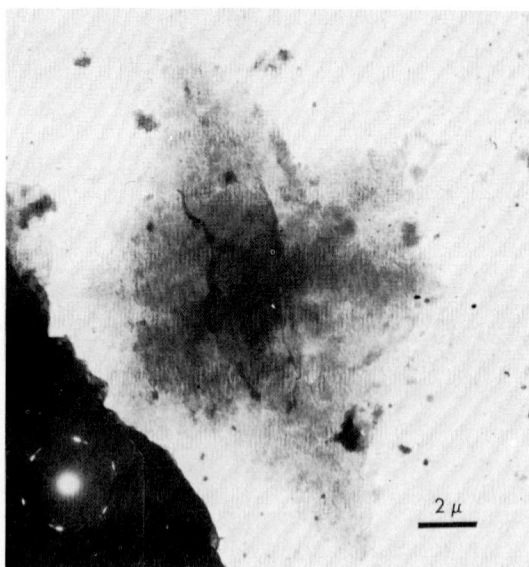

Fig. VIII-2. Electron micrograph and diffraction pattern of a crystal similar to those in Figure VIII-1 except removed after 4 minutes of polymerization (Niegisch (2)).

Fig. VIII-3. Crystals of polyoxymethylene as polymerized from a mixture of concentrated sulfuric acid and formaldehyde solutions over a period of several months. In other cases the length of the crystals is considerably less than the lateral dimensions (incident steep oblique illumination).

(9–12) and by various catalysts (13,14). The resulting crystals of polyoxymethylene yield excellent single crystal x-ray diffraction patterns which indicate that the majority of the chains lie along the c axis of the original monomer crystal. There is also some evidence of a limited number of chains lying at a discrete angle to this axis. The availability of these crystals, which can be on the order of centimeters in length and millimeters in diameter, is expected to be of considerable significance in studies of the physical properties of polymers. Although essentially nothing has been reported, to date, concerning their morphology, it is likely that the molecules are fully extended as polymerized and therefore care is needed in comparing their properties with that of bulk polymer crystallized from the melt.

c. Polytetrafluoroethylene

Polytetrafluoroethylene is generally polymerized to form either particles on the order of 0.2 micron diameter in a dispersion or a granular material of large particle size. The dispersion particles have been investigated more thoroughly than the granular material, primarily because they are easier to observe in the electron microscope. Both materials are highly crystalline as polymerized; the dispersion particles, for instance, are at least 95% crystalline.

As in the case of polymer single crystals, polytetrafluoroethylene dispersion particles deteriorate rapidly in the electron beam. By reducing the beam current it has been possible to follow the deterioration process

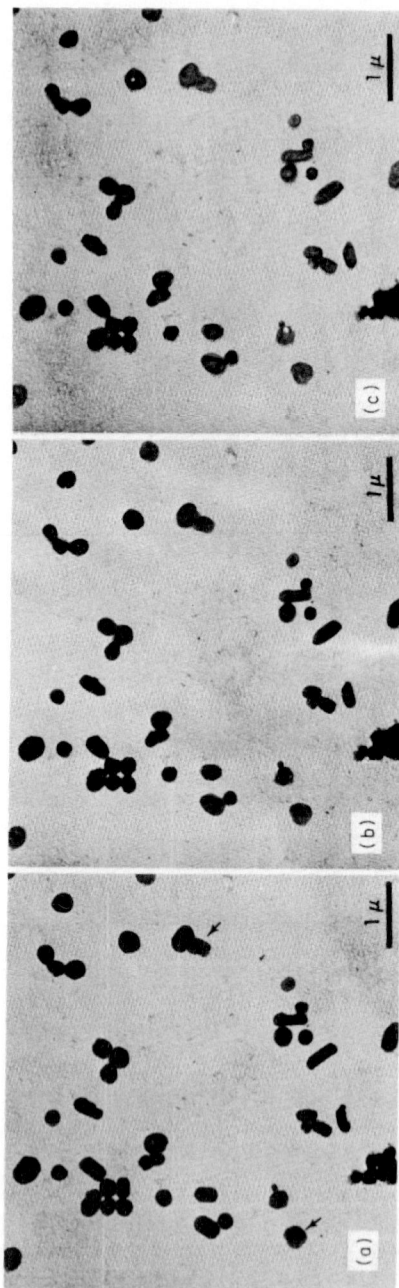

Fig. VIII-4. Deterioration of polytetrafluoroethylene dispersion particles in a low intensity beam of an electron microscope. Micrograph (a) was taken as soon as the sample was inserted in the beam, (b) after 30 seconds, and (c) after $2^{1}/_{2}$ minutes. Electron diffraction, from the particles at this beam intensity, is destroyed in about 15 seconds. Dark portions on the particles in (a) (arrows) are believed due to Bragg extinction effects from crystallites within the particles.

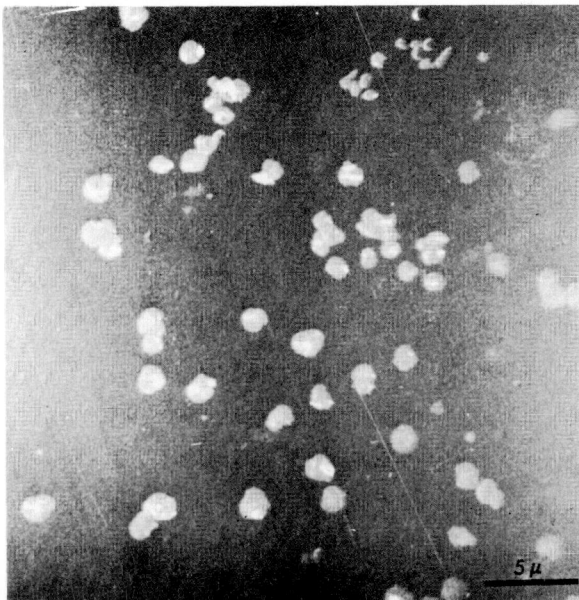

Fig. VIII-5. Dark field electron micrograph of polytetrafluorethylene dispersion particles. The molecules in the bright areas are nearly vertical.

(Fig. VIII-4). The resultant particles often appear hollow with one or more thick appearing bands. The deterioration of the particles has also been noted by Grimaud, Sanlaville, and Troussier (5). Small dark regions can be seen in the original particles which are believed to result from Bragg extinction from individual crystallites in the particles. Similar areas show up illuminated in dark field micrographs (Fig. VIII-5).

Electron diffraction patterns from the dispersion particles are weak and diffuse, usually consisting of only a single ring. If only a few particles are present, individual diffraction spots can be seen (Fig. VIII-6).

A few rod-shaped particles are also found during the dispersion polymerization. Dark field microscopy indicates the molecules are parallel to the axis of the rods.

Shadowed or replicated samples of the dispersion particles often have a particulate surface (6) (Figs. VIII-7 and VIII-8). Although the appearance of the particles in Figure VIII-7 suggests that granulation of the shadowing material has occurred, Symons (6) has shown, using SiO replicas, that the particulate nature is probably real. In both figures there appears to be some regularity or lining up of the smaller particles.

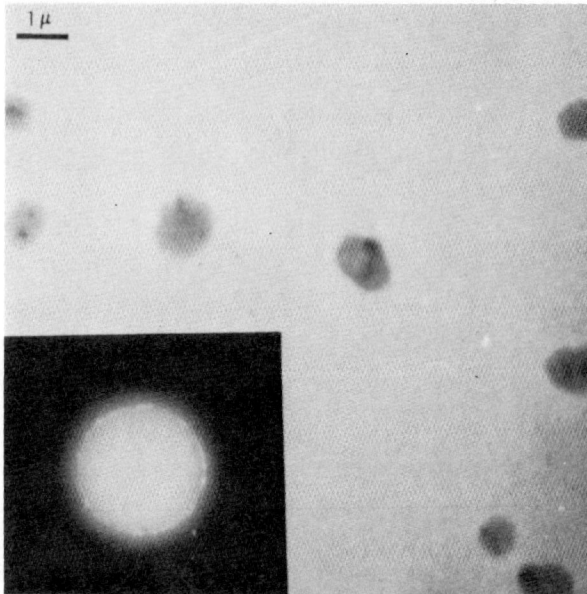

Fig. VIII-6. Electron diffraction pattern (a) from a small number of polytetrafluoro-ethylene dispersion particles (b). Bragg extinction spots can be seen on the dispersion particles.

Granular particles are difficult to observe in the electron microscope because of their size. Replicas of the particles suggest that they are aggregates of considerably smaller particles (Fig. VIII-9). These smaller particles can sometimes be seen in a nearly isolated state.

2. MORPHOLOGY OF AMORPHOUS POLYMERS

The only significant work that has been done on the morphology of amorphous polymers is that which has been done by Kargin and his group. A summary of their ideas was published in 1957 (7) and is quoted below. This summary was published at about the same time as the reports on polyethylene single crystals. The conclusions presented in it, except for the retention of a packet or bundlelike aggregation of chains in the crystals, are still acceptable to this author.

"In our view, amorphous polymers are composed either of chains coiled into globules, or of packets of uncoiled chains. The mechanical and other physical properties of polymers consisting of packets of chains can be

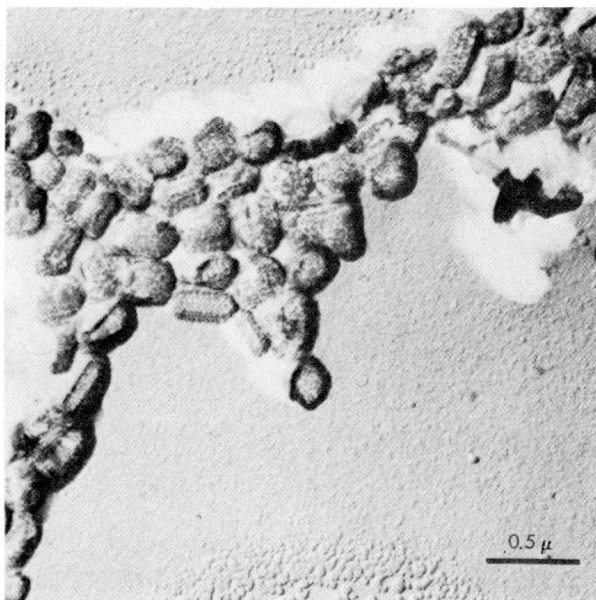

Fig. VIII-7. Chromium shadowed dispersion particles of polytetrafluoroethylene on a carbon substrate. The chromium appears to have aggregated into individual clumps to a greater extent on the particles than on the substrate.

satisfactorily explained by this model because of the inevitable flexibility with several possible structural mechanisms, of the packets themselves. Examples of such mechanisms may be the untwisting of a region of the packet within which the chains are twisted into spirals, or coordinated rotation of one region of the packet relative to another, about C—C bonds. Further structural investigations of this question are, of course, necessary. It is also necessary to develop a statistical theory of deformation of elastic polymers based on this model. It should be pointed out that the above views on the structure of amorphous polymers refer not only to the high-elastic and glass state, but also to the viscofluid state. Indeed, as is known, liquids consisting of rodlike molecules of low molecular weight contain packets of molecules: each molecule is surrounded by other molecules, predominantly arranged with the same axial orientation. Although these packets (sometimes termed 'swarms') are not thermodynamically stable formations, they are nevertheless always present in such liquids. This applies in greater force to polymers. It must be stressed that, in consequence of the well known slow nature of mechanical relaxation processes in poly-

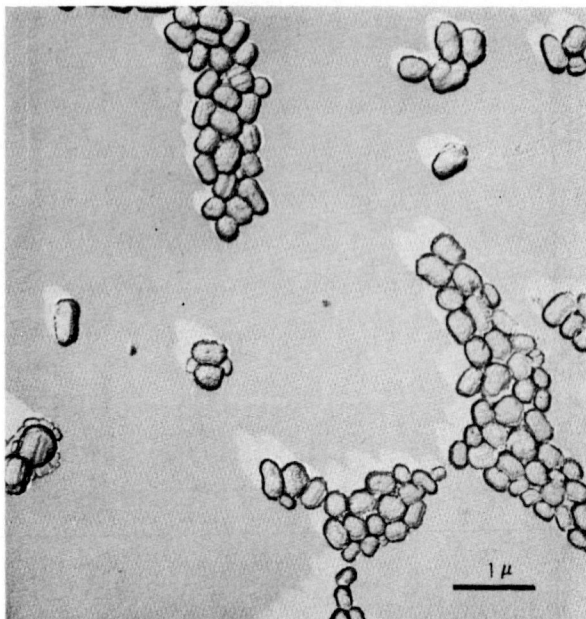

Fig. VIII-8. Silicon monoxide replica of polytetrafluoroethylene dispersion particles. The particles were removed by pyrolysis whereas they are still present in Figure VIII-7 (Symons (6)).

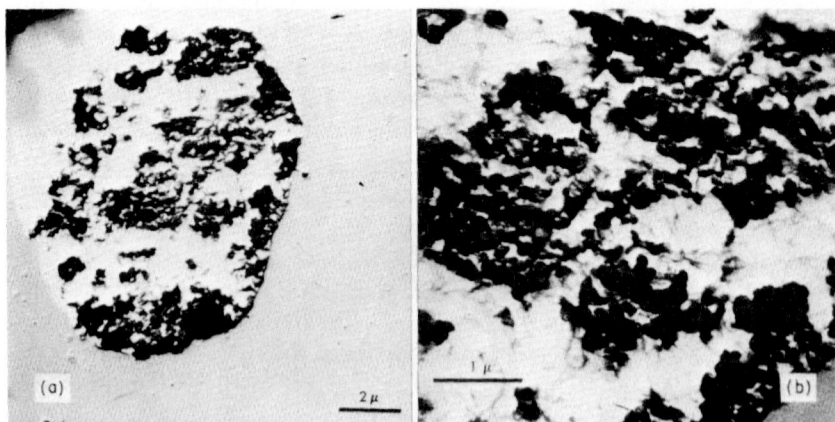

Fig. VIII-9. Granular polytetrafluoroethylene particle removed at an early stage of polymerization. The particle, at higher magnification in (b), is seen to be made up of an aggregate of smaller particles.

mers, the life of a packet of long chain molecules must be exceptionally long, while in the glassy state the bundles will remain practically unchanged.

"These concepts of the structure of amorphous polymers naturally necessitate a revision of the views on the structure of crystalline polymers. The picture, widely accepted at the present time (1957), of the structure of a crystalline polymer as a system of small ordered domains linked by common chains passing successively through regions of oriented and disordered parts of the chains, cannot be valid in the form in which it is usually presented. This follows even on purely geometrical considerations: on the basis of the quite well known dimensions of the ordered domains and the distances between them it is impossible to construct a polymer model in which the chains would leave an ordered domain, become twisted, and then again form an ordered domain. There is no doubt that any one chain passes through several domains of order and disorder. However, the chain does not leave the limits of its packet, and largely retains its same neighbors along its entire length.

"In the proposed model, the degree of order of a polymer is determined by the structure of the packets of chain molecules, since all the packets are of the same type. The order of the packet depends on the degree of lateral order of its molecules and on the alternation of ordered and disordered domains along the packet. The latter is associated with the mechanical stresses which inevitably arise within the packet with growth of the ordered domains. These stresses, which oppose further growth of the ordered domains, can be appreciably diminished only by alternation of the domains of order and disorder along the packet. The proposed model for the molecular structure of amorphous and crystalline polymers must be amplified by structural investigations and analysis of various physical properties of polymers. The structure of polymer systems consisting of individual molecules coiled into globules must be considered separately."

Kargin and Koretskaya (8) have indicated that all variations in structure between globules, packets, and single crystals can be obtained with crystallizable polymers. The globules, similar to those seen in Figure II-1, do not yield crystalline diffraction patterns. This may be either due to their having an amorphous structure, as suggested by Kargin and Koretskaya (8), or due to the improbability of obtaining a diffraction pattern from a crystal with a particular orientation (Chapter I). Similar effects are observed (i.e. no electron diffraction patterns) with polytetrafluoroethylene dispersion particles which are highly crystalline by both density and x-ray diffraction

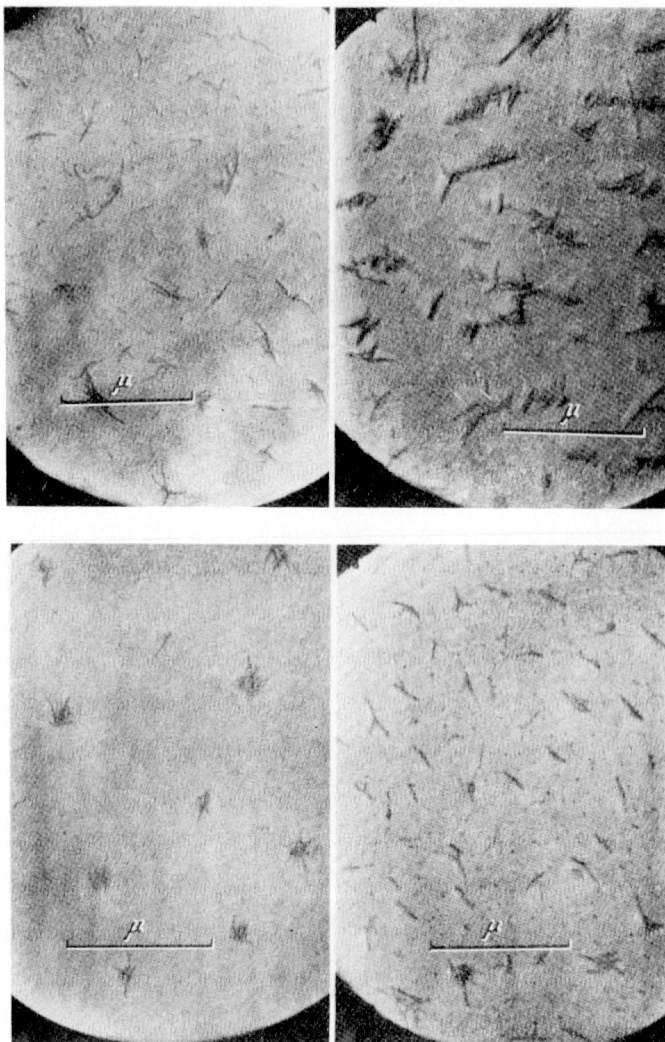

Fig. VIII-10. Packets of polyethylene molecules precipitated from xylene solution onto a nitrocellulose substrate at 100°C (Kargin and Koretskaya (8)).

standards. It should be noted that the globules in Figure II-1 contain many molecules. The globules and single crystals can form from the same solutions. The requirements for the formation of the "packets" (Fig. VIII-10) were not clearly defined nor were diffraction patterns or the orientation of the molecules described. Although it is suggested that the structural unit in lamellar crystals is these globules or packets this feature is doubted by the author. More detailed micrographs and diffraction observations of the globules and the packets are desirable.

REFERENCES

1. W. D. Niegisch, "Electron Diffraction Applied to Single Crystals," paper presented at Elec. Micro. Soc. Am. Meeting, Pittsburgh (1961).
2. W. D. Niegisch, personal communication.
3. W. L. Carrick, R. W. Kluiber, E. F. Bonner, L. H. Wartman, F. M. Rugg, and J. J. Smith, "Transition Metal Catalysts. I. Ethylene Polymerization with a Soluble Catalyst Formed from an Aluminum Halide, Tetraphenyltin and a Vanadium Halide," *J. Am. Chem. Soc.*, **82**, 3883 (1960).
4. E. W. Fisher, "Thermodynamical Explanation of Large Periods in High Polymer Crystals and Drawn Fibers," *Ann. N. Y. Acad. Sci.*, **89**, 620 (1961).
5. E. Grimaud, J. Sanlaville, and M. Troussier, "Dispersions de polytetrafluorethylene," *J. Polymer Sci.*, **31**, 525 (1958).
6. N. K. J. Symons, personal communication.
7. V. A. Kargin, A. I. Kitaigorodsky, and G. L. Slonimsky, "The Structure of Linear Polymers," *Colloid J. (USSR)(Eng. trans.)*, **19**, 141 (1957).
8. V. A. Kargin and T. A. Koretskaya, "Mechanism of the Formation of Polymer Crystals I," *High Mole. Wt. Comp. (USSR)*, **1**, 1721 (1959) (in Russian).
9. K. Hayashi, Y. Kitanishi, M. Nishii, and S. Okamura, "Crystalline Polymers Prepared by the Radiation Induced Solid State Polymerization," *Makromol. Chem.*, **47**, 237 (1961).
10. S. Okamura, K. Hayashi, and Y. Kitanishi, "Radiation Induced Solid-State Polymerization of Ring Compounds," *J. Polymer Sci.*, **58**, 925 (1962).
11. S. Okamura, K. Hayashi, and M. Nishii, "Polymer Crystals Obtained by Radiation Polymerization of Trioxane in Solid State," *J. Polymer Sci.*, **60**, S26 (1962).
12. J. Lando, N. Morosoff, H. Morowetz, and B. Post, "Single Crystal Character of Polyoxymethylene Prepared from Single Crystals of Trioxane," *J. Polymer Sci.*, **60**, S24 (1962).
13. S. Okamura, T. Higashimura, and K. Takeda, "Cationic Polymerization of Trioxane in Solid State," *Makromol. Chem.*, **51**, 217 (1962).
14. S. E. Jamison and H. D. Noether, "Single Polyoxymethylene Crystals by Solid State Polymerization of Trioxane," *Polymer Letters*, **1**, 1 (1963).

IX. PROPERTIES AND MORPHOLOGY

Since the 1920's numerous measurements of the physical properties of numerous polymers have been reported. These properties were often correlated with the density or degree of crystallinity and, in many cases, good correlations were found. On the basis of these correlations, hypotheses and generalizations concerning the effect of the crystallites or the amorphous regions were put forth which appeared logical and could, in fact, often permit the prediction of similar properties in other polymers. Many of these hypotheses will now have to be re-examined. Initial results of this re-examination show convincingly the need to consider the morphology of the polymer in detail in discussing its physical properties. The first four sections of this chapter describe recent experimental results in the areas of diffusion of small molecules in polymers, electron induced crosslinking, chemical reactivity, and mechanical relaxations in which the lamellar nature of crystalline polymers is seen to play a key role. In the concluding section we examine what should probably be considered the first experiments in polymer solid state physics that can lead to new applications of polymers on the basis of their lamellar structure; i.e., experiments in the area of the electronic conduction of polyethylene single crystals.

1. DIFFUSION AND SOLUBILITY

A number of reports have been published relating the swelling of polymers in the presence of solvents and the diffusion and solubility of gas molecules to the amorphous content of the polymer. The crystals are assumed to be impervious to the foreign molecules. The two phase, fringed micelle model would appear to adequately explain the results. The amorphous content of the samples used in these experiments was varied by varying the crystallization conditions, i.e., quenching as compared with slow cooling. A nearly linear relationship is obtained between diffusion and density.

These results, as reported, are difficult to explain in terms of the lamellar structure of polymers. One would expect, for instance, that swelling might easily occur between lamellae but to a lesser extent in defect regions in the crystal, and also that diffusion might take place more readily between lamellae rather than through them. The direct correlation with density and thus "amorphous content" is somewhat surprising.

Since the advent of the lamellar model for crystalline polymers, several publications (1–4) have suggested that the interpretation of diffusion and solubility measurements in terms of a two phase fringed micelle model is probably incorrect. Michaels and Bixler (1,2), for instance, indicate that solubility is a function of density and, therefore, "amorphous" content. However, their diffusion results, which are also density dependent, they suggest, can best be explained in terms of impenetrable, thin, highly aniso-metric sheets of polymer (i.e., lamellae). Michaels and Bixler (2) and Jeschke and Stuart (4) both point out that the polymer segments not in the crystals have a lower degree of chain mobility than in true "amorphous" material; i.e., the permeable portion of the polymer is altered due to the presence of crystals. Jeschke and Stuart suggest that diffusion takes place at im-perfections, imperfectly ordered regions, and at grain boundaries.

Michaels and Bixler worked with polyethylene and found that a con-siderable degree of control of the thermal history of the sample was required to obtain reproducible density and, therefore, diffusion and solubility results. Changes in density were obtained by varying the degree of branch-ing of the polymer. Considerable differences in morphology are probably associated with the changes in density. Jeschke and Stuart, on the other hand, used polyethylene terephthalate and varied the density (and morphology) by varying the crystallization conditions.

Using data obtained by Michaels and co-workers (2,3), it can be shown that, for a given sample of linear polyethylene, the diffusion constant is not proportional to density (Table IX-1). One notes that when a given sample

TABLE IX-1
Diffusion Constant in Linear Polyethylene

Sample	Heat treatment[a]	α[b]	$D \cdot 10^7 cm^2/sec$	
			He	CH_4
Quenched film[c]	None	0.396	61.3	1.58
Quenched film[c]	97°	0.305	59.1	1.25
Quenched film[c]	115°	0.242	57.5	1.12
Quenched film[c]	125°	0.200	65.5	1.55
Slow cooled film[d]	None	0.23	30.7	0.57

[a] Sample was annealed at temperature given for 20 hours, then cooled to room tem-perature. All measurements were made at 25°C.

[b] Amorphous content from density measurements.

[c] Data of de Filippi (3), using a W. R. Grace Company linear polyethylene film.

[d] Data of Michaels and Bixler (2), also using a W. R. Grace Company linear poly-ethylene.

is annealed, the "amorphous content" decreasing by a factor of two, the diffusion constant remains almost constant and, following annealing at 125°C, actually increases. At this temperature it is known that recrystallization with an increase in fold period would occur in a quenched film although it may also occur at the lower annealing temperatures. The diffusion constant in the quenched, annealed films is two to three times larger than in the slow cooled film of equivalent density. The difference between the quenched and slow cooled films, as prepared, is that normally found, i.e., diffusion is faster in the more "amorphous" films. It is obvious from these results that it is not density that is the determining factor in diffusion but rather the morphology of the sample as determined by its thermal history.

2. ELECTRON INDUCED CROSSLINKING

As pointed out in the first volume of this series (5), electron induced crosslinking of crystalline polymers at low doses was generally believed to occur only in amorphous regions as pictured in the two phase, fringed micelle model. In that model it was only in such regions in the solid, or in the melt, that molecules could cross over each other and approach closely enough to react, when free radicals are formed by the ionizing radiation. The molecular orientation and separation in crystalline regions should effectively prohibit crosslinking. Experiments did indeed show that the crosslinking sensitivity of polyethylene, for instance, is inversely proportional to its density. The crosslinking results would thus appear to be difficult to interpret in terms of the lamellar structure of crystalline polymers in which "amorphous" regions are interpreted as being comprised of the folds, defects in the crystal and noncrystallizible molecules or segments.

This apparent conflict has been resolved by Salovey and Keller (6,7); the results are another indication of the lamellar nature of polymers and the necessity of considering this structure when interpreting the details of polymer behavior. They investigated the effect of irradiating with 1 m.e.v. electrons in vacuum: single crystals, either simple or highly dendritic; solution grown "hedrites"; and melt crystallized linear polyethylene (Marlex 50). All samples were fused at 150°C after irradiation while still in the sealed tubes. The amount of crosslinking was measured by extracting the polymer with refluxing xylene in a Soxhlet extraction apparatus. For most of the work a 20 megarad (MR) dose was used, this dose being convenient and producing significantly different results for the various samples.

The solution-crystallized material was precipitated from 0.02 to 2% xylene solutions at 70 and 85°C. At 70°C single crystals were obtained with a fold period of 110A. (The sample probably crystallized while cooling and therefore somewhat above 70°C.) From the more concentrated solutions, as discussed in Chapter II, complex dendritic crystals were obtained in which the individual lamellae diverged. At 85°C compact structures, presumably hedrites, were obtained from the 1% and higher solutions, whereas from the 0.05% and lower solutions, single crystals were obtained. Although not stated, the radiation effects suggest that the dividing line between hedrites and single crystals is at about 0.5% concentration when crystallized at 85°C. All of the samples crystallized at 85°C had a 130A fold period. The single crystals and hedrites were filtered and vacuum dried (presumably at room temperature) before irradiation.

The gel point for the melt crystallized linear polyethylene used in this work is about 2 MR. After a dose of 20 MR, 65–75% of the sample is insoluble in refluxing xylene. However, Salovey and Keller found that the single crystals (crystallized at 70 or 85°C from dilute solution) were fully soluble with the same 20 MR dose (from solutions of 0.1% concentration and higher small amounts of gel (10%) were formed). The polymer was not altered by the precipitation process since a sample of the crystals which was melted and then irradiated resembled the original material; i.e., it was 71% insoluble. The hedrite samples, irradiated with 20 MR dose, were similar to the melt crystallized material, being 50–65% insoluble.

These results, as they stand, indicate that the morphology of the sample strongly affects its crosslinking behavior. At the time of the work of Salovey and Keller, theoretical treatments of crystallization (8) suggested that folded chain lamellae would grow only from dilute solution and that a bundlelike, fringed micelle structure could form during crystallization from concentrated solution and the melt. Thus it appeared that the high irradiation sensitivity of the hedrites and bulk material was due to the presence of two-phase-model amorphous regions.

Two further experiments, however, prove that the morphological difference is a subtle one and that the two phase model is not capable of explaining the results. (1) Samples of the hedrites crystallized from 1% xylene solutions were ultrasonically dispersed, filtered and irradiated with a 20 MR dose in the usual manner. A control sample (not dispersed) and one of the dispersed samples, which had subsequently been fused, were irradiated at the same time. Following irradiation the dispersed sample was fully soluble, the untreated sample had a gel content of 60.9%, and the dispersed, fused sample a gel content of 27.8%. Since the dispersed sample now reacted the same as the single crystals, this experiment indicated that the difference in irradiation sensitivity was not due to the presence of bundlelike crystals in the hedrites and bulk material. (2) If extreme care is used to reduce the solvent content to a minimum, a significant amount of gel is produced in the single crystal mats with a 20 MR dose (9).

The difference between the single crystals and dispersed hedrites on the one hand and the hedrites and bulk material on the other is due, it is believed, to the presence of trapped solvent between the lamellae. On the basis of only the first of the above two experiments Salovey and Keller suggested that the lamellae were held apart by occluded ends of the molecules. The low gel content in the dispersed, fused hedrites was suggested as being due to chain scission during the ultrasonic irradiation. However, we believe it may also be due to fibrillation occurring during the dispersion treatment, these fibrils separating the lamellae and being retained to an extent during fusion.

In either case, the results indicate that crosslinking is taking place between the folds of molecules in adjacent lamellae. These portions of the molecules could be close enough to react. When free radicals are formed within the lamella, they must migrate to the folds. Presumably they can migrate easily at room temperature since the fusion treatment makes only a small difference in the crosslinking efficiency (10). Defects of various types within the lamellae may also be suitable sites for crosslinking or trapping of the free radical.

3. CHEMICAL REACTIVITY

Just as diffusion and crosslinking were generally believed to be restricted to the amorphous regions in "*semi*crystalline" polymers, so also were such chemical reactions as halogenation and oxidation (11). Keller, Matreyek, and Winslow have compared (11) the reactivity of polyethylene single crystals and bulk material to bromine and chlorine at room temperature and at elevated temperatures. The crystals were precipitated from tetrachloroethylene at $70°$ (fold period $= 110$ A).

At room temperature, when reacted with either bromine (in suspension) or chlorine (as a mat), the single crystals take up halogen corresponding to one halogen atom per fold. The low angle x-ray intensity of the treated samples, however, did not increase as would be expected if the halogen addition resulted in layers of heavy atoms on the fold surfaces. When brominated or chlorinated samples are dissolved and reprecipitated at $85°C$ they crystallize more slowly but with the same fold period as untreated polyethylene. Keller *et al.* thus concluded (11) that although the halogen may have been attached predominantly at the folds, and therefore regularly spaced with a spacing determined by the original fold period (110 A), its presence had no effect on the new fold period (135 A). Swan (12) has shown that chlorine atoms can be accommodated to at least a limited extent in the lattice, the unit cell expanding slightly.

The authors indicate that they expected to find a difference between the single crystals and a bulk polymer, i.e. "that uptake by essentially fully crystalline single crystals would be less than uptake by the melt crystallized polymer which contains a larger proportion of disordered regions" (11). However, they found that at room temperature the uptake of chlorine by bulk polymer was essentially the same as that of the single crystal mats.

At elevated temperature (60 and 80°C) the chlorine uptake by both single crystal mats and bulk polymer was considerably higher, but was still essentially the same for equivalent sample thicknesses. After 250 hours at 60°C the single crystals had reacted with chlorine to an amount corresponding to about 1 chlorine atom for every 1.4 carbon atoms. Thus the chlorine must be reacting throughout the crystal lattice.

With increasing chlorine content the x-ray Bragg reflections gradually decrease in intensity while a diffuse, liquid-scatteringlike halo develops. When chlorinated to the extent of 1 chlorine for every 2 carbon atoms considerable Bragg scattering still remains. This plus the fact that this polymer could be reprecipitated during cooling from solution in the form of a mixture of well-formed single crystals and, at a lower temperature, an ill-defined shapeless precipitate, suggested (11) that the polymer does not transform gradually throughout the lattice but rather that certain regions undergo a complete change while others remain unaffected.

Keller *et al.* (11) also chlorinated individual single crystals and observed the results in the electron microscope. A structure on the 200 to 600 A scale can be seen on both shadowed (Fig. IX-1) and unshadowed samples. In the shadowed samples the surface appears puckered, the authors indicating that the puckering is most obvious on the lamella in contact with the substrate. In unshadowed samples, small regions of high electron density are observed which may either be due to the greater thickness in the puckered regions or the higher eléctron density in areas which have completely reacted with the chlorine. Electron diffraction patterns from crystals chlorinated between 40 and 100 hours show the simultaneous presence of Bragg reflections and a broad diffuse ring, and thus are similar to the x-ray patterns from the crystal mats.

On the basis of this work the authors conclude that there is no inherent difference between the texture of solution-grown single crystals and melt-grown bulk material (11). Various crystallization and annealing treatments will probably produce changes in chemical reactivity related to the changes

Fig. IX-1. Electron micrograph of a polyethylene single crystal chlorinated for 72 hours at 60°C. The 200–600 A puckering shows up as dark, electron-dense regions in unshadowed samples. (Keller, Matreyek, and Winslow (11).)

in diffusion and crosslinking and due primarily to difference in the lamellar thickness and perfection in the sample.

4. TRANSITIONS

In this section we shall consider first some present concepts concerning glass transitions, melting, and defects in solids of low molecular weight materials and then consider whether these concepts may also apply to polymers. More extensive treatments of the concepts for solids of low molecular weight materials are available in a number of recent publications (see, for example, references 13 and 14).

As indicated in Chapter I, anharmonicity in the thermal vibrations of atoms in solids results in thermal expansion at temperatures above 0°K. At low temperatures crystalline material and glassy, amorphous solids expand at about the same, nearly uniform rate. With glassy materials it

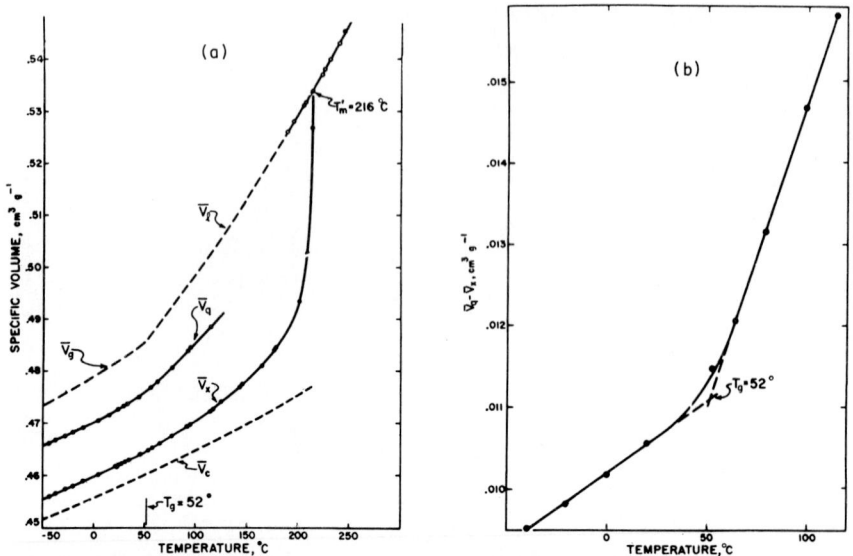

Fig. IX-2. (a) Specific volume-temperature plot for polychlorotrifluoroethylene. $\bar{V}_q$ and $\bar{V}_x$ are experimentally determined values for quenched and slow-cooled samples respectively; $\bar{V}_s$, $\bar{V}_g$, and $\bar{V}_c$ are extrapolated, assumed, and calculated values for the supercooled liquid, glass, and crystalline states respectively. (b) Plot of $\bar{V}_q \bar{V}_x$ from (a). (Hoffman and Weeks (15).)

is well known and accepted that as the temperature is raised a temperature region is reached at which the expansion coefficient increases. There is a change in slope of a specific volume versus temperature curve for the material. Above this temperature region the expansion is again nearly uniform but at a faster rate. The so-called glass transition temperature is obtained by extrapolating the "linear" portions of the volume versus temperature curve and determining the temperature at which they cross. In the case of crystalline polymers containing amorphous regions in which the glass transition temperature, T_g, does not vary significantly with density, a more appropriate method for determining T_g has been introduced by Hoffman and Weeks (15). Specific volume-temperature curves are obtained for samples with as low and high a density as possible and the resulting values subtracted. Nearly linear plots are obtained (Fig. IX-2) from which T_g can easily be determined. The difficulty is determining T_g from the original curves can be seen in Figure IX-2a.

The excess volume formed by this more rapid expansion is usually described as free volume. In glassy materials this free volume, it is suggested

(16), is not associated with any given atom but is distributed at random and can be redistributed without increasing the potential energy of the system. The entropy, however, is changed if the distribution is changed. In glasses of low molecular weight materials one can think of the network as opening up, voids forming which will be significant when they are somewhat smaller in size than the dimensions of the atoms or molecules in the glass.

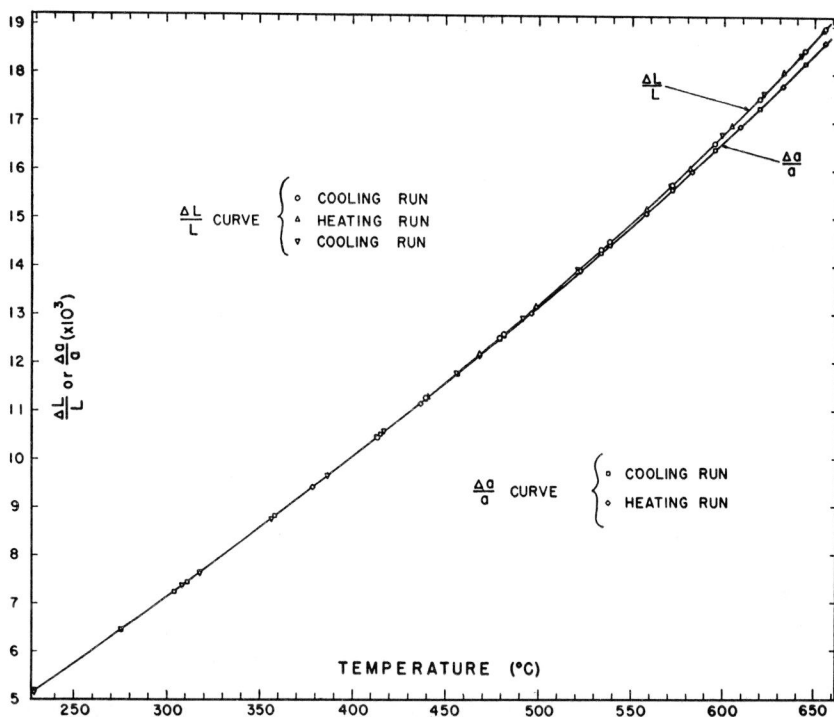

Fig. IX-3. Measured length (L) and lattice parameter (a) expansions of aluminum versus temperature. The fraction of additional lattice sites present in thermal equilibrium at the higher temperatures is 3 ($\Delta L/L$-$\Delta a/a$). These sites correspond to the thermal generation of vacancies and first become measurable at about 425°C. (Simmons and Balluffi (17).)

When a stress is applied the equilibrium distribution of voids is disturbed. If the stress is applied for a sufficient length of time, the voids are redistributed, the network is deformed plastically and the sample flows. The glass transition temperature is often determined by measuring the temperature at which plastic deformation can occur. This temperature varies with the frequency of the applied force, the sample being able to adjust more rapidly at higher temperatures.

In crystalline materials there is also a temperature well below the melting point at which the increases in lattice dimensions, due primarily to thermal expansion, and the increase in macroscopic dimensions diverge. The macroscopic dimensions usually (17,18) (in silver and aluminum, the metals examined to date) increase more rapidly than the lattice dimensions (Fig. IX-3). The vibrational energy, increasing with increasing temperature, is not uniformly distributed among the atoms; some atoms, at any given time, vibrate with considerably great amplitude than others. Occasionally, the frequency increasing with increasing temperature, an atom within the crystal will gain sufficient energy to escape from its lattice site into a neighboring interstitial site. As long as the atom remains within the crystal the density change is the same whether measured from x-ray lattice spacings or by volumetric means. The lattice dimensions and macroscopic dimensions increase at the same rate. It is only if the interstitial atom, in effect, moves to the surface, to a new lattice site, that the volumetric expansion becomes greater than the lattice expansion.

At relatively low temperatures, when only a few interstitial atoms and associated vacancies are formed, the thermal energy is too low and too rapidly redistributed for a newly formed vacancy or interstitial to migrate. Furthermore, the lattice forces are such that the barrier for recombination is considerably lower than that for further separation. Thus if any further motion occurs, the interstitial has a high probablity of returning to its original lattice site and recombining with the vacancy. As the temperature increases both the number of interstitial-vacancy pairs and the probability that one or the other may migrate away from its original position, leading to permanent separation, increases.

The energy barrier for migration of an interstitial is about twice that for migration of a vacancy. Thus one might expect that when migration occurs the macroscopic dimensions would increase less rapidly than the lattice dimensions, vacancies migrating to the surface and destroying lattice sites. The observed results in aluminum and silver (17,18) can be explained, however, on the basis of the generation of vacancies at surfaces and their migration into the lattice. The energy of formation of a vacancy at a surface should be significantly less than that for formation of an interstitial-vacancy pair in the interior of the lattice. These surfaces may include, besides the external surfaces of the sample, internal surfaces such as grain boundaries and dislocations. At an edge dislocation, for instance, the formation of a vacancy may be considered to take place by the transfer of an atom from an adjacent full plane to the edge of the half plane. The resulting vacancy in the full plane may then migrate away from the dislocation line. Repetition of the process would lead to the growing out of the edge dislocation. A large number of vacancies, equal to the number of lattice sites on the originally missing half plane, would be injected into the lattice.

These "excess" vacancies (no corresponding interstitial atom) created at the surface and migrating into the lattice, are in thermal equilibrium at the temperature at which they are formed, the number increasing with increasing temperature. Near the melting point, in silver, for instance, there are more than 10^{19} excess vacancies per cubic centimeter (18). If the temperature is reduced sufficiently slowly, the experiments show that the excess vacancies diffuse out of the lattice and disappear at surfaces. If the samples are rapidly quenched from temperatures near the melting point, some of the excess vacancies are trapped in the lattice. These vacancies, which can be detected by electrical resistivity and nuclear magnetic resonance experiments, can be considered as a type of "free volume," being redistributable under applied stresses at sufficiently high temperatures. The major difference between the voids in a glass and the excess vacancies in a crystal is expected to be the large size and energy of formation of the vacancy.

As the melting point is approached, the thermal vibrations increasing in amplitude and the lattice expanding, more and more interstitial-vacancy pairs (in addition to excess vacancies) are created, separate and begin to diffuse through the lattice. Other defects of a more complicated nature are also formed. At the melting point the combination of decreasing lattice forces due to the lattice expansion and the increasing amplitude of the vibrations results in a catastrophic increase in the number of migrating defects. A liquid with short range order but no long range order is formed. Although in most materials there is a significant decrease in density, the nearest neighbor spacing does not show a correspondingly large increase at the melting point.

Above the melting point the macroscopic dimensions and the "lattice" dimensions (nearest neighbor spacings) diverge at an even greater rate than in the region near the melting point. Since the long range order is nonexistent the excess "vacancies" or voids, i.e., the free volume, resulting in the change in density at the melting point and subsequent rapid change as the temperature is increased, need not be of atomic dimensions. These voids do not result from the removal of an atom from a site in a three dimensional lattice. Rapid redistribution of the voids in a liquid occurs under an applied stress and due to the thermal vibrations since the energy barriers for diffusion are relatively negligible.

In highly imperfect crystals of low molecular weight materials, as in cold worked metals, "localized melting" and recrystallization takes place at temperatures well below the melting point of the perfect crystal. In the vicinity of gross defects such as dislocations and aggregates thereof, forming grain and subgrain boundaries, the formation, separation if necessary, and diffusion of point defects take place more readily than in a perfect lattice. The lattice strains reduce some energy barriers and thus the thermal vibrations are effective at a lower temperature. Repeated formation, diffusion, and annihilation of point defects result in the growth of some grains at the expense of others. In the vicinity of the gross defects one can think of the material as essentially liquidlike in the sense that the lattice forces are not effective in restraining the diffusion of vacancies and interstitials.

One may now ask how these concepts apply to polymers. In this section we have indicated that the so-called glass transition is associated with the formation of redistributable free volume in the glass, that similar effects occur in crystalline materials, that the motion and formation of the free volume is enhanced in imperfect crystals, that atoms and molecules can move in the solid, and that there is an abrupt increase in the free volume and degree of atomic or molecular mobility at the melting point. These ideas should apply to polymers as well as to low molecular weight materials. A major difference arises from the anistropy of the molecular forces in a polymer; the primary bonds along the molecule are much stronger than the secondary bonds between molecules. As indicated in Chapter V, motion along the backbone of the molecule is considerably less restricted than motions normal to the backbone. Furthermore, as suggested by Kargin (19), one would expect alignment of portions of neighboring chains in the liquid and glassy state over larger regions than occur in low molecular weight material. Current theoretical treatments of even low molecular weight liquids (20) (and therefore glasses) are in terms of a quasi-crystalline structure rather than a random gaslike structure.

Molecular motion in polymers can be observed by a number of experimental techniques, including NMR and dielectric and mechanical loss measurements. Changes in the values measured in these experiments occur when the particular motion involved begins to take place, as a function of temperature, at a rate comparable to the measuring frequency. The molecular motions that result in the formation of free volume in low molecular weight glasses and in amorphous polymers, for instance, are observed, at higher temperatures because of the necessarily higher measuring frequency, in mechanical and dielectric loss measurements as a peak in energy absorption and in NMR as a narrowing of the resonance spectrum.

In polymers one finds two or more peaks in the energy loss curves. In amorphous polymers and in the amorphous regions of crystalline polymers these peaks have been attributed to the onset of motions of the appropriate frequency of long and short segments of the main chains and of branches. An additional peak in crystalline polymers has been related to a premelting of the smaller crystals and the accompanying motion of the now liquid segments (for recent reviews of these measurements see references 21 to 23).

In the remainder of this section we shall consider an interpretation of the mechanical loss curves for a polymer in terms of its morphology. Unfortunately there is considerable confusion in the literature at present concerning the detailed interpretation of individual loss peaks in terms of specific molecular motions and we shall in general not attempt to relate the following discussion to specific loss peaks in particular polymers.

Takayanagi et al. (21,22) have indicated that one would expect to obtain only one loss peak for a fringed micelle type crystalline linear polymer whereas two peaks should be observed for a lamellar type structure. In a fringed micelle model, molecules pass through both crystal and amorphous regions and applied stresses will be distributed between the crystallites and the amorphous regions. If the elastic modulus of the crystalline regions is considerably larger than that of the amorphous regions, as would be expected, then the single loss peak would occur at a temperature near that for a wholly amorphous sample but with a magnitude inversely proportional to the volume fraction of amorphous material. On the other hand, in the case of a lamellar type of structure in which any amorphous material is restricted to regions between the lamellae and the majority of the molecules are folded into the lamellae (i.e. stress is not propagated from the crystals to the amorphous regions or vice versa), each region will absorb energy at its characteristic temperature and frequency. The magnitude of the loss at the characteristic temperature would depend on the relative amount of corresponding material. Defects within the crystals would contribute to the energy absorption in the crystal, permitting some variation of the magnitude and temperature of the characteristic loss peak of the crystal with thermal history as well as possibly giving rise to an inherent energy absorption, but would be distinct from loss phenomena in the amorphous regions. If more than one specific type of motion is associated with a given loss peak, as would certainly be the case if energy absorption in crystals is affected by the presence of defects, then one should consider the integrated area of the loss peak rather than its maximum when relating the loss to the fraction of the sample in which the specific motion is taking place (24).

In the subsequent discussion, we shall use the term magnitude to refer to the total area.

In view of our present knowledge of the morphology of crystalline polymers we believe that one should expect to find at least three or four energy loss maxima. With Takayanagi (21,22), and as defined below, we shall label these as the α_c, α_a, β, γ, etc. loss peaks and consider each individually.

At a temperature related to that at which recrystallization during annealing takes place (Chapter V) it is apparent that molecular motion is taking place within the lattice. This motion which, as discussed in Chapter V, may take place through the creation and movement of point defects or by the stepwise motion of an entire segment, is on a scale sufficient to permit mass transport and is thus probably related to that observed in aluminum and silver (Fig. IX-3). Because of the long chain nature of the polymer molecules, the motion is significant only at temperatures closer to the melting point than in the case of the metals. We relate the loss peak associated with this motion to α_c of Takayanagi (21,22) or the α transition as defined by other authors (23). The energy absorption would be associated with the formation and motion of the defects. The creation of these defects in the lattice of polyethylene or any form of rotational motion should result in expansion primarily of the **a** axis lattice spacing, the lattice approaching hexagonal packing. In most crystalline polymers the loss maxima labeled α_c in mechanical loss experiments is found on the order of 50°C below the melting point. At about the same temperature, Li-shen, Andreeva, and Kargin (25) and Takayanagi, Aromoki, Yoskino, and Hoashi (26) find that **a** axis spacing begins to increase more rapidly than at lower temperatures. (There is an abrupt increase in the slope of a lattice expansion-temperature curve.)

The α_c loss peak in polyethylene and other polymers has in the past been related to the melting or "premelting" of the smaller crystals in the sample. Takayanagi, however, has shown (21,26) that the α_c absorption in polyethylene, at least, takes place at a temperature considerably below that at which the density begins to markedly change. He suggests instead that the α_c loss peak is due to the relative motion of neighboring {200} planes in the crystal, this motion being able to occur when the lattice spacing becomes sufficiently large. Scott et al. (24) suggested that a similar peak in polychlorotrifluoroethylene may be related in some fashion to the lamella surfaces, the peak being well defined only in highly crystalline specimens which also have well-developed lamellae (crystallized by slow cooling rather than quenching and annealing). One might expect on the basis of any of the above interpretations, except that of premelting of fringed micelle type crystallites, that the magnitude (not maximum) of the loss peak would be proportional to the density of the sample and that its characteristic temperature would be independent of thermal history. However, the incorporation of branches, end groups, and other configurational defects in the lattice as well as metastable defects due to previous thermal treatments may reduce the restraints on motion in the lattice and result in changes in magnitude-density and temperature relationships. As discussed below additional loss peaks may result from the defects themselves.

Many crystalline polymers, even though lamellar, may contain regions in which the molecules are sufficiently disordered to be considered amorphous (i.e , segregated clusters of low molecular weight molecules, atactic molecules, crosslinked or gellike particles, or even crystallizable molecules in rapidly quenched samples). In such regions and in truly amorphous polymers the lack of lattice restraints will permit molecular motion at considerably lower temperatures than in a crystal. The resulting energy absorption peak will be related to the glass transition of the polymer, occurring at that temperature at which free volume is able to be created and redistributed in the glassy or amorphous regions at the frequency of the measurement. This peak is labeled α_a by Takayanagi (21,22). The magnitude of the loss peak should depend on the proportion of amorphous material while the temperature would stay nearly constant as the proportion is changed. However, again, if the character of the amorphous regions is changed (as often occurs when samples of different "amorphous content" are prepared—for instance by crosslinking or segregation of low moleculear weight polymer) the characteristic temperature as well as the magnitude of the loss peak may change.

In light of Kargin's suggestions (19) for the structure of amorphous polymers (Chapter VIII) and the probably similar structure of most of the amorphous pockets in crystalline polymers (gellike regions may be inherently different depending on the degree of crosslinking), the same type of molecular motions should take place and be of primary concern as those previously suggested as occurring in the lattice; i.e., motion along the molecular backbone. Since the restraints are lower and because the linear translation or angular rotation need not be restricted to lattice spacing steps, one finds, as in comparing low molecular weight glasses and crystals, that the motion takes place at a lower temperature. Lateral translation of the molecule even in amorphous regions is probably greatly restricted because of the long chain nature of polymers. In quenched crystallizable polymers, such as polyethylene terephthalate and polychlorotrifluoroethylene, heating above the characteristic temperature of α_a, i.e. the glass transition temperature, is known to result in sufficient molecular motion in the amorphous regions to permit crystallization (Chapters IV and V). In fact, if it were not for the difficulty in quenching polymers such as polyethylene and polyoxymethylene to a glassy state, such a test might serve as a means of distinguishing the α_a loss peak. At present, for instance, there are suggestions in the literature that mechanical loss peaks at either about $-20°C$ or $-100°C$ may be the α_a peaks of polyethylene (see references 21 and 23).

Additional loss peaks would be expected to be associated with configurational features of the molecules whether they are located in amorphous or crystalline regions. Such factors as branches, either random as in branched polyethylene or regular as in isotactic polyethylene, and ends of the molecules can undergo rotational motion by themselves as well as in conjunction with the main chain. The temperature at which this motion is significant in loss measurements should vary with the environment of the particular factor. It may occur over a broad range of temperatures, related to a more or less continuous range of restraints from that present for randomly coiled chains through the packet structure pictured by Kargin to inclusion within the crystal lattice or it may occur at a specific, different temperature for each environment. These peaks are labeled β, γ, etc. by Takayanagi (21). The magnitude of the loss peaks due to motion of branches etc. should remain constant as the thermal history of a sample (and its density) is changed, but the characteristic temperature may vary. In addition, as previously discussed, the presence of these configurational factors may influence the magnitude and temperature of the α_a and α_c peaks.

Much work has been reported relating loss peak magnitude and temperature to sample density, degree of branching, length of branches, etc. Except for a few cases in which the loss peaks can be directly attributable to the motions of specific branches, we do not believe that the interpretation of the loss peaks in terms of particular types of motion in specific regions of a polymer is on a sound basis at the present time. Besides the sources of energy absorption discussed above, it is possible that relative motions of the lamellae themselves may occur. An interesting step in the direction of unraveling the present situation has been taken by Thurn (27) who has determined the mechanical loss spectrum of polyethylene single crystals, attaching them to a quartz fiber and comparing the damping with that obtained from the fiber coated with polyethylene crystallized from the melt. He finds that the α_c peak ($+80°C$) and a peak at about $-20°C$ are present in both samples but that the peak at -80 to $-100°C$ is absent in the single crystal sample in which there are inherently no pockets of low molecular

weight and other amorphous material. Additional detailed descriptions of
the molecular motions can be expected from NMR studies as a function
of orientation. Initial work in this area has been described by Hyndman
and Origlio (28) and Kedzie (29). The results of Hyndman and Origlio,
for instance, agree well with the librational motion postulated by Clark
and Muus (30) from x-ray diffraction observations as beginning at the 19°
transition in polytetrafluoroethylene (see Chapter I).

5. ELECTRONIC PROPERTIES

Van Roggen has recently reported (31) some extremely interesting scout-
ing studies of the electronic properties of polyethylene single crystals.
The crystals were deposited on metallic substrates with catwhiskers being
used as the second electrode. The crystals were most easily seen, with incident illumination, on copper. Even though he found that the contact
pressure and area of contact could not be kept constant for extended periods of time, the reproducibility of the current-voltage curves from a given
crystal was good. He indicates, however, that considerably larger variations
were found from crystal to crystal which would suggest that there are larger
variations in the thickness of different crystals than within a given crystal.

An I-V curve for a 100 A thick polyethylene crystal using a Pt catwhisker
and Cu substrate (Pt/Cu) is shown in Figure IX-4. Of particular interest

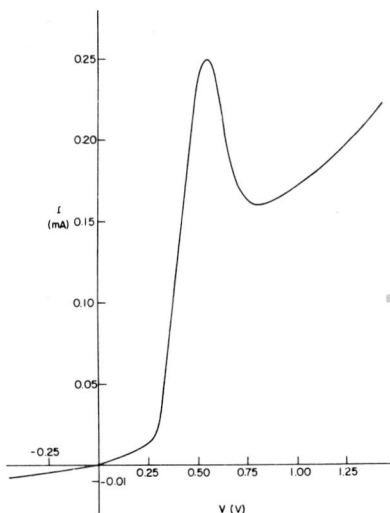

Fig. IX-4. Current-voltage characteristic of a 100 A thick lamella of polyethylene
between a Pt catwhisker and Cu substrate (van Roggen (31)).

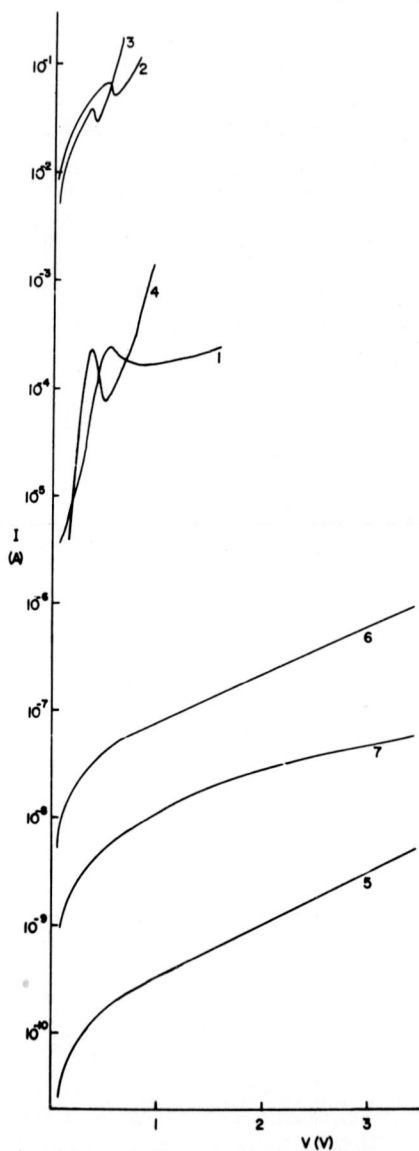

Fig. IX-5. I-V curves for (1) Pt/Cu with 100 A lamella, (2) Mo/Cu with 100 A lamella, (3) W/Cu with 100 A lamella, (4) Mo/Cu with lamella annealed to 300–400 A thickness, (5) Pt/Cu with more than one lamella, (6) Pt/Cu with nitrocellulose film, and (7) Pt/Cu with high energy electron irradiated lamellae (van Roggen (31)).

is the negative resistance region in which the slope is negative. The curves are similar to those obtained with Esaki diodes (32) which find use in, for instance, oscillators and amplifiers. Van Roggen indicates (31) that Mo and W catwhiskers give the same type of I-V curves (Fig. IX-5) but that W/Cu has the opposite rectification as compared with Pt/Cu and Mo/Cu. The magnitude of the current increases with the difference of the cat-whiskers contact potential relative to that of copper in air. Experiments with a Au/Au system gave similar results indicating that oxide layers on the metal are not the cause of the observed effect.

As shown in curve 4 of Figure IX-5, increasing the thickness of the crystals by annealing (Chapter V) results in a decrease in current. The decrease in voltage for the peak on the curve (compared to curve 2), van Roggen indicates (31), is due to a change in contact potential of the catwhisker as a result of a different surface treatment in the sharpening process.

All of the above work was done with single lamellae. When two or more lamellae lying on top of each other are used, not only is the current greatly reduced but the negative resistance is not found (curve 5, Fig. IX-5). The shape of the I-V curve with two lamellae is similar to that observed with thin films of nitrocellulose (curve 6). Similar curves are also obtained (curve 7) if the crystals are irradiated with a low dose of high energy electrons (as used by Salovey and Keller (Section IX-2)).

Van Roggen suggests (31) the following band type model to explain his observations. He considers the interior of the crystal to be a perfect lattice with a large energy gap (between the valence and conduction bands) while the outer layers or fold surfaces are considered as thin imperfect crystals. As a result of ionization by contact with the metal electrodes, energy states become available in the fold surfaces, i.e., one can consider electrons as being introduced to or withdrawn from the fold surfaces by the neighboring electrode. These energy states, which lie in a region differing from the Fermi level of the metal by an amount depending on the contact potential, can act as injector electrodes, i.e. tunneling can take place across the bulk of the crystal between states of equal energy in these regions. The electrons of the tunnel current are supplied and taken up by the adjacent electrodes. For electrodes formed from different metals (i.e. different contact potentials) a voltage will in general have to be applied to equalize the energy of the regions and permit tunneling (see, for instance, reference 33).

Although the model has to be applied with caution, for a number of reasons (31), it does appear to explain the results. As van Roggen points out (31), the current through the crystal will consist of two main parts. There will, as in normal tunnel diodes, be tunneling between the two electrodes through the crystal as a whole. The second and, for this discussion, more interesting part corresponds to tunneling between equal energy states in the fold surfaces. When the applied voltage becomes sufficient to equalize states in the two surfaces, this current will increase, leading to the rapid rise in current seen in Figure IX-4. As the voltage is increased still further, however, the energy levels corresponding to the ionized states will again become unequal and the current will decrease, resulting in the negative resistance portion of the curve. A further increase in voltage may then lead to tunneling between one electrode and the opposite fold surface. The voltage (positive or negative) at which the maximum current occurs will depend on the relative contact potentials of the metal electrodes while the current, assuming efficient contact between fold surface and electrode, will depend on the energy barrier corresponding to the difference in energy levels between the ionized fold surfaces and the conduction levels of the interior of the crystal.

When two or more lamellae are involved the large decrease in current can be attributed to scattering at the intermediate fold surfaces which can be considered as highly imperfect regions and thus as very large energy barriers. The current will be smaller than through a single lamella of equivalent thickness. An increase in thickness of the single lamella will also lead to a decrease in current because of the larger tunneling distance. The observed effects must be related to the morphology of the crystals since disrupting their structure by irradiation or the use of noncrystalline polymer films of equivalent thickness results in effects similar to bulk material under high applied fields.

In this initial study no account is taken of the distortion of the lattice due to the use of crystals consisting of collapsed hollow pyramids. Neither was the effect of known variations in morphology and thickness of the original crystal considered. The consideration of various effects such as these and the possibility of three dimensional electronic anisotropy in crystals of polymers such as nylons suggest that van Roggen's results have opened a fruitful field of research involving the investigation and application of the solid state electronic properties of polymer crystals. This research should be important not only in itself but for its ability to characterize the crystals themselves. It is anticipated that similar research in other areas of polymer solid state physics will be performed in the near future. As indicated in the preface, if the material presented in this book serves as a basis for the development of such research, the book will have served its purpose.

REFERENCES

1. A. S. Michaels and H. J. Bixler, "Solubility of Gases in Polyethylene," *J. Polymer Sci.*, **50**, 393 (1961).
2. A. S. Michaels and H. J. Bixler, "Flow of Gases Through Polyethylene," *J. Polymer Sci.*, **50**, 413 (1961).
3. R. P. de Filippi, "Transport of Liquids in Structurally Modified Polyethylene," Summary of Thesis, Mass. Inst. of Tech., 1961.
4. D. Jeschke and H. A. Stuart, "Diffusion und Permeation von Gasen in Hochpolymeren in Abhangigkeit von Kritallisationsgrad und von der Temperatur," *Z. Naturforsch.*, **16a**, 37 (1961).
5. F. A. Bovey, *Effects of Ionizing Radiation on Natural and Synthetic High Polymers*, Interscience, New York (1958).
6. R. Salovey and A. Keller, "Influence of Crystallization Conditions on Radiation Effects in Polyethylene. Part I Crystallization from Dilute Solution and from the Melt," *Bell Sys. Tech. J.*, **40**, 1397 (1961).
7. R. Salovey and A. Keller, "The Influence of Crystallization Conditions on Radiation Effects in Polyethylene. Part II Crystallization fron Concentrated Solution," *Bell Sys. Tech. J.*, **40**, 1409 (1961).
8. J. I. Lauritzen, Jr., and J. D. Hoffman, "Theory of Formation of Polymer Crystals with Folded Chains in Dilute Solution," *J. Research Natl. Bur. Standards*, **64A**, 73 (1960).
9. R. Salovey, "Irradiation of Single Crystal Polyethylene," *J. Polymer Sci.*, **51**, S1 (1961).
10. A. Keller, personal communication.
11. A. Keller, W. Matreyek and F. H. Winslow, "Halogenation of Polyethylene Single Crystals," *J. Polymer Sci.*, **62**, 291 (1962).
12. P. R. Swan, "Polyethylene Unit Cell Variations with Branching," *J. Polymer Sci.*, **56**, 409 (1962).
13. J. D. MacKenzie, ed., *Modern Aspects of the Vitreous State*, Butterworths, London (1960).
14. F. Seitz and D. Turnbull, eds., *Solid State Physics*, Vols. 1 to 11, Academic Press, New York (1955 to date). (Numerous articles in these volumes review present concepts of the role of defects in solids.)
15. J. D. Hoffman and J. J. Weeks, "Specific Volume and Degree of Crystallinity of Semicrystalline Poly(Chlorotrifluoroethylene), and Estimated Specific Volumes of. the Pure Amorphous and Crystalline Phases," *J. Research Natl. Bur. Standards*, **60**, 465 (1958).
16. D. Turnbull and M. H. Cohen, "A Free Volume Model of the Amorphous Phase: The Glass Transition," *J. Chem. Phys.*, **34**, 120 (1961).
17. R. O. Simmons and R. W. Balluffi, "Measurement of Equilibrium Vacancy Concentrations in Aluminum," *Phys. Rev.*, **117**, 52 (1960).
18. R. O. Simmons and R. W. Balluffi, "Measurement of the Equilibrium Concentration of Lattice Vacancies in Silver near the Melting Point," *Phys. Rev.*, **119**, 600 (1960).
19. V. A. Kargin, A. I. Kitaigorodsky, and G. L. Slonimsky, "The Structure of Linear Polymers," *Colloid J. (USSR) (Eng. Trans.)*, **19**, 141 (1957).
20. J. Frenkl, *Kinetic Theory of Liquids*, Oxford University Press (1946).

21. M. Takayanagi, "Temperature Dispersion of Crystalline High Polymers," *High Polymers (Japan)*, **10**, 289 (1961) (in Japanese).

22. M. Takayanagi, M. Yoshino, and K. Hoashi, "Fine Structure and Viscoelastic Absorption of Crystalline High Polymers," *J. Japan Soc. Testing Materials*, **10**, 418 (1961) (in Japanese, English summary).

23. A. E. Woodward, "Transitions and Segmental Motion in High Polymers," *Trans. N. Y. Acad. Sci.*, **24**, 250 (1962).

24. A. H. Scott, D. J. Scheiber, A. J. Curtis, J. I. Lauritzen, Jr., and J. D. Hoffman, "Dielectric Properties of Semicrystalline Polychlorotrifluoroethylene," *J. Research Natl. Bur. Standards*, **66A**, 269 (1962).

25. L. Li-shen, N. S. Andreeva, and V. A. Kargin, "X-ray Studies of Polyethylene Monocrystals at Various Temperatures," *High Mole. Wt. Comps.*, **3**, 1238 (1961) (in Russian).

26. M. Takayanagi, T. Aramaki, G. Yoshino, and K. Hoashi, "Mechanism of Viscoelastic Absorption in Polyethylene at Higher Temperatures than Room Temperature," *J. Polymer Sci.*, **46**, 531 (1960).

27. H. Thurn, "Eine Methode zur Messung der dynamisch-mechanischen Eigenschaften von dunnen Filmen, Pulvern und Fasern aus hochpolymeren Stoffen," *Kolloid-Z.*, **173**, 72 (1960).

28. D. Hyndman and G. F. Origlio, "N.M.R. Absorption in Teflon Fibers," *J. Appl. Phys.*, **31**, 1849 (1960).

29. R. W. Kedzie, "N.M.R. in Gravitationally Oriented Polyethylene Crystal Mats," to be published.

30. E. S. Clark and L. T. Muus, "Partial disordering and crystal transitions in polytetrafluoroethylene," *Z. Krist.*, **117**, 119 (1962).

31. A. van Roggen, "Electronic Conduction of Polymer Single Crystals," *Phys. Rev Letters*, **9**, 368 (1962).

32. L. Esaki, "New Phenomenon in Narrow Germanium *p-n* Junctions," *Phys. Rev.*, **109**, 603 (1958).

33. I. Giaever, "Electron Tunneling Between Two Superconductors," *Phys. Rev. Letters*, **5**, 464 (1960).

APPENDIX I

Crystallographic Data for Various Polymers. III

The following table has been compiled by R. L. Miller and L. E. Nielsen and is a further revision and extension of their previous data tables.* The various entries are mostly self-explanatory. Unless otherwise noted all entries in a given row were obtained from the reference noted in column 2. The compilers of the table have made no attempt to assess the completeness of the table nor the validity of the data. The references listed in the table are not included in the author index.

References for Crystallographic Data Table

Note that these abbreviations have been used: Atti (Mem) = Atti accad. nazl. Lincei Mem., Classe sci. fis. mat. e nat; Atti = Atti accad. nazl. Lincei Rend., Classe sci. fis. mat. e nat.

1. H. Tadokoro, K. Kozai, S. Siki, and I. Nitta, *J. Polymer Sci.*, **26**, 379 (1957).
2. W. P. Slichter, *J. Polymer Sci.*, **35**, 77 (1959).
3. D. R. Holmes, C. W. Bunn, and D. J. Smith, *J. Polymer Sci.*, **17**, 159 (1955).
4. W. J. Dulmage, *J. Polymer Sci.*, **26**, 277 (1957).
5. A. Prietzschk, *Kolloid-Z.*, **156**, 8 (1958).
6. R. L. Miller, unpublished results.
7. G. Natta and P. Corradini, *J. Polymer Sci.*, **20**, 251 (1956); *Atti*, **19**, 229 (1955).
8. G. Natta, I. W. Bassi, and P. Corradini, *Makromol. Chem.*, **18–19**, 455 (1956).
9. G. Natta, P. Corradini, and I. W. Bassi, *Atti*, **19**, 404 (1955).
10. G. Natta, *SPE Journal*, **15**, 373 (1959).
11. C. W. Bunn and E. R. Howells, *Nature*, **174**, 549 (1954).
12. G. Natta, L. Porri, P. Corradini, and D. Morero, *Atti*, **20**, 560 (1956).
13. G. Natta, P. Corradini, and G. Dall Asta, *Atti*, **20**, 408 (1956).
14. C. W. Bunn, *Trans. Faraday Soc.*, **35**, 482 (1939).
15. E. R. Walter and F. P. Reding, *J. Polymer Sci.*, **21**, 561 (1956).
16. M. L. Miller and C. E. Rauhut, *J. Polymer Sci.*, **38**, 63 (1959).
17. G. Natta, *J. Polymer Sci.*, **16**, 143 (1955).
18. C. C. Price, M. Osgan, R. E. Hughes, and C. Shambelan, *J. Am. Chem. Soc.*, **78**, 690 (1956).
19. G. Natta, P. Corradini, and M. Cesari, *Atti*, **21**, 365 (1956).
20. C. Legrand, *Acta Cryst.*, **5**, 800 (1952).

* R. L. Miller and L. E. Nielsen, "Crystallographic Data for Various Polymers," *J. Polymer Sci.*, **44**, 391 (1960) and **55**, 643 (1961).

Crystallographic Data for Various Polymers

Polymer	Ref.[2]	Crystal system[a]	Space group[b]	a	b	c[c]	Angles[d] α, β, γ	Monomer units in cell	Density, g./cc. Cryst.	Density, g./cc. Amorph.	M.p., °C	Heat of fusion, kcal./monomer unit	Chain conformation[e]
Column No.: 1	2	3	4	5	6	7	8	9	10	11	12	13	14
A. HYDROCARBONS													
Polyethylene *28.05[89]	14	Ortho	D_{2h}^{16}	7.40	4.93	2.534		2	1.00	0.852[57]	110	1.88[85]	Z
	15			7.36	4.92	2.534			1.014		137[85]	1.84[86]	
(second form)	138	Pseudo-mono		4.05	4.85	2.54	$\gamma = 105$	1	0.965	0.855	141[206]	1.86[156]	
(single crystal)	72	Tri		7.84	5.56	120	63, 71, 82	48	1.013[f]				
Polypropylene: Isotactic	19,127	Mono	C_{2h}^{6}	6.65	20.96	6.50	$\beta = 99.3$	12	0.936	0.854[45]	176[10]	2.37[82]	3-1[67]
	6			6.666	20.87	6.488	$\beta = 98.2$		0.937			2.60[83]	
	131			6.64	20.88	6.51	$\beta = 98.7$						
II.	136	Mono	C_{2h}^{5}	6.69	20.98	6.504	$\beta = 99.5$	12	0.9323	0.8535			3-1
	170	Tri	C_i^1	13.36	6.50*	10.99	87, 108, 99	12					
	166	Hex	D_3^4 or D_3^6	12.74	12.74	6.35			0.88[167]				
Syndiotactic	196	Rho	D_2^5	6.38	6.38	6.33		3					
	67	Ortho		14.50	5.81	7.3		8	0.91[f]				
Polyethylidene	169	Ortho		14.5	5.8	7.4		8	0.90				H
	178	Rho		12.38	6.28	2.5							
Polystyrene *104.14*	67	Rho	C_{3h}^{6}	21.90	21.90	6.63		18	1.12[45]	1.04 to	240[10]	2.15[89]	3-1
	128			21.90	21.90	6.65			1.126	1.065[17]	250[90]	2.00[174]	
	6		D_{3d}^{6}	22.08	22.08	6.626			1.111				
	139									1.024			
Polybutene-1 I	35,126	Rho	D_{3d}^{6}	17.7	17.7	6.50		18	0.95[12]	0.87[45]	126[12]	3.33[82]	3-1[67]
II	9	Tet	S_4^1	7.49	7.49	6.85			0.96				4-1
	207	Tet		14.89	14.89	20.87		44	0.89				11-3
Polypentene-1 *70.13*	80					6.60					75[48]		
Polyhexene-1 *84.16*	48										80[9]		
											−55		

Crystallographic Data for Various Polymers (*continued*)

Column No.: Polymer 1	Ref. 2	Crystal system[a] 3	Space group[b] 4	a 5	b 6	c[c] 7	Angles[d] α, β, γ 8	Monomer units in cell 9	Density, g./cc. Cryst. 10	Density, g./cc. Amorph. 11	M.p., °C 12	Heat of fusion, kcal./monomer unit 13	Chain conformation[e] 14
Polyheptene-1 *98.18*	48										−40		
Polyoctene-1 *112.21*	48										−38		
Polydodecene-1 *168.31*	48										45		
Polyoctadecene-1 *252.47*											80[5] 100[90]		
Polyisobutene *56.10*	34 115 139	Ortho	D^4	6.94	11.96	18.63		16	0.937	0.9125[a]	128[9]	2.87[82]	H 8-5
Polyisopropylethylene	82									0.842	310	4.130	
Polyisobutylethylene	82										235	4.710	
Poly-alternating-ethylene-butene-2 *84.16*	191					9.15				0.87	135		
1,2-Polybutadiene, Syndiotactic *54.09*	7	Ortho	D_{2h}^{11}	10.98	6.60	5.14		4	0.963		154		Z
Isotactic	12	Rho	D_{3d}^6	17.3	17.3	6.5		18	0.96		120		3[152]
1,4-trans-Polybutadiene (Above 65°) *54.09*	37 154	Pseudo-hex Pseudo-hex		4.54 4.88	4.54 4.88	4.9 4.68		1	1.02		148[44]		Z
1,4-cis-Polybutadiene *54.09*	36 124	Mono	C_{2h}^6	4.60	9.50	8.60	β = 109	4	1.01				Z
1,4-Poly-2,3-dimethyl-1,3-butadiene *82.14*	104					4.35					1 260		
1,4-trans-Poly-1,1,4,4-tetramethyl-1,3-butadiene	183					4.8					265		
1,4-cis-Polyisoprene *68.11*	40	Mono	C_{2h}^5	12.46	8.89	8.10	β = 92	8	1.00[f]	0.906[3]	28[81]	1.05[81]	Z
1,4-trans-β-Polyisoprene *68.11*	23	Ortho	D_2^4	7.78	11.78	4.72		4	1.11[f]		65[46] 74[81]	3.04[81]	H

(*continued*)

Crystallographic Data for Various Polymers (*continued*)

| Polymer | Ref. | Crystal system[a] | Space group[b] | Unit cell parameters | | | | Monomer units in cell | Density, g./cc. | | M. p., °C. | Heat of fusion, kcal./monomer unit | Chain conformation[e] |
| | | | | a | b | c^c | Angles[d] α, β, γ | | Cryst. | Amorph. | | | |
Column No.: 1	2	3	4	5	6	7	8	9	10	11	12	13	14
1,4-*trans*-Polypentadiene *68.11*	37	Fseudo-hex		5.25	5.25	4.82		1	0.98		95		Z
Poly-3-methylbutene-1 *70.13*	189	Ortho		19.73	4.85	4.8		4			300[48]		4[152]
	9	Mono	S_4^1	9.55	8.54	6.84	$\gamma = 116.5$	4	0.93		310[90]		
Poly-4-methylpentene-1 *84.16*	55	Tet		18.66	18.66	13.80		28	0.813		235[48]		7-2[97]
	67			18.60	18.60	13.84			0.816[f]				
	94								0.828	0.838	250		
Poly-4-methylhexene-1 *98.18*	9					14.00					200		
Poly-5-methylhexene-1 *98.18*	67	Tet		19.64	19.64								7-2
	9	Hex		10.2	10.2	6.50			0.84		130		3-1[67]
Poly-5-methylheptene-1 *112.21*	67					6.40							3-1
	129										52		
Poly-3-cyclopentylpropene *110.19*											225[90]		
Poly-3-cyclohexylpropene *124.22*											230[90]		
Poly-4,4-dimethylpentene *98.18*											350[90]		
Poly-4,4-dimethylhexene *112.21*											350[90]		
Polyacetylene *26.04*	69	Pseudo-hex		4.2	4.2	2.43			1.15				Z
Polyvinylcyclopropane *68.11*	175					6.5					230		H
Polyvinylcyclopentane *180.15*	95	Tri	C_{4a}^6	10.5	7.4	6.6	92, 108, 99	3	0.986				
Polyvinylcyclohexane *110.19*	60	Tet		21.9	21.9	6.5		16	0.95		305[90]		4-1
	67			21.76	21.76	6.50							
Poly-o-methylstyrene *130.18*	95	Tri	C_{4a}^{12}	11.6	7.8	6.6	92, 108, 98	3	0.982				4-1
	67,125	Tet		19.01	19.01	8.10		16	1.071[f]		360[74]		4-1

Crystallographic Data for Various Polymers (continued)

Polymer	Ref.	Crystal system[a]	Space group[b]	Unit cell parameters a	b	c°	Angles[d] α, β, γ	Monomer units in cell	Density Cryst.	Amorph.	M. p., °C	Heat of fusion, kcal./monomer unit	Chain conformation[e]
Column No.: 1	2	3	4	5	6	7	8	9	10	11	12	13	14
Poly-m-methylstyrene *130.18*	67,80 163	Tet	S_4^1	19.81	19.81	21.74 57.1		44	1.010[f]		215[74]		11-3 29-8
Poly-p-methylstyrene *130.18*	103 103					57.0 12.9							
Poly-2,4-dimethylstyrene *144.21*	75										310		
	74										350		
Poly-2,5-dimethylstyrene *144.21*	75										330		
	74										340		
Poly-3,5-dimethylstyrene *144.21*	75										290		
Poly-3,4-dimethylstyrene *144.21*	75										240		
Poly-3-phenylpropene-1 *118.17*	67					6.40					230[90]		3-1
Poly-3-phenylbutene *132.20*											360[90]		
Poly-4-phenylbutene *132.20*	67					6.55					160[90]		3-1
Poly-α-vinylnaphthalene *154.20*	67,122	Tet	$\bar{C}_{4h}^{12}$	21.20	21.20	8.10		16	1.124[f]		360[75]		4-1
Poly-p-xylene *82.14*	142								1.14		375	7.20	
Polyallylbenzene	187										208		
Poly-o-allyltoluene	187										290		
Poly-m-allyltoluene	187										180		
Poly-p-allyltoluene	187										240		
Poly-2-allyl-p-xylene	187										338		
Poly-4-allyl-o-xylene	187										275		
Poly-5-allyl-m-xylene	187										252		
Poly-4-phenyl-1-butene	187										68		

(continued)

Crystallographic Data for Various Polymers (*continued*)

Polymer	Ref.	Crystal system[a]	Space group[b]	a	b	c[c]	Angles[d] α, β, γ	Monomer units in cell	Cryst.	Amorph.	M. p., °C	Heat of fusion, kcal./ monomer unit	Chain conformation[e]
Column No.: 1	2	3	4	5	6	7	8	9	10	11	12	13	14
Poly-4-(o-tolyl)-1-butene	187										235		
Poly-4-(p-tolyl)-1-butene	187										196		
B. VINYL POLYMERS													
Polyvinyl chloride	7	Ortho	D_{2h}^{11}	10.6	5.4	5.1		4[f]	1.44				Z
62.50	64	Ortho		10.11	5.27	5.12			1.522[f]	1.39[41]	190	0.66[145]	
	140	Mono		10.65	5.15*	5.20	β = 90	4	1.455		212[142]		
Polyvinylidene chloride	32	Mono		13.69	4.67*	6.296	β = 55.2	4	1.949	1.66[44]	190[44]		4
96.95	33	Mono		22.54	4.68*	12.53	β = 84.2	16	1.959				
Polytetrafluoroethylene below 20C	11,154	Pseudo-hex		5.54	5.54	16.8	γ = 119.5		2.40[f]				H
100.02	209	Tri					γ = 119.3						
above 20C	11	Hex		5.59	5.59	16.88			2.36[f]				13-6
	209	Hex		5.61	5.61	16.8			2.304[65]		330	0.68[91]	15-2
	21	Hex		5.66	5.66	19.50			2.10		330[101]		15-7
Polychlorotrifluoroethylene	208	Hex		6.5	6.5	35		14	2.194[9]	2.089	220[9]	1.20[88]	14-1
	116			6.34	6.34	35				2.032			
	139									1.925			
	205					43		16					16-1
Polyvinyl alcohol	29	Mono	C_{2h}^{2}	7.81	2.52*	5.51	β = 91.7	2[f]	1.35[f]	1.291[1]			Z
	132			7.805	2.533	5.485	β = 92.2			1.26[3]			
	184								1.345	1.269			
Polymethylmethacrylate Isotactic	30	Pseudo-ortho		21.08	12.17	10.55		20	1.23	1.22[31]	160[31]		5-2
100.11													
Syndiotactic	30									1.19[31]	200[31]		10-4
Polyisopropylacrylate Isotactic	67					6.5					162[120]		3-1
114.14													

Crystallographic Data for Various Polymers (continued)

Polymer	Ref.	Crystal system[a]	Space group[b]	a	b	c°	Angles[d] α, β, γ	Monomer units in cell	Density, g./cc. Cryst.	Amorph.	M.P. °C.	Heat of fusion, kcal./monomer unit	Chain conformation[e]
Column No.: 1	2	3	4	5	6	7	8	9	10	11	12	13	14
Poly-n-butylacrylate	202										47		
Poly-L-butylacrylate	67					6.45					193[16]		3-1
128.17	192					6.48							3-1
Polyisobutylacrylate	202			17.92		6.48			1.04		200		3-1
	192				10.50	6.42							
	202					6.42							
Poly-sec-butylacrylate	192			17.92	17.92	6.49			1.24		81		3-1
	202					6.4							
Poly-n-propyl thiolacrylate	200					6.42							
Polyisopropyl thiolacrylate	200					6.42							
Polyisobutyl thiolacrylate	201					6.42							
Poly-sec-butyl thiolacrylate	201			17.92	10.34	6.35			1.06		130		3-1
Polyvinyl fluoride	62	Hex		4.93	4.93	2.53		1	1.44		200[79]	1.80[79]	Z
46.04													
Polyvinylidene fluoride	171	Mono		5.02	25.4*	4.62	β = 107						H
64.02													
Polyacrylonitrile													
Syndiotactic	76	Hex	C_{2v}^{16}	5.99	5.99			4	1.11		317[77]	1.16[77]	Z
	133	Ortho		10.20	6.10	5.10		8	1.27				
	162	Ortho		18.1	6.12	5.00		4	1.135				
	210	Ortho		10.55	5.80				1.54				
53.06													
Isotactic	133	Tet		4.74	4.74	2.55							
Polyvinylidene bromide	33	Mono		25.88	4.77*	13.87	β = 70.2	16	3.065				Z
185.87													
Polychloroprene	109	Ortho		9.0	8.23	4.79		4	1.657[f]		80[81]	2.00[81]	
88.54													
Poly-1-methoxybutadiene	112										118		Z
84.11													
Polytetrafluoroallene	38	Tet	$C_4^2 - D_4^3$	6.88	6.88	15.4		8	2.02		126		Z
112.03													

(continued)

Crystallographic Data for Various Polymers (*continued*)

Polymer	Ref.	Crystal system[a]	Space group[b]	a	b	c°	Angles[d] α, β, γ	Monomer units in cell	Density Cryst.	Density Amorph.	M.p., °C	Heat of fusion, kcal./monomer unit	Chain conformation[e]
Column No.: 1	2	3	4	5	6	7	8	9	10	11	12	13	14
Poly-*o*-fluorostyrene	70	Rho	C_{3v}^6	20.4	20.4	6.63		18	1.526f		270[75]		
	73	Rho		22.1	22.1	6.65			1.296f				
	123			22.15	22.15	6.63			1.29				
122.14	168												
Poly-*p*-fluorostyrene	70					8.30					265[75]		3-1
136.16	168												
Poly-*o*-methyl-*p*-fluorostyrene	70					8.05							4-1
Polytrimethylsilyl styrene *136.16*	103					60.4					284		3-1[67]
Polytrimethylallylsilane	70					6.50					360[71]		H
Polyallylsilane	70					6.45					128[71]		H
Poly-5-(trimethylsilyl)-pentene-1	70					6.55					133[71]		H
Poly-2-methyl-4-fluorostyrene *136.16*	168										360[75]		4-1
Poly-*n*-isopropyl acrylamide *113.16*	102								1.118	1.070	200		3-1
Poly-(*N,N*-di-*n*-butyl acrylamide) *183.29*	93	Hex		26.3	26.3	6.3		12	1.06				

C. POLYESTERS

Polymer	Ref.	Crystal system[a]	Space group[b]	a	b	c°	Angles[d] α, β, γ	Monomer units in cell	Density Cryst.	Density Amorph.	M.p., °C	Heat of fusion, kcal./monomer unit	Chain conformation[e]
Polyethylene terephthalate	27	Tri	C_i^1	4.56	5.94	10.75	98, 118, 112	1	1.455	1.3357	265	5.82[87]	z
192.16	195			5.54	4.14	10.86	107, 112, 92		1.498	1.337			
Polytrimethylene terephthalate *206.18*	99										233	3.98[188] 5.40[187]	
Polytetramethylene terephthalate *220.21*	99								1.08[111]		232	7.6[111]	
Polypentamethylene terephthalate *234.23*	99										134		
Polyhexamethylene terephthalate	68	Tri		4.57	6.10	15.40	105, 98, 114		1.131f		160[1]	8.3[91]	
248.27	99										154	8.44[189]	

Crystallographic Data for Various Polymers (continued)

Polymer	Ref.	Crystal system[a]	Space group[b]	a	b	c	Angles[d] α, β, γ	Monomer units in cell	Density, g./cc. Cryst.	Density, g./cc. Amorph.	M.p., °C	Heat of fusion, kcal./monomer unit	Chain conformation[e]
Column No.: 1	2	3	4	5	6	7	8	9	10	11	12	13	14
Polyoctamethylene terephthalate *276.31*	99										132		
Polynonamethylene terephthalate *290.34*	99										85		
Polydecamethylene terephthalate *304.37*	68, 99	Tri		4.62	6.30	20.10	107, 96, 113	1	1.012[f]		138[31]	11.00[31]	
Poly-1,4-cyclohexylenedimethylene terephthalate-cis	199	Tri		6.37	6.63	14.2	89, 47, 111	1	1.265		129		
Poly-1,4-cyclohexylenedimethylene terephthalate-trans	199	Tri		6.02	6.01	13.7	89, 53, 112	1	1.303				
Polyethylene isophthalate *192.16*	111								1.358	1.346	240		
Polytrimethylene isophthalate *206.18*	111										132		
Polytetramethylene isophthalate *220.21*	111								1.309	1.268	152.5	10.1	
Polyhexamethylene isophthalate *248.27*	111										140		
Polyethylene succinate *144.12*	105			9.05	11.09	8.32	$\beta = 102.8$	4	1.175[f]				H
	108	Mono		5.0	7.4	8.32							
Polyethylene adipate *172.18*	105	Mono		25.7	30.7	11.71		40	1.274[f]		52[27]		Z
	108			5.0	7.4	11.71	$\beta = 103.8$						
	203	Mono	C_{2A}^{5}	5.47	7.23	11.72	$\beta = 113.5$	2	1.34				
	211	Mono		7.26	5.40	10.85	$\alpha = 67.7$						
Polyethylene suberate *200.23*	108	Mono		5.0	7.4	14.1					55[27]		
	203	Mono	C_{2A}^{5}	5.51	7.25	14.28	$\beta = 114.5$	2					
Polyethylene sebacate *228.28*	105			25.7	30.7	16.67	$\beta = 103.8$	40	1.187[f]			6.11[158]	Z
	100	Mono		5.5	15	16.9	$\beta = 65$	4	1.120[f]				
	108	Mono		5.0	7.4	16.83							
Polyethylene azelate *214.25*	105	Mono		25.7	30.7	31.2	$\beta = 103.8$	80	1.190[f]		76[147]	6.95[147]	Z
	108	Ortho		7.45	4.97	31.5		4	1.220[f]				Z

(continued)

Crystallographic Data for Various Polymers (*continued*)

Polymer	Ref.	Crystal system[a]	Space group[b]	Unit cell parameters a	b	c°	Angles[d] α, β, γ	Monomer units in cell	Density, g./cc. Cryst.	Amorph.	M.p., °C.	Heat of fusion, kcal./ monomer unit	Chain conformation[e]
Column No.: 1	2	3	4	5	6	7	8	9	10	11	12	13	14
Polytrimethylene succinate *158.15*	107	Mono		5.0	7.4	15.2					47		Z
Polytrimethylene glutarate *172.18*	107	Mono		5.0	7.4	15.4					39		Z
Polytrimethylene adipate *186.20*	107	Mono		5.0	7.4	21.5					38		Z
Polytrimethylene pimelate *200.23*	107	Mono		5.0	7.4	23.6					37		Z
Polytrimethylene suberate *214.25*	107	Mono		5.0	7.4	26.1					41		Z
Polytrimethylene azelate *228.28*	107	Mono		5.0	7.4	27.7					50		Z
Polytrimethylene sebacate *242.31*	105	Tet		31.2	31.2	33.5			1.090		53		Z
	107	Mono		5.0	7.4	31.3					59		Z
Polytrimethylene-1,9-dicarboxylate	107	Mono		5.0	7.4	32.4					61		Z
Polytrimethylene-1,10-dicarboxylate	107	Mono		5.0	7.4	35.8					76		Z
Polytrimethylene-1,16-dicarboxylate	107	Mono		5.0	7.4	51.6							Z
Polydecamethylene oxylate *228.38*	106	Mono		5.28	7.00	17.0							Z
Polydecamethylene succinate *256.33*	106	Mono		5.0	7.4	19.6							Z
Polydecamethylene glutarate *270.36*	106	Mono		5.0	7.4	41.6							Z
Polydecamethylene adipate *284.38*	106	Mono		5.0	7.4	22.1					80[81]	10.2[81]	Z
Polydecamethylene suberate *312.44*	106	Mono		5.0	7.4	24.6							Z
Polydecamethylene azelate *326.46*	106	Mono		5.0	7.4	51.7					69[127]	10.0[81]	z
Polydecamethylene sebacate *340.49*	106	Mono		5.0	7.4	27.1					80[137]	12.0[81] 12.3[160]	Z

Crystallographic Data for Various Polymers (continued)

Polymer	Ref.	Crystal system[a]	Space group[b]	Unit cell parameters a	b	c[c]	Angles[d] α, β, γ	Monomer units in cell	Density, g./cc. Cryst.	Amorph.	M.p. °C	Heat of fusion, kcal./monomer unit	Chain conformation[e]
Column No.: 1	2	3	4	5	6	7	8	9	10	11	12	13	14
Polyethylene ω-hydroxydeconoate	108	Ortho		7.45	4.97	27.1		4					
Polydiethylene oxide sebacate	105	Tet		17.6	17.6	38.0			1.128		210		
Poly-trans-methylmethoxy-ethylene *72.10*	141										243		
Poly-trans-methylethoxy-ethylene *86.13*	141										226		
Poly-trans-methylisobutoxy-ethylene *114.18*	141					13.8					217		
Poly-trans-ethoxymethoxy-ethylene *102.13*	141												
Poly-p-xylenediamine sebacate	204	Tri		5.74	4.87	20.6	76, 55, 65	1	1.169				Z
D. POLYAMIDES													
Nylon 4.10	153										254		
Nylon 4.12	153										245		
Nylon 5.9	2					19.5							
Nylon 6.6 α	25 / 54	Tri	C_i^1	4.9	5.4	17.2	48, 77, 63	1	1.24 / 1.220	1.09[2] / 1.069	265[2]	11.10[82] / 9.7[155] / 8.79[216]	Z
β	25	Tri	C_i^1	4.9	8.0	17.2	90, 77, 67	2	1.248[f]				
Nylon 6.10 α *282.42*	25 / 153	Tri	C_i^1	4.95	5.4	22.4	49, 76, 63	1	1.16[f] / 1.17		228[51] / 216	7.32[160]	Z
β	54 / 25	Tri	C_i^1	4.9	8.0	22.4	90, 77, 67	2	1.152 / 1.196[f]	1.041			
Nylon 7.7	63 / 65	Pseudo-hex	C_*^1	4.82	19.0* / 18.95	4.82	β = 60	1	1.108		205[2]		Z
Nylon 8.10	153										207		

(continued)

Crystallographic Data for Various Polymers (*continued*)

Polymer	Ref.	Crystal system[a]	Space group[b]	— Unit cell parameters				Monomer units in cell	Density, g./cc.		M. p., °C	Heat of fusion kcal./monomer unit	Chain conformation[e]
				a	b	c	Angles[d] α, β, γ		Cryst.	Amorph.			
Column No.: 1	2	3	4	5	6	7	8	9	10	11	12	13	14
Nylon 9.9	110	Pseudo-hex				24.0		1			175		
γ	65	Pseudo-hex											
Nylon 10.6	110	Pseudo-hex				20.0					230		
Nylon 10.9	137	Pseudo-hex						2			214	8.76[159]	
γ *324.49*	65												
Nylon 10.10	110	Tri				25.6		1			196	8.29[159]	
338.52	65												
	137												
Nylon 6	3	Mono	C_2^2	9.56	17.2*	8.01	$\beta = 67.5$	8	1.23		216	7.82[160]	
	26	Mono	C_2^2	4.81	17.10	7.61	$\beta = 79.5$	4[f]	1.21		215	4.96[18]	Z
	135			9.65	17.2	8.11	$\beta = 66.3$				223[253]	4.32[216]	
	65							2					
	212	Mono	C_2^2	9.45	8.02	17.08	$\gamma = 68$						
	213								1.23	1.10			
Nylon 7	61	Tri	$C_1–C_i^1$	4.9	5.4	9.85	49, 77, 63	1[f]	1.211[f]		225[146]		
	65							1	1.20[153]		217[153]		
											233[215]		
Nylon 8	61	Hex		4.9	4.9	21.7		2	1.038[f]		185[146]		
	65	Pseudo-hex	C_2^2						1.18[153]		202[153]		
	121	Mono											
Nylon 9	65							1			194[146]		
	177										209		
Nylon 10	61	Hex		4.9	4.9	26.5		1	1.019[f]		177[146]		
	65	Pseudo-hex	C_2^2								192[177]		
Nylon 11	59	Tri		9.6	4.2	15.0	72, 90, 64	2	1.192[f]		194[161]		
	61	Tri	$C_1–C_i^1$	4.9	5.4	14.0	49, 77, 63	2	1.228[f]		182[146]		
	65												
Nylon 12	177										179		
Nylon 13	177										183		
Nylon 22	177										145		

Crystallographic Data for Various Polymers (*continued*)

E. POLYETHERS

Column No.:	Polymer 1	Ref. 2	Crystal system[a] 3	Space group[b] 4	Unit cell parameters			Angles[d] α, β, γ 8	Monomer units in cell 9	Density, g./cc.		M.p., °C. 12	Heat of fusion, kcal./monomer unit 13	Chain conformation[e] 14
					a 5	b 6	c[c] 7			Cryst. 10	Amorph. 11			
	Polyvinylmethyl ether *58.07*	176 67	Rho	D_{3d}^6	16.20	16.20	6.50 6.30		18			144[114]		3-1
	Polyvinylethyl ether *72.10*	114										86		
	Polyvinyl-*n*-propyl ether *86.12*	114										76		
	Polyvinylisopropyl ether *86.12*	114 161	Tet		17.2	17.2	35.5					190		17-5
	Polyvinyl-*n*-butyl ether *100.14*	114										64		
	Polyvinylisobutyl ether *114.17*	8 114	Ortho		16.8	9.70	6.50		6[f]	0.940		115[46] 165		H 3-1
	Polyvinyl-*t*-butyl ether *100.14*	114										260		
	Polyvinylneopentyl ether *100.14*	114 161	Ortho		18.2	10.5	6.50					216		3-1
	Polyvinylbenzyl ether	114										162		
	Polyvinyl-2-chloroethyl ether	114										150		
	Polyvinyl-2-methoxyethyl ether	114										73		
	Polyvinyl-2,2,2-trifluoroethyl ether	114										128		
	Polymethylpropenyl ether	113										287		
	Polyethylpropenyl ether	113										230		
	Poly-*n*-propylpropenyl ether	113										168		
	Polytetrahydrofuran	180										37		
	Polyoxacyclobutane (polytrimethylene oxide) *58.08*	147										35		
	Poly-3,3-bisdimethyl oxacyclobutane *86.13*	147										47		

(*continued*)

Crystallographic Data for Various Polymers (*continued*)

Polymer	Ref.	Crystal system[a]	Space group[b]	a	b	c[c]	Angles[d] α, β, γ	Monomer units in cell	Cryst.	Amorph.	M.p., °C	Heat of fusion, kcal./monomer unit	Chain conformation[e]
Column No.: 1	2	3	4	5	6	7	8	9	10	11	12	13	14
Poly-3,3-bisethoxymethyloxacyclo-butane *174.23*	151										83		
Poly-3,3-bishydroxymethyloxacyclo-butane *118.13*	151										280		
Poly-3,3-bisfluoromethyloxacyclo-butane *122.11*	149										135		
Poly-3,3-bischloromethyloxacyclo-butane	148												
α *155.03*						4.8					180		
β											190[173]		
Poly-3,3-bisbromomethyloxacyclo-butane *243.95*	172	Mono	C_2^1–C_2^d	6.85	11.42	4.75	β = 109.8	2	1.47		220		
Poly-3,3-bisiodomethyloxacyclo-butane 337.98*	150										290		
F. CELLULOSICS													
Cellulose I	22	Mono		8.35	10.3	7.9	β = 84	2	1.592				Z
	28				10.34*								
	20												
Cellulose II	98	Mono	C^2	8.20	10.3*	7.90	β = 83.3		1.625				
	22	Mono		8.14	10.34*	9.14	β = 62	2	1.583				Z
	28												
Cellulose III	28	Mono		8.02	10.3*	9.03	β = 62.8		1.62				
Cellulose IV	98	Mono		8.12	10.3*	7.99	β = 90		1.62				
Cellulose X	98	Mono		8.10	10.3*	8.16	β = 78.3		1.61 / 1.615				
Cellulose trinitrate	24	Ortho		12.25	25.4*	9.0		4[f]	2.07[f]		697[77]	0.9–1.5[77]	Z
											700[81]		
Cellulose 2.4.4-nitrate	81										617	1.35	
Cellulose triacetate	4	Pseudo-ortho	C_2^2	24.5	11.6*	10.43		4	1.30		306[144]		
Cellulose tripropionate	144										234		Z

Crystallographic Data for Various Polymers (*continued*)

Polymer	Ref.[2]	Crystal system[a] [3]	Space group[b] [4]	Unit cell parameters — a [5]	b [6]	c[c] [7]	Angles[d] α, β, γ [8]	Monomer units in cell [9]	Density g./cc. — Cryst. [10]	Amorph. [11]	M.p., °C. [12]	Heat of fusion, kcal./monomer unit [13]	Chain conformation[e] [14]
Cellulose tributyrate	118										207	3.0	
Cellulose tricaprylate	117										116	3.1	
Cellulose trivalerate	144										122		
Cellulose tricaproate	144										94		
Cellulose triheptylate	144										88		
Cellulose tricaprate	144										88		
Cellulose trilaurate	144										91		
Cellulose trimyristate	144										106		
Cellulose tripalmitate	144										105		
G. OTHER POLYMERS													
Polyoxymethylene	42, 134	Hex	$C_3^2 - C_3^3$	4.46	4.46	17.30		9	1.506	1.25	181	0.890[91], 1.59[186]	9-5
Polyethylene oxide	109			9.5	19.5*	12.0		36	1.205[f]				
	188	Mono		7.95	13.11	19.39	$\beta = 101$	28	1.23		66[81]	1.98[81]	7-2
	194	Mono		8.03	13.09	19.52	$\beta = 124.6$		1.47[182]				
Polypropylene oxide	13	Ortho	$C_{2v}^9 - D_2^4$	10.52	4.67	7.16		4	1.096		75[18]		Z
	41	Ortho		10.52	4.68	7.10			1.102[f]				
	78	Ortho	D_2^4	10.40	4.64	6.92	$\beta = 125.1$		1.154[f]				
	139									0.998			
Polytrimethylene oxide	181										34		
Polyhexamethylene oxide	181										58		
Polydecamethylene oxide	181										79		
Poly-2-vinylpyridine	185					6.7					212		
Polystyrene oxide	164										149		
Polybutadiene oxide	164										74		
Polyacetone	214	Tet	S_4^{16}	14.65	14.65	10.22		28			60		7-2
Polyketone alternating Et/CO *56.06*	193	Ortho	D_{2h}^{16}	7.97	4.76	7.57		4	1.296				Z
Poly-t-butylvinylketone	198										150		
Polyepibromohydrin (bromomethyl ethylene oxide)	164										112		

(*continued*)

Crystallographic Data for Various Polymers (continued)

Polymer	Ref.	Crystal system[a]	Space group[b]	Unit cell parameters a	b	c[c]	Angles[d] α, β, γ	Monomer units in cell	Density, g./cc. Cryst.	Amorph.	M. p., °C.	Heat of fusion, kcal./monomer unit	Chain conformation[e]
Column No.: 1	2	3	4	5	6	7	8	9	10	11	12	13	14
Polyepichlorohydrin chloromethyl ethylene	119	Ortho	$D_2^{40r}-C_{2v}^9$	12.14	4.90	7.05					117		
	194					7.07					121		
Polyepifluorohydrin (fluoromethyl ethylene oxide)	119										68		
Poly-trans-2-butene oxide	165										114		
Poly-cis-2-butene oxide	165										162		
Polyacetaldehyde	92	Tet	C_{4h}^6	14.60	14.60	4.79		16	1.14				4-1
	97			14.63	14.63	4.79			1.14				
Polypropionaldehyde	92	Tet	C_{4h}^6	17.52	17.52	4.78		16	1.05				4-1
Poly-n-butyraldehyde	92	Tet	C_{4h}^6	20.01	20.01	4.78		16	1.00				4-1
Polyisobutyraldehyde	96	Tet		20.6	20.6	5.2			0.997				4-1
Polyisovaleraldehyde	96	Tet		20.6	20.6	5.2							
Poly-bisphenol A	5	Ortho	$D_2^2-D_2^3$	11.9	10.1	21.5		16	1.037[f]		207		Z
Polydimethylene oxide *60. 05*	180							8	1.30		62		
Polydecamethylene oxide *172.26*	180												
Polymethylisobutoxyethylene	67					13.77					72		7-2
Polydiketene	190										115		
Poly-β-propiolactone	190										122		
Polytrioxane	190	Hex		9.26	9.26	8.60					198		
Polydimethylsiloxane	197	Mono		13.0	8.3*	7.75	β = 60	6	1.07	0.98			
Polyhexamethylene sulfone	39	Mono		9.88	9.26	18.24	β = 121.7	8	1.39[f]				Z
Polyhexamethylene pentamethylene sulfone	39	Mono		9.88	9.26	34.00	β = 121.7	8	1.42[f]				Z
Polyhexamethylene tetramethylene sulfone	39	Mono		9.88	9.26	15.68	β = 121.7	4	1.46[f]				Z

Crystallographic Data for Various Polymers (*continued*)

Polymer	Ref.	Crystal system[a]	Space group[b]	Unit cell parameters a	b	c[c]	Angles[d] α, β, γ	Monomer units in cell	Density, g./cc. Cryst.	Amorph.	M. p., °C	Heat of fusion, kcal./monomer unit	Chain conformation[e]
Column No.: 1	2	3	4	5	6	7	8	9	10	11	12	13	14
Polypentamethylene sulfone	39	Mono		9.88	9.26	7.76	β = 121.7	4	1.48[f]				Z
Polypentamethylene tetramethylene sulfone	39	Mono		9.88	9.26	28.33	β = 121.7	8	1.53[f]				Z
Polydimethyl ketene													
I	179					8.8		1			250		
	130										255		
II	179										170		
Polymethylene sulfide	181										245		
Polydimethylene sulfide	181										190		
Polytrimethylene sulfide	181										100		
Polytetramethylene sulfide	181										67		
Polyhexamethylene sulfide	181										79		
Polydecamethylene sulfide	181										91		
Polyethylene disulfide	134					8.8							8-2
Polyethylene tetrasulfide	134					4.32							4-1

[a] Tri = triclinic, mono = monoclinic, ortho = orthorhombic, tet = tetragonal, rho = rhombohedral, and hex = hexagonal.

[b] See *International Tables for X-Ray Crystallography*, Kynoch Press, Birmingham, England, 1952.

[c] Unless otherwise noted (by *) c is the fiber axis (chain direction).

[d] Unless otherwise noted, all angles are 90°.

[e] H = helix, Z = zigzag, figures indicate helix pitch.

[f] Calculated from tabulated data.

[g] Molecular weight of monomer unit.

* See footnote c.

21. H. S. Kaufman, *J. Am. Chem. Soc.*, **75**, 1477 (1953).

22. P. H. Hermans, *Physics and Chemistry of Cellulose Fibres*, Elsevier, New York (1949).

23. C. W. Bunn, *Chemical Crystallography*, Clarendon Press, Oxford (1946).

24. H. S. Peiser, H. P. Rooksby, and A. J. C. Wilson, *X-ray Diffraction by Polycrystalline Materials*, Chapman and Hall, London (1955).

25. C. W. Bunn and E. V. Garner, *Proc. Roy. Soc. (London)*, **A189**, 39 (1947).

26. A. Okada, *Chem. High Polymers (Japan)*, **7**, 122 (1950).

27. R. De P. Daubeny, C. W. Bunn, and C. J. Brown, *Proc. Roy. Soc. (London)*, **A226**, 531 (1954).

28. H. J. Wellard, *J. Polymer Sci.*, **13**, 471 (1954).

29. C. W. Bunn, *Nature*, **161**, 929 (1948).

30. J. D. Stroupe and R. E. Hughes, *J. Am. Chem. Soc.*, **80**, 1768 (1958).

31. T. G. Fox, B. S. Garrett, W. E. Goode, S. Gratch, J. F. Kincaid, A. Spell, and J. D. Stroupe, *J. Am. Chem. Soc.*, **80**, 1768 (1958).

32. R. C. Reinhardt, *Ind. Eng. Chem.*, **35**, 422 (1943).

33. S. Narita and K. Okuda, *J. Polymer Sci.*, **38**, 270 (1959).

34. C. S. Fuller, S. J. Frosch, and N. R. Pape, *J. Am. Chem. Soc.*, **62**, 1905 (1940).

35. G. Natta, P. Corradini, and I. W. Bassi, *Makromol. Chem.*, **21**, 240 (1956).

36. G. Natta and P. Corradini, *Angew. Chem.*, **68**, 615 (1956).

37. G. Natta, P. Corradini, and L. Porri, *Atti*, **20**, 728 (1956).

38. J. D. McCullough, R. S. Bauer, and T. L. Jacobs, *Chem. & Ind. (London)*, **1957**, 706.

39. H. D. Noether, *J. Polymer Sci.*, **25**, 217 (1957).

40. C. W. Bunn, *Proc. Roy. Soc. (London)*, **A180**, 40 (1942).

41. C. Shambelan, Ph.D. Thesis, U. of Penna., 1959, *Dissertation Abstr.*, **20**, 120 (1959).

42. C. F. Hammer, T. A. Koch, and J. F. Whitney, *J. Appl. Polymer Sci.*, **1**, 169 (1959).

43. W. Goggin and R. Lowry, *Ind. Eng. Chem.*, **34**, 327 (1942).

44. G. Natta, *Chem. & Ind. (London)*, **1957**, 1520.

45. G. Natta, P. Pino, P. Corradini, F. Danusso, E. Mantica, G. Mazzanti, and G. Moraglio, *J. Am. Chem. Soc.*, **77**, 1708 (1955).

46. R. Boyer, *Compt. Rend. de la 2E Reunion de Chimie Physique*, (June 2–7, 1952, Paris), p. 383.

47. D. E. Roberts and L. Mandelkern, *J. Am. Chem. Soc.*, **80**, 1289 (1958).

48. F. P. Reding, *J. Polymer Sci.*, **21**, 547 (1956).

49. J. D. Hoffman and J. J. Weeks, *J. Research Natl. Bur. Standards*, **60**, 465 (1958).

50. L. Wood, N. Bekkedahl, and R. E. Gibson, *J. Chem. Phys.*, **13**, 475 (1945).

51. R. Beaman and F. Cramer, *J. Polymer Sci.*, **21**, 223 (1956).

52. H. Starkweather, Jr., G. Moore, J. Hansen, T. Roder, and R. Brooks, *J. Polymer Sci.*, **21**, 189 (1956).

53. R. Wiley, *Ind. Eng. Chem.*, **38**, 959 (1946).

54. H. W. Starkweather, Jr., and R. E. Moynihan, *J. Polymer Sci.*, **22**, 363 (1956).

55. F. C. Frank, A. Keller, and A. O'Connor, *Phil. Mag.*, **4**, 200 (1959).

56. G. Natta, P. Corradini, and I. W. Bassi, *Atti*, **23**, 363 (1958).

57. L. E. Nielsen, unpublished results.

58. P. W. Teare and D. R. Holmes, *J. Polymer Sci.*, **24**, 496 (1957).

59. K. Little, *Br. J. Appl. Phys.*, **10**, 225 (1959).

60. G. Natta, P. Corradini, and I. W. Bassi, *Makromol. Chem.*, **33**, 247 (1959).

61. W. P. Slichter, *J. Polymer Sci.*, **36**, 259 (1959).

62. R. C. Golike, *J. Polymer Sci.*, **42**, 583 (1960).

63. Y. Kinoshita, *Makromol. Chem.*, **33**, 21 (1959).

64. P. H. Burleigh, *J. Am. Chem. Soc.*, **82**, 749 (1960).

65. Y. Kinoshita, *Makromol. Chem.*, **33**, 1 (1959).

66. R. E. Moynihan, *J. Am. Chem. Soc.*, **81**, 1045 (1959).

67. G. Natta, *Makromol. Chem.*, **35**, 94 (1960).

68. J. Bateman, R. E. Richards, G. Farrow, and I. M. Ward, *Polymer*, **1**, 63 (1960).

69. P. Corradini, *Atti*, **25**, 517 (1958).

70. G. Natta, P. Corradini, and I. W. Bassi, *Gazz. Chim. Ital.*, **89**, 784 (1959).

71. G. Natta, G. Mazzanti, P. Longi, and F. Bernardini, *Chim. e Ind. (Milan)*, **40**, 813 (1958).

72. W. D. Niegisch and P. R. Swan, *J. Appl. Phys.*, **31**, 1906 (1960).

73. P. Corradini, G. Natta, and I. W. Bassi, *Angew. Chem.*, **70**, 598 (1958).

74. D. Sianesi, G. Natta, and P. Corradini, *Gazz. Chim. Ital.*, **89**, 775 (1959).

75. G. Natta, F. Danusso, and D. Sianesi, *Makromol. Chem.*, **28**, 253 (1958); and D. Sianesi, M. Rampichini, and F. Danusso, *Chim. e Ind. (Milan)*, **41**, 287 (1959).

76. G. Natta, G. Mazzanti, and P. Corradini, *Atti*, **25**, 3 (1958).

77. W. R. Krigbaum and N. Tokita, *J. Polymer Sci.*, **43**, 467 (1960).

78. E. Stanley and M. Litt, *J. Polymer Sci.*, **43**, 453 (1960).

79. D. I. Sapper, *J. Polymer Sci.*, **43**, 383 (1960).

80. P. Corradini and P. Ganis, *J. Polymer Sci.*, **43**, 311 (1960).

81. L. Mandelkern, *Chem. Rev.*, **56**, 903 (1956).

82. J. R. Schaefgen, *J. Polymer Sci.*, **38**, 549 (1959).

83. F. Danusso, G. Moraglio, and E. Flores, *Atti*, **25**, 520 (1958).

84. R. C. Wilhoit and M. Dole, *J. Phys. Chem.*, **57**, 14 (1953).

85. F. A. Quinn, Jr., and L. Mandelkern, *J. Am. Chem. Soc.*, **80**, 3178 (1958).

86. F. W. Billmeyer, Jr., *J. Appl. Phys.*, **28**, 1114 (1957).

87. M. Dole, *J. Polymer Sci.*, **19**, 347 (1956).

88. A. M. Bueche, *J. Am. Chem. Soc.*, **74**, 65 (1952).

89. F. Danusso and G. Moraglio, *Atti*, **27**, 381 (1959).

90. T. W. Campbell and A. C. Haven, Jr., *J. Appl. Polymer Sci.*, **1**, 73 (1959).

91. H. W. Starkweather, Jr., and R. H. Boyd, *J. Phys. Chem.*, **64**, 410 (1960).

92. G. Natta, G. Mazzanti, P. Corradini, and I. W. Bassi, *Makromol. Chem.*, **37**, 156 (1960).

93. D. V. Badami, *Polymer*, **1**, 273 (1960).

94. J. H. Griffith and B. G. Ranby, *J. Polymer Sci.*, **44**, 369 (1960).

95. C. G. Overberger, A. E. Borchert, and A. Katchman, *J. Polymer Sci.*, **44**, 491 (1960).

96. G. Natta, G. Mazzanti, P. Corradini, A. Valvassori, and I. W. Bassi, *Atti*, **28**, 18 (1960).

97. G. Natta, G. Mazzanti, P. Corradini, P. Chini, and I. W. Bassi, *Atti*, **28**, 8 (1960).

98. Ø. Ellefsen, *Norelco Reporter*, **7**, 104 (1960).

99. G. Farrow, J. McIntosh, and I. M. Ward, *Makromol. Chem.*, **38**, 147 (1960).

100. N. G. Esipova, L. Pan-Tun, N. S. Andreeva, and P. V. Kozlov, *Vysokomol. Soed.*, **2**, 1109 (1960).

101. A. G. M. Last, *J. Polymer Sci.*, **39**, 543 (1959).

102. D. J. Shields and H. W. Coover, Jr., *J. Polymer Sci.*, **39**, 532 (1959).

103. S. Murahashi, S. Nozakura, and H. Tadokoro, *Bull. Chem. Soc. Japan*, **32**, 534 (1959).

104. T. F. Yen, *J. Polymer Sci.*, **38**, 272 (1959).

105. C. S. Fuller and C. L. Erickson, *J. Am. Chem. Soc.*, **59**, 344 (1937).

106. C. S. Fuller and C. J. Frosch, *J. Am. Chem. Soc.*, **61**, 2575 (1939).

107. C. S. Fuller, C. J. Frosch, and N. R. Pape, *J. Am. Chem. Soc.*, **64**, 154 (1942).

108. C. S. Fuller and C. S. Frosch, *J. Phys. Chem.*, **43**, 323 (1939).

109. C. S. Fuller, *Chem. Rev.*, **26**, 143 (1940).

110. W. O. Baker and C. S. Fuller, *J. Am. Chem. Soc.*, **64**, 2399 (1942).

111. A. Conix and R. Van Kerpel, *J. Polymer Sci.*, **40**, 521 (1959).

112. R. F. Heck and D. S. Breslow, *J. Polymer Sci.*, **41**, 521 (1960).

113. R. F. Heck and D. S. Breslow, *J. Polymer Sci.*, **41**, 520 (1960).

114. E. J. Vandenberg, R. F. Heck, and D. S. Breslow, *J. Polymer Sci.*, **41**, 519 (1960).

115. A. M. Liquori, *Acta Cryst.*, **8**, 345 (1955).

116. S. Furuya and M. Honda, *J. Polymer Sci.*, **28**, 232 (1958).

117. P. Goodman, *J. Polymer Sci.*, **24**, 307 (1960).

118. L. Mandelkern and P. J. Flory, *J. Am. Chem. Soc.*, **73**, 3206 (1951).

119. S. Ishida and S. Murahashi, *J. Polymer Sci.*, **40**, 571 (1959).

120. B. S. Garrett, W. E. Goode, S. Gratch, J. F. Kincaid, C. L. Levesque, A. Spell, J. D. Stroupe, and W. H. Watanabe, *J. Am. Chem. Soc.*, **81**, 1007 (1959).

121. D. C. Vogelsong and E. M. Pearce, *J. Polymer Sci.*, **45**, 546 (1960).

122. P. Corradini and P. Ganis, *Nuovo Cimento*, **15**, Suppl. 1, 104 (1960).

123. G. Natta, P. Corradini, and I. W. Bassi, *Nuovo Cimento*, **15**, Suppl. 1, 83 (1960).

124. G. Natta and P. Corradini, *Nuovo Cimento*, **15**, Suppl. 1, 111 (1960).

125. P. Corradini and P. Ganis, *Nuovo Cimento*, **15**, Suppl. 1, 93 (1960).

126. G. Natta, P. Corradini, and I. W. Bassi, *Nuovo Cimento*, **15**, Suppl. 1, 52 (1960).

127. G. Natta and P. Corradini, *Nuovo Cimento*, **15**, Suppl. 1, 40 (1960).

128. G. Natta, P. Corradini, and I. W. Bassi, *Nuovo Cimento*, **15**, Suppl. 1, 68 (1960).

129. P. Pino and G. P. Lorenzi, *J. Am. Chem. Soc.*, **82**, 4745 (1960).

130. G. Natta, G. Mazzanti, G. Pregaglia, M. Binaghi, and M. Peraldo, *J. Am. Chem. Soc.*, **82**, 4742 (1960).

131. Z. W. Wilchinsky, *J. Appl. Phys.*, **31**, 1969 (1960).

132. T. Mochizuki, *J. Chem. Soc. Japan*, **81**, 15 (1960).

133. R. Stefani, M. Chevreton, M. Garnier, and C. Eyraud, *Compt. Rend.*, **251**, 2174 (1960).

134. M. L. Huggins, *J. Chem. Phys.*, **13**, 37 (1945).

135. C. Ruscher and H. J. Schroder, *Faserforsch. u. Textiltech.*, **11**, 165 (1960).

136. Z. Mencik, *Chem. prumysl.*, **10**, 377 (1960).

137. M. Dole and B. Wunderlich, *Makromol. Chem.*, **34**, 29 (1960).

138. G. Allen, G. Gee, and G. J. Wilson, *Polymer*, **1**, 456 (1960).

139. G. Allen, G. Gee, D. Mangaraj, D. Sims, and G. J. Wilson, *Polymer*, **1**, 466 (1960).

140. M. Asahina and K. Okuda, *Chem. High Polymers (Japan)*, **17**, 607 (1960).

141. G. Natta, M. Farino, M. Peraldo, P. Corradini, G. Bressan, and P. Ganis, *Atti*, **28**, 442 (1960).

142. L. A. Auspos, C. W. Burnam, L. Hall, J. K. Hubbard, W. Kirk, J. R. Schaefgen, and S. B. Speck, *J. Polymer Sci.*, **15**, 19 (1955).
143. A. T. Walter, *J. Polymer Sci.*, **13**, 207 (1954).
144. C. J. Malm, J. W. Mench, D. L. Kendall, and G. D. Hiatt, *Ind. Eng. Chem.*, **43**, 688 (1951).
145. C. E. Anagnostopoulos, A. Y. Coran, and H. R. Gamrath, *J. Appl. Polymer Sci.*, **4**, 181 (1960).
146. C. F. Horn, B. T. Freure, H. Vineyard, and H. J. Decker, Am. Chem. Soc. Meeting, St. Louis, March, 1961.
147. J. B. Rose, *J. Chem. Soc.*, **1956**, 542, 546.
148. A. C. Farthing and W. J. Reynolds, *J. Polymer Sci.*, **12**, 503 (1954).
149. Y. Etienne, *Ind. Plastiques Mod. (Paris)*, **9**, 37 (1957).
150. T. W. Campbell, *J. Org. Chem.*, **22**, 1029 (1957).
151. A. C. Farthing, *J. Chem. Soc.*, **1955**, 3648.
152. G. Natta, *Angew. Chem.*, **68**, 383 (1956).
153. G. F. Schmidt and H. A. Stuart, *Z. Naturforschung*, **13A**, 222 (1958).
154. G. Natta and P. Corradini, *Nuovo Cimento*, **15**, Suppl. 1, 9 (1960).
155. F. Rybnikar, *Chem. Listy*, **52**, 1042 (1958).
156. B. Wunderlich and M. Dole, *J. Polymer Sci.*, **24**, 201 (1957).
157. C. W. Smith and M. Dole, *J. Polymer Sci.*, **20**, 37 (1956).
158. B. Wunderlich and M. Dole, *J. Polymer Sci.*, **32**, 125 (1958).
159. P. J. Flory, H. D. Bedon, and E. H. Keefer, *J. Polymer Sci.*, **28**, 151 (1958).
160. R. D. Evans, M. R. Mighton, and P. J. Flory, *J. Am. Chem. Soc.*, **72**, 2018 (1950).
161. G. Dall Asta and N. Oddo, *Chim. e Ind. (Milan)*, **42**, 1234 (1960).
162. Z. Mencik, *Vysokomol. Soed.*, **2**, 1635 (1960).
163. Y. Chatani, *J. Polymer Sci.*, **47**, S91 (1960).
164. E. J. Vandenberg, *J. Polymer Sci.*, **47**, 486 (1960).
165. E. J. Vandenberg, *J. Polymer Sci.*, **47**, 489 (1960).
166. H. D. Keith, F. J. Padden, Jr., N. M. Walter, and H. W. Wyckoff, *J. Appl. Phys.*, **30**, 1485 (1959).
167. G. Natta, M. Peraldo, and P. Corradini, *Atti*, **26**, 14 (1959).
168. D. Sianesi, R. Serra, and F. Danusso, *Chim. e Ind. (Milan)*, **41**, 515 (1959).
169. G. Natta, I. Pasquon, P. Corradini, M. Peraldo, M. Pegoraro, and A. Zambelli, *Atti*, **28**, 539 (1960).
170. N. M. Walter, quoted in C. Y. Liang, and F. G. Pearson, *J. Mol. Spectroscopy*, **5**, 290 (1960).
171. S. S. Leshchenko, V. L. Karpov, and V. A. Kargin, *Vysokomol. Soed*, **1**, 1538 (1959).
172. D. J. H. Sandiford, *J. Appl. Chem.*, **8**, 188 (1958).
173. M. Hatano and S. Kambura, *Polymer*, **2**, 1 (1961).
174. R. Dedeurwaerder and J. F. M. Oth, *Bull. Soc. Chim. Belg.*, **70**, 37 (1961).
175. G. Natta, D. Sianesi, D. Morero, I. W. Bassi, and G. Caporiccio, *Atti*, **28**, 551 (1960).
176. I. W. Bassi, *Atti*, **29**, 193 (1960).
177. G. Champetier, M. Laualov, and J. P. Pied, *Bull. Soc. Chem. (France)*, **1958**, 708.
178. A. G. Nasini, L. Trossarelli, and G. Saini, *Makromol. Chem.*, **44-46**, 550 (1961).

POLYMER SINGLE CRYSTALS

179. G. Natta, G. Mazzanti, G. F. Pregaglia, and M. Binaghi, *Makromol. Chem.*, **44–46**, 537 (1961).

180. J. C. Swallow, *Proc. Roy. Soc. (London)*, **A238**, 1 (1956).

181. J. Lal and G. S. Trick, *J. Polymer Sci.*, **50**, 13 (1961).

182. R. J. Wilkinson and M. Dole, International Symposium on Macromolecular Chemistry, Montreal, Canada, July 31, 1961.

183. F. B. Moody, American Chemical Society Meeting, Chicago, September, 1961.

184. I. Sakurada, K. Nukushina, and Y. Sone, *J. Soc. Polymer Sci. Japan*, **12**, 506, 517 (1955).

185. G. Natta, G. Mazzanti, P. Longi, G. Dall Asta, and F. Bernardini, *J. Polymer Sci.*, **51**, 487 (1961).

186. M. Inoue, *J. Polymer Sci.*, **51**, 518 (1961).

187. J. A. Price, M. R. Lytton, and B. G. Ranby, *J. Polymer Sci.*, **51**, 541 (1961).

188. F. P. Price and R. W. Kilb, *J. Polymer Sci.*, **57**, 395 (1962).

189. G. Natta, L. Porri, P. Corradini, G. Zanini, and F. Ciampelli, *J. Polymer Sci.*, **51**, 463 (1961).

190. K. Hayashi, Y. Kitanishi, M. Nishii, and S. Okamura, *Makromol. Chem.*, **47**, 237 (1961).

191. G. Natta, G. Dall Asta, G. Maffanti, I. Pasquon, A. Valvassori, and A. Zambelli, *J. Am. Chem. Soc.*, **83**, 3343 (1961).

192. A. Kawasaki, J. Furukawa, T. Tsuruta, G. Wasai, and T. Makimoto, *Makromol. Chem.*, **49**, 76 (1961).

193. Y. Chatani, T. Takizawa, S. Murahashi, Y. Sakata, and Y. Nishimura, *J. Polymer Sci.*, **55**, 811 (1961).

194. J. R. Richards, Ph.D. Thesis, Univ. of Penna., *Diss. Abstr.* **22**, 1029 (1961).

195. H. G. Kilian, H. Haboth, and E. Jenckel, *Kolloid-Z.*, **172**, 166 (1960).

196. E. J. Addink and J. Beintema, *Polymer*, **2**, 185 (1961).

197. G. Damaschun, *Kolloid-Z.*, **180**, 65 (1962).

198. C. G. Overberger and A. M. Schiller, *J. Polymer Sci.*, **54**, S30 (1961).

199. C. A. Boye, *J. Polymer Sci.*, **55**, 275 (1961).

200. A. Kawasaki, J. Furukawa, T. Tsuruta, Y. Nakayama, and G. Wasai, *Makromol. Chem.*, **49**, 112 (1961).

201. A. Kawasaki, J. Furukawa, T. Tsuruta, Y. Nakayama, and G. Wasai, *Makromol. Chem.*, **49**, 136 (1961).

202. T. Makimoto, T. Tsuruta, and J. Furukawa, *Makromol. Chem.*, **50**, 116 (1961).

203. A. Turner-Jones and C. W. Bunn, *Acta Cryst.*, **15**, 105 (1962).

204. D. C. Vogelsong, *J. Polymer Sci.*, **57**, 895 (1962).

205. C. Y. Liang and S. Krimm, *J. Chem. Phys.*, **25**, 563 (1956).

206. M. G. Broadhurst, *J. Research Natl. Bur. Standards*, **66A**, 241 (1962); *J. Chem. Phys.*, **36**, 2578 (1962).

207. R. L. Miller, In *Crystalline Olefin Polymers*, Raff and Doak, eds., Interscience, New York, 1963.

208. A. V. Ermolina, G. S., Markova, and V. A. Kargin, *Kristallografiia*, **2**, 623 (1957).

209. E. S. Clark and L. T. Muus, *Z. Krist.*, in press.

210. V. F. Holland, J. E. Johnson, and B. B. Bowles, Private Communication.

211. J. J. Point, *Bull. Classe Sci. Acad. Roy. Belge*, **30**, 435 (1953).

212. L. G. Wallner, *Monatshefte fur Chem.*, **79**, 279 (1948).

213. H. Hendus, K. Schmieder, G. Schnell, and Wolf, K. A., *Festschr. Carl Wurster der BASF Vom*, **2**, 12 (1960).

214. J. Furukawa, T. Saegusa, T. Tsuruta, S. Ohta, and G. Wasai, *Makromol. Chem.*, **52**, 230 (1962).

215. G. Champetier and J. P. Pied, *Makromol. Chem.*, **44–46**, 64 (1961).

216. F. Rybnikar, *Coll. Czech. Chem. Comm.*, **24**, 2861 (1959).

AUTHOR INDEX*

* Italic numbers refer to bibliographies of the different chapters.

366, 367 (ref. 56), *373*, 375 (ref. 4), *418*, 422 (ref. 30), *476*, 492 (ref. 4), *511*
Sullivan, P., 74 (ref. 86), *78*, 150, 151, 161 (ref. 71), 176 (ref. 89), *187, 188*
Sutherland, G. B. B. H., 406 (ref. 46), 420
Swan, P. R., 50 (refs. 67, 68), 51, 52 (ref. 68), 77, 85, 94 (ref. 26), 100 (ref. 53), 108, 111, 113, 115, 119, 131, 132, 136–138 (ref. 26), *185, 187*, 496 (ref. 12), *511*
Sweeting, O. J., 422 (refs. 3, 4), 427 (ref. 3), 453 (ref. 4), 454 (ref. 3), 466 (refs. 3, 49), *475, 477*
Symons, N. K. J., 71 (ref. 85), *78*, 81, 89, 98, 100, 163, 179, 180 (ref. 17), *185*, 189, 192, 193, 195 (ref. 4), *220*, 295 (ref. 107), 300–302 (ref. 109), *308*, 483, 486 (ref. 6), *489*
Szigeti, B., 406 (ref. 44), *420*

Takayanagi, M., 288 (ref. 99), *308*, 503 (refs. 21, 22), 504 (refs. 21, 22, 26), 505 (refs. 21, 22), 506 (ref. 26), *512*
Takeda, K., 481 (ref. 13), *489*
Tanaka, K., 52 (ref. 92), *78*, 450 (ref. 52), *477*
Teege, E., 259 (ref. 70), *307*
Thurn, H., 334, 337 (ref. 28), *372*, 506 (ref. 27), *512*
Till, P. H., 81, 83, 94, 95 (ref. 8), *185*
Tiller, W. A., 272 (ref. 80), *307*
Tilley, G. P., 422, 453 (ref. 4), 466 (ref. 49), *475, 477*
Timell, T. E., 183, 184 (ref. 94), *188*
Tordella, J. P., 51 (ref. 90), *78*, 450 (ref. 37), *477*
Tosi, M., 375, 383, 389, 390, 396, 398–400 (ref. 19), *419*
Trapeznikova, O. N., 208 (ref. 20), *220*
Trillat, J. J., 427, 435 (ref. 22), *476*
Troussier, M., 483 (ref. 5), *489*
Tschirch, A., 226 (ref. 22), *305*
Tsurada, M., 38 (ref. 87), *78*, 167 (ref. 80), *188*
Turnbull, D., 375 (ref. 21), 378 (refs. 22, 23), 387 (ref. 22), 395 (refs. 23, 33), 396 (ref. 23), *419*, 497 (ref. 14), 499 (ref. 16), *511*
Turner-Jones, A., 45 (ref. 89), *78*

Vallee, B. L., 303 (ref. 112), *309*
van Bibber, K., 97 (ref. 48), *186*
Vand, V., 81 (ref. 7), *185*
Van Roggen, A., 507–509 (ref. 31), *512*
Veiel, U., 275 (ref. 84), *307*
Vermilyea, D. A., 104 (ref. 62), *187*
Vincent, P. I., 440 (ref. 26), *476*
Vlasov, A. V., 165 (ref. 76), *187*
Vogelsong, D. C., 36, 38, 39 (ref. 52), *76*

Wallerant, F., 225 (ref. 8), *304*
Walter, E. R., 356 (ref. 45), *373*
Walter, N. M., 57, 62 (ref. 76), *306*, 81, 86–88, 181 (ref. 18), 183, 184 (ref. 94), *185, 188*, 231, 266, 268 (ref. 49), *306*, 311, 320, 343 (ref. 2), *371*
Ward, I. M., 48 (ref. 64), 77
Waring, J. R. S., 10 (ref. 24), *75*, 201 (ref. 15), *220*, 226, 243, 275, 289 (ref. 19), *305*
Warren, C. H., 225 (ref. 13), *304*
Wartman, L. H., 479 (ref. 3), *489*
Watase, T., 427 (ref. 17), *476*
Webb, M. B., 22, 67 (ref. 35), *76*
Weeks, J. J., 282, 284 (ref. 96), *308*, 368–370 (ref. 60), *373*, 394 (ref. 27), 395, 396 (refs. 27, 31), 397 (ref. 31), 414 (ref. 31), *419*, 498 (ref. 15), *511*
Weigel, E., 226, 304 (ref. 25), *305*
Westrik, R., 85, 111 (ref. 31), *186*
Whitney, J. F., 242 (ref. 53), *306*, 427 (ref. 21), *476*
Willems, I., 102 (ref. 56), *187*
Willems, J., 102 (refs. 55, 56), *187*
Wilson, F. C., 100 (ref. 52), *186*
Wilson, P. R., 229 (ref. 38), *305*
Wino, H. J., 226, 293 (ref. 24), *305*
Winslow, F. H., 495–497 (ref. 11), *511*
Winter, U., 66 (ref. 80), *78*, 455 (ref. 41), *477*
Woestenenk, J. M., 275 (ref. 86), *307*
Wolf, K. A., 37, 38 (ref. 54), 77
Woodward, A. E., 333, 335–337 (ref. 27), *372*, 503, 504 (ref. 23), *512*
Worthington, C. R., 427 (ref. 18), 455, 456 (ref. 42), *476, 477*
Wunderlich, B., 74 (ref. 86), *78*, 99 (ref. 49), 150 (refs. 70, 71), 151, 161 (ref. 71), 176 (ref. 89), *186–188*

SUBJECT INDEX*

* Page numbers in boldface type refer to chapter or section titles. Page numbers in italics refer to portions of the text in which the subject is discussed or mentioned on several successive pages. Roman numerals refer to individual pages. Although effort has been made to cross-reference the index, it is recognized that reference from one part to another is not always complete and, therefore, when feasible, subjects should be looked up under several different terms.

fibrils, 435
granular, morphology of, 484
granular polymer, 297
nuclear magnetic resonance, 4
polymerization, 481
rolling of, *463*
single crystals, **183**
    from melt, *300*
spherulites, **294**
striations, 295
x-ray diffraction, 4
Polyurethane, annealing of, *366*
annealing of fibers, 455
crystal structure, **66**
crystallization from glassy state, *261*
oligomers, 99
small angle x-ray diffraction, 265, 455
spherulites, 265, **303**, 366
Potential energy of chain in lattice, chains
        at rest, *407*
    librating chains, *409*
    vibrating chains, *408*
Premelting, 4, 504
Pressure, effect on single crystals, 99
Primary lamellae. *See* Lamellae, primary
Proteins, spherulites, **303**

Radioactive tracers, use of, 271
Recrystallization, **311**
    in kinetic theory, 381
    in low molecular weight materials, 502
    in polypropylene spherulites, 268
    in thermodynamic theory, 382
    isothermal, 240, 284
    of hedrites, 204
    See also Annealing
Redrawing of polyethylene, 427, 440
Relaxation of drawn material, 425, 435,
        453
Repeat distance, 16
Repeat unit, 16
Replicas, detachment, 197
Replication, **69**
Ribbons. *See* Single crystal, ribbons;
        Spherulites, ribbons; and ribbons
        under individual polymers
Rolling, **458**

Rubber, crystal structure, **53**
    spherulites, 226, **293**

Screw dislocations, 99, 104, *153*
    in crystal growth, 378
    in hedrites, 206
    in single crystals, 81
    in spherulites, *234*
    See also Spiral growths
Secondary lamellae. *See* Lamellae,
        secondary
Sheafs, 196, 201, 206, 223, 225, 233, 265,
        280, 281, 295
Shear between fold planes, 450; *see also*
        Slip
Single crystals, annealing of, **311**
    Bragg extinction lines, 111
    dark field microscopy, *111*
    dendrites, **139**
        formation of, 400
        splaying of lamellae, *161*
        See also dendrites under individual
                polymers
    description of, *106*, *151*
    drawing of, **441**
    effect of branches, 99
        of concentration, 94
        of pressure, 99
        of molecular weight, *95*, 171
        of solvent 87, 94, 100, 174
        of substrate *101*, 157
    electron diffraction from, 83
    electronic properties, *507*
    free energy of formation, *379*
    freeze-drying, 195
    from solution, **79**
    from the melt; polyamides, **278**
        polypropylene, *211*
        polystyrene, 211
        polytetrafluoroethylene, *300*
    growth by solvent evaporation, 87, 176
    growth faces, *104*
    growth of, 87
    growth rate; effect of defects, 393
        temperature dependence, *391*
        theory of, *389*
    hollow pyramids, 85, **108**
    interatomic forces, 379